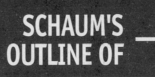

SCHAUM'S
OUTLINE OF

Theory and Problems of

GENETICS

Fourth Edition

SUSAN L. ELROD, Ph.D.

Assistant Professor of Biological Sciences
California Polytechnic State University
at San Luis Obispo

WILLIAM D. STANSFIELD, Ph.D.

Emeritus Professor of Biological Sciences
California Polytechnic State University
at San Luis Obispo

Schaum's Outline Series

McGRAW-HILL

New York   Chicago   San Francisco   Lisbon   London   Madrid   Mexico City
Milan   New Delhi   San Juan   Seoul   Singapore   Sydney   Toronto

SUSAN L. ELROD has degrees in Biological Sciences from California State University, Chico (B.S., 1986) and Genetics from the University of California, Davis (Ph.D., 1995). Graduate work was followed by a Postdoctoral Fellowship at Novozymes Biotech, Inc., in Davis, California. Her research is focused on the molecular genetics of yeast and filamentous fungal systems, with a particular interest in industrial and environmental applications. She holds three patents and has published in the area of gene cloning and protein expression in *Aspergillus oryzae*. She has been an Assistant Professor in the Biological Sciences Department at California Polytechnic State University, San Luis Obispo, since 1997, where she teaches Introductory Microbiology, Industrial Microbiology and Biotechnology, Molecular Genetics, Bioinformatics, and the History of Biology.

WILLIAM D. STANSFIELD has degrees in Agriculture (B.S., 1952), Education (M.A., 1960), and Genetics (M.S., 1962; Ph.D., 1963; University of California at Davis). His published research is in immunogenetics, twinning, and mouse genetics. From 1957 to 1959 he was an instructor in high school vocational agriculture. He was a faculty member of the Biological Sciences Department of California Polytechnic State University from 1963 to 1992 and is now Emeritus Professor. He has written university-level textbooks in evolution and serology/immunology, and has coauthored a dictionary of genetics.

Schaum's Outline of Theory and Problems of
GENETICS

5 6 7 8 9 10 11 12 13 14 15 16 17 18 19 20 CUS CUS 0 9 8 7 6 5

ISBN 0-07-136206-1

Sponsoring Editor: Barbara Gilson
Production Supervisor: Elizabeth J. Shannon
Editing Liaison: Maureen B. Walker
Project Supervisor: Keyword Publishing Services Ltd.

Library of Congress Cataloging-in-Publication Data applied for

McGraw-Hill
A Division of The McGraw-Hill Companies

PREFACE TO THE FOURTH EDITION

It is altogether fitting that this fourth edition of *Schaum's Outline of Theory and Problems of Genetics* should be published at the junction between the first and second centuries of the science of genetics. It was essentially 100 years from the time that Gregor Mendel's pioneering work was rediscovered in 1900 to the announcement of the complete draft sequence of the human genome in 2000. Now that the approximately 30,000 human genes in the human genome have been identified, a major task in this current century will be to discover the protein products of each of these genes in various cell types, then determine all of their structures, functions, and interactions in healthy and diseased states. Concurrent research programs will also be directed to the genes and proteins of many other animals, plants, fungi, bacteria, and viruses of importance in agriculture, medicine, industry, and basic research. We are now at the doorstep of the golden age of applied genetics.

This fourth edition has undergone a vast reorganization of material from the third edition and has had much new information added (especially at the molecular level) by coauthor Susan Elrod. Although this book has traditionally been viewed primarily as a supplement to help students solve problems and pass exams, it also contains at least as much content as could be adequately covered in a one-quarter or a one-semester introductory course when serving as the only course textbook.

An introduction to genetics can be made in one of two equally acceptable ways. One way is by the historical presentation of Mendel's work first, followed by various phenomena discovered by other researchers, such as the chromosome theory of heredity, linkage, gene interactions, maternal inheritance, etc. During the first half of the twentieth century, the physical nature of the gene was essentially unknown, but was suspected to be proteins. In the 1940s, experimental evidence identified DNA as the genetic material of at least some bacteria and viruses. Then in 1953, Watson and Crick published their model for the structure of DNA. This was the key that unlocked an explosion in biological knowledge known as the molecular revolution. Soon it was shown how DNA molecules replicate, mutate, and recombine with one another. Active genes were found to be transcribed into a related class of chemicals called RNA molecules, and these RNAs cooperated in the synthesis of proteins, and the proteins were involved in most of the vital functions of the cell such as photosynthesis, fermentation, oxidative respiration, growth, reproduction, and development. The genetic code was deciphered and techniques were developed to sequence the DNA units known as nucleotides. Rapid, robotic sequencing techniques were later developed that allowed the entire three billion nucleotides of the human genome to be relatively rapidly decoded.

The second way in which genetics can be introduced is by first presenting the molecular knowledge of the second half of the twentieth century, followed by various aspects of transmission genetics. However, there may be a tendency today to overemphasize the molecular concepts of genetics and to relegate the historical aspects of this discipline to the ashcan of forgotten or outdated ideas. In our minds, this would be a great disservice to our students, for without a historical perspective of how we have arrived at our present level of knowledge they will have no general understanding and appreciation of how science advances. That is why in this fourth edition we still present discoveries of the early years of genetics in the first two chapters. This is followed by the biochemical basis of heredity. From this point on, all the rest of genetics and biology can be freely discussed and interrelated at any level of molecules, organelles, cells, tissues, organisms, populations, or species.

S. L. ELROD
W. D. STANSFIELD

CONTENTS

Contents

CHAPTER 1

The Physical Basis of Heredity

Genetics

Genetics is that branch of biology concerned with heredity and variation. The hereditary units that are transmitted from one generation to the next (i.e., inherited) are called **genes**. Genes reside in the long molecules of **deoxyribonucleic acid (DNA)** that exist within all cells. DNA, in conjunction with a protein matrix, forms **nucleoprotein** and becomes organized into structures called **chromosomes** that are found in the **nucleus** or nuclear region of cells. A gene contains coded information for the production of proteins. DNA is normally a stable molecule with the capacity for self-replication. On rare occasions a change may occur spontaneously in some part of DNA. This change, called a **mutation**, may result in an alteration of the code that might lead to the production of a defective protein. The net result of a mutation is sometimes seen as a change in the physical appearance of the individual or a change in some other measurable attribute of the organism called a **character** or **trait**. Through the process of mutation a gene may be changed into two or more alternative forms called **alleles**.

EXAMPLE 1.1
Healthy people have a gene that specifies the normal protein structure of the red blood cell pigment called hemoglobin. Some anemic individuals have an altered form of this gene, i.e., an allele, which makes a defective hemoglobin protein that is unable to carry the normal amount of oxygen to the body cells when oxygen is scarce.

Each gene occupies a specific position on a chromosome, called the gene **locus** (**loci**, plural). All allelic forms of a gene therefore are found at corresponding positions on genetically similar (**homologous**) chromosomes. The word "locus" is sometimes used interchangeably for "gene." When the science of genetics was in its infancy, the gene was thought to behave as a unit particle. These particles were believed to be arranged on the chromosome like beads on a string. This is still a useful concept for beginning students to adopt, but will require considerable modification when we study the biochemical basis of heredity in Chapter 3. All the genes on a chromosome are said to be **linked** to one another and to belong to the same **linkage group**. Wherever the chromosome goes it carries all of the genes associated with it. As we shall see later in this chapter (and in Chapter 6), linked genes are not transmitted independently of one another, but genes in different linkage groups (on different chromosomes) are transmitted independently of one another.

Cells

The smallest unit of life is the **cell**. Each living thing is composed of one or more cells. The most primitive cells alive today are the **bacteria**. Bacteria are generally single-celled organisms that characteristically lack a nucleus. Bacteria therefore belong to a group of organisms called **prokaryotes** (literally, "before a nucleus" had evolved; also spelled **procaryotes**). However, most other life forms (including algae, fungi, plants, and animals) are characterized by the presence of a nucleus and are referred to as **eukaryotes** (literally, "truly nucleated"; also spelled **eucaryotes**). The nucleus is a membrane-bound compartment that isolates the genetic material from the rest of the cell (**cytoplasm**). Most of this book deals with the genetics of eukaryotes. Bacteria will be considered in Chapter 10.

Cells are delimited by a **plasma membrane** and contain all the necessary chemicals and structures for the life of a particular type of cell. The cells of a multicellular organism are often differentiated to perform specific functions (sometimes referred to as a "division of labor"). For example, a neuron is specialized to conduct nerve impulses, a muscle cell contracts, a red blood cell carries oxygen, and so on. Thus, there is no such thing as a typical cell type. Figure 1-1 is a composite diagram of an animal cell showing common subcellular structures. Subcellular structures that are surrounded by a membrane are called **organelles**. Most organelles and other cell structures are too small to be seen with the light microscope, but they can be studied with the electron microscope. The characteristics of organelles and other parts of eukaryotic cells are outlined in Table 1-1.

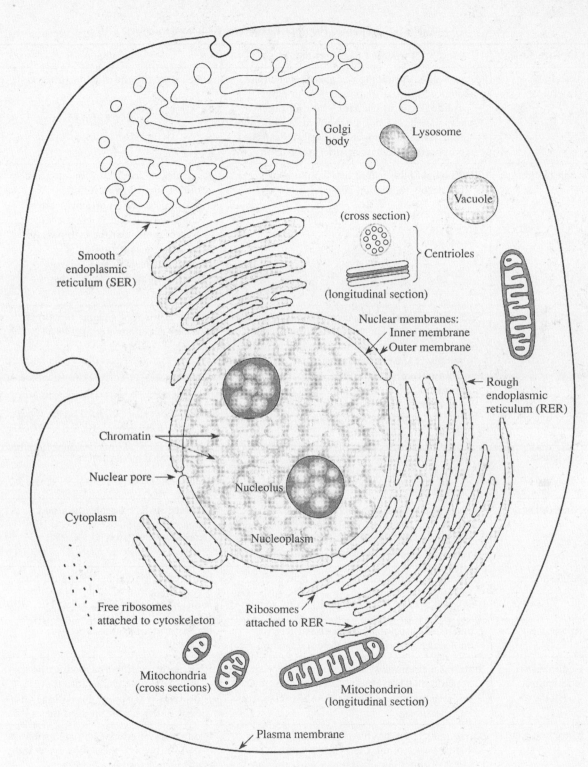

Fig. 1-1.   Diagram of an animal cell.

Table 1-1. Characteristics of Eukaryotic Cellular Structures

Cell Structures	Physical Characteristics	Function(s)
Extracellular structures	A cell wall surrounding the **plasma membrane**; composed primarily of cellulose in plants, peptidoglycan in bacteria, chitin in fungi. Animal cells are not surrounded by cell walls. Some bacteria produce extracellular capsules composed of polysaccharides or glycoproteins	Gives strength and rigidity to the cell
Plasma membrane	**Phospholipid bilayer** that also contains proteins and sterols (in animal cells)	Regulates molecular traffic into and out of the cell; extracellular substances (e.g., nutrients, water) enter the cell and waste substances or secretions exit the cell; passage of substances may require expenditure of energy (active transport) or may be passive (diffusion)
Nucleus	Surrounded by a double membrane (the **nuclear membrane**) that controls the movement of materials between the nucleus and cytoplasm; the membrane contains pores that communicate with the ER. Contains **chromatin**, which is the nucleoprotein component of the **chromosomes**. Only the DNA portion of chromosomes contains the hereditary material	Master control of cellular functions via its genetic material (DNA)
Nucleolus	Site(s) on chromatin where ribosomal RNA (rRNA) is synthesized; disappears from view in light microscope during cellular replication	Ribosomal RNA synthesis
Nucleoplasm	Nonchromatin components of the nucleus containing materials for building DNA and messenger RNA (mRNA molecules serve as intermediates between nucleus and cytoplasm)	Involved in DNA replication and gene expression
Cytoplasm	Contains multiple structures (see below) and enzymatic systems (e.g., glycolysis and protein synthesis)	Involved in providing energy to the cell; execution of the genetic instructions from the nucleus
Ribosome	Consists of three or four ribosomal RNA molecules and over 50 different proteins	Site of protein synthesis
Endoplasmic reticulum (ER)	Internal membrane system; rough endoplasmic reticulum (RER) is studded with ribosomes; smooth endoplasmic reticulum (SER) is free of ribosomes	RER is responsible for modification of polypeptide chains into mature proteins (e.g., by glycosylation) and SER is the site of lipid synthesis
Mitochondria	Surrounded by a double phospholipid bilayer membrane; contains enzymes required for ATP production. Contains a separate set of genes on a circular chromosome that are involved in mitochondrial function	Production of adenosine triphosphate (ATP) through the Krebs cycle and electron transport chain; beta oxidation of long-chain fatty acids; ATP is the main source of energy to power biochemical reactions

Plastid	Plant structure surrounded by a double membrane; contains pigments such as chlorophyll and carotinoids. Example, chloroplasts in plants	Storage and synthesis of food (e.g., starch); photosynthesis occurs in chloroplasts
Golgi body (apparatus)	Organelle composed of flattened, saclike cisternae in close proximity and communication to the ER. Less well-developed cisternae sometimes called dictyosomes in some fungi, protozoans, and plants	Site where sugars, phosphate, sulfate, or fatty acids are added to certain proteins; as membranes bud from the Golgi system they are marked for shipment in transport vesicles to arrive at specific sites (e.g., plasma membrane for secretion outside the cell, lysosome)
Lysosome	Membrane-bound sac of digestive enzymes in all eukaryotic cells	Aids in intracellular digestion of bacteria and other foreign particles; may cause cell destruction if ruptured
Vacuole	Membrane-bound storage organelle	Storage of water and metabolic products (e.g., amino acids, sugars); plant cells often have a large central vacuole that (when filled with fluid to create turgor pressure) makes the cell turgid
Centrioles	Composed of microtubules; capable of being replicated after each cell division; rarely present in plants	Form poles of the spindle apparatus during cell divisions
Cytoskeleton	Consists of microtubules of the protein tubulin (as in the spindle fibers responsible for chromosomal movements during nuclear division or in flagella and cilia), microfilaments of actin and myosin (as occurs in muscle cells), and intermediate filaments (each with a distinct protein such as keratin)	Contributes to shape, division, and motility of the cell and the ability to move and arrange its components
Cytosol	The fluid portion of the cytoplasm exclusive of the formed elements listed above; contains water, minerals, ions, sugars, amino acids, and other nutrients.	Components are involved in building macromolecular biopolymers (nucleic acids, proteins, lipids, and large carbohydrates such as starch and cellulose) and other aspects of cellular metabolism

Chromosomes

1　CHROMOSOME NUMBER

In more complex organisms, such as plants and animals, each **somatic cell** (any cell exclusive of sex cells) contains one set of chromosomes inherited from the maternal (female) parent and a comparable set of chromosomes (homologous chromosomes or **homologues**) from the paternal (male) parent. The number of chromosomes in this dual set is called the **diploid** ($2n$) number. The suffix "-ploid" refers to chromosome "sets." The prefix indicates the degree of **ploidy**. Sex cells, or **gametes**, which contain half the number of chromosome sets found in somatic cells, are referred to as **haploid** cells (n). A **genome** refers to the haploid set of genetic information contained in the cells (i.e., on the chromosomes) of a particular species. The number of chromosomes in each somatic cell is the same for all members of a given species. For example, human somatic cells contain 46 chromosomes, tobacco cells have 48, cattle 60, the garden pea 14, the fruit fly 8, etc.

2　CHROMOSOME MORPHOLOGY

Chromosomes are composed of DNA associated with a variety of proteins. This complex of DNA and proteins can often be seen inside cells as **chromatin** (so called because the complex is readily stained by certain dyes). Members of one group of proteins, called **histones,** help to organize the long strands of DNA into a structure known as a **nucleosome** (see Chapter 13). The structure of chromosomes becomes most easily visible during certain phases of nuclear division when the chromosomes are highly coiled (e.g., metaphase). Each chromosome can generally be distinguished from all others by several criteria, including the relative lengths of the chromosomes, the position of a structure called the **centromere** that divides the chromosome into two arms of varying length, the presence and position of enlarged areas called "knobs" or **chromomeres**, the presence of tiny terminal extensions of chromatin material called "**satellites**," etc. A chromosome with a median centromere (**metacentric**) will have arms of approximately equal size. A **submetacentric** or **acrocentric** chromosome has arms of distinctly unequal size; the size difference is more pronounced in an acrocentric chromosome. The shorter arm is called the **p arm** and the longer arm is called the **q arm**. If a chromosome has its centromere at or very near one end of the chromosome, it is called **telocentric**. Each chromosome of the genome (with the exception of sex chromosomes) is numbered consecutively according to length, beginning with the longest chromosome first.

3　AUTOSOMES VS. SEX CHROMOSOMES

In the males of some species, including humans, sex is associated with a morphologically dissimilar (**heteromorphic**) pair of chromosomes called **sex chromosomes**.

Such a chromosome pair is usually labeled X and Y. Genetic factors on the Y chromosome in humans determine maleness. Females have two morphologically identical X chromosomes. The members of any other homologous pairs of chromosomes (**homologues**) are morphologically indistinguishable, but usually are visibly different from other pairs (**nonhomologous** chromosomes). All chromosomes exclusive of the sex chromosomes are called **autosomes**. Figure 1-2 shows the chromosomal complement of the fruit fly *Drosophila melanogaster* ($2n = 8$), with three pairs of autosomes (2, 3, 4) and one pair of sex chromosomes.

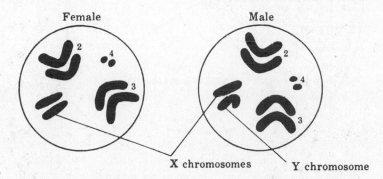

Fig. 1-2.   Diagram of diploid cells in *Drosophila melanogaster*.

Cell Division and Reproduction

1   MITOSIS

All somatic cells in a multicellular organism are descendants of one original cell, the fertilized egg, or **zygote**, through a divisional process called **mitosis**. The function of mitosis is first to construct an exact copy of each chromosome and then to distribute, through division of the original (mother) cell, an identical set of chromosomes to each of the two progeny cells, or **daughter cells**. **Interphase** is the period between successive mitoses and it consists of three phases: G_1, S, and G_2 (Fig. 1-3). During **S** (synthesis) **phase**, the DNA molecules (see Fig. 3-1) of each chromosome are replicated (see Fig. 3-10) producing an identical pair of DNA molecules called **chromatids** (sometimes called "sister" chromatids). Each replicated chromosome thus enters mitosis with two identical DNA molecules. Thin chromatin strands commonly appear as amorphous granular material in the nucleus of stained cells during interphase. Before and after S phase, there are two periods of intense metabolic activity, growth and differentiation, called $\mathbf{G_1}$ (gap 1) and $\mathbf{G_2}$ (gap 2). During G_1, cells are preparing for DNA synthesis (S phase) and during G_2, cell growth and expansion occurs. Cells can leave the cell cycle and enter a resting, or $\mathbf{G_0}$, stage from G_1. G_0 cells are nonproliferative, but viable and metabolically active. Cells may reenter the cell cycle by returning to G_1. Once a cell enters G_1 and the cell cycle, it is committed to completing the cycle. The **M phase**, or mitosis, consists of four major phases detailed below: prophase, metaphase,

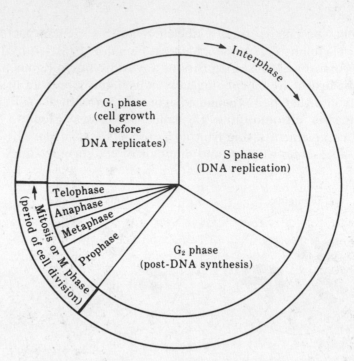

Fig. 1-3. The cell cycle.

anaphase, and telophase (see Fig. 1-4). Mitosis is usually the shortest phase of the cell cycle, taking 1 h of an 18–24 h total cell cycle time in an idealized animal cell. The amount of time spent in the other phases can vary, but a typical G_1 phase lasts 6–12 h, S phase 6–8 h, and G_2 phase 3-4 h. The times spent in each phase of mitosis are quite different. Prophase usually requires far longer than the other phases; metaphase is the shortest.

(a) Prophase. In prophase, the chromosomes **condense** (see Chapter 7), becoming visible under the light microscope first as thin threads, and then becoming progressively shorter and thicker as they coil around histone proteins and then supercoil upon themselves.

EXAMPLE 1.2

A toy airplane can be used as a model to explain the condensation of the chromosomes. A rubber band, fixed at one end, is attached to the propeller at its other end. As the prop is turned, the rubber band coils and supercoils on itself, becoming shorter and thicker in the process. Something akin to this process occurs during the condensation of the chromosomes. However, as a chromosome condenses, the DNA wraps itself around histone proteins to form little balls of nucleoprotein called **nucleosomes**, like beads on a string. At the next higher level of condensation, the beaded string spirals into a kind of cylinder. The cylindrical structure then folds back and forth on itself. Thus, the interphase chromosome becomes condensed several hundred times its length by the end of prophase (see Fig. 7-1).

By late prophase, a chromosome may be sufficiently condensed to be seen under the microscope as consisting of two chromatids connected at their centromeres.

The **centrosome** consists of a pair of **centrioles** and is the site where **microtubules**, composed of two different types of tubulin proteins, organize to form the mitotic spindle. Centrioles are made of microtubules and during prophase, each pair of centrioles is replicated and migrates toward opposite polar regions of the cell. There they establish a **microtubule organizing center (MTOC)** from which a spindle-shaped network of microtubules (called the **spindle**) develops. Microtubules extend from a MTOC to the **kinetochore**, a multiprotein structure attached to centromeric DNA on each chromosome. Most plants, fungi, and some algae lack centrioles but are able to form spindle fibers; thus, centrioles are not required for spindle formation in all organisms. By late prophase, the nuclear membrane has disappeared and the spindle has fully formed. Late prophase is a good time to study and count chromosomes because they are highly condensed and not confined within a nuclear membrane. Mitosis can be arrested at this stage by exposing cells to the alkaloid chemical **colchicine** that interferes with assembly of the spindle fibers. Such treated cells cannot proceed to metaphase until the colchicine is removed.

(b) Metaphase. During metaphase, kinetochore fibers from opposite MTOCs push and pull on the joined centromeres of sister chromatids causing each chromosome to move to a plane usually near the center of the cell, a position designated the **metaphase plate**. The chromosomes are kept in this position by the tension from fibers of opposite MTOCs.

(c) Anaphase. During anaphase, sister chromatids separate at the centromere and are pulled to opposite poles. As each chromatid moves through the viscous cytosol, its arms drag along behind its centromere (attached to spindle fibers via the kinetochore), giving it a characteristic shape depending upon the location of the centromere. Metacentric chromosomes appear V-shaped, submetacentric chromosomes appear J-shaped, and telocentric chromosomes appear rod-shaped.

(d) Telophase. In telophase, each set of separated chromatids is assembled at the two poles of the cell. The chromatids (now, referred to again as chromosomes) begin to uncoil and return to an interphase condition. The spindle degenerates, the nuclear membrane reforms, and the cytoplasm divides in a process called **cytokinesis**. In animals, cytokinesis is accomplished by the formation of a cleavage furrow that deepens and eventually "pinches" the cell in two as shown in Fig. 1-4. Cytokinesis in most plants involves the construction of a cell plate of pectin originating in the center of the cell and spreading laterally to the cell wall. Later, cellulose and other strengthening materials are added to the cell plate (if plant cell), converting it into a new cell wall.

The two products of mitosis are called **daughter cells**, or **progeny cells**, and may or may not be of equal size depending upon where the plane of cytokinesis sections the cell. Thus, while there is no assurance of equal distribution of cytoplasmic components to daughter cells, they do contain exactly the same type and number of chromosomes and hence possess exactly the same genetic constitution.

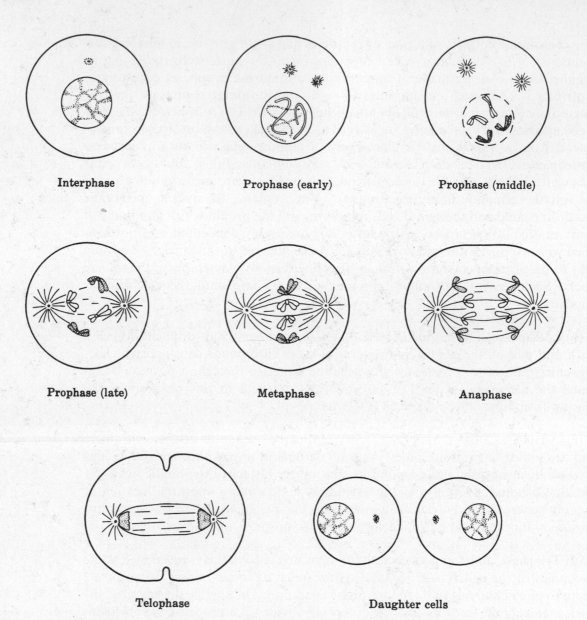

| Interphase | Prophase (early) | Prophase (middle) |

| Prophase (late) | Metaphase | Anaphase |

| Telophase | Daughter cells |

Fig. 1-4. Mitosis in animal cells. Dark chromosomes are of maternal origin; light chromosomes are of paternal origin. One pair of homologues is metacentric, the other pair is submetacentric.

There are three **checkpoints** during the cell cycle to ensure proper progress is being made before proceeding to the next stage: at the G_1/S transition, at the G_2/M transition, and during M phase. For example, if DNA replication has not occurred during S phase (checked at the G_2/M transition), it is pointless for the cell to proceed through mitosis. **Cancers** arise primarily due to unregulated cell division and, in fact, many human cancers are, in part, caused by aberrations (or mutations) in genes that control cell cycle checkpoints. When checkpoints are not properly monitored, abnormal cells that are normally culled from a population may continue to divide and result in a tumor (see Chapter 11 on cancer).

2 SEXUAL REPRODUCTION AND MEIOSIS

Sexual reproduction involves the manufacture of **gametes** (**gametogenesis**) and the union of a male and a female gamete (**fertilization**) to produce a **zygote**. In humans, male gametes are **sperms** and female gametes are **eggs**, or **ova** (**ovum**, singular). Gametogenesis occurs only in the specialized cells (germ line) of the reproductive organs (**gonads**). In animals, the **testes** are male gonads and the **ovaries** are female gonads. Gamete cells are produced through the process of **meiosis**. **Meiosis** (Fig. 1-5) consists of two specialized, consecutive cell divisions in which the chromosome number of the resulting cells is reduced from a diploid ($2n$) to a haploid (n) number. The number of chromosomes must be reduced by half during gametogenesis in order to maintain the chromosome number characteristic of the species after fertilization.

Specifically, meiosis involves a single DNA replication and two divisions of the cytoplasm. The first meiotic division (meiosis I) is a **reductional division** that produces two haploid cells from a single diploid cell. The second meiotic division (meiosis II) is an **equational division** (mitotis-like, in that sister chromatids of the haploid cells are separated). Each of the two meiotic divisions (meiosis I and II) consists of four major phases (detailed below). For meiosis I the four phases are: prophase I, metaphase I, anaphase I, and telophase I; and for meiosis II: prophase II, metaphase II, anaphase II, and telophase II. The DNA replicates during the interphase preceding meiosis I; it does not replicate between telophase I and prophase II.

(a) Meiosis I. In the beginning of meiosis I, replicated chromosomes thicken and condense. **Prophase I** of meiosis differs from the prophase of mitosis in that homologous chromosomes come to lie side by side in a pairing process called **synapsis**. Each pair of synapsed chromosomes is called a **bivalent** (two chromosomes) or a **tetrad** (four chromatids). Each chromosome consists of two identical (replicated) sister chromatids at this stage; the cell contains one set of maternally derived and one set of paternally derived chromosomes. During synapsis, chromatids may cross over and exchange genetic material in a process called crossing over and recombination. The events of prophase I are complex and can be subdivided into five stages.

- **Leptonema** (**Leptotene** or thin-thread stage): The long, thin chromosomes start to condense and, as a consequence, the first signs of threadlike structures begin to appear in the formerly amorphous nuclear chromatin material.
- **Zygonema** (**Zygotene** or joined-thread stage): During this stage, homologous chromosome partners find each other and are joined together by a ribbonlike protein structure called the **synaptonemal complex**. This is the beginning of synapsis. It is thought that synapsis occurs intermittently along the paired chromosomes at sites where the homologues share similar genetic information. A few cases are known in which synaptonemal complexes are not formed, but then synapsis is not as complete and crossing over is markedly reduced or eliminated.

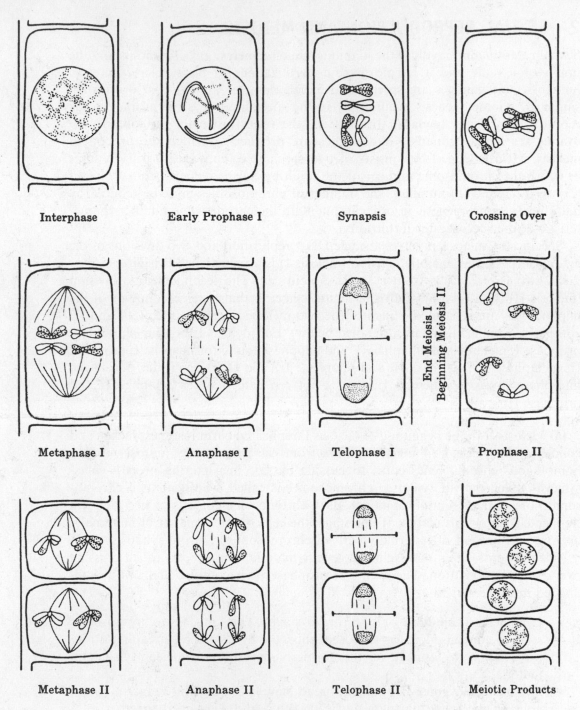

Fig. 1-5. Meiosis in plant cells.

- **Pachynema** (**Pachytene** or thick-thread stage): Synapsis is complete and **recombination nodules** begin to appear along the synapsed chromosomes. At these sites, nonsister chromatids (one from each of the paired chromosomes) of a tetrad **cross over**, break, trade DNA strand and reunite, resulting in an exchange of genetic material (see Fig. 6-1). When viewed under the microscope, the point of exchange appears as a cross-shaped figure called a **chiasma** (**chiasmata**, plural). At a given chiasma, only two of the four chromatids cross over randomly along their paired length. Generally, the number of crossovers per bivalent increases with the length of the chromosome. By chance, a bivalent may experience zero, one, or multiple crossovers. By the breakage and reunion of nonsister chromatids within a chiasma, linked genes become recombined into crossover-type chromatids; the two chromatids within that same chiasma that did not exchange segments maintain the original linkage arrangement of genes as noncrossover or parental-type chromatids. Crossing over is usually a genetic phenomenon that can be inferred only from the results of breeding experiments.

- **Diplonema** (**Diplotene** or double-thread stage): This stage begins when the synaptonemal complex begins to disappear so that individual chromatids and chiasmata can be more readily seen. Chiasmata are also still visible.
- **Diakinesis** (double movement stage): The chromosomes reach their maximal condensation at this stage while the nucleoli and nuclear membrane disappear and the spindle apparatus begins to form.

During **metaphase I**, the bivalents orient at random on the equatorial plane. At **anaphase I**, the centromeres do not separate, but continue to hold sister chromatids together. The chiasmata begin to dissolve, allowing the homologous pairs of chromosomes to separate and move to opposite poles; i.e., whole chromosomes (each consisting of two sister chromatids) move apart. This movement reduces the chromosome number from the diploid ($2n$) to the haploid (n) state. Telophase I occurs when the nuclear membrane reforms and the chromosomes have reached their polar destinations. Cytokinesis follows and results in a division of the diploid mother cell into two haploid daughter cells. Each haploid cell receives a random assortment of maternal and paternal chromosomes; i.e., the chromosomes in one daughter cell will not be uniformly of either maternal or paternal origin. Also, because of crossovers, sister chromatids (still attached at the centromere) may no longer be genetically identical. This ends the first meiotic division.

Genetic aberrations can occur if mistakes are made during the separation of homologous chromosomes at anaphase I. If homologues fail to come apart, or disjoin, and both migrate to the same pole (called **nondisjunction**), the resulting gametes will contain two of those chromosomes, instead of just one. When such a gamete fuses with another during fertilization, the resulting zygote will have three of that particular chromosome. This condition is called a trisomy (see Chapter 7). Most trisomies are lethal; however, trisomy 21 (also called Down syndrome, see Example 7.21), results in an individual who has three copies of chromosome number 21. This trisomy is not lethal, but produces individuals who are mentally and physically disabled. Trisomies of the sex chromosomes also occur without lethality, but also result in genetic abnormalities.

(b) Interkinesis. The period between the first and second meiotic divisions is called interkinesis. Depending on the species, interkinesis may be brief or continue for an extended period of time. During an extensive interkinesis, the chromosomes may uncoil and return to an interphase-like condition with reformation of a nuclear membrane. At some later time, the chromosomes would again condense and the nuclear membrane would disappear. Nothing of genetic importance happens during interkinesis. However, it is important to note one important difference between mitotic interphase and meiotic interkinesis; i.e., no DNA synthesis occurs during interkinesis!

(c) Meiosis II. In **prophase II**, the spindle apparatus reforms and the chromosomes recondense. By **metaphase II**, the individual chromosomes have lined up on the equatorial plane. During **anaphase II**, the centromeres of each chromosome separate, allowing the sister chromatids to be pulled apart in an equational division (mitotis-like) by the attached spindle fibers. During **telophase II**, the chromosomes gather at opposite poles and the nuclear membrane reappears. Each cell then divides by cytokinesis into two progeny cells. Thus, a diploid mother cell becomes four haploid progeny cells as a consequence of a meiotic cycle (meiosis I and meiosis II). The characteristics that distinguish mitosis from meiosis are summarized in Table 1-2.

Table 1-2. Summary of Mitosis and Meiosis

Mitosis	Meiosis
An equational division that separates sister chromatids	The first stage (meiosis I) is a reductional division that separates homologous chromosomes; sister chromatids separate during the second stage (meiosis II)
One division per cycle, i.e., one cytoplasmic division (cytokinesis) per equational chromosomal division	Two divisions per cycle, i.e., two cytoplasmic divisions: one following reductional chromosomal division and one following equational chromosomal division
Homologous chromosomes do not synapse; no chiasmata form	Chromosomes synapse and form chiasmata
Genetic exchange between homologous chromosomes does not occur	Genetic exchange occurs between homologous chromosomes
Two daughter cells produced per cycle	Four daughter cells, called gametes, produced per cycle
Genetic content of mitotic daughter cells is identical to mother cell	Genetic content of meiotic daughter cells is different from each other and from the mother cell
Chromosome number of daughter cells is the same as that of the mother cell	Chromosome number of daughter cells is half that of the mother cell
Mitotic products are usually capable of undergoing additional mitotic divisions	Meiotic products cannot undergo additional meiotic divisions, although they may undergo subsequent mitotic divisions
Normally occurs in almost all somatic cells	Occurs only in specialized cells of the germ line
Begins at the zygote stage and continues through the life of the organism	Occurs only after a higher organism has begun to mature; occurs in the zygote of many algae and fungi

SOLVED PROBLEM 1.1

Consider three pairs of homologous chromosomes with centromeres labeled A/a, B/b, and C/c, where the slash line separates one chromosome from its homologue. How many different kinds of meiotic products can this individual produce?

Solution: For ease in determining all possible combinations, we can use a dichotomous branching system.

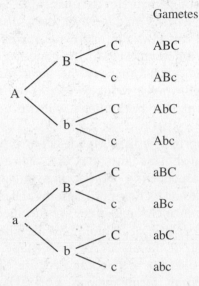

	Gametes
C	ABC
c	ABc
C	AbC
c	Abc
C	aBC
c	aBc
C	abC
c	abc

Eight different chromosomal combinations are expected in the gametes. The solution can also be found by using the formula 2^n, where $n =$ the number of loci. In this case, $n = 3$ so the number of possible combinations is equal to $2^3 = 8$.

SOLVED PROBLEM 1.2

The horse (*Equus caballus*) has a diploid complement of 64 chromosomes including 36 acrocentric autosomes; the ass (*Equus asinus*) has 62 chromosomes including 22 acrocentric autosomes. (*a*) Predict the number of chromosomes to be found in the hybrid offspring (mule) produced by mating a male ass (jack) to a female horse (mare). (*b*) Why are mules usually sterile (incapable of producing viable gametes)?

Solution:

(*a*) The sperm of the jack carries the haploid number of chromosomes for its species ($62/2 = 31$); the egg of the mare carries the haploid number for its species ($64/2 = 32$); the hybrid mule formed by the union of these gametes would have a diploid number of $31 + 32 = 63$.

(*b*) The haploid set of chromosomes of the horse, which includes 18 acrocentric autosomes, is so dissimilar to that of the ass, which includes only 11 acrocentric autosomes, that meiosis in the mule germ line cannot proceed beyond first prophase where synapsis of homologues occurs.

Mendel's Laws of Inheritance

Gregor Mendel published the results of his genetic studies on the garden pea in 1866 and thereby laid the foundation of modern genetics. In his paper, Mendel proposed some basic genetic principles. One of these is known today as the **law of segregation**. Mendel[1] is credited with proposing a model in which any one parent contains two copies of a unit of inheritance (now called a gene) for each trait; however, only one of these two genes (an allele) is transmitted through a gamete to the offspring. For example, a plant that contains two allelic forms of a gene for seed shape, one for round and one for wrinkled seeds, will transmit only one of these two alleles through a gamete to its offspring. Mendel knew nothing of chromosomes or meiosis, as they had not yet been discovered. We now know that the physical basis for this law is in anaphase I, where homologous chromosomes (each containing a different allele of the gene for seed shape, in this case) segregate or separate from each other. If the gene for round seed is on one chromosome and its allelic form for wrinkled seed is on the homologous chromosome, then it becomes clear that both alleles will not normally be found in the same gamete.

The second important principle that Mendel's work helped to establish is the **law of independent assortment**. This law states that the segregation (or separation) of one gene pair occurs independently of any other gene pair. We know that this is true only for unlinked genes or loci on nonhomologous chromosomes. For example, on one homologous pair of chromosomes are the alleles for seed shape (round vs. wrinkled) and on another pair of homologues are alleles for green and yellow seed color. The segregation of the alleles for seed shape occurs independently of the segregation of the alleles for seed color because each pair of homologues behaves as an independent unit during meiosis. Furthermore, because the orientation of bivalents on the first meiotic metaphase plate is essentially random, four combinations of factors could be found in the meiotic products: (1) round-yellow, (2) wrinkled-green, (3) round-green, (4) wrinkled-yellow.

Gametogenesis

Usually, the immediate end products of meiosis are not fully developed gametes. A period of maturation commonly follows meiosis. In plants, one or more mitotic divisions are required to produce reproductive spores, whereas in animals the meiotic products develop directly into gametes through growth and/or differentiation. The entire process of producing mature gametes or spores, of which meiotic division is the most important part, is called **gametogenesis**. Refer back to Figs. 1-3 and 1-5 for details of mitotic and meiotic divisions as needed.

---

[1]For an alternative viewpoint, see Floyd V. Monaghan and Alain Corcos, "The True Mendelian Laws", *The Journal of Heredity*, 1994, 75:321–323.

1 ANIMAL GAMETOGENESIS (AS REPRESENTED IN MAMMALS)

Gametogenesis in the male animal is called **spermatogenesis** [Fig. 1-6(*a*)]. Mammalian spermatogenesis originates in the germinal epithelium of the seminiferous tubules of the male gonads (testes) from diploid primordial cells. These cells undergo repeated mitotic divisions to form a population of **spermatogonia**. By growth, a spermatogonium may differentiate into a diploid primary **spermatocyte** with the capacity to undergo meiosis. The first meiotic division occurs in these primary spermatocytes, producing haploid secondary spermatocytes. From these cells, the second meiotic division produces four haploid meiotic products called **spermatids**. Almost the entire amount of cytoplasm then extrudes into a long whiplike tail during maturation and the cell becomes transformed into a mature male gamete called a **sperm cell** or **spermatozoan** (spermatozoa, plural).

Gametogenesis in the female animal is called **oogenesis** [Fig. 1-6(*b*)]. Mammalian oogenesis originates in the germinal epithelium of the female gonads

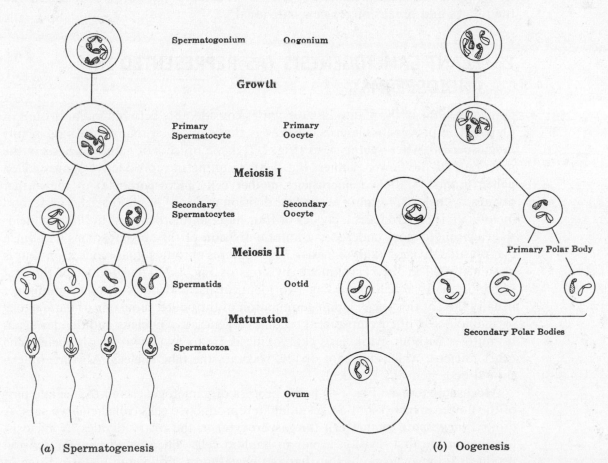

(a) Spermatogenesis   **(b) Oogenesis**

Fig. 1-6. Animal gametogenesis. The number of chromatids in each chromosome at each stage may not be accurately represented. Also, crossovers have been deleted from this figure for the sake of simplicity; thus, if two gamete cells appear to contain identical chromosomes, they are probably dissimilar because of crossovers.

(ovaries) in diploid primordial cells called **oogonia** (**oogonium**, singular). By growth and storage of much cytoplasm or yolk (to be used as food by the early embryo), the oogonium is transformed into a diploid primary **oocyte** with the capacity to undergo meiosis. The first meiotic division reduces the chromosome number by half and also distributes vastly different amounts of cytoplasm to the two products by a grossly unequal cytokinesis. The larger cell thus produced is called a secondary oocyte and the smaller is a primary **polar body**. In some cases, the first polar body may undergo the second meiotic division, producing two secondary polar bodies. All polar bodies degenerate, however, and take no part in fertilization. The second meiotic division of the oocyte again involves an unequal cytokinesis, producing a large yolky **ootid** and a secondary polar body. By additional growth and differentiation, the ootid becomes a mature female gamete called an **ovum** or **egg cell**.

The union of male and female gametes (sperm and egg) is called fertilization and reestablishes the diploid number in the resulting cell called a zygote. During fertilization, the head of the sperm enters the egg, but the tail piece (the bulk of the cytoplasm of the male gamete) remains outside and degenerates. Subsequent mitotic divisions produce the numerous cells of the **embryo** that become organized into the tissues and organs of the new individual.

2 PLANT GAMETOGENESIS (AS REPRESENTED IN ANGIOSPERMS)

Gametogenesis in the plant kingdom varies considerably between major groups of plants. The process as described below is that typical of many flowering plants (**angiosperms**). **Microsporogenesis** (Fig. 1-7) is the process of gametogenesis in the male part of the flower (**anther**, Fig. 1-8), resulting in reproductive spores called **pollen grains**. A diploid **microspore mother cell** (**microsporocyte**) in the anther divides by meiosis, forming at the first division a pair of haploid cells. The second meiotic division produces a cluster of four haploid microspores. Following meiosis, each microspore undergoes a mitotic division of the chromosomes without a cytoplasmic division (**karyokinesis**). This requires chromosomal replication that is not illustrated in the karyokinetic divisions of Fig. 1-7. The product of the first karyokinesis is a cell containing two identical haploid nuclei. Pollen grains are usually shed at this stage. Upon germination of the pollen tube, one of these nuclei (or haploid sets of chromosomes) becomes a generative nucleus and divides again by mitosis without cytokinesis (karyokinesis II) to form two **sperm nuclei**. The other nucleus, which does not divide, becomes the **tube nucleus**. All three nuclei should be genetically identical.

Megasporogenesis (Fig. 1-9) is the process of gametogenesis in the female part of the flower (**ovary**, Fig. 1-8), resulting in reproductive cells called **embryo sacs**. A diploid **megaspore mother cell** (**megasporocyte**) in the ovary divides by meiosis, forming in the first division a pair of haploid cells. The second meiotic division produces a linear group of four haploid **megaspores**. Following meiosis, three of the megaspores degenerate. The remaining megaspore undergoes three mitotic divisions of the chromosomes without intervening cytokineses (karyokineses),

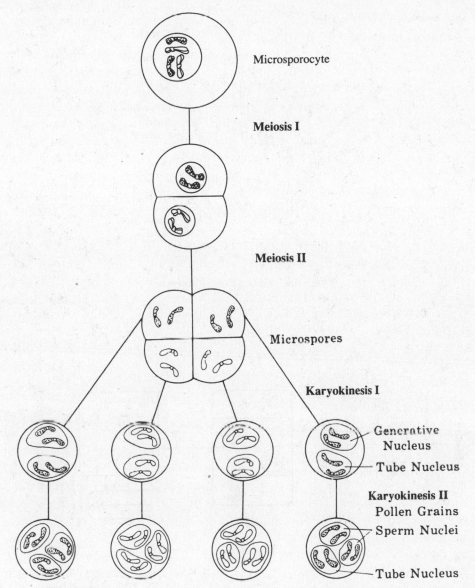

Microsporocyte

Meiosis I

Meiosis II

Microspores

Karyokinesis I

Generative
Nucleus

Tube Nucleus

Karyokinesis II
Pollen Grains

Sperm Nuclei

Tube Nucleus

Fig. 1-7.   Microsporogenesis. The number of chromatids in each chromosome at each stage may not
be accurately represented. Also, crossovers have been deleted from this figure for the sake
of simplicity; thus, if two gamete cells appear to contain identical chromosomes, they are
probably dissimilar because of crossovers.

producing a large cell with eight haploid nuclei (immature embryo sac).
Remember that chromosomal replication must precede each karyokinesis, but
this is not illustrated in Fig. 1-9. The sac is surrounded by maternal tissues of
the ovary called **integuments** and by the megasporangium (**nucellus**). At one end of
the sac there is an opening in the integuments (**micropyle**) through which the pollen
tube will penetrate. Three nuclei of the sac orient themselves near the micropylar
end and two of the three (**synergids**) secrete products that attract the pollen tube.
The third nucleus develops into an **egg nucleus**. Another group of three nuclei

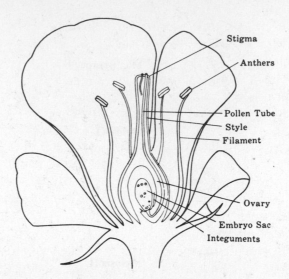

Fig. 1-8. Diagram of a flower.

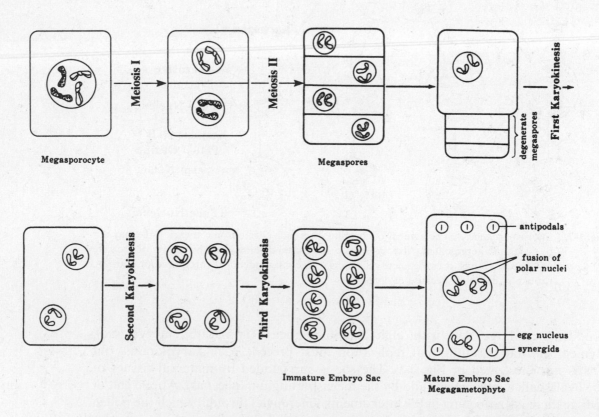

Fig. 1-9. Megasporogenesis. The number of chromatids in each chromosome at each stage may not be accurately represented. Also, crossovers have been deleted from this figure for the sake of simplicity; thus, if two gamete cells appear to contain identical chromosomes, they are probably dissimilar because of crossovers.

moves to the opposite end of the sac and degenerates (**antipodals**). The two remaining nuclei (**polar nuclei**) unite near the center of the sac, forming a single diploid **fusion nucleus**. The mature embryo sac (**megagametophyte**) is now ready for fertilization.

Pollen grains from the anthers are carried by wind or insects to the **stigma**. The pollen grain germinates into a pollen tube that grows down the **style**, presumably under the direction of the tube nucleus. The pollen tube enters the ovary and makes its way through the micropyle of the ovule into the embryo sac (Fig. 1-10). Both sperm nuclei are released into the embryo sac. The pollen tube and the tube nucleus, having served their function, degenerate. One sperm nucleus fuses with the egg nucleus to form a diploid zygote, which will then develop into the embryo. The other sperm nucleus unites with the fusion nucleus to form a triploid (3n) nucleus, which, by subsequent mitotic divisions, forms a starchy nutritive tissue called **endosperm**. The outermost layer of endosperm cells is called **aleurone**. The embryo, surrounded by endosperm tissue, and in some cases such as corn and other grasses where it is also surrounded by a thin outer layer of diploid maternal tissue called **pericarp**, becomes the familiar seed. Since two sperm nuclei are involved, this process is termed double fertilization. Upon germination of the seed, the young seedling (the next sporophytic generation) utilizes the nutrients stored in the endosperm for growth until it emerges from the soil, at which time it becomes capable of manufacturing its own food by photosynthesis.

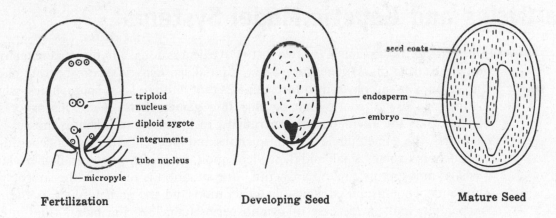

Fig. 1-10.   Fertilization and development of a seed.

SOLVED PROBLEM 1.3
When a plant of chromosomal type *aa* pollinates a plant of type *AA*, what chromosomal type of embryo and endosperm is expected in the resulting seeds?

Solution: The pollen parent produces two sperm nuclei in each pollen grain of type *a*, one combining with the *A* egg nucleus to produce a diploid zygote (embryo) of type *Aa*, and the other combining with the maternal fusion nucleus *AA* to produce a triploid endosperm of type *AAa*.

SOLVED PROBLEM 1.4
Given the first meiotic metaphase orientation shown on the right in the diagram below, and keeping all products in sequential order as they would be formed from

left to right, diagram the embryo sac that develops from the meiotic product at the left and label the chromosomal constitution of all its nuclei.

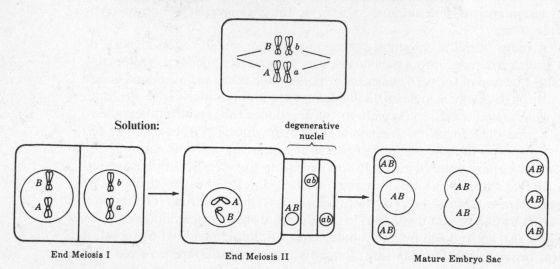

Life Cycles and Genetic Model Systems

Life cycles of most plants have two distinctive generations: a haploid **gametophytic** (gamete-bearing plant) generation and a diploid **sporophytic** (spore-bearing plant) generation. Gametophytes produce gametes that unite to form **sporophytes**, which in turn give rise to spores that develop into **gametophytes**, etc. This process is referred to as the **alternation of generations**. In lower plants, such as mosses and liverworts, the gametophyte is a conspicuous and independently living generation, the sporophyte being small and dependent upon the gametophyte. In higher plants (ferns, gymnosperms, and angiosperms), the situation is reversed; the sporophyte is the independent and conspicuous generation and the gametophyte is the less conspicuous and, in the case of gymnosperms (cone-bearing plants) and angiosperms (flowering plants), completely dependent generation. We have just seen in angiosperms that the male gametophytic generation is reduced to a pollen tube and three haploid nuclei (**microgametophyte**); the female gametophyte (**megagametophyte**) is a single multinucleated cell called the embryo sac that is surrounded and nourished by ovarian tissue. *Arabidopsis thaliana* is a plant commonly used as a genetic model system.

Many simpler organisms such as one-celled animals (protozoa), algae, yeast (*Saccharomyces cerevisiae* and *Schizosaccharomyces pombe*), and other fungi are useful in genetic studies and have interesting life cycles that exhibit considerable variation. Some of these life cycles, as well as those of bacteria and viruses, are presented in later chapters. Other genetic model systems such as the nematode *Caenorhabditis elegans* and the fruit fly *Drosophila melanogaster* will be discussed throughout the book.

Supplementary Problems

1.5. There are 40 chromosomes in somatic cells of the house mouse. (a) How many chromosomes does a mouse receive from its father? (b) How many autosomes are present in a mouse gamete? (c) How many sex chromosomes are there in a mouse ovum? (d) How many autosomes are there in somatic cells of a female?

1.6. Identify the mitotic stage represented in each of the following diagrams of isolated cells from an individual with a diploid chromosome complement of one metacentric pair and one acrocentric pair of chromosomes.

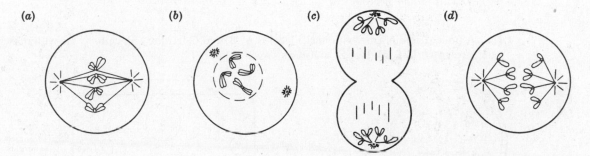

1.7. Identify the meiotic stage represented in each of the following diagrams of isolated cells from the germ line of an individual with one acrocentric pair and one metacentric pair of chromosomes.

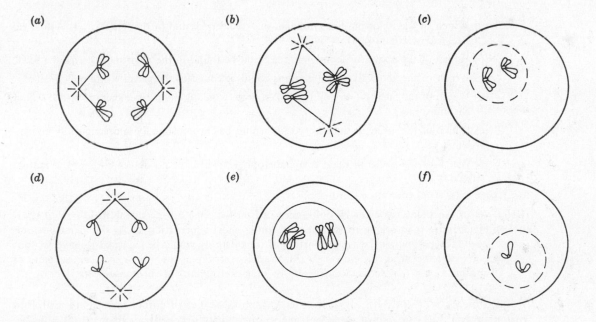

1.8. How many different types of gametic chromosomal combinations (paternal vs. maternal) can be formed in the garden pea ($2n = 14$)? (*Hint*: See Solved Problem 1.1.)

1.9. What animal cells correspond to the three megaspores that degenerate following meiosis in plants?

1.10. What plant cell corresponds functionally to the primary spermatocyte?

1.11. What is the probability of a sperm cell of a man ($n = 23$) containing only replicas of the centromeres that were received from his mother? (*Hint*: See Solved Problem 1.1.)

1.12. How many chromosomes of humans ($2n = 46$) will be found in (*a*) a secondary spermatocyte, (*b*) a spermatid, (*c*) a spermatozoan, (*d*) a spermatogonium, (*e*) a primary spermatocyte?

1.13. How many human egg cells (ova) are produced by (*a*) an oogonium, (*b*) a primary oocyte, (*c*) an ootid, (*d*) a polar body?

1.14. Corn (*Zea mays*) has a diploid number of 20. How many chromosomes would be expected in (*a*) a meiotic product (microspore or megaspore), (*b*) the cell resulting from the first nuclear division (karyokinesis) of a megaspore, (*c*) a polar nucleus, (*d*) a sperm nucleus, (*e*) a microspore mother cell, (*f*) a leaf cell, (*g*) a mature embryo sac (after degeneration of nonfunctional nuclei), (*h*) an egg nucleus, (*i*) an endosperm cell, (*j*) a cell of the embryo, (*k*) a cell of the pericarp, (*l*) an aleurone cell?

1.15. A pollen grain of corn with nuclei labeled A, B, and C fertilized an embryo sac with nuclei labeled D, E, F, G, H, I, J, and K as shown below.

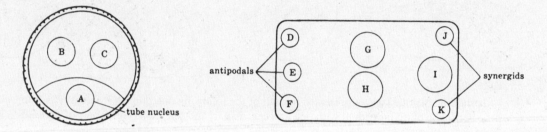

(*a*) Which of the following five combinations could be found in the embryo: (1) ABC   (2) BCI   (3) GHC   (4) AI   (5) CI?

(*b*) Which of the above five combinations could be found in the aleurone layer of the seed?

(*c*) Which of the above five combinations could be found in the germinating pollen tube?

(*d*) Which of the nuclei, if any, in the pollen grain would contain genetically identical sets of chromosomes?

(*e*) Which of the nuclei in the embryo sac would be chromosomally and genetically equivalent?

(*f*) Which of the nuclei in these two gametophytes will have no descendants in the mature seed?

1.16. A certain plant has eight chromosomes in its root cells: a long metacentric pair, a short metacentric pair, a long telocentric pair, and a short telocentric pair. If this plant fertilizes itself (self-pollination), what proportion of the offspring would be expected to have (*a*) four pairs of telocentric chromosomes, (*b*) one telocentric pair and three metacentric pairs of chromosomes, (*c*) two metacentric and two telocentric pairs of chromosomes?

For Problems 1.17 and 1.18, diagram the designated stages of gametogenesis in a diploid organism that has one pair of metacentric and one pair of acrocentric chromosomes. Label each of the chromatids assuming that the locus of gene *A* is on the metacentric pair (one of which carries the *A* allele and its homologue carries the *a* allele) and that the locus of gene *B* is on the acrocentric chromosome pair (one of which carries the *B* allele and its homologue carries the *b* allele).

1.17. Oogenesis: (*a*) first metaphase; (*b*) first telophase resulting from part (*a*); (*c*) second metaphase resulting from part (*b*); (*d*) second telophase resulting from part (*c*).

1.18. Spermatogenesis: (*a*) anaphase of a dividing spermatogonium, (*b*) anaphase of a dividing primary spermatocyte, (*c*) anaphase of a secondary spermatocyte derived from part (*b*), (*d*) four sperm cells resulting from part (*b*).

Review Questions

Matching Questions   Choose the one best match for each organelle (in the left column) with its function or description (in the right column). No organelle has more than one match.

Cell Organelle	Function or Description
1. Mitochondria	A. Establishes polar region
2. Centrioles	B. May contain a photosynthetic system
3. Chromosome	C. Site of protein synthesis
4. Nucleus	D. Location of genes
5. Nucleolus	E. Organelle involved in protein modification and subcellular localization
6. Ribosome	F. Storage of excess water and other metabolic products
7. Endoplasmic reticulum	G. Site of major ATP synthesis
8. Plastid	H. Organelle containing linear chromosomes
9. Golgi body	I. Cytoplasmic membrane network
10. Vacuole	J. RNA-rich region in nucleus

Vocabulary   For each of the following definitions, give the appropriate term and spell it correctly. Terms are single words unless indicated otherwise.

1. Any chromosome other than a sex chromosome.
2. Site on a chromosome to which spindle fibers attach.
3. Adjective applicable to a chromosome with arms of about equal length.
4. Adjective referring to the number of chromosomes in a gamete.
5. Specialized cell division process consisting of a reductional and equational division.
6. Division of the cytoplasm.
7. The first phase of mitosis.
8. The cytological structure on paired chromosomes with which genetic exchange (crossing over) is correlated.
9. Chromosomes that contain enough similar genetic material to pair in meiosis.
10. The period between mitotic division cycles.

Multiple-Choice Questions   Choose the single most appropriate answer.
1. An organelle present in animal cells but missing from plant cells is   (*a*) a nucleolus   (*b*) a centriole   (*c*) a vacuole   (*d*) a mitochondrion   (*e*) more than one of the above

2. During the meiosis I division, chromosomes divide (a) equationally (b) reductionally.

3. Humans normally have 46 chromosomes in skin cells. How many autosomes would be expected in a kidney cell? (a) 46 (b) 23 (c) 47 (d) 44 (e) none of the above

4. During mitosis, synapsis occurs in the phase called (a) telophase (b) anaphase (c) prophase (d) metaphase (e) none of the above

5. If the genetic endowments of two nuclei that unite to produce the plant zygote are labeled A and B, and the other product of fertilization within that same embryo sac is labeled ABB, then the tube nucleus that was in the pollen tube that delivered the fertilizing male gametes must be labeled (a) A (b) AB (c) B (d) BB (e) none of the above

6. A locus is best defined as (a) an alternative form of a gene (b) the position of a gene on a chromosome (c) a cellular structure composed of DNA and protein (d) a fertilized egg (e) the place on a chromosome where spindle fibers attach

7. The different forms of a gene are called (a) gametes (b) loci (c) homologues (d) alleles (e) ribosomes

8. Which of the following cells is normally diploid? (a) primary polar body (b) spermatid (c) primary spermatocyte (d) spermatozoan (e) secondary polar body

9. Upon which two major features of chromosomes does their cytological identification depend? (a) length of chromosome and position of centromere (b) amount of DNA and intensity of staining (c) numbers of nucleoli and centromeres (d) number of chromatids and length of arms (e) chromosome thickness and length

10. In oogenesis, the cell that corresponds to a spermatid is called a(an) (a) ovum (b) egg (c) secondary oocyte (d) oogonium (e) secondary polar body

Answers to Supplementary Problems

1.5. (a) 20 (b) 19 (c) 1 (d) 38

1.6. (a) Metaphase (b) prophase (c) telophase (d) anaphase

1.7. (a) 1st anaphase (b) 1st metaphase (c) 2nd prophase or end of 1st telophase (d) 2nd anaphase (e) 1st prophase (f) 2nd telophase (meiotic product)

1.8. $2^7 = 128$

1.9. Polar bodies

1.10. Microspore mother cell (microsporocyte); both are diploid cells with the capacity to divide meiotically

1.11. $(1/2)^{23}$, less than one chance in 8 million

1.12. (a) 23 (b) 23 (c) 23 (d) 46 (e) 46

1.13. (a) 1 (b) 1 (c) 1 (d) 0

1.14. (a) 10 (b) 20 (c) 10 (d) 10 (e) 20 (f) 20 (g) 30 (h) 10 (i) 30 (j) 20 (k) 20 (l) 30

1.15. (a) 5 (b) 3 (c) 1 (d) A, B, C (e) D, E, F, G, H, I, J, K (f) A, D, E, F, J, K

1.16. (a) 0 (b) 0 (c) All

Only one of several possible solutions is shown for each of Problems 1.17 and 1.18.

1.17.

(a) 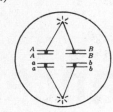   (b)   (c) 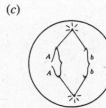   (d)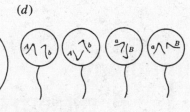

1.18.

(a) 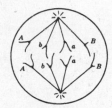   (b)   (c)   (d)

Answers to Review Questions

Matching Questions

1. G   2. A   3. D   4. H   5. J   6. C   7. I   8. B   9. E   10. F

Vocabulary

1. autosome
2. centromere or kinetochore
3. metacentric
4. haploid
5. meiosis

6. cytokinesis
7. prophase
8. chiasma or recombination nodule
9. homologues
10. interphase

Multiple-Choice Questions

1. *b*   2. *b*   3. *d*   4. *e* (synapsis occurs in meiosis, not mitosis)   5. *a*   6. *b*   7. *d*   8. *c*   9. *a*   10. *e*

CHAPTER 2

Patterns of Inheritance

Terminology

1 PHENOTYPE

A **phenotype** is any measurable characteristic or distinctive trait possessed by an organism. The trait may be visible to the eye, such as the color of a flower or the texture of hair, or it may require special tests for its identification, as in a serological test for blood type. The phenotype is the result of gene products brought to expression in a given environment.

> **EXAMPLE 2.1**
> Rabbits of the Himalayan breed will, in the usual range of environments, develop black pigment (a phenotype) at the tips of the nose, tail, feet, and ears. If raised at very high temperatures, an all-white rabbit is produced. The gene for Himalayan color pattern specifies a temperature-sensitive enzyme that is inactivated at high temperature, resulting in a loss of pigmentation.

The kinds of traits that we shall encounter in the study of simple Mendelian inheritance, i.e., those that follow Mendel's two laws (Chapter 1), will be considered to be relatively unaffected by the normal range of environmental conditions in which the organism is found. It is important, however, to remember that genes establish boundaries within which the environment may modify the phenotype.

2 GENOTYPE

All of the alleles possessed by an individual constitute its **genotype**.

 (a) Homozygous. The union of gametes carrying identical alleles produces a **homozygous** genotype. A **homozygote** contains the same alleles at a single locus and produces only one kind of gamete.

EXAMPLE 2.2

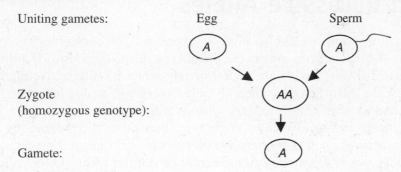

Uniting gametes: Egg   Sperm

Zygote
(homozygous genotype):   AA

Gamete: A

(b) Pure Line. A group of individuals with similar genetic background (breeding) is often referred to as a line or strain or variety or breed. Self-fertilization or mating closely related individuals for many generations (inbreeding) usually produces a population that is homozygous at nearly all loci. Matings between the homozygous individuals of a **pure line** produce only homozygous offspring like the parents. Thus, we say that a pure line "breeds true."

EXAMPLE 2.3

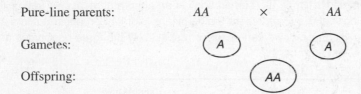

Pure-line parents: AA   ×   AA

Gametes: A   A

Offspring: AA

(c) Heterozygous. The union of gametes carrying different alleles produces a **heterozygous** genotype. A **heterozygote** contains two different alleles at a single locus and produces different kinds of gametes.

EXAMPLE 2.4

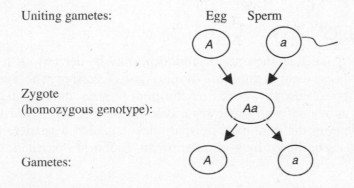

Uniting gametes: Egg   Sperm

                              A   a

Zygote
(homozygous genotype):   Aa

Gametes: A   a

(d) Hybrid. The term **hybrid,** as used in the problems of this book, is synonymous with the heterozygous condition. It can also refer to the progeny of interspecies crosses. **Monohybrids** involve a single-factor hybrid. Later in this chapter we will consider heterozygosity at two (**dihybrids**) or more loci (**polyhybrids**).

Dominant and Recessive Alleles

Allelic forms of a gene are almost always "expressed" by coding for the synthesis of a protein, which, in turn, affects the phenotype of the organism. If a distinctive phenotype is associated with an allele (*a*) only when its alternative allele (*A*) is absent from the genotype, the allele is called **recessive**. A **dominant** allele (*A*) is observed phenotypically in the heterozygote as well as in the homozygote. In some cases, dominance and recessiveness might be conceived as the presence or absence of a trait, protein, or gene product; however, there is no general mechanism that applies to all cases of dominance in molecular or cellular terms. Dominance is not a causal property inherent in the trait or allele itself, but rather it is a relationship between pairs of alleles (see Example 2.18). Other forms of allelic relationships, such as codominance or incomplete dominance, are more common in nature than complete dominance and are discussed later in this chapter. It is important to remember that the frequency of any allele in a natural population of individuals is ultimately determined by evolutionary processes such as natural selection (Chapter 9), not by its relationship to other alleles (i.e., dominance).

EXAMPLE 2.5

Lack of pigment deposition in the human body is an abnormal recessive trait called "albinism." Using *A* and *a* to represent the dominant (pigment-producing) allele and the recessive (albino) allele, we can describe respectively the three genotypes and two phenotypes that are possible:

Genotypes	Phenotypes
AA (homozygous dominant)	Pigmented
Aa (heterozygote)	Pigmented
aa (homozygous recessive)	Albino (no pigment)

1   CARRIERS

Recessive alleles (such as the one for albinism) may be deleterious to those individuals who possess them in duplicate (homozygous recessive genotype). A heterozygote may appear just as normal as the homozygous dominant genotype. A heterozygous individual who possesses a deleterious recessive allele hidden from phenotypic view by the dominant normal allele is called a **carrier.** Most of the deleterious alleles harbored by a population are found in carrier individuals.

2   WILD-TYPE AND MUTANT ALLELES

An allele that is very common in a population is referred to as **wild type**. Alleles that are less common are referred to as **mutant** alleles. An organism exhibiting the phenotype associated with the wild-type allele is termed a wild-type organism; an organism exhibiting the phenotype associated with the rare allele is termed a mutant. Often, wild-type alleles are dominant and mutant alleles are recessive; however, this is not always true and should never be assumed. Genetic testing is always required to determine the relationships between alleles.

3   GENETIC SYMBOLS

Different systems for symbolizing genotypes (dominant and recessive alleles) and phenotypes (wild type and mutant) are used by geneticists for numerous organisms, from plants and animals to yeast, bacteria, and viruses. Students should become familiar with the different types of symbolic representations and be able to work genetic problems regardless of the symbolic system used. The base letter for the gene usually is taken from the name of the mutant or abnormal trait and can be a single letter (e.g., *a*), an abbreviation (e.g., *cdc*, for cell division cycle), or the first few letters of the gene name (e.g., *pro*, for a gene involved in proline amino acid biosynthesis). Generally, dominant alleles are denoted with an uppercase letter (e.g., *A*), first letter upper case (e.g., *Pb*), or all letters upper case (e.g., *HIS4*), and a recessive allele is denoted with all lowercase letters (e.g., *a*, *pb*, *his4*). However, there are other systems. Wild-type alleles in fruit flies are additionally designated with a + symbol; they can be either dominant or recessive. The case of the symbol indicates the dominance or recessiveness of the mutant allele to which the superscript + for wild type must be referred. The wild-type allele in this system can also just be referred to as "+" with no letter designation (e.g., $+/a$). Also, since most organisms that we will use as examples are diploid, both alleles are represented in a genotype either as two symbols side-by-side (e.g., *Aa*) or two symbols with a slash in between (e.g., A/a). Also, gene symbols are usually *italicized*.

EXAMPLE 2.6
Ebony (black) body color in *Drosophila* (fruit fly) is governed by a recessive, mutant allele *e*, while wild-type body color is gray and is governed by a dominant allele e^+. Thus, a fly with both e^+ alleles (e^+/e^+) or with a dominant e^+ and a recessive *e* allele (e^+/e) will have a gray body. However, a fly with both recessive *e* alleles (e/e) will be black.

EXAMPLE 2.7
Lobed-shaped eyes in *Drosophila* are governed by a mutant allele *L* that is dominant, *L*, while wild-type (oval-shaped) eyes are governed by a recessive allele L^+. Thus, a fly with both dominant *L* alleles (L/L) or a fly with one of each allele (L/L^+) will have lobed eyes, while a fly with both recessive alleles (L^+/L^+) will have wild-type, oval-shaped eyes.

In bacteria, mutant alleles are additionally designated with a superscript minus sign after the gene symbol (e.g., $proA^-$). In some systems, specific alleles are often designated by a number or letter following the allele symbol. For example, a specific mutant allele of the yeast gene might be designated as *his4-11*, suggesting that this is the eleventh allele of the fourth wild-type *HIS* gene (genes involved in the synthesis of histidine) to be identified. In yeast, wild-type alleles are designated by all upper-case letters.

Monofactorial (Single Gene) Crosses

The **monofactorial cross** is a mating in which only a single gene is analyzed or monitored in the cross. This cross demonstrates **Mendel's law of segregation** (see Chapter 1). There are six types of monofactorial crosses that can be performed to uncover various genetic relationships; these are described in Section 1 below.

1   SIX MONOFACTORIAL CROSSES

This begins with a cross between a pair of homozygous, pure-breeding parents in the first round (P in cross #1 below) to yield the **first filial generation (F$_1$)**. The resulting offspring are all **monohybrids** (heterozygous for a single pair of alleles). A cross is then carried out between these F$_1$ individuals (cross #2 below) to yield a **second filial generation (F$_2$)**. Unless otherwise specified in the problem, the F$_2$ generation is produced by crossing the F$_1$ individuals among themselves randomly. If plants are normally self-fertilized, they can be artificially cross-pollinated in the parental generation and the resulting F$_1$ progeny may then be allowed to pollinate themselves to produce the F$_2$ progeny. In this generation, the alleles are observed to segregate from one another and four possible gametic combinations

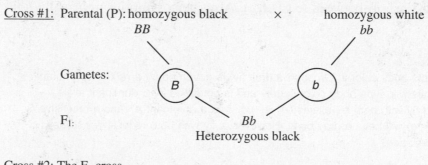

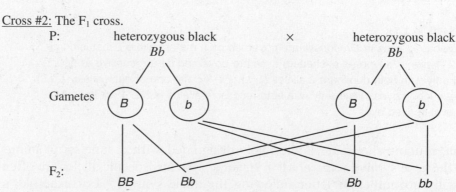

result in three classes of genotypes. Dominance and recessiveness relationships between the alleles for each specific gene set will determine the phenotypic outcomes.

EXAMPLE 2.8

A pair of alleles governs coat color in the guinea pig; a dominant allele B produces black and its recessive allele b produces white. The results of an F_1 cross ($Bb \times Bb$) are summarized in the table below.

Genotypes	Number (fraction)	Phenotypes
BB (homozygous dominant)	1 (1/4)	Black
Bb (heterozygous)	2 (1/2)	Black
bb (homozygous recessive)	1 (1/4)	White

The numbers shown in the table above represent the expected F_2 ratio of offspring from this cross: 1 : 2 : 1. Note the phenotypic ratio is 3 : 1 black to white. These expected results can also be shown as fractions (in parentheses).

Crosses 1 and 2 are solved using a **branch diagram** or a **forked-line** method to help solve the offspring of the cross. This procedure was introduced in Chapter 1 as a means for determining all possible ways in which any number of chromosome pairs could orient themselves on the first meiotic metaphase plate. It can also be used to find all possible genotypic or phenotypic combinations. Another method to use is called the **Punnett square**, named after its inventor, Reginald C. Punnett. The next example shows how to solve cross #2 and Example 2.8 using a Punnett square.

EXAMPLE 2.9

A Punnett square uses a checkerboard or table format to show the genotypes of possible gametes from each parent, one on the top row (bold) and the other on the left-most column (bold). The possible offspring genotypic combinations of each gamete combination are shown in the "remaining" squares. For production of the F_2 in Example 2.8, the same genotypic combinations (BB, Bb, bb) and ratios result (1 BB: 2 Bb: 1 bb) from this method, as do the same phenotypic ratios (3 black : 1 white).

	B	b
B	BB	Bb
b	Bb	bb

There are four other types of monofactorial crosses:

3. homozygous black (BB) × homozygous black (BB)
4. homozygous white (bb) × homozygous white (bb)
5. heterozygous black (Bb) × homozygous black (BB)
6. heterozygous black (Bb) × homozygous white (bb)

The results from each of the six monofactorial crosses are summarized in Table 2-1. Note that crosses 3 and 4 show pure breeding lines.

Table 2-1. Summary of Monofactorial Cross Results

Cross No.	Mating	Expected F_1 Ratios	
		Genotypes	Phenotypes
1	$BB \times bb$	All Bb	All black
2	$Bb \times Bb$	1 BB : 2 Bb : 1 bb	3 black : 1 white
3	$BB \times BB$	All BB	All black
4	$bb \times bb$	All bb	All white
5	$Bb \times BB$	1 BB : 1 Bb	All black
6	$Bb \times bb$	1 Bb : 1 bb	1 black : 1 white

SOLVED PROBLEM 2.1

Black coat of guinea pigs is a dominant trait; white is the alternative recessive trait. When a pure black guinea pig is crossed to a white one, what fraction of the black F_2 is expected to be heterozygous?

Solution: As shown in Example 2.9, the F_2 genotypic ratio is 1 BB : 2 Bb : 1 bb. Considering only the black F_2, we expect 1 BB : 2 Bb or two out of every three black pigs to be heterozygous; the fraction is 2/3.

2 TESTCROSS

A **testcross** is used to determine the genotype of an individual exhibiting a dominant phenotype because this individual could have either a homozygous or heterozygous genotype. The testcross parent is always homozygous recessive for all of the genes under consideration. The purpose of a testcross is to discover how many different kinds of gametes are being produced by the individual whose genotype is in question. A homozygous dominant individual will produce only one kind of gamete; a heterozygous individual will produce two kinds of gametes with equal frequency.

EXAMPLE 2.10

Consider the case in which a testcross is performed with a white-coat male guinea pig and a black-coat female of unknown genotype.

Scenario A: The black female is homozygous.

P:	black female	$\times$	white male	
	BB	$\times$	bb	
Gametes:	only B		only b	
F_1:		Bb		
		all offspring black		

Conclusion: The black female parent must be producing only one kind of gamete and therefore she is homozygous dominant BB.

<u>Scenario B:</u> The black female is heterozygous.

P:	black female	×	white male
	Bb	×	bb

Gametes: B and b only b

Punnett square:

	b
B	Bb
b	bb

F$_1$:   1 Bb : 1 bb
1 black : 1 white

Conclusion: The black female parent must be producing only two kinds of gametes and therefore she is heterozygous dominant *Bb.*

SOLVED PROBLEM 2.2

A dominant gene b^+ is responsible for the wild-type body color of *Drosophila;* its recessive allele b produces black body color. A testcross of a wild-type female gave 52 black and 58 wild type in the F$_1$. If the wild-type F$_1$ females are crossed to their black F$_1$ brothers, what genotypic and phenotypic ratios would be expected in the F$_2$? Diagram the results using the appropriate genetic symbols.

Solution:

P:	$b^+ -$	×	bb
	wild-type female		black male
F$_1$:	52 bb (black) : 58 b^+b (wild type)		

The dash in the genotype of the parental female indicates that one allele is unknown. Since the recessive black phenotype appears in the F$_1$ in approximately a 1 : 1 ratio, we know that the female parent must be heterozygous b^+b. Furthermore, we know that the wild-type F$_1$ progeny must also be heterozygous. The wild-type F$_1$ females are then crossed with their black brothers:

F$_1$ cross:	b^+b	×	bb
	wild-type female		black male
F$_2$:	1/2 b^+b wild type : 1/2 bb black		

The expected F$_2$ ratio is therefore the same as that observed in the F$_1$, namely, 1 wild type : 1 black.

SOLVED PROBLEM 2.3

If a black female guinea pig is testcrossed and produces two offspring in each of three litters, all of which are black, what is her probable genotype? With what degree of confidence may her genotype be specified?

Solution:

P:	$B -$	×	bb
	black female		white male
F$_1$:	all Bb = all black		

The female parent could be homozygous BB or heterozygous Bb and still be phenotypically black; hence, the symbol $B-$. If she is heterozygous, each offspring from this testcross has a 50% chance of being black. The probability of six offspring being produced, all of which are black, is $(1/2)^6 = 1/64 = 0.0156 = 1.56\%$. In other words, we expect such results to occur by chance less than 2% of the time. Since it is chance that operates in the union of gametes, she might actually be heterozygous and thus far only her B gametes have been the "lucky ones" to unite with the b gametes from the white parent. Since no white offspring have appeared in six of these chance unions, we may be approximately 98% confident ($1 - 0.0156 = 0.9844$ or 98.44%), on the basis of chance, that she is of homozygous genotype (BB). It is possible, however, for her very next testcross offspring to be white, in which case we would then become certain that her genotype was heterozygous Bb and not BB.

3 BACKCROSS

In the **backcross**, the F_1 progeny are mated back to one of their parents (or to individuals with a genotype identical to that of their parents). Sometimes "backcross" is used synonymously with "testcross" in genetic literature. The testcross is different in that a recessive homozygote is always used as the testcross parent; this is not necessarily true in a backcross.

EXAMPLE 2.11

A homozygous black female guinea pig is crossed to a white male. An F_1 son is backcrossed to his mother. This cross and backcross is diagrammed as follows, using the symbols ♀ for female and ♂ for male.

P:	BB ♀	×	bb ♂
	black-coat female		white-coat male
F_1:		Bb	
		black males and females	
F_1 backcross:	Bb ♂	×	BB ♀
	black-coat son		black-coat mother
Backcross progeny:	$1/2\ BB : 1/2\ Bb$		All-black offspring

Other Allelic Relationships

1 CODOMINANT ALLELES

Alleles that lack dominant and recessive relationships and are both observed phenotypically are called **codominant.** This means that the phenotypic effect of each allele is observable in the heterozygous condition. Hence, the heterozygous genotype gives rise to a phenotype distinctly different from either of the homozygous genotypes, but possesses characteristics of each. For codominant alleles, all uppercase base symbols with different superscripts are used. The uppercase letters call attention to the fact that each allele can be detected phenotypically to some degree even when in the presence of its alternative allele (heterozygous).

EXAMPLE 2.12

The alleles governing the M-N blood group system in humans are codominant and may be represented by the symbols L^M and L^N, the base letter (L) being assigned in honor of its codiscoverers (Landsteiner and Levine). Two antisera (anti-M and anti-N) are used to distinguish three genotypes and their corresponding phenotypes (blood groups). Agglutination of red blood cells results from a reaction between the antisera and a specific protein antigen (i.e., anti-M reacts only with protein M) and is represented by +; nonagglutination (0) occurs when the specific antigen is not present (i.e., anti-M will not agglutinate cells with protein N if M is absent). Similarly, cells possessing the N antigen will only agglutinate with the anti-N antibodies.

Genotype	Reaction with:		Blood Group (Phenotype)
	Anti-M	Anti-N	
$L^M L^M$	+	0	M
$L^M L^N$	+	+	MN
$L^N L^N$	0	+	N

SOLVED PROBLEM 2.4

Coat colors of the Shorthorn breed of cattle represent a classical example of codominant alleles. Red is governed by the genotype $C^R C^R$, roan (mixture of red and white) by $C^R C^W$, and white by $C^W C^W$. (a) When roan Shorthorns are crossed among themselves, what genotypic and phenotypic ratios are expected among their progeny? (b) If red Shorthorns are crossed with roans, and the F_1 progeny are crossed among themselves to produce the F_2, what percentage of the F_2 will probably be roan?

Solution:

(a)   P: $C^R C^W$ × $C^R C^W$
                                    roan   roan

    F_1:   1/4 $C^R C^R$ red : 1/2 $C^R C^W$ roan : 1/4 $C^W C^W$ white

Since each genotype produces a unique phenotype, the phenotypic ratio 1 : 2 : 1 corresponds to the same genotypic ratio.

(b)   P: $C^R C^R$ × $C^R C^W$
                                    red   roan

    F_1:   1/2 $C^R C^R$ red : 1/2 $C^R C^W$ roan

There are three types of matings possible for the production of the F_2. Their relative frequencies of occurrence may be calculated by preparing a mating table.

	1/2 $C^R C^R$ male	1/2 $C^R C^W$ male
1/2 $C^R C^R$ female	(1) 1/4 $C^R C^R$ female × $C^R C^R$ male	(2) 1/4 $C^R C^R$ female × $C^R C^W$ male
1/2 $C^R C^W$ female	(2) 1/4 $C^R C^W$ female × $C^R C^R$ male	(3) 1/4 $C^R C^W$ female × $C^R C^W$ male

1. The mating $C^R C^R \times C^R C^R$ (red × red) produces only red ($C^R C^R$) progeny. But only one-quarter of all matings are of this type. Therefore, four of all the F_2 should be red from this source.

2. The matings $C^R C^W \times C^R C^R$ (roan female × red male or roan male × red female) are expected to produce 1/2 $C^R C^R$ (red) and 1/2 $C^R C^W$ (roan) progeny. Half of all matings are of this kind. Therefore, (1/2)(1/2) = 1/4 of all the F_2 progeny should be red and four should be roan from this source.

3. The mating $C^R C^W \times C^R C^W$ (roan × roan) is expected to produce 1/4 $C^R C^R$ (red), 1/2 $C^R C^W$ (roan), and 1/4 $C^W C^W$ (white) progeny. This mating type constitutes 1/4 of all crosses. Therefore, the fraction of all F_1 progeny contributed from this source is (1/4)(1/4) = 1/16 $C^R C^R$, (1/4)(1/2) = 1/8 $C^R C^W$, (1/4)(1/4) = 1/16 $C^W C^W$.

The expected F_2 contributions from all three types of matings are summarized in the following table.

Type of Mating	Frequency of Mating	F₂ Progeny		
		Red	Roan	White
(1) Red × red	1/4	1/4	0	0
(2) Red × roan	1/2	1/4	1/4	0
(3) Roan × roan	1/4	1/16	1/8	1/16
Totals		9/16	6/16	1/16

The fraction of roan progeny in the F_2 is 3/8, or approximately 38%.

2 INCOMPLETE DOMINANCE

Alleles that lack dominance relationships and result in heterozygotes that have an intermediate phenotype that is distinct from either homozygous parent are called incompletely or partially dominant alleles. The phenotype may appear to be a "blend" in heterozygotes, but each allele maintains its individual identity and alleles will segregate from each other in the formation of gametes.

EXAMPLE 2.13

Flower color in flowering plants, such as snapdragons, is a good example of incomplete dominance. A cross between pure-breeding red-flowered plants ($R^1 R^1$) and pure-breeding white flowered plants ($R^2 R^2$) results in pink plants ($R^1 R^2$). The F_2 offspring produce red, pink and white progeny in the ratio 1 : 2 : 1, respectively.

Parents: $R^1 R^1$ (red)   × $R^2 R^2$ (white)

Gametes: R^1 R^2

F_1 offspring: $R^1 R^2$ (pink)

F_2 offspring: 1/4 red : 1/2 pink : 1/4 white

3 LETHAL ALLELES

The phenotypic manifestation of some genes is the death of the individual organism prior to sexual maturity. The factors that cause such a manifestation are called

lethal alleles. A fully dominant lethal allele (i.e., one that kills in both the homozygous and heterozygous conditions) occasionally arises by mutation from a wild-type allele. Individuals with a dominant lethal die before they can reproduce. Therefore, the mutant dominant lethal is removed from the population in the same generation in which it arose. Lethals that kill only when homozygous may be of two kinds: (1) one that has no obvious phenotypic effect in heterozygotes, and (2) one that exhibits a distinctive phenotype when heterozygous.

EXAMPLE 2.14
The following phenotypes are associated with the possible genotypes that involve a completely recessive lethal (l) allele.

Genotype	Phenotype
LL, Ll	Normal viability
ll	Lethal

EXAMPLE 2.15
The amount of chlorophyll in snapdragons is controlled by a pair of alleles C^1 and C^2, one of which, C^2, exhibits a lethal effect when homozygous and a distinctive color phenotype when heterozygous. Thus, with regard to color, these alleles are incompletely dominant. However, with regard to viability, the C^2 allele is fully recessive; i.e., the C^2 allele only causes death when C^1 is absent. Note that these genes have at least two phenotypic manifestations (color and viability). This phenomenon of a single gene producing more than one phenotype manifestation is called pleiotropism (Chapter 4).

Genotype	Phenotype
C^1C^1	Green (wild type)
C^1C^2	Pale green
C^2C^2	White (lethal)

SOLVED PROBLEM 2.5
The absence of legs in cattle ("amputated") has been attributed to a completely recessive lethal gene. A normal bull is mated with a normal cow and they produce an amputated calf (usually dead at birth). The same parents are mated again. (*a*) What is the chance of the next calf being amputated? (*b*) What is the chance of these parents having two calves, both of which are amputated? (*c*) Bulls carrying the amputated allele (heterozygous) are mated to noncarrier cows. The F_1 is allowed to mate at random to produce the F_2. What genotypic ratio is expected in the adult F_2? (*d*) Suppose that each F_1 female in part (*c*) rears one viable calf, i.e., each of the cows that throws an amputated calf is allowed to remate to a carrier sire until she produces a viable offspring. What genotypic ratio is expected in the adult F_2?

Solution:
(*a*) If phenotypically normal parents produce an amputated calf, they must both be genetically heterozygous.

P: Aa × Aa
          normal   normal

F_1:   Genotypes   Phenotypes

          1/2 AA
                              }   3/4 normal
          1/2 Aa

          1/4 aa =   1/4 amputated (dies)

Thus, there is a 25% chance of the next offspring being amputated.

(b) The chance of the first calf being amputated and the second calf also being amputated is the product of the separate probabilities: $(1/4)(1/4) = 1/16$ (see Probability Theory later in this chapter).

(c) The solution to part (c) is analogous to that of Problem 2.4(b). A summary of the expected F_2 is as follows:

	F_2 Genotypes		
Type of Mating	AA	Aa	aa
$AA \times AA$	4/16		
$AA \times Aa$	4/16	4/16	
$Aa \times Aa$	1/16	2/16	1/16
Totals	9/16	6/16	1/16

All aa genotypes die and fail to appear in the adult progeny. Therefore, the adult progeny has the genotypic ratio $9AA : 6Aa$ or $3AA : 2Aa$.

(d) The results of matings $AA \times AA$ and $AA \times Aa$ remain the same as in part (c). The mating of $Aa \times Aa$ now is expected to produce 1/3 AA and 2/3 Aa adult progeny. Correcting for the frequency of occurrence of this mating, we have $(1/4)(1/3) = 1/12$ AA and $(1/4)(2/3) = 2/12$ Aa.

Summary of the F_2:

	F_2 Genotypes	
Type of Mating	AA	Aa
$AA \times AA$	3/12	
$AA \times Aa$	3/12	3/12
$Aa \times Aa$	1/12	2/12
Totals	7/12	5/12

The adult F_2 genotypic ratio is expected to be $7AA : 5Aa$.

4   PENETRANCE AND EXPRESSIVITY

Differences in environmental conditions or in genetic backgrounds may cause individuals that are genetically identical at a particular locus to exhibit different phenotypes. The percentage of individuals with a particular gene combination that exhibit the corresponding character to any degree represents the **penetrance** of the trait.

EXAMPLE 2.16
One type of polydactyly (extra fingers and/or toes) in humans can be produced by a dominant gene (P). The wild-type condition with five digits on each limb is produced by the recessive genotype (pp). However, some heterozygous individuals (Pp) are not polydactylous. If 20% of Pp individuals do not show polydactyly (i.e., are wild type), the gene has a penetrance of 80%.

A trait, although penetrant, may be quite variable in its level of expression. The degree of effect produced by a penetrant genotype is termed **expressivity.**

EXAMPLE 2.17
The polydactylous condition may be penetrant in the left hand (six digits) and not in the right (five digits), or it may be penetrant in the feet but not in the hands.

A recessive lethal gene that lacks complete penetrance and expressivity will kill less than 100% of the homozygotes before sexual maturity. The terms **semilethal,** **sublethal,** or **subvital** apply to such genes. The effects that various kinds of lethals have on the reproduction of the next generation form a broad spectrum from complete lethality to sterility in completely viable genotypes. Problems in this book, however, will consider only those lethals that become completely penetrant, usually during the embryonic stage. Genes other than lethals will likewise be assumed completely penetrant.

5   MULTIPLE ALLELES

The genetic systems proposed thus far have been limited to a single pair of alleles. The maximum number of alleles at a gene locus that any individual possesses is two, with one on each of the homologous chromosomes. But since a gene can be changed to alternative forms by the process of mutation, a large number of alleles is theoretically possible in a population of individuals. Whenever more than two alleles are identified at a gene locus in a population, we have a **multiple allelic series.** The dominance hierarchy should be defined at the beginning of each problem involving multiple alleles. A capital letter is commonly used to designate the allele that is dominant to all others in the series. The corresponding lowercase letter designates the allele that is recessive to all others in the series. Other alleles, intermediate in their degree of dominance between these two extremes, are usually assigned the lowercase letter with some suitable superscript.

EXAMPLE 2.18

The color of *Drosophila* eyes is governed by a series of alleles that cause the hue to vary from red or wild type (w^+ or W) through coral (w^{co}), blood (w^{bl}), eosin (w^e), cherry (w^{ch}), apricot (w^a), honey (w^h), buff (w^{bf}), tinged (w^t), pearl (w^p), and ivory (w^i) to white (w). Each allele in the system except w can be considered to produce pigment, but successively less is produced by alleles as we proceed down the hierarchy: $w^+ > w^{co} > w^{bl} > w^e > w^{ch} > w^a > w^h > w^{bf} > w^t > w^p > w^i > w$. The wild-type allele ($w^+$) is completely dominant and w is completely recessive to all other alleles in the series. **Compounds** are heterozygotes that contain unlike members of a multiple allelic series. The compounds of this series that involve alleles other than w^+ tend to be phenotypically intermediate between the eye colors of the parental homozygotes

EXAMPLE 2.19

A classical example of multiple alleles is found in the ABO blood group system of humans, where the allele I^A for the A antigen is codominant with the allele I^B for the B antigen. Both I^A and I^B are completely dominant to the allele i, which fails to specify any detectable antigenic structure. The hierarchy of dominance relationships is symbolized as $(I^A = I^B) > i$. Two antisera (anti-A and anti-B) are required for the detection of four phenotypes.

	Reaction with:		
Genotypes	**Anti-A**	**Anti-B**	**Blood Groups (Phenotype)**
$I^A I^A$, $I^A i$	+	−	A
$I^B I^B$, $I^B i$	−	ǀ	B
$I^A I^B$	+	+	AB
ii	−	−	O

EXAMPLE 2.20

A slightly different kind of multiple allelic system is encountered in the coat colors of rabbits: C allows full color to be produced (typical gray rabbit); c^{ch}, when homozygous, removes yellow pigment from the fur, making a silver-gray color called chinchilla; c^{ch}, when heterozygous with alleles lower in the dominance hierarchy, produces light-gray fur; c^h produces a white rabbit with black extremities, called "Himalayan"; c fails to produce pigment, resulting in albino. The dominance hierarchy may be symbolized as follows: $C > c^{ch} > c^h > c$.

Phenotypes	Possible Genotypes
Full color	CC, Cc^{ch}, Cc^h, Cc
Chinchilla	$c^{ch} c^{ch}$
Light gray	$c^{ch} c^h$, $c^{ch} c$
Himalayan	$c^h c^h$, $c^h c$
Albino	cc

SOLVED PROBLEM 2.6

The genetics of rabbit coat colors is given in Example 2.20. Determine the genotypic and phenotypic ratios expected from mating full-colored males of genotype Cc^{ch} to light-gray females of genotype $c^{ch}c$.

Solution:

P:
$$Cc^{ch} \times c^{ch}c$$
full color   light gray

F_1:

	C	c^{ch}
c^{ch}	Cc^{ch} full color	$c^{ch}c^{ch}$ chinchilla
c	Cc full color	$c^{ch}c$ light gray

Thus, we have a 1 : 1 : 1 : 1 genotypic ratio, but a phenotypic ratio of 2 full color : 1 chinchilla : 1 light gray.

SOLVED PROBLEM 2.7

A man is suing his wife for divorce on the grounds of infidelity. Their first child and second child, whom they both claim, are blood groups O and AB, respectively. The third child, whom the man disclaims, is blood type B. (*a*) Can this information be used to support the man's case? (*b*) Another test was made in the M-N blood group system. The third child was group M, the man was group N. Can this information be used to support the man's case?

Solution:

(*a*) The genetics of the ABO blood group system was presented in Example 2.19. Because the group O baby has the genotype *ii*, each of the parents must have been carrying the recessive allele. The AB baby indicates that one of the parents had the dominant I^A allele and the other had the codominant allele I^B. Any of the four blood groups can appear among the children whose parents are $I^A i \times I^B i$. The information given on ABO blood groups is of no use in supporting the man's claim.

(*b*) The genetics of the M-N blood group system was presented in Example 2.12. The M-N blood groups are governed by a pair of codominant alleles, where groups M and N are produced by homozygous genotypes. A group N father must pass the L allele to his offspring; they all would have the N antigen on their red blood cells, and would all be classified serologically as either group MN or N depending upon the genotype of the mother. This man could not be the father of a group M child.

Probability Theory

There are two laws of probability that are used for genetic analysis. The first law, the **law of products** (or **product rule**), is used to predict the probability of two or more independent events occurring together. Two or more events are said to be **independent** if the occurrence or nonoccurrence of any one of them does not affect the probability of occurrence of any of the others. When two independent events occur with the probabilities p and q, respectively, then the probability of their joint occurrence is $(p)(q)$. That is, the combined probability is the product of the probabilities of the independent events. If the word "and" is used or implied in the phrasing of a problem solution, a *multiplication* of independent probabilities is usually required.

EXAMPLE 2.21
Theoretically, there is an equal opportunity for a tossed coin to land on either heads or tails. Let p = probability of heads = 1/2, and q = probability of tails = 1/2. In two tosses of a coin the probability of two heads appearing (i.e., a head on the first toss *and* a head on the second toss) is $p \times p = p^2 = (1/2)^2 = 1/4$.

EXAMPLE 2.22
In test crossing a heterozygous black guinea pig ($Bb \times bb$), let the probability of a black (Bb) offspring be p = 1/2 and of a white (bb) offspring be q = 1/2. The combined probability of the first two offspring being white (i.e., the first offspring is white *and* the second offspring is white) = $q \times q = q^2 = (1/2)^2 = 1/4$.

There is only one way in which two heads may appear in two tosses of a coin, i.e., heads on the first toss and heads on the second toss. The same is true for two tails. There are two ways, however, to obtain one head and one tail in two tosses of a coin. The head may appear on the first toss and the tail on the second *or* the tail may appear on the first toss and the head on the second. If a head occurs on the first toss and on the second toss, a tail cannot also occur on the second toss. Therefore, these events are mutually exclusive. **Mutually exclusive events** are those in which the occurrence of any one of them excludes the occurrence of the others. The **law of the sum** (or **sum rule**) is used to predict the probability that two mutually exclusive events will occur and states that the probability is the sum of their individual probabilities. The word "or" is usually required or implied in the phrasing of problem solutions involving mutually exclusive events, signaling that an *addition* of probabilities is to be performed. That is, whenever alternative possibilities exist for the satisfaction of the conditions of a problem, the individual probabilities are combined by addition.

EXAMPLE 2.23
In two tosses of a coin, there are two ways to obtain a head and tail.

	First Toss		Second Toss		Probability
First alternative:	Head (p)	(and)	Tail (q)	=	$(p)q$
					(or)
Second alternative:	Tail (q)	(and)	Head (p)	=	$(q)(p)$
Combined probability:					$2(p)(q)$

If $p = q = 1/2$, then the combined probability = $2(1/2)(1/2) = 1/2$.

EXAMPLE 2.24
In test crossing heterozygous black guinea pigs ($Bb \times bb$), there are two ways to obtain one black (Bb) and one white (bb) offspring in a litter of two animals. Let p = probability of black = 1/2 and q = probability of white = 1/2.

	First Offspring		Second Offspring		Probability
First alternative:	Black (p)	(and)	White (q)	=	(p)(q)
					(or)
Second alternative:	White (q)	(and)	Black (p)	=	(q)(p)
Combined probability:					2(p)(q)

Many readers will recognize that the application of the above two laws for combining probabilities is the basis of the binomial distribution, which will be considered in the next section.

SOLVED PROBLEM 2.8

Heterozygous black guinea pigs (*Bb*) are crossed among themselves. (*a*) What is the probability of the first three offspring being alternately black-white-black or white-black-white? (*b*) What is the probability among three offspring of producing two black and one white in any order?

Solution:

(*a*)　　P:　　　　　　　　　　　*Bb*　　×　　*Bb*
　　　　　　　　　　　　　　　　black　　　　　black

　　　　F$_1$:　　　　　　　　　3/4 black : 1/4 white

Let p = probability of black = 3/4, q = probability of white = 1/4.
Probability of black *and* white *and* black = $p \times q \times p = p^2q$, or
Probability of white *and* black *and* white = $q \times p \times q = pq^2$
Combined probability = $p^2q + pq^2$ = 3/16
(*b*) Consider the number of ways that two black and one white offspring could be produced.

Offspring Order			Probability
1st	2nd	3rd	

Black *and* black *and* white = (3/4)(3/4)(1/4) = 9/64, or
Black *and* white *and* black = (3/4)(1/4)(3/4) = 9/64, or
White *and* black *and* black = (1/4)(3/4)(3/4) = 9/64
Combined probability = 27/64

Once we have ascertained that there are three ways to obtain two black and one white, the total probability becomes $3(3/4)^2(1/4)$ = 27/64.

Statistical Distributions

1 THE BINOMIAL DISTRIBUTION

In $(p + q)^n$, the p and q represent the probabilities of alternative independent events, and the power n to which the binomial is raised represents the number of trials. The sum of the factors in the binomial must add to unity; thus,

$$p + q = 1$$

Recall that when two independent events are occurring with the probabilities p and q, then the probability of their joint occurrence is pq. That is, the combined probability is the product of the independent events. When alternative possibilities exist for the satisfaction of the conditions of the problem, the probabilities are combined by addition.

EXAMPLE 2.25

In two tosses of a coin, with $p = $ heads $= $ and $q = $ tails $=$, there are four possibilities.

First Toss		Second Toss		Probability
Heads (p)	(and)	Heads (p)	=	p^2
Heads (p)	(and)	Tails (q)	=	pq
Tails (q)	(and)	Heads (p)	=	pq
Tails (q)	(and)	Tails (q)	=	q^2

which may be expressed as follows:

$$p^2 \quad + \quad 2pq \quad + \quad q^2 \quad = \quad 1.0$$
$$\text{(2 heads)} \qquad \text{(1 head : 1 tail)} \qquad \text{(2 tails)}$$

EXAMPLE 2.26

Expanding the binomial $(p + q)^2$ produces the same expression as in the previous example. Thus, $(p + q)^2 = p^2 + 2pq + q^2$.

EXAMPLE 2.27

When a coin is tossed three times, the probabilities for any combination of heads and/or tails can be found from $(p + q)^3 = p^3 + 3p^2q + 3pq^2 + q^3$. Let $p = $ probability of heads $= 1/2$ and $q = $ probability of tails $= 1/2$.

No. of Heads	No. of Tails	Term	Probability	
3	0	p^3	$(1/2)^3$	$= 1/8$
2	1	$3p^2q$	$3(1/2)^2(1/2)$	$= 3/8$
1	2	$3pq^2$	$3(1/2)(1/2)^2$	$= 3/8$
0	3	q^3	$(1/2)^3$	$= 1/8$

The expansion of $(p + q)^3$ is found by multiplying $(p^2 + 2pq + q^2)$ by $(p + q)$. This process can be extended for higher powers, but obviously becomes increasingly laborious. A short method for expanding $(p + q)$ to any power (n) may be performed by following these rules. (1) The coefficient of the first term is 1. The power of the first factor (p) is n, and that of (q) is 0. (*Note*: Any factor to the zero power is 1.) (2) Thereafter in each term, multiply the coefficient by the power of p and divide by the number of that term in the expansion. The result is the coefficient of the next term. (3) Also thereafter, the power of p will decrease by 1 and the power of q will increase by 1 in each term of the expansion. (4) The fully expanded binomial will have ($n + 1$) terms. The coefficients are symmetrical about the middle term(s) of the expansion.

Summary:

Term	Coefficient	Powers	
		p	q
1	1	n	0
2	$n(1)/1$	$n - 1$	1
3	$n(n - 1)/(1)(2)$	$n - 2$	2
4	$n(n - 1)(n - 2)/(1)(2)(3)$	$n - 3$	3
.	.	.	.
.	.	.	.
.	.	.	.
$n + 1$	1	0	n

2   SINGLE TERMS OF THE EXPANSION

The coefficients of the binomial expansion represent the number of ways in which the conditions of each term may be satisfied. The number of combinations (C) of n different things taken k at a time is expressed by

$$_nC_k = \frac{n!}{(n - k)!k!} \tag{2.1}$$

where $n!$ (called "factorial n") = $n(n - 1)(n - 2) \ldots 1$. ($0! = 1$ by definition.)

EXAMPLE 2.28
If $n = 4$, then $n! = 4(4 - 1)(4 - 2)(4 - 3) = (4)(3)(2)(1) = 24$.

EXAMPLE 2.29
The number of ways to obtain two heads in three tosses of a coin is

$$_3C_2 = \frac{3!}{(3 - 2)!2!} = \frac{3 \cdot 2 \cdot 1}{(1) \cdot 2 \cdot 1} = 3$$

These three combinations are HHT, HTH, and THH.

Formula (2.1) can be used for calculating the coefficients in a binomial expansion,

$$(p + q)^n = \sum_{k=0}^{n} {_nC_k} p^{n-k} q^k = \sum_{k=0}^{n} \frac{n!}{(n-k)!k!} p^{n-k} q^k \tag{2.2}$$

where $\sum$ means to sum what follows as k increases by one unit in each term of the expansion from $k = 0$ to n. This method is obviously much more laborious than the short method presented previously. However, it does have utility in the calculation of one or a few specific terms of a large binomial expansion. To represent this formula in another way, we can let p = probability of the occurrence of one event (e.g., a success) and q = probability of the occurrence of the alternative event (e.g., a failure); then the probability that in n trials a success will occur s times and a failure will occur f times is given by

$$\left(\frac{n!}{s!f!}\right)(p^s)(q^f) \tag{2.3}$$

SOLVED PROBLEM 2.9

Expand the binomial $(p + q)^5$.

Solution: 1st term: $1p^5q^0$;   coefficient of 2nd term $= (5)(1)/1$
2nd term: $5p^4q^1$;   coefficient of 3rd term $= (4)(5)/2$
3rd term: $10p^3q^2$;   coefficient of 4th term $= (3)(10)/3$
4th term: $10p^2q^3$;   coefficient of 5th term $= (2)(10)/4$
5th term: $5p^1q^4$;   coefficient of 6th term $= (1)(5)/5$
6th term: $1p^0q^5$

Summary: $(p + q)^5 = p^5 + 5p^4q + 10p^3q^2 + 10p^2q^3 + 5pq^4 + q^5$

SOLVED PROBLEM 2.10

Find the middle term of the expansion $(p + q)^{10}$ by application of formula (2.2).

Solution: The middle term of the expansion $(p + q)^{10}$ is the sixth term since there are $(n + 1)$ terms in the expansion. The power of q starts at zero in the first term and increases by 1 in each successive term so that the sixth term would have q^5, and so $k = 5$. Then the sixth term is

$$\frac{n!}{(n-k)!k!}p^{n-k}q^k = \frac{10 \cdot 9 \cdot 8 \cdot 7 \cdot 6 \cdot 5!}{(10-5)!5!}p^{10-5}q^5 = \frac{10 \cdot 9 \cdot 8 \cdot 7 \cdot 6}{5 \cdot 4 \cdot 3 \cdot 2 \cdot 1}p^5q^5 = 252\,p^5q^5$$

3   THE MULTINOMIAL DISTRIBUTION

The binomial distribution may be generalized to accommodate any number of variables. If events $e_1, e_2, \ldots, e_k$ will occur with probabilities $p_1, p_2, \ldots, p_k$, respectively, then the probability that $e_1, e_2, \ldots, e_k$ will occur $k_1, k_2, \ldots, k_n$ times, respectively, is

$$\frac{N!}{k_1!k_2!\ldots k_n!}p_1^{k_1}p_2^{k_2}\ldots p_n^{k_n} \tag{2.4}$$

where $k_1 + k_2 + \ldots + k_n = N$.

SOLVED PROBLEM 2.11

A multiple allelic series is known with seven alleles. How many kinds of matings are possible?

Solution:

No. of genotypes possible = No. of different allelic   +   No. of genotypes with two
combinations (heterozygotes)   of the same alleles
(homozygotes)

$$\frac{n!}{(n-k)!k!} + n = \frac{7!}{(5)!2!} + 7 = \frac{7 \cdot 6 \cdot 5!}{2 \cdot 5!} + 7$$

$$= 21 + 7 = 28 \text{ genotypes}$$

No. of different matings = No. of matings between   +   No. of matings between
                                unlike genotypes   identical genotypes

$$\frac{28!}{26!2!} + 28 = \frac{28 \cdot 27 \cdot 26!}{2 \cdot 26!} + 28 = 406$$

SOLVED PROBLEM 2.12

The MN blood types of humans are under the genetic control of a pair of codominant alleles as explained in Example 2.12. In families of six children where both parents are blood type MN, what is the chance of finding three children of type M, two of type MN, and one of type N? [*Hint*: use formula (2.4).]

Solution:

P: $L^M L^N \times L^M L^N$
F_1: $1/4\ L^M L^M$ = type M
            $1/2\ L^M L^N$ = type MN
            $1/4\ L^N L^N$ = type N

Let p_1 = probability of child being type M = 1/4
      p_2 = probability of child being type MN = 1/2
      p_3 = probability of child being type N = 1/4
Let k_1 = number of children of type M required   = 3
      k_2 = number of children of type MN required = 2
      k_3 = number of children of type N required   = 1
                              $N = \sum k_i = 6$

$$(p_1 + p_2 + p_3)^N = \frac{N!}{k_1!k_2!k_3!} p_1^{k_1} p_2^{k_2} p_3^{k_3}$$

$$(p_1 + p_2 + p_3)^6 = \frac{6!}{3!2!1!}\left(\frac{1}{4}\right)^3\left(\frac{1}{2}\right)^2\left(\frac{1}{4}\right) = \frac{6 \cdot 5 \cdot 4 \cdot 3!}{2 \cdot 3!}\left(\frac{1}{4}\right)^4\left(\frac{1}{2}\right)^2 = \frac{15}{256}$$

Crosses Involving Two or More Genes

1 THE DIHYBRID CROSS

In this section the simultaneous inheritance of two or more traits, each specified by a different pair of independently assorting autosomal genes (i.e., genes on different chromosomes other than the sex chromosomes) will be considered. A cross that involves the analysis of two independent traits is termed a **dihybrid** cross. This type of cross demonstrates Mendel's second **law of independent assortment** (see Chapter 1). In the conventional dihybrid cross, two true-breeding parents are mated (cross #1 in Example 2.30) to yield an F_1 generation. The F_1 hybrids are then crossed to yield an F_2 generation (cross #2).

EXAMPLE 2.30

In addition to the coat-color locus of guinea pigs introduced earlier in this chapter ($B-$ = black, bb = white), another locus on a different chromosome (independently assorting) is known to govern length of hair, such that $L-$ = short hair and ll =

long hair. Any of four different genotypes exist for the black, short-haired phenotype: *BBLL*, *BBLl*, *BbLL*, *BbLl*. Two different genotypes produce a black, long-
haired pig: *BBll* or *Bbll*; likewise two genotypes for a white, short-haired pig: *bbLL*
or *bbLl*; and only one genotype specifies a white, long-haired pig: *bbll*. A dihybrid
genotype is heterozygous at two loci. Dihybrids form four genetically different
gametes with approximately equal frequencies because of the random orientation
of nonhomologous chromosome pairs on the first meiotic metaphase plate
(Chapter 1).

Cross #1:

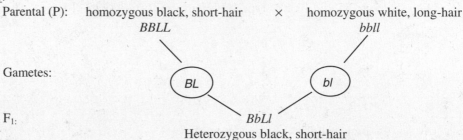

Parental (P):   homozygous black, short-hair   ×   homozygous white, long-hair
                 BBLL   *bbll*

Gametes:

                        BL   bl

F_1:   *BbLl*
                    Heterozygous black, short-hair

Cross # 2: The F_1 cross.
        P: heterozygous black, short-hair   ×   heterozygous black, short-hair
                     BbLl   *BbLl*

When two dihybrids are crossed, four kinds of gametes are produced in equal frequencies in both the male and the female.

Gametes from a *BbLl* hybrid: (Branch diagram)

Allele 1	Allele 2	Gametes	Frequency*
1/2 *B*	1/2 *L*	*BL*	1/4
	1/2 *l*	*Bl*	1/4
1/2 *b*	1/2 *L*	*bL*	1/4
	1/2 *l*	*bl*	1/4

* The product of the probabilities of one allele *and* another (product rule).

These gametes then randomly combine in the F_1 cross. The use of a Punnett
square helps demonstrate all 16 possible combinations of these gametes.

	BL	**Bl**	**bL**	**bl**
BL	*BBLL*	*BBLl*	*BbLL*	*BbLl*
Bl	*BBLl*	*BBll*	*BbLl*	*Bbll*
bL	*BbLL*	*BbLl*	*bbLL*	*bbLl*
bl	*BbLl*	*Bbll*	*bbLl*	*bbll*

The four dark squares are homozygous *BB* genotypes, the eight light squares are heterozygous *Bb*, and the four open squares are homozygous *bb* in a 1 : 2 : 1 ratio, respectively. The same ratio occurs for the *LL*, *Ll*, and *ll* genotypes.

Summary of F_1 cross results:

Genotypes	Number (fraction)	Phenotypes
BBLL	1 (1/16)	Black coat, short hair
BBLl	2 (2/16 or 1/8)	Black coat, short hair
BBll	1 (1/16)	Black coat, long hair
BbLL	2 (2/16 or 1/8)	Black coat, short hair
BbLl	4 (4/16 or 1/4)	Black coat, short hair
Bbll	2 (2/16 or 1/8)	Black coat, long hair
bbLL	1 (1/16)	White coat, short hair
bbLl	2 (2/16 or 1/8)	White coat, short hair
bbll	1 (1/16)	White coat, long hair

The phenotypes in this type of conventional dihybrid cross give a 9 : 3 : 3 : 1 ratio. This ratio is derived from grouping the phenotypes in the table above as follows:

9/16 Black coat, short hair (*B−L−*)

3/16 Black coat, long hair (*B− ll*)

3/16 White coat, short hair (*bbL−*)

1/16 White coat, long hair (*bbll*)

The following examples demonstrate several other methods for determining the ratios in a dihybrid cross.

EXAMPLE 2.31

Determination of genotypic ratios and frequencies.

   Considering only the *B* locus, *Bb* × *Bb* produces 1/4 *BB*, 1/2 *Bb*, and 1/4 *bb*. Likewise for the *L* locus, *Ll* × *Ll* produces 1/4 *LL*, 1/2 *Ll*, and 1/4 *ll*. Place these genotypic probabilities in a Punnett square and combine independent probabilities by multiplication (law of products).

F_2:

	1/4 *LL*	1/2 *Ll*	1/4 *ll*
1/4 *BB*	1/16 *BBLL*	1/8 *BBLl*	1/16 *BBll*
1/2 *Bb*	1/8 *BbLL*	1/4 *BbLl*	1/8 *Bbll*
1/4 *bb*	1/16 *bbLL*	1/8 *bbLl*	1/16 *bbll*

Another way to arrive at the frequencies seen in the above Punnett square is to use the branch diagram. Again, the final frequencies are determined by using the law of products because traits are being combined (e.g. *BB* <u>and</u> *LL*).

B locus		L locus		Ratio	Genotypes	
		1/4 LL	=	1/16	$BB\ LL$	
1/4 BB		1/2 Ll	=		1/8	$BB\ Ll$
		1/4 ll	=	1/16	$BB\ ll$	
		1/4 LL	=	1/8	$Bb\ LL$	
1/2 Bb		1/2 Ll	=	1/4	$Bb\ Ll$	
		1/4 ll	=	1/8	$Bb\ ll$	
		1/4 LL	=	1/16	$bb\ LL$	
1/4 bb		1/2 Ll	=	1/8	$bb\ Ll$	
		1/4 ll	=	1/16	$bb\ ll$	

EXAMPLE 2.32

Determination of phenotypic ratios and frequencies.

Considering the B locus, $Bb \times Bb$ produces 3/4 black and 1/4 white. Likewise at the L locus, $Ll \times Ll$ produces 3/4 short and 1/4 long. Place these independent phenotypic probabilities in a Punnett square and combine them by multiplication.

F_2:

	3/4 black	1/4 white
3/4 short	9/16 black, short	3/16 white, short
1/4 long	3/16 black, long	1/16 white, long

Another way to arrive at the frequencies seen in the above Punnett square is to use the branch diagram. Again, the final frequencies are determined by using the law of products because traits are being combined (e.g., black coat *and* long hair).

Black or white	Short or long		Probability	Phenotypes
	3/4 short	=	9/16	black, short
3/4 black				
	1/4 long	=	3/16	black, long
	3/4 short	=	3/16	white, short
1/4 white				
	1/4 long	=	1/16	white, long

If only one of the genotypic frequencies or phenotypic frequencies is required, there is no need to be concerned with any other genotypes or phenotypes. A mathematical solution can be readily obtained by combining independent probabilities.

EXAMPLE 2.33

Determination of the frequency of only one genotype, *BBLl*, in the offspring of dihybrid parents.

First consider each locus separately: $Bb \times Bb = 1/4\ BB$; $Ll \times Ll = 1/2\ Ll$. Combining these independent probabilities, $1/4 \times 1/2 = 1/8\ BBLl$.

EXAMPLE 2.34

To find the frequency of only one phenotype, white coat, short hair, in the offspring of dihybrid parents.

First consider each trait separately: $Bb \times Bb = 1/4$ white (bb); $Ll \times Ll = 3/4$ short (L–). Combining these independent probabilities, $1/4 \times 3/4 = 3/16$ white, short.

SOLVED PROBLEM 2.13

In the garden pea, Mendel found that yellow seed color was dominant to green ($Y > y$) and round seed shape was dominant to shrunken ($S > s$). (a) What phenotypic ratio would be expected in the F_2 from a cross of a pure yellow, round × green, shrunken? (b) What is the F_2 ratio of yellow: green and of round: shrunken?

Solution:

(a)　　　P:　　　　　　　　　　　$YY\ SS$　　　×　　　　　$yy\ ss$
　　　　　　　　　　　　　　yellow, round　　　　　green, shrunken

　　　　F_1:　　　　　　　　　　　　　$Yy\ Ss$
　　　　　　　　　　　　　　　　　yellow, round

　　　　F_2:　　　　　　　9/16 Y- S- yellow, round
　　　　　　　　　　　　3/16 Y- ss yellow, shrunken
　　　　　　　　　　　　3/16 yy S- green, round
　　　　　　　　　　　　1/16 yy ss green, shrunken

(b) The ratio of yellow : green = (9/16 yellow, round + 3/16 yellow, shrunken) : (3/16 green, round + 1/16 green, shrunken) = 12 : 4 = 3 : 1. The ratio of round : shrunken = (9/16 yellow, round + 3/16 green, round) : (9/16 yellow, shrunken + 1/16 green, shrunken) = 12 : 4 = 3 : 1. Thus, at each of the individual loci an F_2 phenotypic ratio of 3 : 1 is observed, just as would be expected for a monohybrid cross.

SOLVED PROBLEM 2.14

How many different crosses may be made (a) from a single pair of factors, (b) from two pairs of factors, and (c) from any given number of pairs of factors (n)?

Solution:

(a) All possible matings of the three genotypes produced by a single pair of factors may be represented in a Punnett square.

	AA	Aa	aa
AA	$AA \times AA$ (1)	$AA \times Aa$ (2)	$AA \times aa$ (3)
Aa	$Aa \times AA$ (2)	$Aa \times Aa$ (4)	$Aa \times aa$ (5)
aa	$aa \times AA$ (3)	$aa \times Aa$ (5)	$aa \times aa$ (6)

The symmetry of matings above and below the squares on the diagonal becomes obvious. The number of different crosses may be counted as follows: 3 in the first column, 2 in the second, and 1 in the third: $3 + 2 + 1 = 6$ different types of matings.

(b) There are $3^2 = 9$ different genotypes possible with two pairs of segregating factors. If a 9×9 Punnett square were constructed, the same symmetry would exist above and below the squares on the diagonal as was shown in part (a). Again, we may count the different types of matings as an arithmetic progression from 9 to 1; $9 + 8 + 7 + 6 + 5 + 4 + 3 + 2 + 1 = 45$.

(c) The sum of any arithmetic progression of this particular type may be found by the formula $M = 1/2(g^2 + g)$, where M = number of different types of matings, and g = number of genotypes possible with n pairs of factors.

2 TESTCROSS WITH TWO TRAITS

A testcross of a homozygote (BB) gives all one phenotype, while a testcross of a heterozygote (Bb) gives a 1 : 1 phenotypic ratio, indicating that one pair of factors is segregating (see section on Monohybrid Crosses). A dihybrid testcross with a dihybrid (i.e., heterozygote; $BbLl$) gives 1 : 1 : 1 : 1 genotypic and phenotypic ratios, indicating that two pairs of factors are segregating and assorting independently. Testcrosses with individuals that are homozygous for one trait and heterozygous for the second trait give a 1 : 1 phenotypic ratio.

EXAMPLE 2.35

A testcross of a black-coated, short-haired individual with an incompletely known genotype (B-L-) is performed with an individual whose genotype is homozygous recessive at all of the loci under consideration ($bbll$).

Scenario A: The individual is a dihybrid ($BbLl$). A 1 : 1 : 1 : 1 phenotypic ratio results.

P:	$BbLl$	×	$bbll$
	Black coat, short-haired		white coat, long-haired

F_1:	1/4 $BbLl$	black, short-haired
	1/4 $Bbll$	black, long-haired
	1/4 $bbLl$	white, short-haired
	1/4 $bbll$	white, long-haired

Scenario B. The individual has the genotype $BBLl$ (heterozygous at only one locus). A 1 : 1 phenotypic ratio results.

P:	$BBLl$	×	$bbll$
	Black coat, short-haired		white coat, long-haired

F_1:	1/2 $BbLl$	black coat, short-haired
	1/2 $Bbll$	black coat, long-haired

A 1 : 1 phenotypic ratio would also result from the testcross of a $BbLL$ individual; however the genotypes and phenotypes would be as follows:

F_1:	1/2 $BbLl$	black coat, short-haired
	1/2 $bbLl$	white coat, short-haired

A knowledge of the monohybrid probabilities presented earlier in Chapter 2 may be applied in a simplified genotypic or phenotypic Punnett square.

SOLVED PROBLEM 2.15

Tall tomato plants are produced by the action of a dominant allele *D*, and dwarf plants by its recessive allele *d*. Hairy stems are produced by a dominant gene *H*, and hairless stems by its recessive allele *h*. A dihybrid tall, hairy plant is testcrossed. The F_1 progeny were observed to be 118 tall, hairy : 121 dwarf, hairless : 112 tall, hairless : 109 dwarf, hairy. (*a*) Diagram this cross. (*b*) What is the ratio of tall : dwarf; of hairy : hairless? (*c*) Are these two loci assorting independently of one another?

Solution:

(*a*)   P: *Dd Hh*   ×   *dd hh*

                     tall, hairy   dwarf, hairless

Gametes: *DH, Dh, dH, dh*   *dh*

F_1:

Genotypes	Number	Phenotypes
Dd Hh	118	Tall, hairy
Dd hh	112	Tall, hairless
dd Hh	109	Dwarf, hairy
dd hh	121	Dwarf, hairless

Note that the observed numbers approximate a 1 : 1 : 1 : 1 phenotypic ratio.

(*b*) The ratio of tall : dwarf = (118 + 112) : (109 + 121) = 230 . 230 or 1 . 1 ratio. The ratio of hairy : hairless = (118 + 109) : (112 + 121) = 227 : 233 or approximately 1 : 1 ratio. Thus, the testcross results for each locus individually approximate a 1 : 1 phenotypic ratio.

(*c*) Whenever the results of a testcross approximate a 1 : 1 : 1 : 1 ratio, it indicates that the two gene loci are assorting independently of each other in the formation of gametes. That is to say, all four types of gametes have an equal opportunity of being produced through the random orientation that nonhomologous chromosomes assume on the first meiotic metaphase plate. If the testcross does not approximate a 1 : 1 : 1 : 1 ratio, the two genes are probably on the same chromosome (linked). A statistical method for testing genetic hypotheses (e.g., independent assortment) is provided later in this chapter.

3   MODIFIED DIHYBRID RATIOS

The classical phenotypic ratio resulting from the mating of dihybrid genotypes is 9 : 3 : 3 : 1. This ratio appears whenever the alleles at both loci display complete dominant and recessive relationships. The classical dihybrid ratio may be modified if one or both loci have codominant alleles or lethal alleles. A summary of these modified phenotypic ratios in adult progeny is shown in Table 2-2.

Table 2-2. Modified Dihybrid Ratios

| Allelic Relationships in Dihybrid Parents | | Expected Adult Phenotypic Ratio |
First Locus	Second Locus	
Dominant-recessive	Codominant	$3:6:3:1:2:1$
Codominant	Codominant	$1:2:1:2:4:2:1:2:1$
Dominant-recessive	Codominant lethal*	$3:1:6:2$**
Codominant	Codominant lethal*	$1:2:1:2:4:2$**
Lethal*	Codominant lethal*	$4:2:2:1$**

* See Example 2.15.
** Phenotypic ratio in viable adults.

SOLVED PROBLEM 2.16

A dominant allele L governs short hair in guinea pigs and its recessive allele l governs long hair. Codominant alleles at an independently assorting locus specify hair color, such that $C^Y C^Y$ = yellow, $C^Y C^W$ = cream, and $C^W C^W$ = white. From matings between dihybrid short, cream pigs ($Ll C^Y C^W$) predict the phenotypic ratio expected in the progeny.

Solution:

$3/4\ L\text{-}$

$1/4\ C^Y C^Y\quad =\quad 3/16\ L\text{-}\ C^Y C^Y\quad$ short, yellow
$1/2\ C^Y C^W\quad =\quad 6/16\ L\text{-}\ C^Y C^W\quad$ short, cream
$1/4\ C^W C^W\quad =\quad 3/16\ L\text{-}\ C^W C^W\quad$ short, white

$1/4\ ll$

$1/4\ C^Y C^Y\quad =\quad 1/16\ ll\ C^Y C^Y\quad$ long, yellow
$1/2\ C^Y C^W\quad =\quad 2/16\ ll\ C^Y C^W\quad$ long, cream
$1/4\ C^W C^W\quad =\quad 1/16\ ll\ C^W C^W\quad$ long, white

Thus, six phenotypes appear in the offspring in the ratio $3:6:3:1:2:1$. The dash (–) in the genotypes indicates that either allele L or l may be present, with both combinations resulting in a short-haired phenotype.

SOLVED PROBLEM 2.17

Normal leg size, characteristic of Kerry-type cattle, is produced by the homozygous genotype DD. Short-legged Dexter-type cattle possess the heterozygous genotype Dd. The homozygous genotype dd is lethal, producing grossly deformed stillbirths called "bulldog calves." The presence of horns in cattle is governed by the recessive allele of another gene locus p, the polled condition (absence of horns) being produced by its dominant allele P. In matings between polled Dexter cattle of genotype $DdPp$, what phenotypic ratio is expected in the adult progeny?

Solution:

P: *DdPp*   ×   *DdPp*
          Dexter, polled   Dexter, polled

F_1:

                        3/4 *P*-   =   3/16 *DD P*-   polled, Kerry
        1/4 *DD*
                        1/4 *pp*   =   1/16 *DD pp*   horned, Kerry

                        3/4 *P*-   =   6/16 *Dd P*-   polled, Dexter
        1/2 *Dd*
                        1/4 *pp*   =   2/16 *Dd pp*   horned, Dexter

                        3/4 *P*-   =   3/16 *dd P*-   lethal
        1/4 *dd*
                        1/4 *pp*   =   1/16 *dd pp*   lethal

The phenotypic ratio of viable offspring thus becomes: 3 polled : Kerry : 1 horned, Kerry : 6 polled, Dexter : 2 horned, Dexter.

4   HIGHER COMBINATIONS

The methods for solving two-factor (dihybrid) crosses may easily be extended to solve problems involving three (trihybrid) or more pairs of independently assorting autosomal factors. Given any number of heterozygous pairs of factors (n) in the F_1, the following general formulas apply:

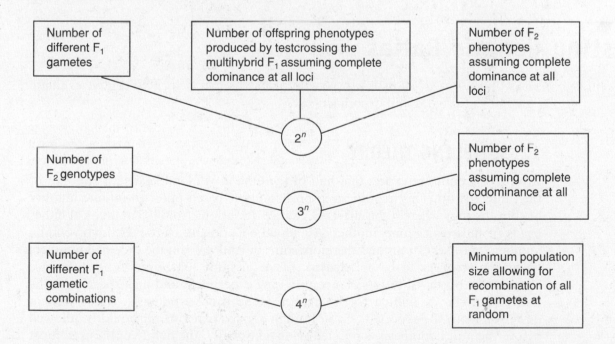

EXAMPLE 2.36

In the following cross, $AaBb \times AaBb$, $n = 2$ (both loci are heterozygous). Then, 2^n or 4 equals the number of different gametes that are possible from each parent, 3^n or 9 equals the number of different genotypes possible, and 4^n or 16 equals the possible gametic combinations.

For a conventional trihybrid F_2 cross ($AaBbCc \times AaBbCc$), eight phenotypic classes result and the phenotypic ratio is $27:9:9:9:3:3:3:1$. The branch diagram below helps illustrate this point, again using the law of products to combine probabilities.

A locus	B locus	C locus	Probability	Phenotypic class
		3/4 C-	$(3/4)(3/4)(3/4) = 27/64$	A-B-C-
	3/4 B-	1/4 cc	$(3/4)(3/4)(1/4) = 9/64$	A-B-cc
3/4 A-		3/4 C-	$(3/4)(1/4)(3/4) = 9/64$	A-bbC-
	1/4 bb	1/4 cc	$(3/4)(1/4)(1/4) = 3/64$	A-$bbcc$
		3/4 C-	$(1/4)(3/4)(3/4) = 9/64$	aaB-C-
	3/4 B-	1/4 cc	$(1/4)(3/4)(1/4) = 3/64$	aaB-cc
1/4 aa		3/4 C-	$(1/4)(1/4)(3/4) = 3/64$	$aabbC$-
	1/4 bb	1/4 cc	$(1/4)(1/4)(1/4) = 1/64$	$aabbcc$

Testing Genetic Ratios

The application of statistics to genetic problems can help predict and evaluate outcomes of crosses and experiments.

1 SAMPLING THEORY

If we toss a coin, we expect that half of the time it will land heads up and half of the time tails up. This hypothesized probability is based upon an infinite number of coin tossings wherein the effects of chance deviations from 0.5 in favor of either heads or tails cancel one another. All actual experiments, however, involve finite numbers of observations and therefore some deviation from the expected numbers (sampling error) is to be anticipated. Let us assume that there is no difference between the observed results of a coin-tossing experiment and the expected results that cannot be accounted for by chance alone (null hypothesis). How great a deviation from the expected 50 : 50 ratio in a given experiment should be allowed before the null hypothesis is rejected? Conventionally, the null hypothesis in most biological experiments is rejected when the deviation is so large that it could be accounted for by chance less than 5% of the time. Such results are said to be significant. When the null hypothesis is rejected at the 5% level, we take 1 chance in 20 of discarding a valid hypothesis. It must be remembered that statistics can never render absolute proof of the hypothesis, but merely sets limits to our uncertainty. If we wish to be even more certain that the rejection of the hypothesis is warranted we could use the 1% level, often called highly significant, in which case the experimenter would be taking only 1 chance in 100 of rejecting a valid hypothesis.

2   SAMPLE SIZE

If our coin-tossing experiment is based on small numbers, we might anticipate relatively large deviations from the expected values to occur quite often by chance alone. However, as the sample size increases, the deviation should become proportionately less, so that in a sample of infinite size the plus and minus chance deviations cancel each other completely to produce the 50 : 50 ratio.

3   DEGREES OF FREEDOM

Assume a coin is tossed 100 times. We may arbitrarily assign any number of heads from 0 to 100 as appearing in this hypothetical experiment. However, once the number of heads is established, the remainder is tails and must add to 100. In other words, we have $n - 1$ degrees of freedom (df) in assigning numbers at random to the n classes within an experiment.

EXAMPLE 2.37
In an experiment involving three phenotypes ($n = 3$), we can fill two of the classes at random, but the number in the third class must constitute the remainder of the total number of individuals observed. Therefore, we have $3 - 1 = 2$ degrees of freedom.

EXAMPLE 2.38
A 9 : 3 : 3 : 1 dihybrid ratio has four phenotypes ($n = 4$). There are $4 - 1 = 3$ degrees of freedom. The number of degrees of freedom in these kinds of problems is the number of variables (n) under consideration minus 1. For most genetic problems, the degrees of freedom will be 1 less than the number of phenotypic classes. Obviously, the more variables involved in an experiment the greater the total deviation may be by chance.

4   CHI-SQUARE TEST

In order to evaluate a genetic hypothesis, we need a test that can convert deviations from expected values into the probability of such inequalities occurring by chance. Furthermore, this test must also take into consideration the size of the sample and the number of variables (degrees of freedom). The chi-square test (pronounced ki-square; symbolized χ^2 includes all of these factors.

$$\chi^2 = \sum_{i=1}^{n} \frac{(o_i - e_i)^2}{e_i} = \frac{(o_1 - e_1)^2}{e_1} + \frac{(o_2 - e_2)^2}{e_2} + \ldots + \frac{(o_n - e_n)^2}{e_n} \qquad (2.5)$$

where $\sum_{i=1}^{n}$ means to sum what follows it as the i classes increase from 1 to n, o represents the number of observations within a class, e represents the number expected in the class according to the hypothesis under test, and n is the number of classes. The value of chi-square may then be converted into the probability that the deviation is due to chance by entering Table 2-3 at the appropriate number of degrees of freedom.

Table 2-3.   Chi-Square Distribution

Degrees of Freedom	Probability										
	0.95	0.90	0.80	0.70	0.50	0.30	0.20	0.10	0.05	0.01	0.001
1	0.004	0.02	0.06	0.15	0.46	1.07	1.64	2.71	3.84	6.64	10.83
2	0.10	0.21	0.45	0.71	1.39	2.41	3.22	4.60	5.99	9.21	13.82
3	0.35	0.58	1.01	1.42	2.37	3.66	4.64	6.25	7.82	11.34	16.27
4	0.71	1.06	1.65	2.20	3.36	4.88	5.99	7.78	9.49	13.28	18.47
5	1.14	1.61	2.34	3.00	4.35	6.06	7.29	9.24	11.07	15.09	20.52
6	1.63	2.20	3.07	3.83	5.35	7.23	8.56	10.64	12.59	16.81	22.46
7	2.17	2.83	3.82	4.67	6.35	8.38	9.80	12.02	14.07	18.48	24.32
8	2.73	3.49	4.59	5.53	7.34	9.52	11.03	13.36	15.51	20.09	26.12
9	3.32	4.17	5.38	6.39	8.34	10.66	12.24	14.68	16.92	21.67	27.88
10	3.94	4.86	6.18	7.27	9.34	11.78	13.44	15.99	18.31	23.21	29.59
	Nonsignificant								Significant		

Source: R. A. Fisher and F. Yates, *Statistical Tables for Biological, Agricultural and Medical Research*, 6th ed., Table IV, Oliver & Boyd, Ltd., Edinburgh, 1963, by permission of the authors and publishers.

An alternative method for computing chi-square in problems involving only two phenotypes will give the same result as the conventional method and often makes computation easier:

$$\chi^2 = \frac{(a - rb)^2}{r(a + b)} \qquad (2.6)$$

where a and b are the numbers in the two phenotypic classes and r is the expected ratio of a to b.

(a) Chi-Square Limitations. The chi-square test as used for analyzing the results of genetic experiments has two important limitations: (1) it must be used only on the numerical data itself, never on percentages or ratios derived from the data; (2) it cannot properly be used for experiments wherein the expected frequency within any phenotypic class is less than 5.

(b) Corrections for Small Samples. The formula from which the chi-square table is derived is based upon a continuous distribution, namely, that of the "normal" curve (see Chapter 8). Such a distribution might be expected when we plot the heights of a group of people. The most frequent class would be the average height and successively fewer people would be in the taller or shorter phenotypes. All sizes are possible from the shortest to the tallest, i.e., heights form a continuous distribution. However, the kinds of genetic problems in the previous chapters of this book involve separate or discrete phenotypic classes such as blue eyes vs. brown eyes. A correction should therefore be applied in the calculation of chi-square to correct for this lack of continuity. The "Yates Correction for Continuity" is applied as follows, where $|o_i - e_i|$ is an absolute (positive) value:

$$\chi^2(\text{corrected}) = \sum_{i=1}^{n} \frac{[|o_i - e_i| - 0.5]^2}{e_i}$$

$$= \frac{[|o_1 - e_1| - 0.5]^2}{e_1} + \frac{[|o_2 - e_2| - 0.5]^2}{e_2} + \cdots + \frac{[|o_n - e_n| - 0.5]^2}{e_n}$$

$$(2.7)$$

This correction usually makes little difference in the chi-square of most problems, but may become an important factor near the critical values. The Yates correction should be routinely applied whenever only 1 degree of freedom exists, or in small samples where each expected frequency is between 5 and 10. If the corrected and uncorrected methods each lead to the same conclusion, there is no difficulty. However, if these methods do not lead to the same conclusion, then either more data need to be collected or a more sophisticated statistical test should be employed. Do not apply the Yates correction for problems in this book unless requested to do so.

SOLVED PROBLEM 2.18

(a) A coin is tossed 10 times and lands heads up 6 times and tails up 4 times. Are these results consistent with the expected 50 : 50 ratio? (b) If the coin is tossed 100 times with the same relative magnitude of deviation from the expected ratio, is the hypothesis still acceptable? (c) What conclusion can be drawn from the results of parts (a) and (b)?

Solution:

(a)

Phenotypic Classes	Observed (o)	Expected (e)	Deviations ($o-e$)	$(o-e)^2$	$(o-e)^2/e$
Heads	6	1/2(10) = 5	1	1	1/5 = 0.2
Tails	4	1/2(10) = 5	−1	1	1/5 = 0.2
Total	10	10	0		$\chi^2 = 0.4$

Two mathematical checkpoints are always present in the chi-square calculations: (1) the total of the expected column must equal the total observations, and (2) the sum of the deviations should equal 0. The squaring of negative deviations converts all values to a positive scale. The number of degrees of freedom is the number of phenotypes minus 1

$(2-1=1)$. We enter Table 2-3 on the first line (df = 1) and find the computed value of 0.4 lying in the body of the table between the values 0.15 and 0.46, corresponding to the probabilities 0.7 and 0.5 shown at the top of the respective columns. This implies that the magnitude of the deviation in our experimental results could be anticipated by chance alone in more than 50% but less than 70% of an infinite number of experiments of comparable size. This range of values is far above the critical probability value of 0.05 or 5%. Therefore, we accept the null hypothesis and conclude that our coin is conforming to the expected probabilities of heads = 1/2 and tails = 1/2.

(b) In part (a), heads appeared in 60% and tails in 40% of the tosses. The same relative magnitude of deviations will now be considered in a sample of size of 100. In problems such as this, where expected values are equivalent in all the phenotypic classes, chi-square may be calculated more rapidly by adding the squared deviations and making a single division by the expected number.

Phenotypes	o	e	$o-e$	$(o-e)^2$
Heads	60	1/2(100) = 50	10	100
Tails	40	1/2(100) = 50	−10	100
Total	100	100	0	$\chi^2 = 200/50 = 4.0$

With df = 1, this χ^2 value lies between 6.64 and 3.84, corresponding to the probabilities 0.01 and 0.05, respectively. This means that a deviation as large as or larger than the one observed in this experiment is to be anticipated by chance alone in less than 5% of an infinite number of trials of similar size. This is in the "critical region," and we are therefore obliged to reject the null hypothesis and conclude that our coin is not conforming to the expected 50 : 50 ratio. Either of two explanations may be involved: (1) this is not a normal well-balanced coin, or (2) our experiment is among the 1 in 20 (5%) expected to have a large deviation produced by chance alone.

(c) The results of parts (a) and (b) demonstrate the fact that large samples provide a more critical test of a hypothesis than small samples. Proportionately larger deviations have a greater probability of occurring by chance in small samples than in large samples.

SOLVED PROBLEM 2.19

In the garden pea, yellow cotyledon color is dominant to green, and inflated pod shape is dominant to the constricted form. When both of these traits were considered jointly in self-fertilized dihybrids, the progeny appeared in the following numbers : 193 green, inflated : 184 yellow, constricted : 556 yellow, inflated : 61 green, constricted. Test the data for independent assortment.

Solution:

P:	*Gg Cc*	×	*Gg Cc*
	yellow, inflated		yellow, inflated

F_1 (expectations):   9/16 *G- C-*   yellow, inflated
                           3/16 *G- cc*   yellow, constricted
                           3/16 *gg C-*   green, inflated
                           1/16 *gg cc*   green, constricted

Phenotypes	Observed	Expected	Deviation d	d^2	d^2/e
yellow, inflated	556	$9/16(994) = 559.1$	-3.1	9.61	0.017
yellow, constricted	184	$3/16(994) = 186.4$	-2.4	5.76	0.031
green, inflated	193	$3/16(994) = 186.4$	6.6	43.56	0.234
green, constricted	61	$1/16(994) = 62.1$	-1.1	1.21	0.019
Total	994	994.0	0		$\chi^2 = 0.301$

$$df = 4 - 1 = 3 \qquad p > 0.95$$

This is not a significant chi-square value, and thus we accept the null hypothesis, i.e., the magnitude of the deviation $(o - e)$ is to be expected by chance alone in greater than 95% of an infinite number of experiments of comparable size. This is far above the critical value of 5% necessary for acceptance of the hypothesis. We may therefore accept the data as being in conformity with a 9 : 3 : 3 : 1 ratio, indicating that the gene for cotyledon color assorts independently of the gene for pod form.

SOLVED PROBLEM 2.20

A total of 160 families with 4 children each were surveyed with the following results:

Girls	4	3	2	1	0
Boys	0	1	2	3	4
Families	7	50	55	32	16

Is the family distribution consistent with the hypothesis of equal numbers of boys and girls?

Solution: Let $a =$ probability of a girl $= 1/2$, $b =$ probability of a boy $= 1/2$.

$$
\begin{array}{cccccccccc}
(a+b)^4 = & a^4 & + & 4a^3b & + & 6a^2b^2 & + & 4ab^3 & + & b^4 \\
& \text{4 girls} & & \text{3 girls} & & \text{2 girls} & & \text{1 girl} & & \text{0 girls} \\
& \text{0 boys} & & \text{1 boy} & & \text{2 boys} & & \text{3 boys} & & \text{4 boys} \\[4pt]
& (1/2)^4 & + & 4(1/2)^3(1/2) & + & 6(1/2)^2(1/2)^2 & + & 4(1/2)(1/2)^3 & + & (1/2)^4 \\[4pt]
& 1/16 & + & 4/16 & + & 6/16 & + & 4/16 & + & 1/16
\end{array}
$$

Expected number with:
$$
\begin{array}{lcl}
\text{4 girls and 0 boys} & = & 1/16(160) = 10 \\
\text{3 girls and 1 boy} & = & 4/16(160) = 40 \\
\text{2 girls and 2 boys} & = & 6/16(160) = 60 \\
\text{1 girl and 3 boys} & = & 4/16(160) = 40 \\
\text{0 girls and 4 boys} & = & 1/16(160) = 10
\end{array}
$$

Then

$$\chi^2 = \frac{(7-10)^2}{10} + \frac{(50-40)^2}{40} + \frac{(55-60)^2}{60} + \frac{(32-40)^2}{40} + \frac{(16-10)^2}{10} = 9.02$$
$$df = 5 - 1 = 4 \qquad p = 0.05 - 0.10$$

This value is close to, but less than, the critical value 9.49. We may therefore accept the hypothesis, but the test would be more definitive if it could be run on a larger sample. It is a well-known fact that a greater mortality occurs in males than in females and therefore an attempt should be made to ascertain family composition on the basis of sex of all children at birth, including prematures, aborted fetuses, etc.

SOLVED PROBLEM 2.21

In Problem 2.18 (*b*), it was shown that observations of 60 : 40 produced a significant chi-square at the 5% level when uncorrected for continuity. Apply the Yates correction for continuity and retest the data.

Solution:

o	e	$[|o - e| - 0.5]$	$[|o - e| - 0.5]^2$
60	50	$10 - 0.5 = 9.5$	90.25
40	50	$10 - 0.5 = 9.5$	90.25
			$\chi^2 = 180.50/50 = 3.61$

Notice that the correction 0.5 is always applied to the *absolute value* $|o - e|$ of the deviation of expected from observed numbers. This is not a significant chi-square value. Because the data are discrete (jumping from unit to unit) there is a tendency to underestimate the probability, causing too many rejections of the null hypothesis. The Yates correction removes this bias and produces a more accurate test near the critical values (column headed by a probability of 0.05 in Table 2-3).

Pedigree Analysis

A **pedigree** is a systematic listing (either as words or as symbols) of the ancestors of a given individual, or it may be the "family tree" for a large number of individuals. It is customary to represent females as circles and males as squares. Matings are shown as horizontal lines between two individuals. The offspring of a mating are connected by a vertical line to the mating line. Different shades or colors added to the symbols can represent various phenotypes. Each generation is listed on a separate row labeled with Roman numerals. Individuals within a generation receive Arabic numerals. Pedigree analysis is used in place of breeding studies, particularly in humans, where experimental matings are not possible. A pedigree can help determine the genetic basis of a particular trait or disease.

EXAMPLE 2.39
Let solid symbols represent black-coated guinea pigs and open symbols represent white-coated guinea pigs; circles represent females and squares represent males.

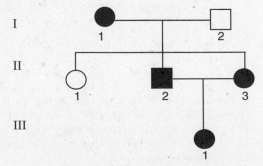

Summary of pedigree analysis:

Individuals	Phenotype	Genotype
I1	Black, female	Bb
I2	White, male	bb
II1	White, female	bb
II2	Black, male	Bb
II3	Black, female	Bb
III1	Black, female	$B\text{-}^*$

* The dash indicates that the genotype could
be either homozygous or heterozygous.
How would you determine this? (*Hint*: see Example 2.10.)

SOLVED PROBLEM 2.22

The black hair of guinea pigs is produced by a dominant gene B and white by its
recessive allele b. Unless there is evidence to the contrary, assume that III1 and II4 do
not carry the recessive allele. Calculate the probability that an offspring of III1 $\times$
III2 will have white hair.

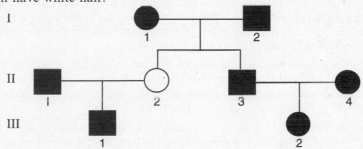

Solution: Both I1 and I2 must be heterozygous (Bb) in order to have the white (bb) offspring
II2. If III1 or III2 had been white, this would constitute evidence that II1 or II4 were heterozygous. In the absence of this evidence, the problem tells us to assume that II1 and II4 are
homozygous (BB). If the offspring of III1 $\times$ III2 is to be white, then both III1 and III2 would
have to be heterozygous (Bb). In this case, II3 would also have to be heterozygous in order to
pass the recessive allele on to III2. Under the conditions of the problem, we are certain that
III1 is heterozygous because his parents (II1 $\times$ II2) are $BB \times bb$. We notice that II3 is black.
The probability that *black* progeny from I1 $\times$ I2 are heterozygous is 2/3. If II3 is heterozygous, the probability that III2 is heterozygous is 1/2. If III2 is heterozygous, there is a 25%
chance that the offspring of III1 $\times$ III2 will be white (bb). Thus, the combined probability that
II3 is heterozygous *and* III2 is heterozygous *and* producing a white offspring is the product of
the independent probabilities $= (2/3)(1/2)(1/4) = 2/24 = 1/12$.

Supplementary Problems

DOMINANT AND RECESSIVE ALLELES

2.23. Several black guinea pigs of the same genotype were mated and they produced 29 black and 9 white offspring. What would you predict the genotypes of the parents to be?

2.24. If a black female guinea pig is testcrossed and produces at least one white offspring, determine (*a*) the genotype and phenotype of the sire (male parent) that produced the white offspring, (*b*) the genotype of this female.

2.25. In *Drosophila,* sepia-colored eyes are due to a recessive allele s and wild type (red eye color) to its dominant allele s^+. If sepia-eyed females are crossed to pure wild-type males, what phenotypic and genotypic ratios are expected if the F_2 males are backcrossed to the sepia-eyed parental females?

2.26. The lack of pigmentation, called albinism, in humans is the result of a recessive allele a and normal pigmentation is the result of its dominant allele A. Two normal parents have an albino child. Determine the probability that (*a*) the next child is albino, (*b*) the next two children are albinos. (*c*) What is the chance of these parents producing two children, one albino and the other normal?

2.27. Short hair is due to a dominant gene L in rabbits, and long hair to its recessive allele l. A cross between a short-haired female and a long-haired male produces a litter of one long-haired and seven short-haired bunnies. (*a*) What are the genotypes of the parents? (*b*) What phenotypic ratio was expected in the offspring generation? (*c*) How many of the eight bunnies were expected to be long-haired?

2.28. Black wool of sheep is due to a recessive allele b and white wool to its dominant allele B. A white ram (male) is crossed to a white ewe (female), both animals carrying the allele for black. They produce a white male lamb that is then backcrossed to the female parent. What is the probability of the backcross offspring being black?

2.29. In foxes, silver-black coat color is governed by a recessive allele b and red color by its dominant allele B. Determine the genotypic and phenotypic ratios expected from the following matings: (*a*) pure red × carrier red, (*b*) carrier red × silver-black, (*c*) pure red × silver-black.

CODOMINANCE AND INCOMPLETE DOMINANCE

2.30. When chickens with splashed white feathers are crossed with black-feathered birds, their offspring are all slate blue (Blue Andalusian). When Blue Andalusians are crossed among themselves, they produce splashed white, blue, and black offspring in the ratio of $1 : 2 : 1$, respectively. (*a*) How are these feather traits inherited? (*b*) Using any appropriate symbols, indicate the genotypes for each phenotype.

2.31. The shape of radishes may be long ($S^L S^L$), round ($S^R S^R$), or oval ($S^L S^R$). If long radishes are crossed to oval radishes and the F_1 then allowed to cross at random among themselves, what phenotypic ratio is expected in the F_2?

2.32. A palomino horse is a hybrid exhibiting a golden color with lighter mane and tail. A pair of codominant alleles (D^1 and D^2) is known to be involved in the inheritance of these coat colors. Genotypes homozygous for the D^1 allele are chestnut-colored (reddish), heterozygous genotypes are palomino-colored, and genotypes homozygous for the D^2 allele are almost white and called *cremello.* (*a*) From matings between palominos, determine the expected palomino : nonpalomino ratio among the offspring. (*b*) What percentage of the nonpalomino offspring in part (*a*) *will* breed true? (*c*) What kind of mating will produce only palominos?

LETHAL ALLELES

2.33. Chickens with shortened wings and legs are called "creepers." When creepers are mated to normal birds they produce creepers and normals with equal frequency. When creepers are mated to creepers they produce two creepers to one normal. Crosses between normal birds produce only normal progeny. How can these results be explained?

2.34. In the Mexican Hairless breed of dogs, the hairless condition is produced by the heterozygous genotype (Hh). Normal dogs are homozygous recessive (hh). Puppies homozygous for the H allele are usually born dead, with abnormalities of the mouth and absence of external ears. If the average litter size at weaning is six in matings between hairless dogs, what would be the average expected *number* of hairless and normal offspring at weaning from matings between hairless and normal dogs?

2.35. A pair of codominant alleles are known to govern cotyledon leaf color in soybeans. The homozygous genotype $C^G C^G$ produces dark-green leaves, the heterozygous genotype $C^G C^Y$ produces light-green leaves, and the other homozygous genotype $C^Y C^Y$ produces yellow leaves so deficient in chloroplasts that seedlings do not grow to maturity. If dark-green plants are pollinated only by light-green plants and the F_1 crosses are made at random to produce an F_2, what phenotypic and genotypic ratios would be expected in the mature F_2 plants?

2.36. Thalassemia is a hereditary disease of the blood of humans that results in anemia. Severe anemia (thalassemia major) is found in homozygotes ($T^M T^M$) and a milder form of anemia (thalassemia minor) is found in heterozygotes ($T^M T^N$). Normal individuals are homozygous $T^N T^N$. If all individuals with thalassemia major die before sexual maturity, (*a*) what proportion of the adult F_1 from marriages of thalassemia minors to normals would be expected to be normal, (*b*) what fraction of the adult F_1 from marriages of minors to minors would be expected to be anemic?

MULTIPLE ALLELES

2.37. A multiple allelic series is known in the Chinese primrose where A (Alexandria type = white eye) > a^n (normal type = yellow eye) > a (Primrose Queen type = large yellow eye). List all of the genotypes possible for each of the phenotypes in this series.

2.38. Plumage color in mallard ducks is dependent upon a set of three alleles: M^R for restricted mallard pattern, M for mallard, and m for dusky mallard. The dominance hierarchy is $M^R > M > m$. Determine the genotypic and phenotypic ratios expected in the F_1 from the following crosses: (*a*) $M^R M^R \times M^R M$, (*b*) $M^R M^R \times M^R m$, (*c*) $M^R M \times M^R m$, (*d*) $M^R m \times Mm$, (*e*) $Mm \times mm$.

2.39. A number of self-incompatibility alleles are known in clover such that the growth of a pollen tube down the style of a diploid plant is inhibited when the latter contains the same self-incompatibility allele as that in the pollen tube. Given a series of self-incompatibility alleles S^1, S^2, S^3, S^4, what genotypic ratios would be expected in embryos and in endosperms of seeds from the following crosses?

	Seed Parent	Pollen Parent
(*a*)	$S^1 S^4$	$S^3 S^4$
(*b*)	$S^1 S^2$	$S^1 S^2$
(*c*)	$S^1 S^3$	$S^2 S^4$
(*d*)	$S^2 S^3$	$S^3 S^4$

2.40. The coat colors of many animals exhibit the "agouti" pattern, which is characterized by a yellow band of pigment near the tip of the hair. In rabbits, a multiple allelic series is known where the genotypes $E^D E^D$ and $E^D e$ produce only black (nonagouti), but the heterozygous genotype $E^D E$ produces black with a trace of agouti. The genotypes EE or Ee produce full color, and the recessive genotype ee produces reddish-yellow. What phenotypic and genotypic ratios would be expected in the F_1 and F_2 from the cross (a) $E^D E^D \times Ee$, (b) $E^D e \times ee$?

2.41. The inheritance of coat colors of cattle involves a multiple allelic series with a dominance hierarchy as follows: $S > s^h > s^c > s$. The S allele puts a band of white color around the middle of the animal and is referred to as a Dutch belt; the s^h allele produces Hereford-type spotting; solid color is a result of the s^h allele; and Holstein-type spotting is due to the s allele. Homozygous Dutch-belted males are crossed to Holstein-type spotted females. The F_1 females are crossed to a Hereford-type spotted male of genotype $s^h s^c$. Predict the genotypic and phenotypic frequencies in the progeny.

2.42. The genetics of the ABO human blood groups was presented in Example 2.19. A man of blood group B is being sued by a woman of blood group A for paternity. The woman's child is blood group O. (a) Is this man the father of this child? Explain. (b) If this man actually is the father of this child, specify the genotypes of both parents. (c) If it was impossible for this group B man to be the father of a type O child, regardless of the mother's genotype, specify his genotype. (d) If a man was blood group AB, could he be the father of a group O child?

STATISTICAL DISTRIBUTIONS

2.43. Black hair in the guinea pig is dominant to white hair. In families of five offspring where both parents are heterozygous black, with what frequency would we expect to find (a) three whites and two blacks, (b) two whites and three blacks, (c) one white and four blacks, (d) all whites?

2.44. Assuming that boys and girls are equally frequent in a population, in families of five children, what is the probability of finding (a) three or more boys, (b) three or more boys or three or more girls?

2.45. A dozen strains of corn are available for a cross-pollination experiment. How many different ways can these strains be paired?

2.46. Five coat colors in mice are agouti, cinnamon, black, chocolate, and albino. (a) List all of the possible crosses between different phenotypes. (b) Verify the number of different crosses by applying formula (2.1).

2.47. In mice litters of size eight, determine (a) the most frequently expected number of males and females, (b) the term of the binomial that part (a) represents, (c) the percentage of all litters of size eight expected to have four males and four females.

DIHYBRID CROSSES WITH DOMINANT AND RECESSIVE ALLELES

2.48. The position of the flower on the stem of the garden pea is governed by a pair of alleles. Flowers growing in the axils (upper angle between petiole and stem) are produced by the action of a dominant allele T, those growing only at the tip of the stem by its recessive allele t. Colored flowers are produced by a dominant gene C, and white flowers by its recessive allele c. A dihybrid plant with colored flowers in the leaf axils is crossed to a pure strain of the same phenotype. What genotypic and phenotypic ratios are expected in the F_1 progeny?

2.49. In summer squash, white fruit color is governed by a dominant allele W and yellow fruit color by its recessive allele w. A dominant allele S at another locus produces disk-shaped fruit and its recessive allele s yields sphere-shaped fruit. If a homozygous white disk variety of genotype $WWSS$ is crossed with a homozygous yellow sphere variety $wwss$, the F_1 are all white disk dihybrids of genotype $WwSs$. If the F_1 is allowed to mate at random, what would be the phenotypic ratio expected in the F_2 generation?

2.50. In *Drosophila*, ebony body color is produced by a recessive gene a and wild-type (gray) body color by its dominant allele a^+. Vestigial wings are governed by a recessive gene vg, and normal wing size (wild type) by its dominant allele vg^+. If wild-type dihybrid flies are crossed and produce 256 progeny, how many of these progeny flies are expected in each phenotypic class?

2.51. Short hair in rabbits is governed by a dominant gene L and long hair by its recessive allele l. Black hair results from the action of the dominant genotype $B-$ and brown from the recessive genotype *bb*. (*a*) In crosses between dihybrid short-haired, black and homozygous short-haired, brown rabbits, what genotypic and phenotypic ratios are expected among their progeny? (*b*) Determine the expected genotypic and phenotypic ratios in progeny from the cross *LlBb × Llbb*.

2.52. The genetic information for the following eight parts is found in Problem 2.51. (*a*) What phenotypic ratio is expected among progeny from crosses of *LlBb × LlBb*? (*b*) What percentage of the F_1 genotypes in part (*a*) breeds true (i.e., what percentage is of homozygous genotypes)? (*c*) What percentage of the F_1 genotypes is heterozygous for only one pair of genes? (*d*) What percentage of the F_1 genotypes is heterozygous at both loci? (*e*) What percentage of the F_1 genotypes could be used for testcross purposes (i.e., homozygous double-recessive)? (*f*) What percentage of the F_1 progeny could be used for testcross purposes at the *B* locus (i.e., homozygous recessive *bb*)? (*g*) What percentage of all short-haired F_1 individuals is expected to be brown? (*h*) What percentage of all black F_1 individuals will breed true for both black and short hair?

2.53. The presence of feathers on the legs of chickens is due to a dominant allele *F* and clean legs to its recessive allele *f*. Pea-comb shape is produced by another dominant allele *P* and single comb by its recessive allele *p*. In crosses between pure feathered-leg, single-combed individuals and pure pea-combed, clean-leg individuals, suppose that only the single-combed, feathered-leg F_2 progeny are saved and allowed to mate at random. What genotypic and phenotypic ratios would be expected among the progeny (F_3)?

2.54. List all the different gametes produced by the following individuals: (*a*) *AA BB Cc* (*b*) *aa Bb Cc*, (*c*) *Aa Bb cc Dd*, (d) *AA Bb Cc dd Ee Ff*.

2.55. The normal cloven-footed condition in swine is produced by the homozygous recessive genotype *mm*. A mule-footed condition is produced by the dominant genotype *M-*. White coat color is governed by the dominant allele of another locus *B* and black by its recessive allele *b*. A white, mule-footed sow (female) is mated to a black, cloven-footed boar (male) and produces several litters. Among 26 offspring produced by this mating, all were found to be white with mule feet. (*a*) What is the most probable genotype of the sow? (*b*) The next litter produced eight white, mule footed offspring and one white, cloven-footed pig. Now, what is the most probable genotype of the sow?

2.56. A white, mule-footed boar (see Problem 2.55) is crossed to a sow of the same phenotype. Among the F_1 offspring there were found to be 6 white, cloven-footed : 7 black, mule-footed : 15 white, mule-footed : 3 black, cloven-footed pigs. (*a*) If all the black, mule-footed F_1 offspring from this type of mating were to be testcrossed, what phenotypic ratio would be expected among the testcross progeny? (*b*) If the sow were to be testcrossed, what phenotypic ratio of progeny would be expected?

MODIFIED DIHYBRID RATIOS

2.57. In peaches, the homozygous genotype $G^O G^O$ produces oval glands at the base of the leaves, the heterozygous genotype $G^O G^A$ produces round glands, and the homozygous genotype $G^A G^A$ results in the absence of glands. At another locus, a dominant gene *S* produces fuzzy peach skin and its recessive allele *s* produces smooth (nectarine) skin. A homozygous variety with oval glands and smooth skin is crossed to a homozygous variety with fuzzy skin and lacking glands at the base of its leaves. What genotypic and phenotypic proportions are expected in the F_2?

2.58. In Shorthorn cattle, coat colors are governed by a codominant pair of alleles C^R and C^W. The homozygous genotype $C^R C^R$ produces red, the other homozygote produces white, and the heterozygote produces roan (a mixture of red and white). The presence of horns is produced by the homozygous recessive genotype *pp* and the polled condition by its dominant allele *P*. If roan cows heterozygous for the horned gene are mated to a horned, roan bull, what phenotypic ratio is expected in the offspring?

2.59. A gene locus with codominant alleles is known to govern feather color in chickens such that the genotype $F^B F^B$ = black, $F^W F^W$ = splashed white, and $F^B F^W$ = blue. Another locus with codominant alleles governs feather morphology such that $M^N M^N$ = normal feather shape, $M^N M^F$ = slightly abnormal feathers called "mild frizzle," and $M^F M^F$ = grossly abnormal feathers called "extreme frizzle." If blue, mildly frizzled birds are crossed among themselves, what phenotypic proportions are expected among their offspring?

2.60. In the above problem, if all the blue offspring with normal feathers and all the splashed-white, extremely frizzled offspring are isolated and allowed to mate at random, what phenotypic ratio would be expected among their progeny?

2.61. The shape of radishes may be long (LL), round ($L'L'$), or oval (LL'). Color may be red (RR), white ($R'R'$), or purple (RR'). If a long, white strain is crossed with a round, red strain, what phenotypic proportions are expected in the F_1 and F_2?

2.62. Suppose that two strains of radishes are crossed (see the above problem) and produce a progeny consisting of 16 long white, 31 oval purple, 16 oval white, 15 long red, 17 oval red, and 32 long purple. What would be the phenotypes of the parental strains?

2.63. A dominant gene K in mice produces a kinked tail; recessive genotypes at this locus kk have normal tails. The homozygous condition of another locus AA produces a gray color called agouti; the heterozygous condition $A^y A$ produces yellow color; the homozygous genotype $A^y A^y$ is lethal. (a) If yellow mice, heterozygous for kinky tail, are crossed together, what phenotypic proportions are expected in their offspring? (b) What proportion of the offspring is expected to be of genotype $A^y A Kk$? (c) If all the yellow offspring were allowed to mate at random, what would be the genotypic and phenotypic ratios among their adult progeny?

2.64. An incompletely dominant gene N in the Romney Marsh breed of sheep causes the fleece of homozygotes to be "hairy," i.e., containing fibers lacking the normal amount of crimp. Normal wool is produced by the homozygous genotype $N'N'$. Heterozygotes NN' can be distinguished at birth by the presence of large, medulated fibers called "halo-hairs" scattered over the body. A gene known as "lethal gray" causes homozygous gray fetuses ($G^l G^l$) to die before 15 weeks in gestation. The heterozygous genotype $G^l G$ produces gray fleece, and the homozygous genotype GG produces black. If heterozygous halo-haired, gray individuals are mated together, (a) what would be the phenotypic proportions expected in the live progeny, (b) what proportion of the live progeny would carry the lethal gene, (c) what proportion of the live progeny with halo-hairs would carry the lethal gene, (d) what proportion of all the zygotes would be expected to be of genotype $NN'G^l G^h$?

2.65. Tay-Sachs disease is a recessive hereditary abnormality causing death within the first few years of life only when homozygous (ii). The dominant condition at this locus produces a normal phenotype (I-). Abnormally shortened fingers (brachyphalangy) is thought to be due to a genotype heterozygous for a lethal gene (BB^L), the homozygote (BB) being normal, and the other homozygote ($B^L B^L$) being lethal. What are the phenotypic expectations among teenage children from parents who are both brachyphalangic and heterozygous for Tay-Sachs disease?

2.66. In addition to the gene governing infantile amaurotic idiocy in the above problem, the recessive genotype of another locus (jj) results in death before age 18 due to a condition called "juvenile amaurotic idiocy." Only individuals of genotype I-J- will survive to adulthood. (a) What proportion of the children from parents of genotype $IiJj$ would probably not survive to adulthood? (b) What proportion of the adult survivors in part (a) would not be carriers of either hereditary abnormality?

2.67. A genetic condition on chromosome 2 in the fruit fly *Drosophila melanogaster* is lethal when homozygous (Pm/Pm), but when heterozygous (Pm/Pm^+) produces a purplish eye color called "plum." The other homozygous condition (Pm^+/Pm^+) produces wild-type eye color. On chromosome 3, a gene called "stubble" produces short, thick bristles when heterozygous (Sb/Sb^+) but is lethal when homozygous (Sb/Sb). The homozygous condition of its alternative allele (Sb^+/Sb^+), produces bristles of normal size (wild type). (a) What phenotypic ratio is expected among progeny from crosses between plum, stubble parents? (b) If the offspring of part (a) are allowed to mate at random to produce an F_2, what phenotypic ratio is expected?

2.68. Feather color in chickens is governed by a pair of codominant alleles such that $F^B F^B$ produces black, $F^W F^W$ produces splashed white, and $F^B F^W$ produces blue. An independently segregating locus governs the length of leg; CC genotypes possess normal leg length, CC^L genotypes produce squatty, short-legged types called "creepers," but homozygous $C^L C^L$ genotypes are lethal. Determine the kinds of progeny phenotypes and their expected ratios that crosses between dihybrid blue creepers are likely to produce.

HIGHER COMBINATIONS

2.69. The seeds from Mendel's tall plants were round and yellow, all three characters due to a dominant gene at each of three independently assorting loci. The recessive genotypes dd, ww, and gg produce dwarf plants with wrinkled and green seeds, respectively. (*a*) If a pure tall, wrinkled, yellow variety is crossed with a pure dwarf, round, green variety, what phenotypic ratio is expected in the F_1 and F_2? (*b*) What percentage of the F_2 is expected to be of genotype $DdWWgg$? (*c*) If all the dwarf, round, green individuals in the F_2 are isolated and artificially crossed at random, what phenotypic ratio of offspring is expected?

2.70. The coat colors of mice are known to be governed by several genes. The presence of a yellow band of pigment near the tip of the hair is called "agouti" pattern and is produced by the dominant allele A. The recessive condition at this locus (aa) does not have this subapical band and is termed nonagouti. The dominant allele of another locus B produces black and the recessive genotype bb produces brown. The homozygous genotype $c^h c^h$ restricts pigment production to the extremities in a pattern called Himalayan, whereas the genotype C- allows pigment to be distributed over the entire body. (*a*) In crosses between pure brown, agouti, Himalayan, and pure black mice, what are the phenotypic expectations of the F_1 and F_2? (*b*) What proportion of the black-agouti, full-colored F_2 would be expected to be of genotype $AaBBCc$? (*c*) What percentage of all the Himalayans in the F_2 would be expected to show brown pigment? (*d*) What percentage of all the agoutis in the F_2 would be expected to exhibit black pigment?

2.71. In addition to the information given in the problem above, a fourth locus in mice is known to govern the density of pigment deposition. The genotype D- produces full color, but the recessive genotype dd produces a dilution of pigment. Another allele at this locus, d^l is lethal when homozygous, produces a dilution of pigment in the genotype dd^l, and produces full color when in heterozygous condition with the dominant allele Dd^l. (*a*) What phenotypic ratio would be expected among the live F_2 progeny if the F_1 from the cross $aabbCCDd \times AABBccdd^l$ were allowed to mate at random? (*b*) What proportion of the live F_2 would be expected to be of genotype $AABbccdd^l$?

2.72. In the parental cross $AABBCCDDEE \times aabbccddee$, (*a*) how many different F_1 gametes can be formed, (*b*) how many different genotypes are expected in the F_2, (*c*) how many squares of a Punnett square would be necessary to accommodate the F_2?

2.73. A pure strain of Mendel's peas, dominant for all seven of his independently assorting genes, was testcrossed. (*a*) How many different kinds of gametes could each of the parents produce? (*b*) How many different gametes could the F_1 produce? (*c*) If the F_1 was testcrossed, how many phenotypes would be expected in the offspring and in what proportions? (*d*) How many genotypes would be expected in the F_2? (*e*) How many combinations of F_1 gametes are theoretically possible (considering, e.g., $AABBCCDDEEFFGG$ sperm nucleus $\times$ $aabbccddeeffgg$ egg nucleus, a different combination than $AABBCCDDEEFFGG$ egg nucleus $\times$ $aabbccddeeffgg$ sperm nucleus)? (*f*) How many different kinds of matings could theoretically be made among the F_2? [*Hint*: See solution to Problem 2.14(*c*)].

TESTING GENETIC RATIOS

2.74. Determine the number of degrees of freedom when testing the ratios (*a*) $3:1$ (*b*) $9:3:3:1$ (*c*) $1:2:1$ (*d*) $9:3:4$. Find the number of degrees of freedom in applying a chi-square test to the results from (*e*) testcrossing a dihybrid, (*f*) testcrossing a trihybrid, (*g*) trihybrid $\times$ trihybrid cross.

2.75. Two phenotypes appear in an experiment in the ratio 16 : 4. (*a*) How well does this sample fit a 3 : 1 ratio? Would a sample with the same proportional deviation fit a 3 : 1 ratio if it were (*b*) 10 times larger than (*a*), (*c*) 20 times larger than (*a*)?

2.76. The flowers of four o'clock plants may be red, pink, or white. Reds crossed to whites produced only pink offspring. When pink-flowered plants were crossed they produced 113 red, 129 white, and 242 pink. It is hypothesized that these colors are produced by a single-gene locus with codominant alleles. Is this hypothesis acceptable on the basis of a chi-square test?

2.77. A heterozygous genetic condition called "creeper" in chickens produces shortened and deformed legs and wings, giving the bird a squatty appearance. Matings between creepers produced 775 creeper : 388 normal progeny. (*a*) Is the hypothesis of a 3 : 1 ratio acceptable? (*b*) Does a 2 : 1 ratio fit the data better? (*c*) What phenotype is probably produced by the gene for creeper when in homozygous condition?

2.78. Among fraternal (nonidentical, dizygotic) twins, the expected sex ratio is 1 MM : 2 MF : 1 FF (M = male, F = female). A sample from a sheep population contained 50 MM, 142 MF, and 61 FF twin pairs. (*a*) Do the data conform within statistically acceptable limits to the expectations? (*b*) If identical (monozygotic) twin pairs = total pairs − (2 × MF pairs), what do the data indicate concerning the frequency of monozygotic sheep twins?

2.79. A total of 320 families with six children each were surveyed with the results shown below. Does this distribution indicate that boys and girls are occurring with equal frequency?

No. of girls	6	5	4	3	2	1	0
No. of boys	0	1	2	3	4	5	6
No. of families	6	33	71	99	69	37	5

2.80. In guinea pigs, it is hypothesized that a dominant allele *L* governs short hair and its recessive allele *l* governs long hair. Codominant alleles at an independently assorting locus are assumed to govern hair color, such that $C^y C^y$ = yellow, $C^y C^w$ = cream, and $C^w C^w$ = white. From the cross $Ll\ C^y C^w \times Ll\ C^y C^w$, the following progeny were obtained: 50 short cream : 21 short yellow : 23 short white : 21 long cream : 7 long yellow : 6 long white. Are the data consistent with the genetic hypothesis?

2.81. Observations of 30 : 3 in a genetic experiment are postulated to be in conformity with a 3 : 1 ratio. Is a 3 : 1 ratio acceptable at the 5% level on the basis of (*a*) an uncorrected chi-square test, (*b*) a corrected chi-square test? [*Hint*: Corrected indicates use of Yates correction factor of continuity.]

PEDIGREE ANALYSIS

2.82. The phenotypic expression of a dominant gene in Ayrshire cattle is a notch in the tips of the ears. In the pedigree below, where solid symbols represent notched individuals, determine the probability of notched progeny being produced from the matings (*a*) III1 × III3 (*b*) III2 × III3 (*c*) III3 × III4 (*d*) III1 x III5 (*e*) III2 × III5

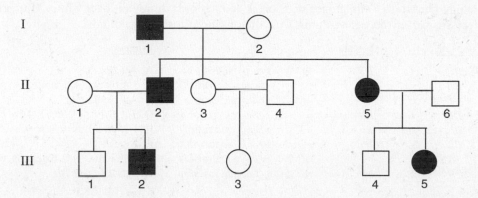

2.83. A multiple allelic series in dogs governs the distribution of coat-color pigments. The allele A^s produces an even distribution of dark pigment over the body; the allele a^y reduces the intensity of pigmentation and produces sable or tan-colored dogs; the allele a^t produces spotted patterns such as tan and black, tan and brown, etc. The dominance hierarchy is $A^s > a^y > a^t$. Given the following family pedigree, (*a*) determine the genotypes of all the individuals insofar as possible, (*b*) calculate the probability of spotted offspring being produced by mating III1 by III2, (*c*) find the fraction of the dark-pigmented offspring from I1 × III3 that is expected to be heterozygous.

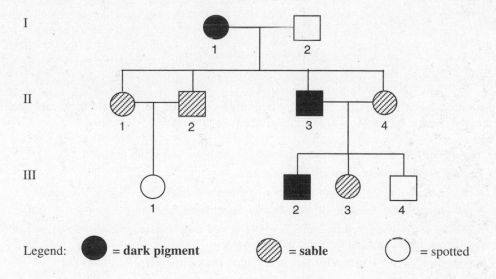

Legend: ● = **dark pigment**   ▨ = **sable**   ○ = spotted

Review Questions

Matching Questions I   In guinea pigs, black coat color is dominant over white. Match the correct answer in the right column with the question in the left column.

	Male		Female			Progeny
1.	*BB*	×	*BB*	=	*genotypic ratio?*	A. All *Bb*
2.	*BB*	×	*Bb*	=	*phenotypic* ratio?	B. 3/4 black : 1/4 white
3	*BB*	×	*bb*	=	*genotypic ratio?*	C. All black
4.	*BB*	×	*Bb*	=	*genotypic ratio?*	D. 1/4 *BB* : 1/2 *Bb* : 1/4 *bb*
5.	*Bb*	×	*Bb*	=	*phenotypic* ratio?	E. 1/2 *Bb* : 1/2 *bb*
6.	*Bb*	×	*Bb*	=	*genotypic ratio?*	F. All *BB*
7.	*bB*	×	*bb*	=	*phenotypic* ratio?	G. 1/2 white : 1/2 black
8.	*bB*	×	*bb*	=	*genotypic ratio?*	H. 1/2 *BB* : 1/2 *Bb*

Matching Questions II   In guinea pigs, black (*B-*) is dominant to white (*bb*). Assorting independently on a different pair of homologues, short hair (*S-*) is dominant to long hair (*ss*). In the parental generation, pure (homozygous) black, short-haired pigs are crossed to white, long-haired pigs. In the F_2 we expect the following:

1. Black, short		A. 1/16	
2. Black, long		B. 1/8	
3. White, short		C. 3/16	
4. White, long		D. 1/2	
5. *BbSs*		E. 1/3	
6. *BBLL*		F. 1/4	
7. *Bbss*		G. 3/4	
8. *bbSs*		H. 9/16	

Vocabulary   For each of the following definitions, give the appropriate term and spell it correctly. Terms are single words unless indicated otherwise.

1. Any measurable or distinctive characteristic or trait possessed by an organism.
2. The genetic endowment of an individual or cell.
3. A cell produced by the union of gametes carrying identical alleles.
4. A cell produced by the union of gametes carrying different alleles.
5. Adjective descriptive of an allele that is not expressed in a heterozygote; also descriptive of the phenotype produced when the allele is homozygous.
6. Any phenotype that is extremely rare in a natural population. (One or two words.)
7. Adjective describing any pair of alleles that interact in the heterozygous condition to produce a phenotype different from those of the respective homozygotes. (One or two words.)
8. Any gene that when homozygous results in the death of the individual during development.
9. The proportion of individuals of a specified genotype that show the expected phenotype.
10. The degree of effect produced by a given genotype under a given set of environmental conditions or over a range of environmental conditions.

Multiple-Choice Questions   Choose the one best answer.

Questions 1–5 use the following information. In guinea pigs, black coat color (governed by gene B) is a dominant trait, and white (attributed to allele b) is a recessive trait.

1. A black female is testcrossed, producing six black offspring. The probability that a heterozygous black female would do this by chance alone is approximately   (*a*) 50%   (*b*) 25%   (*c*) 1%   (*d*) cannot be determined from the information given   (*e*) none of the above

2. A mating that is expected to produce 50% homozygotes and 50% heterozygotes is   (*a*) $BB \times Bb$ (*b*) $Bb \times Bb$ (*c*) $bb \times Bb$ (*d*) two of the above   (*e*) matings (*a*), (*b*), and (*c*) above

3. When heterozygous black pigs are intercrossed, approximately what fraction of the black progeny are expected to be homozygous?   (*a*) 4   (*b*) 2/3   (*c*) 2   (*d*) 3   (*e*) none of the above

4. How many genetically different kinds of matings can be made in a population containing these two alleles (ignoring reciprocal crosses)?   (*a*) 4   (*b*) 6   (*c*) 8   (*d*) more than 8   (*e*) none of the above

5. When heterozygous black pigs are intercrossed the chance of the first two offspring being black is   (*a*) more than 75%   (*b*) 56%   (*c*) 44%   (*d*) 6%   (*e*) none of the above

6. The ABO blood groups of humans are determined by three alleles (Example 2.19). How many genotypes are possible for these phenotypes?   (*a*) 3   (*b*) 4   (*c*) 6   (*d*) 8   (*e*) none of the above

7. A mother of blood group O has a group O child. The father could be   (*a*) A or B or O   (*b*) O only   (*c*) A or B   (*d*) AB only   (*e*) none of the above

8. How many different genotypes can exist in a population with the dominance hierarchy $g^a > g^b > g^c > g^d$?   (*a*) 6   (*b*) 8   (*c*) 16   (*d*) more than 16   (*e*) none of the above

9. If an individual of genotype $AaBbCcDd$ is testcrossed, how many different phenotypes can appear in the progeny?   (*a*) 4   (*b*) 8   (*c*) 12   (*d*) 16   (*e*) none of the above

10. If individuals of genotype $AaBbCc$ are intercrossed, how many different phenotypes can appear in their offspring?   (*a*) 3   (*b*) 6   (*c*) 8   (*d*) 16   (*e*) none of the above

11. If individuals of genotype $AaBbCc$ are intercrossed, how many different genotypes can occur in their progeny?   (*a*) 6   (*b*) 8   (*c*) 16   (*d*) 21   (*e*) none of the above

12. The minimum progeny population size allowing for random union of all kinds of gametes from $AaBbCc$ parents is   (*a*) 9   (*b*) 27   (*c*) 64   (*d*) more than 100   (*e*) none of the above

Questions 13 and 14 use the following information. Given that A^1A^1 = lethal, A^1A^2 = gray, A^2A^2 = black, B^1B^1 = long hair, B^1B^2 = short hair, B^2B^2 = very short hair (fuzzy), and parents that are $A^1A^2B^1B^2$:

13. The fraction of the adult offspring that is expected to be gray, fuzzy is   (*a*) 1/4   (*b*) 1/2   (*c*) 2/3   (*d*) 3/4   (*e*) none of the above

14. If fuzzy is lethal shortly after birth, the fraction of the adult progeny expected to be black, short is   (*a*) 1/4   (*b*) 1/3   (*c*) 1/2   (*d*) 2/3   (*e*) none of the above

Answers to Supplementary Problems

2.23. $Bb \times Bb$

2.24. (*a*) bb = white   (*b*) Bb

2.25. 1/2 wild-type s^+s : 1/2 sepia ss

2.26. (*a*) 1/4   (*b*) 1/16   (*c*) $2(3/4 \times 1/4) = 3/8$

2.27. (*a*) Ll female $\times$ ll male   (*b*) 1 short : 1 long   (*c*) 4

2.28. 1/6

2.29. (*a*) 1/2 BB : 1/2 Bb = all red   (*b*) 1/2 Bb = red : 1/2 bb = silver-black   (*c*) all Bb = red

2.30. (*a*) Single pair of codominant alleles (*b*) $F^S F^S$ = splashed-white : $F^S F^B$ = Blue Andalusian : $F^B F^B$ = black

2.31. 9/16 long : 6/16 oval : 1/16 round

2.32. (*a*) 1 palomino : 1 nonpalomino   (*b*) 100%, $D^1 D^1 \times D^1 D^1$ = all $D^1 D^1$ (chestnut); similarly $D^2 D^2 \times D^2 D^2$ = all $D^2 D^2$ (cremello)   (*c*) $D^1 D^1$ (chestnut) $\times$ $D^2 D^2$ (cremello)

2.33. Creepers are heterozygous. Normal birds and lethal zygotes are homozygous for alternative alleles. One of the alleles is dominant with respect to the creeper phenotype; the other allele is dominant with respect to viability.

2.34. 4 normal : 4 hairless

2.35. 9/15 dark-green $C^G C^G$: 6/15 light-green $C^G C^Y$

2.36. (*a*) 1/2   (*b*) 2/3

2.37. Alexandria type (white eye) = AA, Aa^n, Aa; normal type (yellow eye) = $a^n a^n$, $a^n a$; Primrose Queen type (large yellow eye) = aa

2.38. (*a*) 1/2 $M^R M^R$: 1/2 $M^R M$; all restricted   (*b*) 1/2 $M^R M^R$: 1/2 $M^R m$; all restricted   (*c*) 1/4 $M^R M^R$: 1/4 $M^R m$: 1/4 $M^R M$: 1/4 Mm; 3/4 restricted : 1/4 mallard   (*d*) 1/4 $M^R M$: 1/4 $M^R m$: 1/4 Mm : 1/4 mm; 1/2 restricted : 1/4 mallard : 1/4 dusky   (*e*) 1/2 Mm = mallard : 1/2 mm = dusky

2.39. (*a*) Embryos = 1/2 $S^1 S^3$: 1/2 $S^3 S^4$
Endosperms = 1/2 $S^1 S^1 S^3$: 1/2 $S^4 S^4 S^3$

(*b*) None

(*c*) Embryos = 1/4 $S^1 S^2$: 1/4 $S^1 S^4$: 1/4 $S^3 S^2$: 1/4 $S^3 S^4$
Endosperms = 1/4 $S^1 S^1 S^2$: 1/4 $S^1 S^1 S^4$: 1/4 $S^3 S^3 S^2$: 1/4 $S^3 S^3 S^4$

(*d*) Embryos = 1/2 $S^2 S^4$: 1/2 $S^3 S^4$,
Endosperm = 1/2 $S^2 S^2 S^4$: 1/2 $S^3 S^3 S^4$

2.40. (*a*) F_1 = 1/2 $E^D E$ (black with trace of agouti) : 1/2 $E^D e$ (nonagouti black)
F_2 = 1/4 $E^D E^D$: 1/4 $E^D e$: 1/4 $E^D E$: 1/16 EE : 1/8 Ee : 1/16 ee
1/2 nonagouti black : 1/4 black with trace of agouti : 3/16 full color : 1/16 reddish-yellow

(*b*) F_1 = 1/2 $E^D e$ (nonagouti black) : 1/2 ee (reddish-yellow)
F_2 = 1/16 $E^D E^D$: 3/8 $E^D e$: 9/16 ee; 7/16 nonagouti black : 9/16 reddish-yellow

2.41. 1/4 Ss^h : 1/4 Ss^c : 1/4 $s^h s$: 1/4 $s^c s$; 1/2 Dutch-belted : 1/4 Hereford-type spotting : 1/4 solid color

2.42. (*a*) The man could be the father, but paternity cannot be proved by blood type. In certain cases, a man may be excluded as a father of a child [see part (*d*)]. (*b*) $I^B i$ man $\times$ $I^A i$ woman   (*c*) $I^B I^B$ (*d*) no

2.43. (*a*) 90/1024   (*b*) 270/1024   (*c*) 405/1024   (*d*) 1/1024

2.44. 1/4

2.45. 66

2.46. (*a*) (1) agouti × cinnamon (2) agouti × black (3) agouti × chocolate (4) agouti × albino (5) cinnamon × black (6) cinnamon × chocolate (7) cinnamon × albino (8) black × chocolate (9) black × albino (10) chocolate × albino

2.47. (*a*) 4 males : 4 females (*b*) 5th (*c*) 27.34%

2.48. 1/4 *CCTT* : 1/4 *CCTt* : 1/4 *CcTT* : 1/4 *CcTt*; all axial, colored

2.49. 9/16 white, disk : 3/16 white, sphere : 3/16 yellow, disk : 1/16 yellow, sphere

2.50. 144 wild type : 48 vestigial : 48 ebony : 16 ebony, vestigial

2.51. (*a*) 1/4 *LLBb* : 1/4 *LlBb* : 1/4 *LLbb* : 1/4 *Llbb*; 1/2 short, black : 1/2 short, brown (*b*) 1/8 *LLBb* : 1/4 *LIBb* : 1/8 *LLbb* : 1/4 *Llbb* : 1/8 *llBb* : 1/8 *llbb*; 3/8 short, black : 3/8 short, brown : 1/8 long, black : 1/8 long, brown

2.52. (*a*) 9/16 short, black : 3/16 short, brown : 3/16 long, black : 1/16 long, brown (*b*) 25% (*c*) 50% (*d*) 25% (*e*) 6.25% (*f*) 25% (*g*) 25% (*h*) 8.33%

2.53. 4 *FFpp* : 4 *Ffpp* : 1 *ffpp*; 8 feathered leg, single comb : 1 clean leg, single comb

2.54. (*a*) *ABC, ABc* (*b*) *aBC, aBc, abC, abc* (*c*) *ABcD, ABcd, AbcD, Abcd, aBcD, aBcd, abcD, abcd* (*d*) *ABCDEF, ABCDdEf, ABCdeF, ABCdef, ABcdEF, ABcdEf, ABcdeF, ABcdef, AbCdEF, AbCdEf, AbCdeF, AbCdef, AbcdEF, AbcdEf, AbcdeF, Abcdef*

2.55. (*a*) *BBMM* (*b*) *BBMm*

2.56. (*a*) 2 black, mule-foot : 1 black, cloven-foot (*b*) 1/4 white, mule-foot : 1/4 white, cloven-foot : 1/4 black, mule-foot : 1/4 black, cloven-foot

2.57. 1/16 G^AG^ASS : 2/16 G^AG^ASs : 1/16 G^AG^Ass : 2/16 G^AG^OSS : 4/16 G^AG^OSs : 2/16 G^AG^Oss : 1/16 G^OG^OSS : 2/16 G^OG^OSs : 1/16 G^OG^Oss; 3/16 fuzzy, glandless : 1/16 smooth, glandless : 6/16 round gland, fuzzy : 2/16 round gland, smooth : 3/16 oval gland, fuzzy: 1/16 oval gland, smooth

2.58. 1 red, polled : 1 red, horned : 2 roan, polled : 2 roan, horned : 1 white, polled : 1 white, horned

2.59. 1/16 black : 1/8 black, mildly frizzled : 1/16 black, extremely frizzled : 1/8 blue : 1/4 blue, mildly frizzled : 1/8 blue, extremely frizzled : 1/16 splashed-white : 1/8 splashed-white, mildly frizzled : 1/16 splashed-white, extremely frizzled

2.60. 1 black : 2 blue : 1 splashed-white : 2 blue, mildly frizzled : 2 splashed-white, mildly frizzled : 1 splashed white, extremely frizzled

2.61. F_1 is all oval, purple; F_2 is 1/16 long, red : 1/8 long, purple : 1/16 long, white : 1/8 oval, red : 1/4 oval, purple : 1/8 oval, white : 1/16 round, red : 1/8 round, purple: 1/16 round, white

2.62. Long, purple × oval, purple

2.63. (*a*) 1/2 yellow, kinky : 1/6 yellow : 1/4 agouti, kinky : 1/12 agouti (*b*) 1/3 (*c*) 1/6 A^yAKK : 1/3 A^yAKk : 1/6 A^yAkk : 1/12 $AAKK$: 1/6 $AAKk$: 1/12 $AAkk$; 1/2 yellow, kinky : 1/6 yellow : 1/4 agouti, kinky : 1/12 agouti

2.64. (*a*) 1/12 black, hairy : 1/6 black, halo-haired : 1/12 black : 1/6 gray, hairy : 1/3 gray, halo-haired : 1/6 gray (*b*) 2/3 (*c*) 2/3 (*d*) 1/8

2.65. 1/3 normal : 2/3 brachyphalangic

2.66. (*a*) 7/16 (*b*) 1/9

2.67. (*a*) 4/9 plum, stubble : 2/9 plum : 2/9 stubble : 1/9 wild type (*b*) 1 plum, stubble : 1 plum : 1 stubble : 1 wild type

2.68. 1/12 black : 1/6 blue : 1/12 splashed-white : 1/6 black, creeper : 1/3 blue, creeper : 1/6 splashed-white, creeper

2.69. (*a*) F_1 is all tall, round, yellow; F_2 is 27 tall, round, yellow : 9 tall, round, green : 9 tall, wrinkled, yellow : 9 dwarf, round, yellow : 3 tall, wrinkled, green : 3 dwarf, round, green : 3 dwarf, wrinkled, yellow : 1 dwarf, wrinkled, green (*b*) 3.12% (*c*) 8 round : 1 wrinkled

2.70. (a) F_1 is all agouti, black; F_2 is 27 agouti, black : 9 agouti, black, Himalayan : 9 agouti, brown : 9 black : 3 agouti, brown, Himalayan : 3 black, Himalayan : 3 brown : 1 brown, Himalayan   (b) 4/27   (c) 25%   (d) 75%

2.71. (a) 189 agouti, black : 216 agouti, black, dilute : 63 agouti, black, Himalayan : 72 agouti, black, Himalayan, dilute : 63 agouti, brown : 72 agouti, brown, dilute : 63 black : 72 black, dilute : 21 agouti, brown, Himalayan : 24 agouti, brown, Himalayan, dilute: 21 black, Himalayan : 24 black, Himalayan, dilute : 21 brown : 24 brown, dilute: 7 brown, Himalayan : 8 brown, Himalayan, dilute   (b) 1/120

2.72. (a) $2^5 = 32$ (b) $3^5 = 243$ (c) $4^5 = 1024$

2.73. (a) One each   (b) 128   (c) 128, each with equal frequency   (d) 2187   (e) 16,384   (f) 2,392,578

2.74. (a) 1   (b) 3   (c) 2   (d) 2   (e) 3   (f) 7   (g) 7

2.75. (a) $\chi^2 = 0.27$; p = 0.5−0.7; acceptable   (b) $\chi^2 = 2.67$; $p = 0.1$-0.2; acceptable   (c) $\chi^2 = 5.33$; $p = 0.01 − 0.05$; not acceptable.

2.76. Yes; $\chi^2 = 1.06$; $p = 0.5−0.7$; acceptable

2.77. (a) $\chi^2 = 43.37$; $p < 0.001$; not acceptable   (b) $\chi^2 = 0.000421$; $p > 0.95$; a 2 : 1 ratio fits the data almost perfectly   (c) lethal

2.78. (a) $\chi^2 = 4.76$; $0.10 > p > 0.05$; hypothesis acceptable   (b) Monozygotic twins are estimated to be −31; the negative estimate indicates that identical sheep twins are rare provided that unlike-sex twins do not have a survival advantage over like-sex twins.

2.79. Yes; $\chi^2 = 2.83$; $p = 0.8−0.9$; the distribution is consistent with the assumption that boys and girls occur with equal frequency.

2.80. Yes; $\chi^2 = 2.69$; $p = 0.7−0.8$

2.81. (a) No; $\chi^2 = 4.45$; $p < 0.05$ (b) Yes; $\chi^2 = 3.64$; $p > 0.05$

2.82. (a) 0   (b) 1/2   (c) 0   (d) 1/2   (e) 3/4

2.83. (a) $I1 = A^s a^y$, $I2 = a^t a^t$, $II1 = a^y a^t$, $III2 = a^y a^t$, $III3 = A^s a^t$, $II4 = a^y a^t$, $III1 = a^t a^t$, $III2 = A^{s-}$ ($A^s a^y$ or $A^s a^t$), $III3 = a^y a^t$, $III4 = a^t a^t$ (b) 1/4   (c) 2/3

Answers to Review Questions

Matching Questions I

1. F   2. C   3. A   4. H   5. B   6. D   7. G   8. E

Matching Questions II

1. H   2. C   3. C   4. A   5. F   6. A   7. B   8. B

Vocabulary

1. phenotype
2. genotype
3. heterozygote (heterozygous cell)
4. homozygote (homozygous cell)
5. recessive
6. mutant type
7. codominant, incompletely dominant, partially dominant, semidominant
8. lethal gene
9. penetrance
10. expressivity

Multiple-Choice Questions

1. *c*   2. *e*   3. *d*   4. *b*   5. *b*   6. *c*   7. *a*   8. *e* (10 genotypes)   9. *d*   10. *c*   11. *e* (27)   12. *c*   13. *e* (1/6)   14. *e* (2/9)

CHAPTER 3

The Biochemical Basis of Heredity

Nucleic Acids

1 DEOXYRIBONUCLEIC ACID (DNA)

Genes are made of the nucleic acid **deoxyribonucleic acid (DNA)**. This molecule serves as the carrier of genetic information in all organisms other than some viruses. The double-helical structure of this long molecule is shown in Fig. 3-1. The backbone of the helix is composed of two chains with alternating sugar (S)-phosphate (P) units. The sugar is a pentose (five-carbon molecule) called **deoxyribose**, differing from its close relative **ribose** by one oxygen atom in the 2′ position (Fig. 3-2). The phosphate group (PO_4) connects adjacent sugars by a 3′ to 5′ **phosphodiester** linkage. In one chain the linkages are polarized 3′ to 5′; in the other chain, read in the same direction, they are in the reverse order 5′ to 3′. All nucleic acid chains pair in this **antiparallel** fashion, whether the pairing is DNA with DNA chains, DNA with RNA chains, or RNA with RNA chains. The steps in the spiral staircase (i.e., the units connecting one strand of DNA to its polarized complement) consist of paired organic bases of four kinds: adenine, cytosine, guanine, and thymine (symbolized A, C, G, T, respectively), classified into two groups, the **purines** and the **pyrimidines**. Purines only pair with pyrimidines and vice versa, thus producing a symmetrical double helix. A **hydrogen bond** forms between a covalently bound donor hydrogen atom (e.g., an imino group, NH) with some positive charge and a negatively charged covalently bound acceptor atom (e.g., a keto group, CO) by sharing of a hydrogen atom. Adenine pairs with thymine by two hydrogen bonds; guanine and cytosine pair by three hydrogen bonds (Fig. 3-3). A base-sugar complex is called a **nucleoside**; a nucleoside plus a phosphate is called a **nucleotide**. Thus, DNA is a long polymer (i.e., a macromolecule composed of a number of similar or identical subunits, called monomers, covalently bonded) of thousands of nucleotide base pairs (bp).

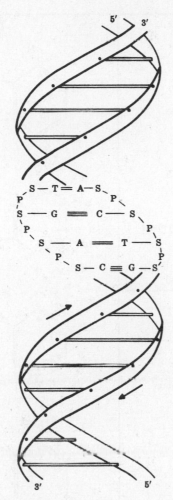

Fig. 3-1. Diagram of the Watson-Crick model of DNA.

EXAMPLE 3.1
Because A pairs with T, and G pairs with C, the ratio of A : T in one DNA strand is 1.0.
The same is true for the ratio of G : C.

SOLVED PROBLEM 3.1
The percentage of nucleotide A in DNA isolated from human liver is observed to be 30.7%. What is the expected percentage of (*a*) T   (*b*) G   (*c*) C?

Solution: Because the amount of A always equals T, the percentage of T is expected to be very close to 30.7%. G and C together would make up the remainder, or $100 - (30.7 + 30.7) = 38.6$; thus, the percentage of C and G separately would be expected to equal half of 38.6, or 19.3.

2   RIBONUCLEIC ACID (RNA)

Another class of nucleic acids, called ribonucleic acid (RNA), is slightly different from DNA in the following respects:

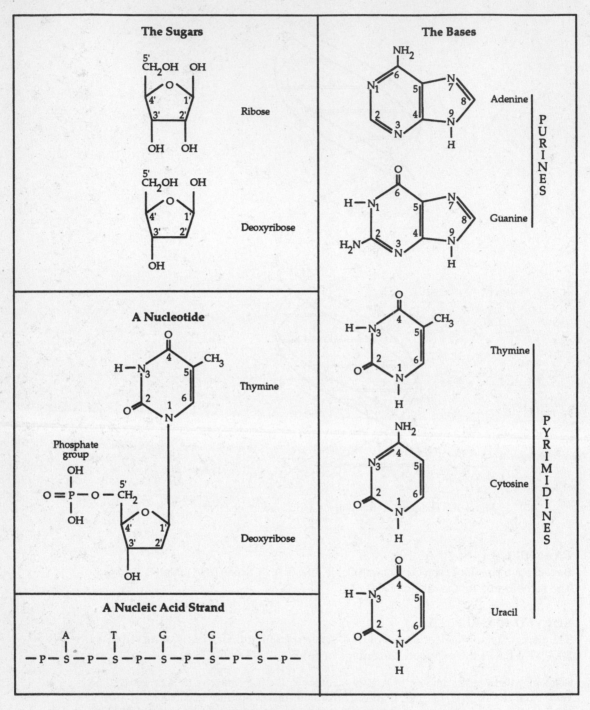

Fig. 3-2. Structural components of nucleic acids. (From W. D. Stansfield, J. S. Colomé, R. J. Cano, *Schaum's Molecular and Cell Biology*, McGraw-Hill, New York, 1996.)

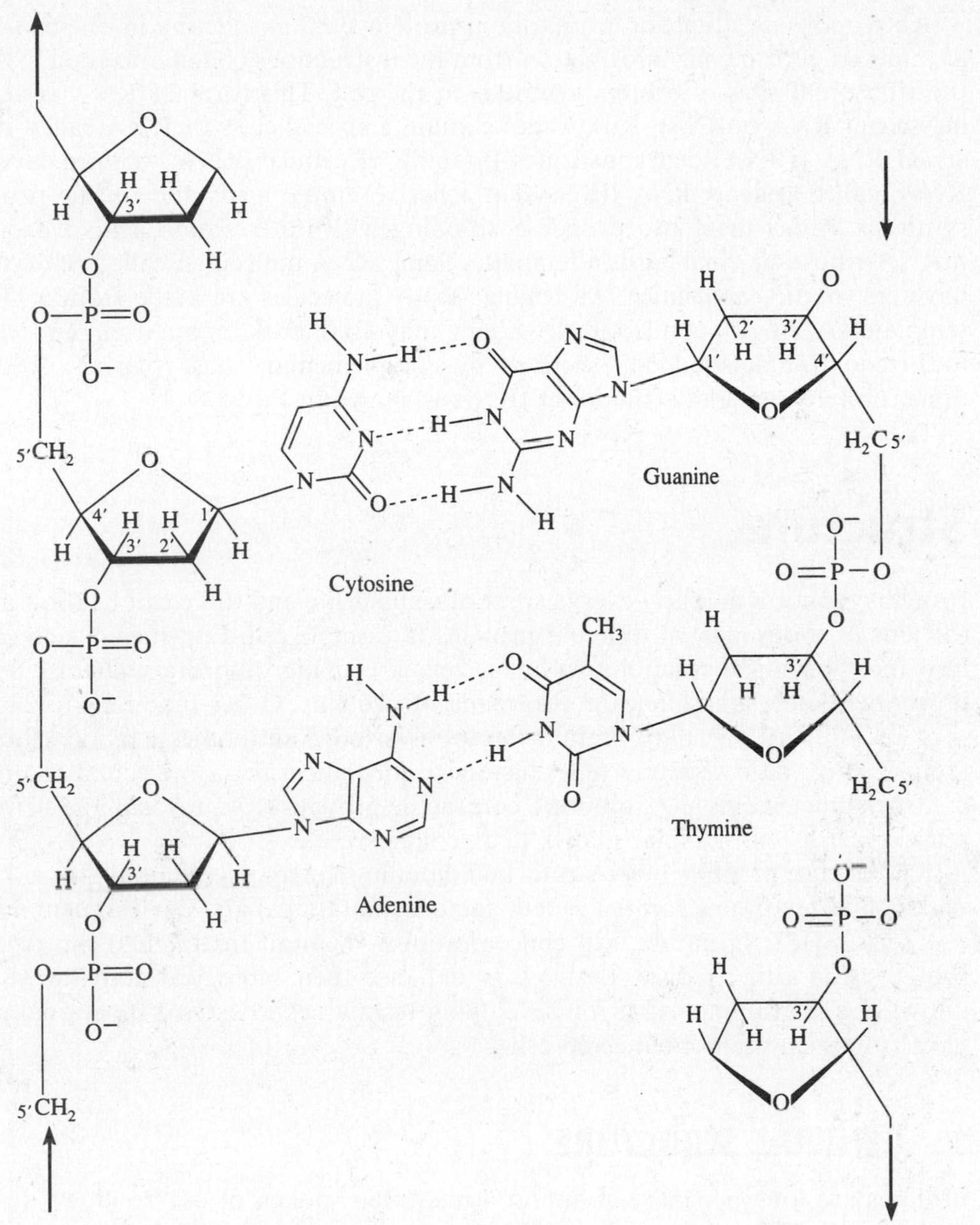

Fig. 3-3. Base pairing in DNA.

1. Cellular RNA is single-stranded; DNA is double-stranded. However, some viruses have a single-stranded DNA genome; a few viruses have a single-stranded RNA genome; very few have a double-stranded RNA genome.
2. RNA contains ribose sugars instead of the deoxyribose sugars that are found in DNA.
3. RNA contains the pyrimidine uracil (U) instead of thymine (T), and U pairs with A.
4. RNA molecules are much shorter than DNA molecules.

RNA molecules function primarily in protein synthesis, acting in one capacity as a messenger carrying information from the instructions coded into the DNA to the ribosomal sites of protein synthesis in the cell. This form of RNA is called **messenger RNA (mRNA)**. Ribosomes contain a special class of RNA called **ribosomal RNA (rRNA)** that constitutes the bulk of cellular RNA. A third kind of RNA, called **transfer RNA (tRNA)**, attaches to amino acids and during protein synthesis brings them into proper positioning with other amino acids using the mRNA-ribosome complex as a template. Some RNA molecules, called **ribozymes**, have enzymatic capabilities. All cellular RNA molecules are made from a DNA template. A single-stranded RNA chain may fold back upon itself and form localized "double-stranded" sections by complementary base pairing. A two-dimensional "cloverleaf" model of tRNA is shown in Fig. 3-8.

Protein Structure

Proteins function in nearly every aspect of cellular life and there can be thousands (or tens of thousands) of different proteins in a single cell. Enzymes, which catalyze most chemical reactions within a cell, are made of protein chains. Some hormones, such as insulin, are also made of proteins. Other functions involving proteins are cell signaling, immune responses (e.g., antibodies), blood clotting factors, chromatin structure (e.g., histones), movement (e.g., molecular motors), cytoskeletal elements (e.g., tubulin), contractile proteins (e.g., myosin and actin of muscle fibers), extracellular matrix (e.g., collagen), etc.

Knowledge of protein structure and bonding forces is essential for a keen understanding of how various genetic factors (mutations) and environmental factors (e.g., pH, temperature, salt concentrations, chemical treatments) can modify proteins and either reduce, destroy, or enhance their biological activities. Such knowledge is also important for developing techniques to extract functional proteins from genetically engineered cells.

1 GENERAL STRUCTURE

Proteins are long polymers of amino acids, often spoken of as "residues" (especially during degradation of proteins to ascertain their amino acid sequences), covalently held together by **peptide bonds**. Twenty different kinds of amino acids occur naturally in proteins. All completely ionized biological amino acids except proline have the general structure shown in Fig. 3-4. The α-carbon is the central atom to which an amino (NH^{3+}) and a carboxyl (COO^-) group are attached. As

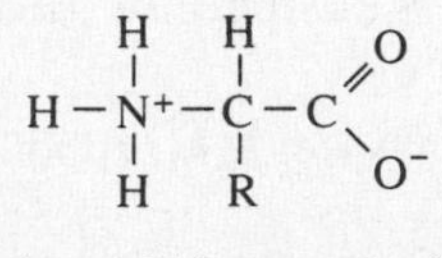

Fig. 3-4. General structure of an amino acid in completely ionized form. R represents a side chain.

pH increases above neutrality (pH 7), the more basic nature of the environment tends to neutralize the acidic carboxyl groups of proteins. As pH decreases below neutrality, the more acidic nature of the environment tends to neutralize the basic amino groups. **Polar molecules** are those with separate positive and negative charges at each end, as exemplified by an amino acid at pH 7. Water is also a polar molecule because the two positive hydrogen atoms are near one end of the molecule and the more negative oxygen atom is at the other end. **Nonpolar molecules** (such as methane, CH_4) are uncharged.

2   THE PEPTIDE BOND

The peptide bond that joins adjacent amino acids during protein synthesis is a strong covalent bond, in which atoms are coupled by sharing an electron. By the removal of water, the carboxyl group of one amino acid becomes joined to the amino group of an adjacent amino acid as shown in Fig. 3-5. This union, which is accompanied by the removal of water, is an example of dehydration synthesis. Each complete polypeptide chain thus has an uncomplexed ("free") amino group at one end and a free carboxyl group at its other end. The amino end of the polypeptide corresponds to the $5'$ end of its respective mRNA. The carboxyl end of the polypeptide corresponds to the $3'$ terminus of the same mRNA. An enzymatic component of the ribosome, called **peptidyl transferase**, is responsible for peptide bond formation.

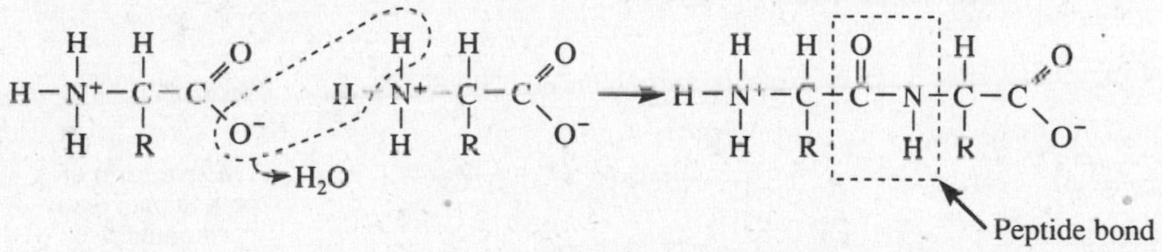

Fig. 3-5.   Dehydration synthesis of a dipeptide and formation of a peptide bond.

3   SIDE CHAINS

Each kind of amino acid differs according to the nature of the side chain or radical attached to the α-carbon. Glycine has the simplest side chain, consisting of a hydrogen atom. Other amino acids have hydrocarbon side chains of various lengths; some of these chains are ionized positively (basic proteins such as lysine and arginine), others are negatively charged (acidic amino acids such as aspartic and glutamic acids), and still others are nonionized (e.g., valine, leucine). Different proteins can be separated on the basis of their net electrical charges by a technique known as **electrophoresis**. Closely related proteins differing by a single amino acid can sometimes be resolved in this way. Some amino acids, such as phenylalanine and tyrosine, have aromatics (ring structures) in their side chains. The amino acid proline does not contain a free imino group (NH) because its nitrogen atom is involved in a ring structure with its side chain. Only two amino acids (cysteine and methionine) contain sulfur in their side chains. The sulfurs of different cysteines can be covalently linked into a disulfide bond (S–S) that is responsible for helping to stabilize the tertiary and quaternary shapes of proteins containing them.

4   STRUCTURAL LEVELS

The linear sequence of amino acids forms the **primary structure** of proteins (Fig. 3-6). Some portions of many proteins have a **secondary structure** in the form of an **alpha helix** in which the carbonyl group (C=O) next to one peptide bond forms a hydrogen bond with an imino group (NH) flanking a peptide bond a few amino acids further along the polypeptide chain. The protein chain may fold back upon itself, forming weak internal bonds (e.g., hydrogen bonds, ionic bonds) as well as strong covalent disulfide bonds that stabilize its **tertiary structure** into a precisely and often intricately folded pattern. Two or more tertiary structures may unite

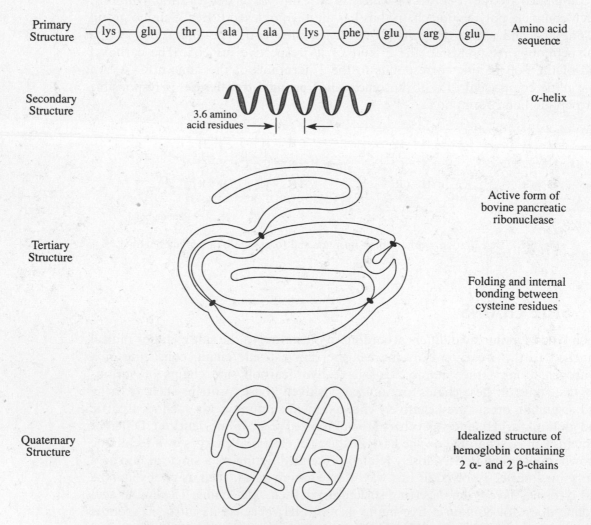

Fig. 3-6.   Stages in the development of a functional protein.

into a functional **quaternary structure**. For example, hemoglobin consists of four polypeptide chains (two identical α-chains and two identical β-chains). A protein cannot function until it has assumed its full tertiary or quaternary configuration. Any disturbance of its normal configuration may inactivate the function of the protein. For example, if the protein is an enzyme, heating may destroy its catalytic activity because weak bonds that hold the protein in its secondary or higher structural forms are ruptured. The shape of an active enzyme molecule fits its **substrate** (the substance that is catalyzed by the enzyme) in a manner analogous to the way a key fits a lock (Fig. 3-7). An enzyme that is altered, either genetically (by mutation of the respective gene), physically (e.g., heat), or chemically (e.g., pH change) may not fit the substrate and therefore would be incapable of catalyzing the conversion of substrate to normal product.

Two degenerative brain diseases in humans (kuru and Creutzfeldt-Jakob disease) and a similar disease in sheep (scrapie) and cattle [bovine spongiform encephalopathy (BSE) or "mad cow disease"] seem to be caused by a proteinaceous infectious particle (**prion**) devoid of nucleic acid. Proponents of this "protein-only hypothesis" suggest that exposure of a mutant protein (**PrP** for prion protein) to normal proteins can cause the latter to change into mutant forms. The mutant forms aggregate and cause neurological disease. There are examples of one protein affecting the structure and function of another (i.e., proteolytic enzymes), but prions seem to interfere with the way that nascent proteins fold into their three-dimensional shapes, and the misfolded proteins can do the same to other proteins in a way that mimics an infection. There is still much controversy over this idea and scientists are actively working to gather more data to solve the mystery.

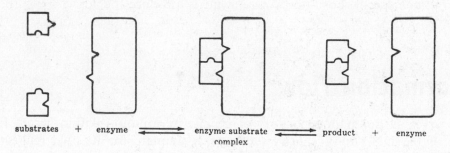

substrates   +   enzyme ⇌ enzyme substrate ⇌ product   +   enzyme
complex

Fig. 3-7.   Diagram of enzymatic action.

5   FACTORS GOVERNING STRUCTURAL LEVELS

Relatively weak bonds such as hydrogen bonds and **ionic bonds** (attraction of positively and negatively charged ionic groups) are mainly responsible for the secondary and higher structural levels of protein organization. Enzymes are not involved in the formation of weak bonds. The extent to which a protein contains alpha-helical regions is dependent upon at least three factors. The most important factor governing tertiary protein structure involves formation of the most favorable energetic interactions between atomic groupings in the side chains of the amino acids. A second factor is the presence of proline, which cannot participate in alpha-helical formation because it is an imino acid rather a true amino acid. Proline is therefore often found at the "corners," or "hairpin turns," of polypeptide chains. Finally, the formation of intrastrand (on the same chain) disulfide bridges tends to distort the alpha helix.

6   FORMATION OF QUATERNARY STRUCTURE

The ionized side chains of some amino acids readily interact with water and therefore are called **hydrophilic** ("water-loving") amino acids. **Hydrophobic** ("water-fearing") amino acids contain nonionized side chains that tend to avoid contact with water. When a polypeptide chain folds into its tertiary shape, these forces cause amino acids with hydrophilic groups to predominate on the outside and hydrophobic segments of the chain to predominate in the interior of globular proteins. The multiple polypeptide chains of quaternary proteins are usually joined by hydrophobic forces. Nonpolar groups of the individual polypeptide chains come together as a way of excluding water. Hydrogen bonds, ionic bonds, and possibly interstrand (between chains) disulfide bonds may also participate in forming quaternary protein structures. Some quaternary proteins consist of two or more identical polypeptide chains (e.g., the bacterial enzyme β-galactosidase consists of four identical polypeptide chains). Such proteins are called **homopolymers**. Other quaternary proteins (such as hemoglobin) consist of nonidentical chains and are called **heteropolymers**. In order to become functional proteins, some polypeptide chains must be subject to modifications after they have been synthesized. For example, the protein chymotrypsinogen must be cleaved at one specific position by an enzyme to produce the active split-product chymotrypsin.

Genetic Information Flow

DNA serves as the main repository of genetic information within a cell. Each strand of the DNA double helix serves as a template for its own **replication**. This activity precedes all cell division and is thus how genetic information is transmitted, or "handed down," to new generations of cells. All cellular RNA molecules are synthesized from DNA templates in a process called **transcription**. Within a transcriptional unit, only one of the strands of DNA serves as a template for the synthesis of RNA molecules. Different transcriptional units may reside on the same or on different DNA strands. Genes are said to be active, or expressed, when they are being transcribed into RNA. Proteins are synthesized from mRNA templates by a process called **translation**. This generalized flow of genetic information from DNA to protein is often referred to as the **central dogma of molecular biology** (see figure below). This oversimplification emphasizes the central idea that DNA does not serve as the direct template for protein synthesis and that nucleic acids are not synthesized from proteins. In the 1990s, the discovery that DNA in telomeric regions of at least some chromosomes could be synthesized from an RNA template served to demonstrate that, in special cases, RNA could be used as a template for DNA synthesis. Furthermore, certain viruses (viruses are not cells) use RNA as genetic material that is copied into a DNA intermediate (**reverse transcription**) by an enzyme called **reverse transcriptase** (dashed arrow in figure below). Such viruses are called "retroviruses" (see Chapter 11) because they reverse the cellular dogma that RNA strands are made from DNA templates.

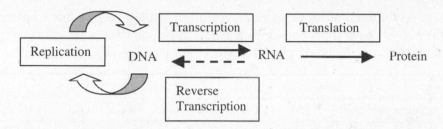

THE GENETIC CODE

Each protein consists of a certain number of amino acids in a precisely ordered sequence. The blueprint that specifies this amino acid sequence is encoded in the nucleotide sequence of a region of DNA called a **gene**. Genes are composed of codons that each specify a specific amino acid through the intermediate of mRNA molecules (see Table 3-1). A **codon** is made up of an adjacent group of three nucleotides, or a triplet of nucleotides, in either the DNA or in its mRNA transcript. If the transcription of a functional genetic unit of DNA into mRNA is always read from a fixed position, then the first six codons in one chain of the corresponding DNA might be as follows. (Note: DNA and RNA sequences, including the codons in Table 3-1, are conventionally written starting with the 5′ end on the left.)

Codon #:	1	2	3	4	5	6
DNA template sequence:	3′ TAC	CCG	ATA	TCA	GCC	AAG 5′
mRNA codon sequence:	5′ AUG	GGC	UAU	AGU	CGG	UUC 3′
Amino acid*:	Met	Gly	Tyr	Ser	Arg	Leu

* See Table 3-1.

One of the two DNA strands acts as the template for synthesis of an mRNA strand (see Transcription section later in this chapter). The specific order of these codons is called the **reading frame**. An **open reading frame (ORF)** is one that starts with an initiator mRNA codon (usually AUG for the amino acid methionine) followed by codons that specify the remaining amino acid sequence, and ends with at least one mRNA stop codon (UAA, UAG, or UGA). The addition of a single base (e.g., **G**) at the end of the second codon below would shift all other codons one nucleotide out of register and prevent the correct reading of all codons to the right of the base addition. The insertion event results in a mutation (see Mutations section, p.99). In fact, in this case a different codon followed by a stop codon is created.

Codon #:	1	2	3	4	5	6	
DNA template sequence:	3′ TAC	CCG	**GAT**	ATC	AGC	CAA	G 5′
mRNA codon sequence:	5′ AUG	GGC	**CUA**	UAG	UCG	GUU	C 3′
Amino acid:	Met	Gly	**Leu**	**STOP**			

Nucleotides and amino acids in bold are new as a result of base addition.

By successively adding bases in a nearby region, it should be possible to place the reading frame of the codons back into register, although the codon changes result in different amino acids being placed in the chain (see below).

Codon #:	1	2	3	4	5	6	7
DNA template sequence:	3′ TAC	CCG	**GAT**	**GAT**	TCA	GCC	AAG 5′
mRNA codon sequence:	5′ AUG	GGC	**CUA**	**CUA**	AGU	CGG	UUC 3′
Amino acid:	Met	Gly	**Leu**	**Leu**	Ser	Pro	Leu

Nucleotides and amino acids in bold are new as a result of base additions.

High concentrations of ribonucleotides in the presence of the enzyme polynucleotide phosphorylase can generate synthetic mRNA molecules *in vitro* without a template by forming an internucleotide 3′-5′ phosphodiester bond. In this way, a number of uracil molecules can become linked together to form a synthetic poly-U with mRNA activity. The addition of poly-U to bacterial cell extracts results in the limited synthesis of polypeptides containing only the amino acid phenylalanine. Thus, one or more uracils codes for phenylalanine. Mixtures of different ribonucleotides can also form synthetic mRNA molecules with the nucleotides in random order (see Solved Problem 3.3). A combination of organic chemical and enzymatic techniques can be used to prepare synthetic polyribonucleotides with known repeating sequences such as, for example, AUAUAUAU . . ., which alternately codes for the amino acids isoleucine and tyrosine; CUCUCUCU . . ., which codes for leucine and serine alternately, etc. In historic experiments, it was found that one base addition altered the reading frame (i.e., changed the amino acid specified) and only two more added bases restored the reading frame. The same has also been found to be true for single nucleotide deletions. Three deletions or multiples thereof can correct the reading frame in the synthesis of an active protein. Supplementary Problem 3.19 explores additions and deletions in more depth. Several other lines of evidence indicate that the codon is a sequence of three nucleotides and the genetic code is generally referred to as a "triplet code." The simplest explanation for these results is that the genetic code consists of a triplet of contiguous nucleotides.

Even in the absence of mRNA and protein synthesis, an RNA trinucleotide will bind to a ribosome. Chemically synthesized trinucleotides of known sequence can thus be made to bind to ribosomes, and this complex will specifically bind one out of a mixture of 20 different tRNA-amino acid complexes *in vitro*. By radioactively labeling only one kind of amino acid in such a mixture, the specificity of the codon can be established. For example, UUG binds only leucine-tRNA complexes to ribosomes, and UGU binds only cysteine-tRNA complexes.

The genetic code is **degenerate** because more than one codon exists for most amino acids. The genetic code is generally the same in all organisms, and thus is said to be **universal**. However, a few codons in some organelle DNAs have different meanings than those in nuclear DNAs (see Chapter 12). Three of the 64 possible three-letter codons do not specify an amino acid. These triplets are called **nonsense codons** or **stop codons**, and they serve as part of the translation termination signal. The AUG codon that specifies methionine is referred to as the **initiator codon,** as well as a sense codon, as it is most likely the first codon in a mRNA message. The mRNA codons for the 20 amino acids are listed in Table 3-1. These codons are conventionally written from the 5′ end (at the left) toward the 3′ end (at the right) because translation (protein synthesis) begins at the 5′ end of an mRNA molecule.

Table 3-1. mRNA Codons

		Second Letter				
		U	C	A	G	
First Letter	U	UUU Phe	UCU Ser	UAU Tyr	UGU Cys	U
		UUC Phe	UCC Ser	UAC Tyr	UGC Cys	C
		UUA Leu	UCA Ser	**UAA Stop**	**UGA Stop**	A
		UUG Leu	UCG Ser	**UAG Stop**	UGG Trp	G
	C	CUU Leu	CCU Pro	CAU His	CGU Arg	U
		CUC Leu	CCC Pro	CAC His	CGC Arg	C
		CUA Leu	CCA Pro	CAA Gln	CGA Arg	A
		CUG Leu	CCG Pro	CAG Gln	CGG Arg	G
	A	AUU Ile	ACU Thr	AAU Asn	AGU Ser	U
		AUC Ile	ACC Thr	AAC Asn	AGC Ser	C
		AUA Ile	ACA Thr	AAA Lys	AGA Arg	A
		AUG Met	ACG Thr	AAG Lys	AGG Arg	G
	G	GUU Val	GCU Ala	GAU Asp	GGU Gly	U
		GUC Val	GCC Ala	GAC Asp	GGC Gly	C
		GUA Val	GCA Ala	GAA Glu	GGA Gly	A
		GUG Val	GCG Ala	GAG Glu	GGG Gly	G

The three-letter symbols, names, and one-letter symbols (in parentheses) of the amino acids are as follows: ala = alanine (A), arg = arginine (R), asn = asparagine (N), asp = aspartic acid (D), cys = cysteine (C), glu = glutamic acid (E), gln = glutamine (Q), gly = glycine (G), his = histidine (H), ile = isoleucine (I), leu = leucine (L), lys = lysine (K), met = methionine (M), phe = phenylalanine (F), pro = proline (P), ser = serine (S), thr = threonine (T), trp = tryptophan (W), tyr = tyrosine (Y), val = valine (V). The start codon, AUG, and its corresponding amino acid is shown in italics. The three stop or nonsense codons, UAA, UAG, and UGA, are shown in bold.

SOLVED PROBLEM 3.2

How many triplet codons can be made from the four ribonucleotides A, U, G, and C containing (*a*) no uracils, (*b*) one or more uracils?

Solution:

(*a*) Since uracil represents 1 among 4 nucleotides, the probability that uracil will be the first letter of the codon is 1/4; and the probability that U will not be the first letter is 3/4. The same reasoning holds true for the second and third letters of the codon. The probability that none of the three letters of the codon are uracils is $(3/4)^3 = 27/64$.

(*b*) The number of codons containing at least one uracil is $1 - (27/64) = 37/64$.

SOLVED PROBLEM 3.3

A synthetic polyribonucleotide is produced from a mixture containing U and C in the relative frequencies of 5 : 1, respectively. Assuming that the ribonucleotides form in a random linear array, predict the relative frequencies in which the various triplets are expected to be formed.

Solution: The frequencies with which the different triplets are expected to be formed by chance associations can be predicted by combining independent probabilities through multiplication as follows:

UUU should occur with a frequency of $(5/6)(5/6)(5/6) = 125/216$.
Codons with 2U and IC $= (5/6)^2(1/6) = 25/216$ each (UUC, UCU, CUU)
Codons with 1U and 2C $= (5/6)(1/6)^2 = 5/216$ each (UCC, CUC, CCU)
CCC $= (1/6)^3 = 1/216$

Protein Synthesis

1 TRANSCRIPTION

There are two major steps in protein synthesis: transcription and translation. The first step in the production of proteins is the transcription of DNA to an mRNA molecule. This process is carried out by the enzyme **RNA polymerase**. This enzyme attaches to the DNA at a specific nucleotide sequence called a **promoter** ahead of (upstream from) the gene to be translated. A number of enzymes stimulate the local unwinding of DNA, and this allows RNA polymerase to begin transcription of one of the DNA strands. Within a gene, only one of the DNA strands is transcribed into mRNA. This DNA strand is called the **anticoding strand** or **antisense strand**; the DNA strand that is not transcribed is called the **coding strand** or **sense strand**. Some other gene on that same DNA molecule may use the other strand as a template for RNA synthesis. Within a gene, however, RNA polymerase does not jump from one DNA strand to another to transcribe the RNA molecule. Termination of transcription occurs when RNA polymerase encounters a "terminator nucleotide sequence" at the end of a structural gene. In some bacterial genes, an accessory protein binds to the terminator sequence and thereby aids in dislodging RNA polymerase from the DNA. The mechanism of transcription termination in eukaryotes is still unknown.

In eukaryotic cells, primary mRNA transcripts are "processed" before they are released from the nucleus as mature mRNA molecules. Initially, most eukaryotic primary transcripts (pre-mRNAs) are mosaics of coding regions (**exons**) and noncoding regions (**introns**). Before the mRNA leaves the nucleus to become mature cytoplasmic mRNA, the noncoding regions must be precisely removed and the exons must be **spliced** together. In addition, an unusual guanine nucleotide (called a **cap**) is attached to the 5′ end, and a string of adenine nucleotides (called a **poly-A tail**) is attached to the 3′ end of the mRNA. In prokaryotic cells, however, there is no nuclear membrane, and mRNA processing does not occur. Except for the archaebacteria (see Chapter 13), bacterial genes do not contain introns. Thus, translation of mRNA into protein can commence in bacteria even before the mRNA has been completely transcribed from the DNA.

2 TRANSLATION

In the second major step of protein synthesis, ribosomes and tRNA-methionine complexes (called "charged" methionyl tRNAs) attach near the 5' end of the mRNA molecule at the first start codon or initiation codon (AUG) and begin to translate its ribonucleotide sequence into the amino acid sequence of a protein. **Ribosomes** consist of three different rRNA molecules and about 50 different proteins. Each amino acid is coded for by at least one tRNA molecule. Because the genetic code is so degenerate, many more than 20 tRNAs are actually involved in protein synthesis. Each amino acid becomes attached or loaded (at its carboxyl terminus) to the 3' end of its own species of tRNA (Fig. 3-8) by a specific enzyme (**amino-acyl synthetase**). Thus, there are at least 20 different synthetases. The "loaded" tRNA is said to be **activated** or **charged**. A loop of unpaired bases near the middle of the tRNA carries a triplet of adjacent bases called the **anticodon**. Other parts of the tRNA are thought to form complementary base pairs with rRNA of the ribosome during protein synthesis or to act as recognition sites for a specific amino-acyl synthetase.

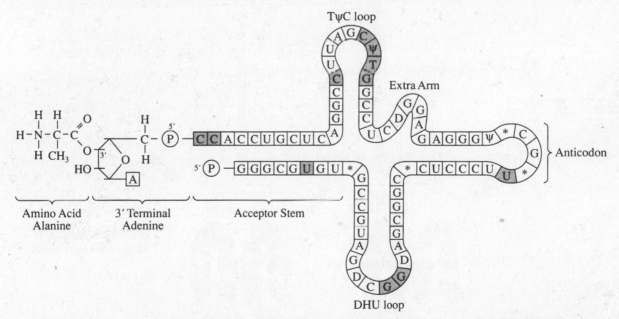

Fig. 3-8. A "cloverleaf model" of the yeast alanine transfer RNA molecule. All species of tRNA are about 75 ribonucleotides in length, with three major loops of unpaired bases. The middle loop contains the anticodon that can base pair with a codon in mRNA. The loop nearest the 3' end (called the TψC loop) is thought to interact with rRNA in the ribosome. There are several unusual bases (*) in tRNAs; pseudouridine (ψ) is one of them. The loop nearest the 5' end is called the DHU loop because it contains another unusual base, dihydrouridine (D). Enzymes in the nucleus modify the normal bases (A, U, C, G) in the preformed tRNA to create these unusual (rare) bases. All tRNAs end with CCA3'; the proper species of amino acid is attached to the terminal A by its cognate amino-acyl-tRNA synthetase enzyme. Some of the positions bearing identical bases in almost all tRNA species are indicated by shading.

Translation of most proteins begins with the start codon 5′ AUG 3′, which specifies the amino acid methionine (refer to Fig. 3-9 throughout the following discussion). Two sites exist on a ribosome for activated tRNAs: the **peptidyl site (P site)** and the **amino-acyl site (A site)**. The initiating methionine-loaded tRNA enters the P site (perhaps by passing through the A site). The 3′ UAC 5′ anticodon of the tRNA pairs with the complementary 5′ AUG 3′ codon in the mRNA. The ribosome holds all of the reactants in the proper alignment during translation. A second activated tRNA (e.g., one loaded with threonine) enters the A site (again by specific codon-anticodon base pairing). A peptide bond is formed between the two adjacent amino acids by the action of an enzymatic portion of the ribosome called **peptidyl transferase**. In bacteria, it appears that ribosomal RNA is responsible for peptide bond formation (Chapter 10). This is an example of **ribozyme** activity. The amino-acyl bond that held the methionine to its tRNA is broken when the peptide bond forms. The now "unloaded" methionyl-tRNA in the P site leaves (usually to become activated again). The ribosome shifts (translocates) three nucleotides along the mRNA to position a new open codon in the vacant A site while at the same time moving the thr-loaded tRNA (now attached to a dipeptide)

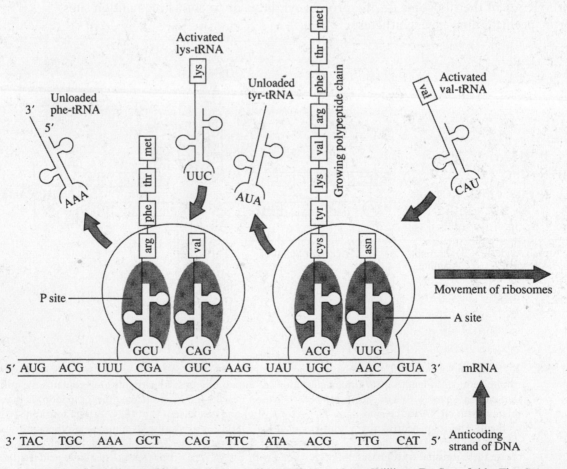

Fig. 3-9. Diagram of protein synthesis. (Reprinted, by permission, from William D. Stansfield, *The Science of Evolution*, ©1977 by Macmillan Publishing Co., Inc.)

from the A to the P site. The third tRNA (e.g., one loaded with phenylalanine) enters the A site; a peptide bond forms between the second and the third amino acids; the second tRNA exits the P site; translocation of the ribosome along the mRNA displays the next codon for arginine in the A site while shifting the phe-loaded tRNA (now carrying a tripeptide) from the A to the P site; and so on. Eventually the system reaches one or more nonsense or stop codons (UAA, UAG, or UGA) causing the polypeptide chain to be released from the last tRNA, the last tRNA to be released from the ribosome, and the ribosome to be released from the mRNA. Thus, the 5′ end of mRNA corresponds to the amino terminus of the polypeptide chain; the 3′ end of the mRNA corresponds to the carboxyl terminus of the polypeptide chain.

The preceding description of protein synthesis is only a broad outline of the process. Some important aspects of this process are performed differently in bacteria and in eukaryotes, details of which are presented in separate chapters dealing with these two major life forms (Chapters 10 and 13, respectively).

SOLVED PROBLEM 3.4
Using the information in Table 3-1, convert the following mRNA segment (shown in register as triplets) into a polypeptide chain. Does the number of codons equal the number of amino acids?

$$...5′ \text{ AUG GAA GCA UCA CCC UAG } 3′ ...$$

Solution:

$$met - glu - ala - ser - pro - (stop)$$

The number of codons does not equal the number of amino acids because the final codon is a stop codon and does not specify an amino acid.

DNA Replication

The hydrogen bonds linking base pairs together are relatively weak bonds. During DNA replication, the two strands separate along this line of weakness in zipperlike fashion (Fig. 3-10). Each strand of the DNA molecule can serve as a template against which a complementary strand can form (according to the rules of specific base pairing) by the catalytic activity of enzymes known as **DNA polymerases**. This mode of replication, in which each replicated double helix contains one original (parental strand) and one newly synthesized daughter strand, is referred to as **semiconservative replication** (see Solved Problem 12.1). At least three forms of DNA polymerase have been identified in prokaryotes and at least four in eukaryotes. All DNA polymerase enzymes add free nucleotides only to the 3′ ends of existing chains, so that the chains will grow from their 5′ ends toward their 3′ ends. All three kinds of DNA polymerases can also degrade DNA in the 3′ to 5′ direction. Enzymes that degrade nucleic acids are called **nucleases**. If the enzyme cleaves nucleotides from the end of the chain it is called an **exonuclease**; if it makes cuts in the interior of the molecule it is termed an **endonuclease**. As long as deoxyribonucleotide precursors are present in even moderate amounts, the synthetic activity of

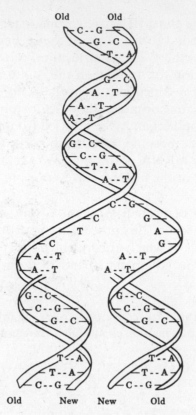

Fig. 3-10. Replication of DNA.

a DNA polymerase is greatly favored over its degradation activity. During replication, incorrectly paired bases have a high probability of being removed by the exonuclease activity of the DNA polymerases before the next nucleotide is added. This is part of the "proofreading system" that protects the DNA from errors (mutations).

All DNA polymerases can extend existing polynucleotide chains only from their 3′ ends; they cannot initiate new chains from their 5′ ends. A special kind of RNA polymerase, called **primase**, constructs a short (about 10 bp) segment of ribonucleotides (as an RNA primer) complementary to the DNA template at a specific DNA sequence called an **origin of replication (ori)** site. This **RNA primer** has a free 3′ end to which additional deoxyribonucleotides can be added by DNA polymerases. Primers are later removed by a DNA nuclease enzyme. There may be one or many ori sites, depending on the type of DNA being replicated. For example, bacterial circular chromosomes have a single ori (Chapter 10), while linear eukaryotic chromosomes have multiple ori sites. Each unit of replication is called a **replicon**. When multiple ori sites are present, they may be 50–250 kb apart and result in the formation of multiple replication bubbles during DNA replication (Figure 3-11).

Replication begins at the 3′ end of a template (parental) strand (Fig. 3-12). A primer RNA is synthesized from 5′ to 3′ toward the replication fork, and the primer is extended by DNA polymerase III, forming the **leading strand**. The

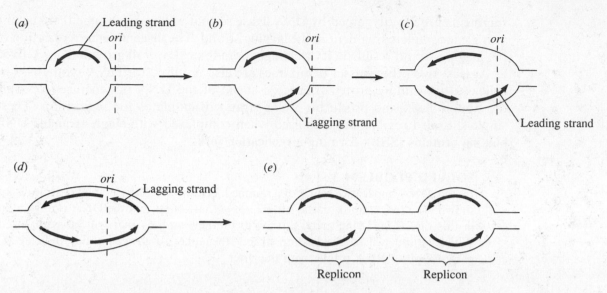

Fig. 3-11. Bidirectional DNA replication. (*a*) From any origin of replication site (*ori*) a leading strand is synthesized. (*b*) Initiation of a lagging strand synthesis. (*c*) The first precursor (Okazaki) fragment has passed the *ori* site and has become the rightward leading strand. (*d*) Initiation of leftward lagging strand begins with establishment of a second complete replication fork. (*e*) Two replication bubbles about to coalesce.

opposite template strand has a 5′ end, so no complementary primer can be formed from 3′ to 5′. Instead, a **lagging strand** is replicated (5′ to 3′) in short segments (a few hundred nucleotides each) in a direction opposite to the movement of the replication fork. These segments are called **Okazaki fragments** (named after their discoverer, Reji Okazaki). A gap is said to exist where one or more adjacent nucleotides are missing from one strand of a duplex DNA molecule. **DNA polymerase I** temporarily creates gaps by removing RNA primers, but quickly fills the gaps with replacement deoxyribonucleotides. Nicks between adjacent Okazaki

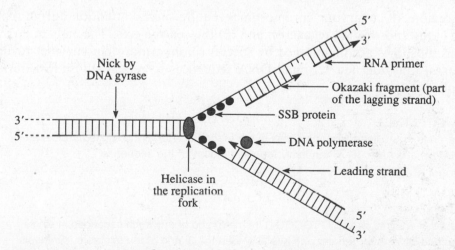

Fig. 3-12. Production of Okazaki fragments on the lagging strand during DNA replication.

fragments are rapidly joined by **DNA ligase** so that at any given time there is only a single incomplete fragment in the lagging strand. The **discontinuous** replication of the lagging strand results in its seemingly paradoxical overall growth from 3′ to 5′.

At least two other classes of enzymes are also required for DNA synthesis. The **helicases** (unwinding proteins) proceed ahead of the DNA polymerases, opening the double helix and producing single-stranded templates for replication. These single-stranded regions are stabilized when complexed with **single-stranded DNA binding proteins (SSB)**, forming a **replication fork**.

SOLVED PROBLEM 3.5

In the DNA replication fork diagrammed below, the old (template) strands are shown. The two strands on the left are not yet unwound, but the DNA is unwinding in this direction. Using arrows, draw in the new strands that will be synthesized. Which strand will be the leading strand, the upper or lower strand? Indicate the presence of an RNA primer with the letter "P."

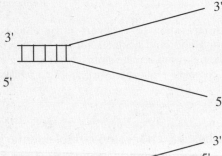

Solution:

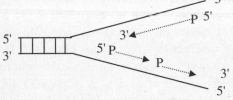

The upper strand will be the leading strand, as its synthesis can occur from 5′ to 3′ continuously in the direction that the replication fork is moving. The lower strands are Okazaki fragments that will be joined together to form the lagging strand.

Two regions of eukaryotic chromosomes require special attention during replication: (1) the ends (called **telomeres**) and (2) the **centromeres**. The ends of eukaryotic chromosomes are maintained by special ribonucleoprotein enzymes (called **telomerases**) that add new terminal DNA sequences to replace those lost during each replication cycle.

EXAMPLE 3.2

The ends of the linear chromosomes in the macronucleus of the ciliated protozoan *Tetrahymena* have 30-70 tandemly repeated blocks of the sequence

$$3' - AACCCC - 5'$$
$$5' - TTGGGG - 3'$$

A special enzyme adds 5′-TTGGGG-3′ to the 3′ end of any such sequence in single-stranded DNA. After the enzyme has extended the 3′ end of the telomere, synthesis of the repeats in the complementary DNA strand could be primed by a primase.

Similarly, all yeast chromosomes end with approximately 100 bp of the irregularly repeated sequence

$$5' - C_{1-3}A\ldots$$
$$3' - G_{1-3}T\ldots$$

The mechanism that determines the length of such telomeres is not well understood.

In the ciliate *Euplotes crassus*, the RNA component of the telomerase serves a template function for synthesis of telomeric T_4G_4 repeats. Thus, at least some portions of certain eukaryotic DNA molecules are known to be replicated from RNA templates. Enzymes that make DNA from RNA templates are called reverse transcriptases; this kind of telomerase represents a specialized kind of reverse transcriptase.

Little is presently known about the centromeric regions where sister chromatids are joined. The centromeric sequences in yeast are about 130 bp long and very rich in A-T. It is thought that the chromatin material is not organized into nucleosomes in the centromeric region, either because the histones have become modified or because a complex of proteins other than the normal histones are specifically bound to the centromeric sequences. This proteinaceous region of the centromere (called the kinetochore) somehow attaches to microtubule bundles of the spindle. Each sister chromatid has its own kinetochore. The centromeric sequence is generally bordered (flanked) by heterochromatin containing repetitive DNA sequences. The centromeric sequences replicate during the contracted metaphase state, and thus are the last DNA segments to do so in eukaryotic cells prior to normal cell division.

Mutations

Mutations are changes in the sequence of DNA. Mutations can occur in any region of the genome; however, phenotypic changes are only observed in the organism if a mutation occurs in the sequence of a gene (see Table 3-2 for specific examples). Recall that there can be different alleles at each genetic locus and the wild-type allele is the one causing the most common phenotype in a population. Alleles are distinguished from each other based on their DNA sequence; the wild-type allele having one sequence and all **mutant** alleles having slightly different sequences.

EXAMPLE 3.3

The normal (wild-type) human hemoglobin protein (Hb A) has about 140 amino acid residues in each of its α- and β-chains. The sequence of the β-chain has been determined to be

Hb A amino acid squence:	val	his	leu	thr	pro	glu	glu	lys	etc.
Amino acid #:	1	2	3	4	5	6	7	8	

An abnormal (mutant) hemoglobin protein (Hb S) is produced by individuals with a mutant allele, resulting in a deformity of the red blood cell called "sickling." In a heterozygous condition this allele produces a mild anemia; in a homozygous condition the severity of the anemia may be lethal. The difference between Hb A and Hb S is that the latter has valine substituted for glutamic acid in the sixth position of the β^s-chain.

Hb S amino acid sequence: val – his – leu – thr – pro – **val** – glu – lys – etc.
Amino acid #:   1   2   3   4   5   6   7   8

Another potentially lethal abnormal hemoglobin (Hb C) is known in which the glutamic acid of the sixth position is replaced by lysine.

Hb A amino acid squence   val – his – leu – thr – pro – **lys** – glu – lys – etc.
Amino acid #:   1   2   3   4   5   6   7   8

One of the codons for glutamic acid is GAA. If a mutation occurred that changed the first A to a U, then the codon GUA (a missense mutation, see Table 3-2) would be translated as valine. The substitution of A for G would produce the missense codon AAA, which codes for lysine. Thus, a change in a single nucleotide in the hemoglobin gene can produce a substitution of one amino acid in a chain of about 140 residues with profound phenotypic consequences!

Alleles that differ at the same nucleotide site are referred to as **homoalleles**. Intragenic recombination between homoalleles cannot result in production of new alleles. Alleles that differ due to nucleotide mutations at different sites are called **heteroalleles**. Intragenic recombination between two heteroalleles can result in production of new alleles, one wild-type and one containing two different nucleotide mutations.

EXAMPLE 3.4

An individual with a mutant phenotype has two mutant sites, m_1, and m_2, within homologous genes (diagrammed as boxes). These heteroalleles can recombine and be transmitted to the progeny as either a doubly mutant gene or as a wild-type gene.

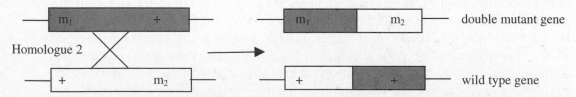

Fortunately, most genes are relatively stable and mutation is a rare event. The great majority of genes have mutation rates of 1×10^{-6}; i.e., 1 gamete in 100,000 to 1 gamete in a million would contain a mutation at a given locus. However, in a higher organism containing 10,000 genes, 1 gamete in 10 to 1 gamete in 100 would be expected to contain at least one mutation. The rate at which a given gene mutates under specified environmental conditions is as much a characteristic of the gene as is its phenotypic expression. The mutation rate of each gene is probably dependent to some extent upon the residual genotype. The only effect that some genes seem to exhibit is to increase the mutation rate of another locus. These kinds of genes are called **mutator genes**.

EXAMPLE 3.5
A dominant gene called "dotted" (*Dt*) on chromosome 9 in corn causes a recessive gene *a* governing colorless aleurone, on chromosome 3, to mutate quite frequently to its allele *A* for colored aleurone. Plants that are *aaDt-* often have kernels with dots of color in the aleurone produced by mutation of *a* to *A*. The size of the dot will be large or small depending on how early or late, respectively, during the development of the seed the mutational event occurred.

Mutations that are deleterious to the organism are kept at a low frequency in the population by the action of natural selection. Organisms harboring these types of mutation are generally unable to compete equally with wild-type individuals. Even under optimal environmental conditions, many mutants appear less frequently than expected. Mendel's laws of heredity assume equality in survival and/or reproductive capacity of different genotypes. Observed deviations from the expected Mendelian ratios would be proportional to the decrease in survival and/or reproductive capacity of the mutant type relative to wild type. The ability of a given mutant to survive and reproduce in competition with other genotypes is an extremely important phenotypic characteristic from an evolutionary point of view.

EXAMPLE 3.6
In *Drosophila*, white-eyed flies are produced by a sex-linked recessive gene (see Example 5.6). White-eyed flies may be only 60% as viable as wild-type flies. Among 100 zygotes from w^+w (wild-type females) crossed to wY (white-eyed males), the Mendelian zygotic expectation is 50 wild type : 50 white eyed (see Example 5.7). If only 60% of white-eyed flies survive, then we would observe $50 \times 0.6 = 30$ white : 50 wild type.

Ionizing radiation such as X-rays are known to cause mutations in genes in direct proportion to the radiation dosage. A linear relationship between dosage (in roentgen units) and the induction of sex-linked recessive lethal mutations in *Drosophila* is shown in Fig. 3-13. This indicates that there is no level of dosage

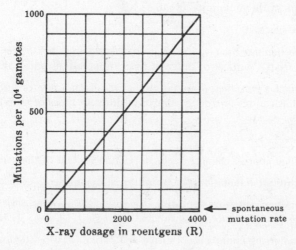

Fig 3-13.   X-chromosome lethal mutations induced by X-rays in *Drosophila*.

that is safe from the genetic standpoint. If a given amount of radiation is received gradually in small amounts over a long period of time (chronic dose) the genetic damage is sometimes less than if the entire amount is received in a short time interval (acute dose). Treatment with ionizing radiation produces mutations most frequently by inducing small deletions in the DNA of the chromosome. Other types of mutagenic agents, such as the chemicals or other types of radiation, are also known to induce mutations (see Table 3-2).

Often, the first encounter with the terminology of mutations is a source of confusion. Mutations can be classified on the basis of several criteria. The outline in Table 3-2 may be useful in showing the relationships among concepts and terms.

Table 3-2.   A Classification of Mutations

I. Size

 A. Point mutation—a change in a very small segment of DNA; usually considered to involve a single nucleotide or nucleotide pair

 B. Gross (large) mutation—changes involving more than one or a few nucleotides of DNA; may involve megabases of DNA that may encompass the entire gene, the entire chromosome, or sets of chromosomes (polyploidy)

II. Effect on protein (codons)

 A. Silent mutation—change in a codon (usually in the third position) that does not change the amino acid coded for

 B. Nonsense mutation—change in a codon from amino acid specificity to a stop codon; results in premature amino acid chain termination during translation

 C. Missense mutation—change in a codon that changes the specificity to a different amino acid; changes the primary sequence of the polypeptide chain and alters the function of the protein

 D. Neutral mutation—change in the codon such that a different amino acid is specified however, the new amino acid behaves similarly to the original one (e.g., has a similar functional group) and does not alter the function of the protein

 E. Frameshift mutation—a shift of the reading frame caused by a deletion or insertion of one or a few nucleotides; creates numerous missense and nonsense codons downstream of the mutational event

III. Effect on gene function

 A. Loss-of-function mutation—a mutation that results in a lack of gene function; this can result from a number of different types of mutations and is recessive in nature

 B. Gain-of-function mutation—a mutation that results in a new or different gene function; this can result from a number of different types of mutations and is dominant in nature

IV. Effect on DNA

 A. Structural mutations—changes in the nucleotide content of the gene

  1. **Base substitution mutations**—substitution of one nucleotide for another

   (*a*) **Transition mutations** substitute one purine for another purine or one pyrimidine for another pyrimidine

   (*b*) **Transversion mutations** substitute one purine for a pyrimidine or vice versa

(*continued*)

Table 3-2   (*continued*)

2. **Deletion mutations**—loss of some portion of DNA

3. **Insertion mutations**—addition of one or more extra nucleotides

B. **Chromosomal rearrangements**—changing the location of a piece of DNA within the genome can result in large structural changes (translocations or inversions) in genes or may change the expression of a gene by placing it under the control of a different promoter (called a "**position effect**")

1. **Translocations**—movement of DNA to a nonhomologous chromosome; usually an exchange occurs between two nonhomologous chromsomes

2. **Inversions**—movement of DNA within the same chromosome; a 180° rotation or "flip"

V. **Origin**

A. **Spontaneous mutation**—mutation occurs during normal cellular activities, primarily DNA replication and repair

B. **Induced mutation**—mutation occurs as a result of treatment with a mutagenic agent or environment; mutation rate is usually higher than background levels

1. **Ionizing radiation**—α-, β-, γ-, or X-rays; usually results in deletions or insertions of DNA

2. **Non-ionizing radiation—UV light;** causes adjacent thymines on one DNA strand to bond together (**thymine dimer**) resulting in a structure that must be repaired in order for DNA replication to proceed; inefficient repair can lead to point mutations

3. **Chemicals**—chemical substances that interact with DNA to create base changes

(*a*) **Base analogs**—chemicals that are structurally similar to bases in DNA, but may have different base pairing properties; bromouracil (BrdU) is structurally similar to thymine so will be incorporated in a growing DNA strand in place of T, but due to its properties it base pairs more frequently with G than with A. The mutagenic affect is mostly due to incorrect base pairing with G, leading to GC-AT transitions

(*b*) **Base modifiers**—chemicals that make changes to a specific base changing its ability to base pair properly; e.g., deamination of cytosine creates a uracil base that will pair with an A instead of the G previously designated by the original C, or alkylating agents that add a methyl group causing guanine to mispair with thymine

(*c*) **Intercalating agents**—chemicals that insert themselves into the DNA helix causing DNA replication and transcription problems; usually results in deletions or insertions

C. **Mutator mutations**—mutations that influence the mutability of other genes

1. Specific mutators—limited to one locus

2. Nonspecific mutators—effect is not specific to one locus; these mutations are generally in genes that control DNA repair

VI. **Magnitude of phenotypic effect**

A. **Change in mutation rate**—alleles mutate at different rates; some can be distinguished based on their rate of mutation.

(*continued*)

Table 3-2 *(continued)*

B. **Isoalleles**—produce identical phenotypes in homozygous or heterozygous combinations with each other, but prove to be distinguishable when in combination with other alleles

C. **Mutants affecting viability:**

1. **Subvitals**—relative viability is greater than 10% but less than 100% compared with wild type

2. **Semilethals**—cause more than 90% but less than 100% mortality

3. **Lethals**—kill all individuals before adult stage

VII. **Direction**

A. **Forward mutation**—creates a change from wild type to abnormal phenotype

B. **Reverse or back mutation**—changes an altered nucleotide sequence back to its original sequence

C. **Supressor mutations**—produces a change from abnormal (i.e., mutated) phenotype back to wild type. There are two types of suppressor mutations:

1. **Intragenic suppressor**—a mutation in the same gene as was originally mutated, but at a different site, that results in restoration of wild-type function (e.g., if an arginine codon CGU was originally mutated to serine codon AGU, the suppression causes a change back to an arginine codon, AGA—see Table 3-1 for codons; also, restoration of a reading frame by additions or deletions)

2. **Intergenic suppressor**—a mutation in another gene that results in restoration of wild-type function (e.g., a nonsense mutation may be suppressed by a mutation in the tRNA for that codon so that it now inserts an amino acid). These are sometimes referred to as **suppressor genes** or extragenic suppressors.

VIII. **Cell type**

A. **Somatic cell mutation**—occurs in cells of the body, excluding sex cells, often producing a mutant phenotype in only a sector of the organism (mosaic or chimera); not a heritable change

B. **Germ cell (gametic) mutation**—occurs in the sex cells, producing a heritable change

Mutations are often clustered to specific regions of the genome called **hot spots**. Hot spots have been observed to accumulate 10 × or even 100 × the number of mutations expected by a random distribution. There are hot spots for spontaneous mutations and potentially different hot spots for induced mutations. Spontaneous mutation hot spots in the bacterium *E. coli* are localized to places in the DNA where a special, modified base called 5-methylcytosine occurs. This base is susceptible to spontaneous rearrangement of a hydrogen atom (called a **tautomeric shift**) that alters its base-pairing capacity, i.e., a rearranged 5-methylcytosine will more likely pair with A, rather than G. This results in a change in base pairs from AT to GC.

Large genomic rearrangements can lead to a phenomenon called **position effect** (for more on position effects, see Chapter 7). By moving a gene out of its chromosomal context, its expression may be altered. The most common type of alteration is a decrease in the production of the encoded gene product that, in turn, leads to an abnormal phenotype. However, some genes, when relocated, can be inappropriately produced. For example, some **oncogenes** (cancer-causing genes) are wild-type genes that have been relocated to a new chromosomal position that allows them to be produced at times during the cell cycle when they are normally turned off. Their inappropriate expression leads to increased cell division and growth of a tumor (for more on cancer, see Chapter 11).

SOLVED PROBLEM 3.6

If 54 mutations are detected among 723 progeny of males that received a gamma ray dosage of 2500 roentgen (r) units and 78 mutations among 649 progeny of males that received 4000 roentgens, how many mutants would be expected to appear among 1000 progeny of males that received 6000 roentgens?

Solution: The number of mutations induced by ionizing radiation is directly proportional to the dosage.

$$78/649 \quad = 12.02\% \text{ mutations at 4000 roentgens}$$

$$54/723 \quad = 7.47\% \text{ mutations at 2500 roentgens}$$

$$\text{Difference} = 4.55\% \text{ mutations for 1500 roentgens}$$

Among 1000 progeny at 6000 roentgens we expect $1000(6000/1500)(0.0455) = 182$ mutants.

DNA Repair

Cells employ DNA repair mechanisms to correct mistakes in the base sequence of DNA molecules. Mistakes can occur spontaneously during normal cellular actitities, or be induced (see Table 3-2). There are several different types of repair mechanisms.

1. **Photoreactivation**. This mechanism involves the removal or reversal of the damaged DNA by a single, light-dependent enzyme. In the bacterium *E. coli*, the enzyme is encoded by the *phr* gene.
2. **Excision or Dark Repair**. This mechanism involves four steps.
   (*a*) A single-strand break is made on the 5′ side near the dimer by a specific endonuclease called UV endonuclease.
   (*b*) The 5′ to 3′ exonuclease activity of DNA polymerase I removes nucleotides near the cut, including the damage.
   (*c*) One of the DNA polymerases (possibly pol I) synthesizes a correct replacement strand from 5′ to 3′ using information from the intact complementary strand.
   (*d*) Polynucleotide ligase seals the break. In the bacterium *E. coli*, the enzymes involved in excision repair are encoded by the *uvr* genes.
3. **Mismatch Repair.** This repair mechanism detects base pairs that are not "matched" or not paired properly. Mismatches can occur during DNA replication or recombination. Mismatches are repaired using excision repair (point 2, above). There are several genes, such as the *mut* genes, known to participate in mismatch repair in *E. coli*.
4. **SOS Repair.** This is a form of error-prone replication that repairs lesions in DNA without regard for restoring the original base sequence. This type of

repair may be triggered by high levels of DNA damage. In the bacterium *E. coli*, this system is governed by *recA* and *umu* genes that are hypothesized to alter the fidelity of the native DNA polymerase. In the process, the polymerase replicates past DNA damage, thus allowing the cell the possibility of survival. If the cell survives the overall DNA damage, it has a high probability of containing one or more mutations.

Defining the Gene

Mendel's work suggested that a **gene** was a discrete "factor" that controlled a given phenotype (e.g., a gene for tall vs. short pea vine growth). Although the physical nature of the gene was not understood until the middle of the 20th century, the work of geneticists established it as the basic biological unit of heredity. Later work showed that genes were composed of DNA, not protein. One of the earliest concepts of gene action to explain human metabolic disorders proposed that each such gene was responsible for a specific enzymatic reaction; hence, the **"one gene–one enzyme hypothesis"** was born. Then it was discovered that some enzymes consist of more than one kind of polypeptide chain. For example, the bacterial enzyme tryptophan synthetase is a tetramer of two α-chains and two β-chains; each of the two types of chains is specified by a different genetic locus. This type of protein is known as a **heterotetramer**, since it is made up of different polypeptide chains. A tetramer composed of four of the same polypeptide chains would be a **homotetramer**. The homo- and hetero- prefixes can be applied to protein dimers, octomers, etc. The paradigm then became **"one gene–one polypeptide chain."** More recently, it has been found that, in some cases, a single gene can give rise to more than one different polypeptide chain by a process called **alternative splicing** (see Example 13.8). It is also known that many genes may contribute to a single character or trait (**polygenic traits** or traits exhibiting continuous variation; Chapter 8) and that each gene may have multiple phenotypic effects (**pleiotropy**). The terms **structural gene** or **protein-coding gene** are often used for genes whose code specifies the sequence of a polypeptide chain. Because tRNA and rRNA molecules are also encoded by DNA sequences, it can be argued that the definition of a gene should be extended to a **"transcription unit"** or a region of DNA between the sites of initiation and termination of transcription. However, some transcription units may contain more than one structural or RNA gene. Some genes are **overlapping** in the sense that the DNA sequence of one gene begins (usually in a different reading frame) in the DNA sequence of another. In summary, a gene has the following characteristics: (1) the physical unit of heredity, (2) a sequence of DNA that occupies a particular locus on a chromosome, and (3) codes for a functional product, such as a protein or RNA molecule.

There are other important genetic elements in the genome of an organism. One major class of such elements is the "control region." Control regions, such as **promoters** and **transcription termination signals**, are not transcribed or translated so they do not specify proteins or functional RNA molecules. However, they contain sequence information that is involved in controlling when a gene is transcribed (see Chapters 10 and 13).

The term **cistron** was originally equated with a region of DNA that specifies a complete polypeptide chain; thus, a cistron is equivalent to a structural gene. However, the term "cistron" was originally given to a functional genetic unit defined by the **complementation test (cis-trans test)**. This test is used to determine whether two different mutations lie in the same or different genes.

EXAMPLE 3.7

Two point mutants (m_1 and m_2) in the same gene are functionally allelic.

(*a*) Cis position (both mutant nucleotides on one homologue)

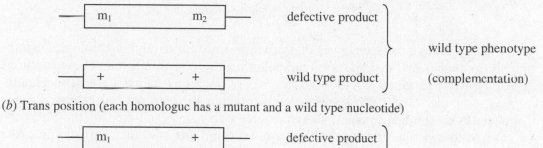

(*b*) Trans position (each homologue has a mutant and a wild type nucleotide)

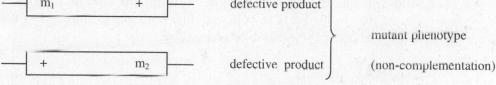

Thus, two mutations are considered to be **functionally allelic** (in the same gene, or cistron) if they complement in the cis position, but not in the trans position. However, if a heterozygote contains two mutations in either the cis or the trans position in different functional units (i.e., different genes, or cistrons), **complementation** can produce a normal phenotype because for each mutant functional unit on one DNA molecule there is a corresponding normal functional unit on the other DNA molecule. Thus, two mutants are **functionally nonallelic** if they complement in either the cis or the trans position.

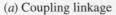

EXAMPLE 3.8

Two point mutants in different genes are functionally nonallelic.

(*a*) Coupling linkage

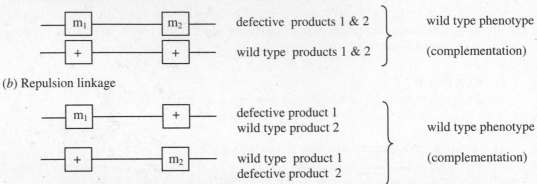

(*b*) Repulsion linkage

Many genes encode enzymes that are important for catalyzing biological synthesis (**anabolic**) and degradation (**catabolic**) reactions within a cell. These reactions are typically grouped together into a series of reactions, called a **biochemical pathway**, starting with a **substrate** that is converted through several chemical **intermediates** to the **end product**. Enzymes help carry out each step of the pathway; thus, wild-type enzyme function is required of all enzymes in a pathway in order for it to function properly. If any one enzyme is missing or mutated so that it is not functional, the pathway will not be completed.

EXAMPLE 3.9

A biochemical pathway is known that converts substrate A into product E, through three intermediate compounds B, C, and D. Enzymes 1–4 are required for each of the indicated steps.

$$A \xrightarrow{\ 1\ } B \xrightarrow{\ 2\ } C \xrightarrow{\ 3\ } D \xrightarrow{\ 4\ } E$$

If the gene encoding enzyme 3 contains a nonsense mutation so that a non-functional, truncated protein is produced, products D and E will not be produced and intermediate C will most likely accumulate inside the cell. This type of mutation can be "rescued" in one of several ways: (1) provide intermediate D to the cell, (2) provide product E to the cell, (3) provide a wild-type copy of the enzyme 3 gene to the cell.

In bacteria and other microorganisms, it is relatively simple to isolate a number of mutant strains that are defective in the synthesis of a particular end product, such as an amino acid. Presumably, each mutant contains a mutation in one of the enzyme genes required for the synthesis of the end product. By analysis of the nutrient requirements of these mutants, the order of the pathway can be established.

SOLVED PROBLEM 3.7

Four single mutant strains of *Neurospora* are unable to grow on minimal medium unless supplemented by one or more of the substances A–F. In the following table, growth is indicated by + and no growth by 0. In addition, both strains 2 and 4 grow if E and F, or C and F, are added to minimal medium. Diagram a biochemical pathway consistent with the data involving all six metabolites, indicating where the mutant block occurs in each of the four strains.

Strain	A	B	C	D	E	F
1	+	0	0	0	0	0
2	+	0	0	+	0	0
3	+	0	+	0	0	0
4	+	0	0	0	0	0

Solution: Strain 1 will grow only if given substance A. Therefore, the defective enzyme produced by the mutant gene in this strain must act sometime prior to the formation of substance A and after the formation of substances B, C, D, E, and F. In other words, this mutation is probably causing a metabolic block in the last step of the biochemical sequence in the synthesis of substance A.

$$(B, C, D, E, F) \xrightarrow{\quad 1 \quad} A$$
order unknown

Strain 2 grows if supplemented by either A or D, but not by B. Therefore, the metabolic block in strain 2 must occur after B but before A. Furthermore, since the dual addition of substances E and F, or C and F, allows strain 2 to grow, we can infer that the intervenient pathway is split, with E and C in one line and F in the other. Substance D could be at one of two positions as shown below.

$$B \xrightarrow{\quad 2 \quad} D? \left\langle \begin{array}{c} E, C \\ \text{(unknown order)} \\ F \end{array} \right\rangle D? \xrightarrow{\quad 1 \quad} A$$

Strain 3 grows if supplemented by A or C, but not by D. Therefore, D cannot immediately precede A (as shown above), and the metabolic block in strain 3 must precede the formation of C but not E.

$$B \xrightarrow{\quad 2 \quad} D \left\langle \begin{array}{c} E \xrightarrow{\quad 3 \quad} C \\ F \end{array} \right\rangle \xrightarrow{\quad 1 \quad} A$$

Strain 4 can grow if given dual supplementation of E and F, or C and F, but not if given D alone. The mutation in strain 4 apparently cannot split D into E and F.

$$B \xrightarrow{\quad 2 \quad} D \xrightarrow{\quad 4 \quad} \left\langle \begin{array}{c} E \xrightarrow{\quad 3 \quad} C \\ F \end{array} \right\rangle \xrightarrow{\quad 1 \quad} A$$

Supplementary Problems

DNA AND PROTEIN SYNTHESIS

3.8. Given a single strand of DNA . . . 3′-TACCGAGTACTGACT-5′ . . . , construct (*a*) the complementary DNA chain, (*b*) the mRNA chain that would be made from the given strand, and (*c*) the polypeptide encoded (use Table 3-1). (*d*) Which strand is the sense strand?

3.9. In 1928, before DNA was recognized as the hereditary material, F. Griffith performed a series of experiments in which he infected mice with two different strains of pneumonia-causing bacteria (*S. pneumonia*). He had in his laboratory two strains of this bacterium; the S strain was virulent (disease-causing) and had a smooth colony appearance and the R strain was not virulent and had a rough colony appearance.

　　Experiment 1: Mice infected with the S strain became sick and died.
　　Experiment 2: Mice infected with the R strain did not become sick.
　　Experiment 3: Mice infected with heat-killed S bacteria did not become sick.
　　Experiment 4: Mice infected with a mixture of heat-killed S bacteria and live R bacteria became sick and died.

(*a*) How can the results of Experiment 4 be explained? (*b*) Do these results provide evidence that DNA is the hereditary material?

3.10. If the ratio (A + G)/(T + C) in one strand of DNA is 0.7, what is the same ratio in the complementary strand?

3.11. The bacterial virus, phage T4, is a simple infectious agent that is composed of a DNA core and a protein coat. It infects a bacterial cell by attaching to its surface and then injecting material into the cell. This material then directs the synthesis of new viral particles. Using radioactive phosphorous (^{32}P) and radioactive sulfur (^{35}S), design an experiment using this virus to show that DNA is the hereditary material. (*Hint*: see Example 12.1.)

3.12. How many different mRNAs could specify the amino acid sequence met-phe-ser-pro?

3.13. If the coding region of a gene is estimated to consist of 450 nucleotide base pairs (bp), how many amino acids would the corresponding polypeptide chain contain?

3.14. Given the hypothetical enzyme below with regions A, B, C, and D (* = disulfide bond; hatched area = active site), explain the effect of each of the following mutations in terms of the biological activity of the mutant enzyme: (*a*) nonsense in DNA coding for region A, (*b*) silent in region D, (*c*) deletion of one complete codon in region C, (*d*) missense in region B, (*e*) nucleotide addition in region C.

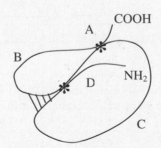

3.15. A large dose of ultraviolet irradiation can kill a wild-type cell even if the DNA repair system is unsaturated. How?

3.16. A single base addition and a single base deletion approximately 15 bases apart in the mRNA specifying the protein lysozyme from the bacterial virus T4 caused a change in the protein from its wild-type composition . . . lys-ser-pro-ser-leu-asn-ala-ala-lys . . . to the mutant form . . . lys-val-his-his-leu-met-ala-alalys . . . . (a) From the mRNA codons listed in Table 3.1, decipher the segment of mRNA for both the original protein and the double mutant. (b) Which base was added? Which was deleted?

3.17. If the DNA of an *E. coli* has 4.2×10^6 nucleotide pairs in its DNA, and if an average gene contains 1500 nucleotide pairs, how many genes does it potentially possess?

3.18. The DNA of bacterial phage lambda has 1.2×10^5 nucleotides. How many proteins of molecular weight 40,000 could be coded by this DNA? Assume a molecular weight of 100 for the average amino acid.

MUTATIONS

3.19. Acridine dyes can apparently cause a mutation in the bacteriophage T4 by the addition or deletion of a base in the DNA chain. A number of such mutants have been found in the r_{II} region of T4 to be single base addition type (+) or deletion type (−) mutants. A normal or wild-type r_{II} region produces a normal lytic period (small plaque size) in the host bacterium *Escherichia coli* strain B. Phage T4 mutants in the r_{II} region rapidly lyse strain B, producing a larger plaque. Several multiple mutant strains of T4 have been developed. Determine the lytic phenotypes (large plaque or small plaque) produced by the following r_{II}, single base mutations in *E. coli* B assuming a triplet codon (suppose the mutant sites are close together):
(a) (+)   (b) (+)(−)   (c) (+)(+)   (d) (−)(−)   (e) (+)(−)(+)   (f) (+)(+)(+)
(g) (−)(−)(+)   (h) (−)(−)(−)   (i) (+)(+)(+)(+)   (j) (+)(−) (+)(+)(+).

3.20. The "dotted" gene in maize (*Dt*) is a "mutator" gene influencing the rate at which the gene for colorless aleurone (a) mutates to its dominant allele (*A*) for colored aleurone. An average of 7.2 colored dots (mutations) per kernel was observed when the seed parent was *dt/dt, a/a* and the pollen parent was *Dt/Dt, a/a*. An average of 22.2 dots per kernel was observed in the reciprocal cross. How can these results be explained?

3.21. Assuming no intensity effect is operative, which individual would carry fewer mutations: an individual who receives 25 roentgens in 5 h or an individual who receives only 0.5 roentgen per year for his or her normal lifetime (60 years)? In terms of percentage, how many more mutations would be expected in the individual with the higher total dosage?

3.22. If the mutation rate of a certain gene is directly proportional to the radiation dosage, and the mutation rate of *Drosophila* is observed to increase from 3% at 1000 roentgens to 6% at 2000 roentgens, what percentage of mutations would be expected at 3500 roentgens?

3.23. A number of nutritional mutant strains were isolated from wild-type *Neurospora* that responded to the addition of certain supplements in the culture medium by growth (+) or no growth (0). Given the following responses for single-gene mutants, diagram a metabolic pathway that could exist in the wild-type strain consistent with the data, indicating where the chain is blocked in each mutant strain.
(a)

Strain	Citrulline	Glutamic Semialdehyde	Arginine	Ornithine	Glutamic Acid
1	+	0	+	0	0
2	+	+	+	+	0
3	+	0	+	+	0
4	0	0	+	0	0

Mutant Supplements Added to Minimal Culture Medium

(b)

Strain	Growth Factors			
	A	B	C	D
1	0	0	+	+
2	0	0	0	+
3	+	0	+	+
4	0	+	+	+

(c)

Strain	Nutrients			
	E	F	G	H
1	+	0	0	+
2	0	0	+	+
3	0	0	0	+
4	0	+	0	+

3.24. Point mutations correlated with amino acids in the active site in widely separated regions of a gene can render its enzymatic or antibody product inactive. What inference can be made concerning the three-dimensional structure of the active sites in such proteins?

3.25. A nonsense point mutation in one gene can sometimes be at least partially suppressed in its phenotypic manifestation by a point mutation in a different gene. Offer an explanation for this phenomenon of second-site suppression.

3.26. In addition to the kind of mechanism accounting for second-site suppression of nonsense mutations (see previous problem), give two other possible mechanisms for this type of suppression of missense mutations.

3.27. Interallelic *in vitro* complementation has been observed in alkaline phosphatase enzymes and other proteins. How can a diploid heterozygote bearing two point mutations within homologous genes result in progeny with normal or nearly normal phenotypes (complementation)?

3.28. Why are most mutations in protein-coding genes recessive to their wild-type alleles?

3.29. Bacterial cells that are sensitive to the antibiotic streptomycin (str^s) can mutate to a resistant state (str^r). Such "gain of function" mutations, however, occur much less frequently than "loss of function" mutations such as mutation from the ability to make the amino acid histidine (his^+) to the inability to do so (his^-), or mutation from the ability to metabolize the sugar lactose (lac^+) to the inability to do so (lac^-). Formulate a hypothesis that explains these observations.

DNA REPAIR

3.30. Exposure to UV light results in the formation of thymine dimers between adjacent thymines in a single DNA strand. Describe how repair of this damage might lead to a mutation.

3.31. A new mutant strain of *E. coli* has been identified that is observed to produce progeny containing spontaneous mutations in many different genes at a higher rate than wild-type *E. coli*. The mutant is described as having a "mutator" phenotype. Hypothesize about the wild-type function of the mutated gene and why it has this mutant phenotype.

Review Questions

Vocabulary For each of the following definitions, give the appropriate term and spell it correctly. Terms are single words unless indicated otherwise.

1. The method of DNA replication in which each strand of the double-helical molecule serves as a template against which a complementary new strand is synthesized.
2. A genetic locus that serves as a recognition site for RNA polymerase attachment.
3. A group of three nucleotides in mRNA that together specify an amino acid.
4. A short RNA sequence onto which DNA polymerase III adds deoxyribonucleotides during bacterial DNA replication.
5. Development of a wild-type (normal) trait in an organism or cell containing two different mutations combined in a hybrid diploid or a heterokaryon.
6. A spiral secondary structure in parts of many peptide chains, constituting the secondary level of organization. (One or two words.)
7. The process whereby RNA is synthesized from a DNA template.
8. The single-stranded pieces of DNA produced by discontinuous replication of double-stranded DNA. (Two words.)
9. Regions within an eukaryotic primary transcript that are removed during processing of mRNA.
10. Removal or reversal of damaged DNA by a light-dependent enzyme.

Multiple-Choice Questions Choose the one best answer.

1. A genetic unit that codes for the amino acid sequence of a complete polypeptide chain is most closely related to   (*a*) an anticodon   (*b*) a promoter   (*c*) a gene   (*d*) a codon   (*e*) a homotrimer
2. Without referring to a table of mRNA codons, solve the following problem. Given a hypothetical segment of antisense strand DNA 3'-GGC AAC CTT GGC 5', the corresponding polypeptide segment could be   (*a*) H$_2$N-gly-asn-leu-pro-COOH   (*b*) HOOC-his-arg-ser-tyr-NH$_2$ (*c*) HOOC-asp-val-ile-gln-NH$_2$ (*d*) H$_2$N-met-thr-phe-cys-COOH   (*e*) H$_2$N-pro-leu-glu-pro-COOH
3. Given the antisense strand DNA codon 3' TAC 5', the anticodon that pairs with the corresponding mRNA codon could be   (*a*) 3' CAT 5'   (*b*) 5' AUG 3'   (*c*) 3' UAC 5'   (*d*) 5' GUA 3'   (*e*) none of the above
4. Which of the following is not a characteristic of cellular RNA?   (*a*) contains uracil   (*b*) is single-stranded   (*c*) is much shorter than DNA   (*d*) serves as template for its own synthesis   (*e*) contains ribose
5. A mutation in the codon UCG to UAG (Table 3-1) is be described as   (*a*) a missense mutation   (*b*) a neutral mutation   (*c*) a silent mutation   (*d*) a nonsense mutation   (*e*) a frameshift mutation
6. An amino acid that cannot participate in alpha-helical formation is   (*a*) proline   (*b*) histidine   (*c*) phenylalanine   (*d*) threonine   (*e*) more than one of the above
7. A coding system in which each word may be coded by a variety of symbols or groups of letters (e.g., the genetic code) is said to be   (*a*) archaic   (*b*) redundant   (*c*) degenerate   (*d*) polysyllabic   (*e*) amplified
8. An amino-acyl synthetase is responsible for   (*a*) formation of a peptide bond   (*b*) attaching an amino group to an organic acid   (*c*) causing a peptide chain to form secondary and higher structural organizations   (*d*) movement of tRNA molecules from A to P sites on a ribosome   (*e*) joining an amino acid to a tRNA
9. The step in the flow of genetic information wherein DNA is copied into mRNA is called   (*a*) translation   (*b*) reverse transcription   (*c*) replication   (*d*) transcription   (*e*) polymerization
10. The growing polypeptide chain is first attached to a tRNA molecule in which site of the ribosome?   (*a*) A site   (*b*) B site   (*c*) P site   (*d*) anticodon loop   (*e*) active site

Answers to Supplementary Problems

3.8. (*a*) 5′-ATGGCTCATGACTGA 3′ (*b*) 5′-AUGGCUCAUGACUGA 3′ (*c*) met-ala-his-asp-(stop) (*d*) The complementary strand shown in the answer for part (*a*) is the sense or untranscribed strand.

3.9. (*a*) The R bacteria have become "transformed" into virulent bacteria by obtaining material from the dead S strain cells. (*b*) No, the material from the dead S strain could be DNA, RNA, protein, or any other cellular material.

3.10. 1.43

3.11. In one population of a bacteriophage, the ^{32}P radioisotope can be incorporated into DNA; and in a second phage population, the ^{35}S radioisotope can be incorporated into proteins (some amino acids have sulfur in their side chains). These two populations of phage can be used to infect *E. coli* cells in separate experiments. If DNA is the hereditary material, then it should be injected into the cell. Thus, the ^{32}P radioisotope label should be found within the bacterial cells, once the empty phage particles are removed from the outside of the cell. If protein is the hereditary material, the ^{35}S radioisotope label will be found within the bacterial cells.

3.12. $1 \times 2 \times 6 \times 4 = 48$

3.13. Approximately 150 amino acids

3.14. (*a*) The protein would be slightly shorter than normal. Since region A does not seem to interact with other portions of the polypeptide chain, the mutant enzyme should still function normally (barring unpredicted interaction of the side chain of the mutant amino acid with other parts of the molecule). If a nonsense mutation had occurred in region D, however, a very small chain would have been produced that would be devoid of a catalytic site, because proteins are synthesized beginning at the NH$_2$ end. (*b*) Same sense mutants produce no change in their polypeptide products from normal. (*c*) The polypeptide would be one amino acid shorter than normal. Since region C does not seem to be critical to the tertiary shape of the molecule, the mutant enzyme would probably function normally. (*d*) An incorrect amino acid would be present in region B. As long as its side chain did not alter the tertiary shape of the molecule, the mutant enzyme would be expected to function normally. (*e*) A frameshift mutant in region C is bound to create many missense codons (or perhaps a nonsense codon) from that point on through the carboxyl terminus, including the enzymatic site. Such a protein would be catalytically inactive.

3.15. Nonfunctional DNA fragments might be produced if DNA replication occurs before all of the critical repairs have been made.

3.16. (*a*)

	lys	ser	pro	ser	leu	asn	ala
normal mRNA	AA?	AGU	CCA	UCA	CUU	AAU	GC?
mutant mRNA	AA?	GUC	CAU	CAC	UUA	AUG	GC?
	lys	val	his	his	leu	met	ala

(*b*) G was added, A was deleted (bold)

3.17. 2800 genes

3.18. 50 proteins, assuming only one strand of the DNA is transcribed into mRNA

3.19. (*a*), (*c*), (*d*), (*e*), (*g*), (*i*) = large plaque, (*b*), (*f*), (*h*), (*j*) = small plaque

3.20.   Seed parent contributes two sets of chromosomes to triploid endosperm; one *Dt* gene gives 7.2 mutations per kernel, two *Dt* genes increase mutations to 22.2 per kernel

3.21.   20% more mutations in the individual receiving 0.5 roentgen per year.

3.22. $10\frac{1}{2}\%$.

3.23.

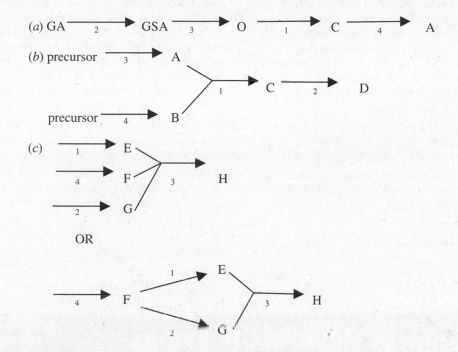

OR

3.24.   The polypeptide chain folds into a configuration such that noncontiguous regions form portions of the catalytic or antibody-combining sites.

3.25.   The suppressing mutation could be in that portion of a gene specifying the anticodon region of a tRNA molecule. For example, a tyrosine suppressor gene changes the anticodon of tRNAtyr from 3'-AUG-5' to 3'-AUC–5', thereby allowing it to recognize UAG mRNA nonsense codons. If the genes for tRNAtyr exists in multiple copies and only one of the tRNAtyr genes was mutated to a suppressor form, there would still be other normal (nonsuppressor) tRNAtyr genes to make some normal proteins. The efficiency of suppression must be low to be compatible with survival of the organism.

3.26.   (1) A change in one of the ribosomal proteins in the 30S subunit could cause misreading of the codon-anticodon alignment, resulting in substitution of an "acceptable" (although perhaps not the normal) amino acid in a manner analogous to the misreading induced by the antibiotic streptomycin. In a cell-free system with synthetic poly-U mRNA, streptomycin causes isoleucine tRNA to be substituted for that of phenylalanine tRNA. (2) A mutation in a gene coding for an amino-acid-activating enzyme (amino-acyl synthetase) causes a different amino acid to occasionally be attached to a given species of tRNA. For example, if AUU (isoleucine mRNA codon) is mutated to UUU (phenylalanine mRNA codon), its effect may be suppressed by the occasional misattachment of isoleucine to tRNAphe by a mutant amino-acyl synthetase that is less than 100% specific in its normal action of attaching phenylalanine to tRNAphe.

3.27.   Such proteins are normally homopolymers (quaternary complexes consisting of two or more identical polypeptide chains). If two mutant polypeptide chains contain compensating amino acid substitutions, they may aggregate into a heterodimer that exhibits at least partial enzymatic activity.

3.28. Wild-type alleles usually code for complete, functional enzymes or other proteins. One active wild-type allele can often cause enough enzyme to be produced so that normal or nearly normal phenotypes result (dominance). Mutations of normally functioning genes are more likely to destroy the biological activities of proteins. Only in the complete absence of the wild-type gene product would the mutant phenotype be expressed (recessiveness).

3.29. Loss of function can potentially occur by point mutations at a number of sites within a cistron coding for a given fermentation enzyme or in any gene coding for one of the multiple enzymes in a common biosynthetic pathway such as those in histidine synthesis. The loss of such an indispensable function is lethal. Streptomycin distorts ribosomes, causing misreading of the genetic code. Only a limited number of changes in the ribosomal proteins or rRNA could render the ribosome immune from interference by streptomycin and still preserve the way these components normally interact with mRNA, tRNA, initiation factors, etc., during protein synthesis.

3.30. Repair of thymine dimers occurs using excision repair. This type of repair involves using the remaining template (after excision) to synthesize a new, replacement strand. This synthesis, especially when there is a lot of DNA damage, can be error-prone, thus leading to mutations.

3.31. The wild-type function of this gene may be in a DNA repair function. When mutated, the cell cannot repair its DNA efficiently, leading to a higher rate of spontaneous mutations across the genome.

 # Answers to Review Questions

Vocabulary

1. semiconservative
2. promoter
3. codon
4. primer
5. complementation
6. alpha helix
7. transcription
8. Okazaki fragments
9. introns
10. photoreactivation

Multiple-Choice Questions

1. *c*　2. *e*　3. *c*　4. *d*　5. *d*　6. *a*　7. *c*　8. *e*　9. *d*　10. *c*

CHAPTER 4

Genetic Interactions

Two-Factor Interactions

The phenotype is a result of many gene products expressed in a given environment. The environment includes not only external factors such as temperature and the amount or quality of light, but also internal factors such as hormones and enzymes. One important type of protein gene product is enzymes. **Enzymes** perform catalytic functions, causing the splitting or union of various molecules. **Metabolism** is the sum of all the physical and chemical processes by which cells are produced and maintained and by which energy is made available for the uses of the organism. These biochemical reactions occur as stepwise conversions of one substance into another, each step being mediated by a specific enzyme. All of the steps that transform a precursor substance to its end product constitute a **biosynthetic pathway**.

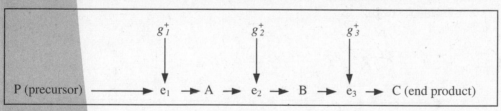

Legend: e = enzyme, g = gene, A, B = metabolites, and C = end product.

Several genes are usually required to specify the enzymes involved in even the simplest pathways. Each metabolite (A, B, C) is produced by the catalytic action of different enzymes (e_x) specified by different wild-type genes (g_x^+). One common example of **genetic interaction** occurs whenever two or more genes specify enzymes that catalyze steps in a common pathway. If substance C is essential for the production of a normal phenotype, and the recessive mutant alleles g_1, g_2, and g_3 produce defective enzymes, then a mutant (abnormal) phenotype would result from a genotype that is homozygous recessive at any of the three loci. If g_3 contains a mutation, the conversion of B to C does not occur and substance B tends to accumulate in excessive quantity; if g_2 contains a mutation, substance B will not be produced and substance A will accumulate. Thus, gene mutations are said to produce "metabolic blocks." An organism with a mutation in only g_2 could produce a normal phenotype if it were given either substance B or C, but an organism with a mutation in g_3 has a specific requirement for C. Thus, gene g_3^+ becomes dependent upon gene g_2^+ for its expression as a normal phenotype. If the genotype is homozygous for the recessive g_2 allele, then the pathway ends with substance A. Neither g_3^+ nor its recessive allele g_3 has any effect on the phenotype. Thus, the genotype g_2g_2 can hide or mask the phenotypic expression of alleles at the g_3 locus. It is said that g_2 is epistatic to g_3, i.e., it masks any expression that may or may not occur from g_3^+ alleles. Originally, a gene or locus that suppressed or masked the action of a gene at another locus was termed **epistatic**. The gene or locus suppressed was **hypostatic**. Later it was found that both loci could be mutually epistatic to one another. Now the term "epistasis" has come to be synonymous with almost any type of gene interaction that involves the masking of one of the gene effects. Dominance involves intra-allelic gene suppression, or the masking effect that one allele has upon the expression of another allele at the same locus. Epistasis involves interallelic gene suppression, or the masking effect that one gene locus has upon the expression of another. The classical phenotypic ratio of 9 : 3 : 3 : 1 observed in the progeny of dihybrid parents becomes modified by epistasis into ratios that are various combinations of the 9 : 3 : 3 : 1 groupings.

EXAMPLE 4.1
A particularly illuminating example of gene interaction occurs in white clover. Some strains have a high cyanide content; others have a low cyanide content. Crosses between two strains with low cyanide have produced an F_1 with a high concentration of cyanide in their leaves. The F_2 shows a ratio of 9 high cyanide : 7 low cyanide. Cyanide is known to be produced from the substrate cyanogenic glucoside by enzymatic catalysis. One strain of clover has the enzyme but not the substrate. The other strain makes substrate but is unable to convert it to cyanide. The pathway may be diagrammed as follows where G_x produces an enzyme and g_x results in a metabolic block.

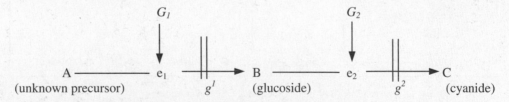

Tests on leaf extracts have been made for cyanide content before and after the addition of either glucoside or the enzyme e_2.

F_2 Ratio	Genotype	Leaf Extract Alone	Leaf Extract Plus Glucoside	Leaf Extract Plus e_2
9	$G^1\text{-}G^2\text{-}$	+	+	+
3	$G^1\text{-}g^2g^2$	0	0	+
3	$g^1g^1G^2$	0	+	0
1	$g^1g^1g^2g^2$	0	0	0

Legend: + = cyanide present, 0 = no cyanide present.

If the leaves are phenotypically classified on the basis of cyanide content of extract alone, a ratio of 9 : 7 results. If the phenotypic classification is based either on extract plus glucoside or on extract plus e_2, a ratio of 12 : 4 is produced. If all of these tests form the basis of phenotypic classification, the classical 9 : 3 : 3 : 1 ratio emerges.

Genetic interactions can also occur during embryogenesis in developmental pathways as well as other complex biological processes (Chapter 13).

Epistatic Interactions

When **epistasis** is operative between two gene loci, the number of phenotypes appearing in the offspring from dihybrid parents will be less than four. There are six types of epistatic ratios commonly recognized, three of which have three phenotypes and the other three have only two phenotypes.

1 DOMINANT EPISTASIS (12 : 3 : 1)

When the dominant allele at one locus, for example, the A allele, produces a certain phenotype regardless of the allelic condition of the other locus, then the A locus is said to be epistatic to the B locus. Furthermore, since the dominant allele A is able to express itself in the presence of either B or b, this is a case of dominant epistasis. Only when the genotype of the individual is homozygous recessive at the epistatic locus (aa) can the alleles of the hypostatic locus (B or b) be expressed. Thus, the genotypes $A-B-$ and $A-bb$ produce the same phenotype, whereas $aaB-$ and $aabb$ produce two additional phenotypes. The classical 9 : 3 : 3 : 1 ratio becomes modified into a 12 : 3 : 1 ratio.

SOLVED PROBLEM 4.1
Coat colors of dogs depend upon the action of at least two genes. At one locus a dominant epistatic inhibitor of coat color pigment ($I-$) prevents the expression of color alleles at another independently assorting locus, producing white coat color. When the recessive condition exists at the inhibitor locus (ii), the alleles of the hypostatic locus may be expressed, $iiB-$ producing black and $iibb$ producing brown. When dihybrid white dogs are mated together, determine (a) the phenotypic proportions expected in the progeny, (b) the chance of choosing, from among the white progeny, a genotype that is homozygous at both loci.

Solution:

(a) P:
$IiBb$ × $IiBb$
white   white

F_1:

$$9/16 \; I\text{-} \; B\text{-} \; \Big\} = 12/16 \text{ white}$$
$$3/16 \; I\text{-}bb \; \Big\}$$
$$3/16 \; iiB\text{-} \quad = \quad 3/16 \text{ black}$$
$$1/16 \; iibb \quad = \quad 1/16 \text{ brown}$$

(b) The genotypic proportions among the white progeny are as follows:

			Proportion of Total F$_1$	Proportion of White F$_1$
1/4 II		1/4 BB	1/16 IIBB	1/12
		1/2 Bb	2/16 IIBb	2/12
		1/4 bb	1/16 IIbb	1/12
1/2 Ii		1/4 BB	2/16 IiBB	2/12
		1/2 Bb	4/16 IiBb	4/12
		1/4 bb	2/16 Iibb	2/12
Totals:			12/16	12/12

The only homozygous genotypes at both loci in the above list are 1/12 *IIBB* and 1/2 *IIbb* = 2/12 or 1/6 of all the white progeny. Thus there is 1 chance in 6 of choosing a homozygous genotype from among the white progeny.

2 RECESSIVE EPISTASIS (9 : 3 : 4)

If the recessive genotype at one locus (e.g., *aa*) suppresses the expression of alleles at the *B* locus, the *A* locus is said to exhibit recessive epistasis over the *B* locus. Only if the dominant allele is present at the *A* locus can the alleles of the hypostatic B locus be expressed. The genotypes *A-B-* and *A-bb* produce two additional phenotypes. The 9 : 3 : 3 : 1 ratio becomes a 9 : 3 : 4 ratio.

SOLVED PROBLEM 4.2

Matings between black rats of identical genotype produced offspring as follows: 14 cream-colored, 47 black, and 19 albino. (*a*) What epistatic ratio is approximated by these offspring? (*b*) What are the genotypes of the parents and the offspring (use your own symbols)?

Solution:

(*a*) The total number of offspring is 80 (14 + 47 + 19) and 80/16 = 5. So 1/16 of 80 = 5. In order to figure out what the ratio is, use this factor. So, 5 × 3 = 15, which is close to 14; 5 × 4 = 20, which is close to 19; and 5 × 9 = 45, which is close to 47. Thus, the ratio is 9 : 3 : 4 (the numbers are not given in this order). This is an example of recessive epistasis.

(*b*)
P: *BbCc*   ×   *BbCc*
         black males   black females

F$_1$:
9/16 *B-C-* = 9/16 black
3/16 *bbC-* = 3/16 cream
3/16 *B-cc* $\Big\}$
1/16 *bbcc* $\Big\}$ = 4/16 albino

This translates to a 9 black : 3 cream : 4 albino ratio; *cc*, the recessive condition, is epistatic to alleles at locus *B*.

3 DUPLICATE GENES WITH CUMULATIVE EFFECT (9 : 6 : 1)

If the dominant condition (either homozygous or heterozygous) at either locus (but not both) produces the same phenotype, the F$_2$ ratio becomes 9 : 6 : 1. For example, where the epistatic genes are involved in producing various amounts of a substance such as pigment, the dominant genotypes of each locus may be considered to produce one unit of pigment independently. Thus, genotypes A-bb and aaB- produce one unit of pigment each and therefore have the same phenotype. The genotype $aabb$ produces no pigment, but in the genotype A-B- the effect is cumulative and two units of pigment are produced.

SOLVED PROBLEM 4.3

Red color in wheat kernels is produced by the genotype R-B-, white by the double-recessive genotype ($rrbb$). The genotypes R-bb and rrB- produce brown kernels. A homozygous red variety is crossed to a white variety. (a) What phenotypic results are expected in the F_1 and F_2? (b) If the brown F_2 is artificially crossed at random (wheat is normally self-fertilized), what phenotypic and genotypic proportions are expected in the offspring?

Solution:

(a)

P:	$RRBB$	$\times$	$rrbb$
	red		white
F_1:		$RrBb$	
		red	

F_2

9/16 R-B-	=	9/16 red
3/16 R-bb $\big\}$	=	6/16 brown
3/16 rrB-		
1/16 $rrbb$	=	1/16 white

(b) The proportion of genotypes represented among the brown F_2 must first be determined.

	Proportion of Total F_2	Proportion of Brown F_2
$(\frac{1}{4}RR)(\frac{1}{4}bb)$	1.16 $RRbb$	1/6
$(\frac{1}{2}Rr)(\frac{1}{4}bb)$	2/16 $Rrbb$	2/6
$(\frac{1}{4}rr)(\frac{1}{4}BB)$	1/16 $rrBB$	1/6
$(\frac{1}{4}rr)(\frac{1}{2}Bb)$	2/16 $rrBb$	2/6
Totals:	6/16	6/6

Next, the relative frequencies of the various matings may be calculated in a Punnett square.

	1/6 $RRbb$	2/6 $Rrbb$	1/6 $rrBB$	2/6 $rrBb$
1/6 $RRbb$	1/36 $RRbb \times RRbb$ (1)	2/36 $RRbb \times Rrbb$ (2)	1/36 $RRbb \times rrBB$ (3)	2/36 $RRbb \times rrBb$ (4)
2/6 $Rrbb$	2/36 $Rrbb \times RRbb$ (2)	4/36 $Rrbb \times Rrbb$ (5)	2/36 $Rrbb \times rrBB$ (6)	4/36 $Rrbb \times rrBb$ (7)
1/6 $rrBB$	1/36 $rrBB \times RRbb$ (3)	2/36 $rrBB \times Rrbb$ (6)	1/36 $rrBB \times rrBB$ (8)	2/36 $rrBB \times rrBb$ (9)
2/6 $rrBb$	2/36 $rrBb \times RRbb$ (4)	4/36 $rrBb \times Rrbb$ (7)	2/36 $rrBb \times rrBB$ (9)	4/36 $rrBb \times rrBb$ (10)

Matings	Progeny	Genotypic Proportions (f)	Mating Frequency (m)	m · f
(1) $RRbb \times RRbb$	$RRbb$	100%	1/36	1/36
(2) $RRbb \times Rrbb$	$RRbb$	1/2	4/36	4/72
	$Rrbb$	1/2		4/72
(3) $RRbb \times rrBB$	$RrBb$	100%	2/36	2/36
(4) $RRbb \times rrBb$	$RrBb$	1/2	4/36	4/72
	$Rrbb$	1/2		4/72
(5) $Rrbb \times Rrbb$	$RRbb$	1/4	4/36	4/144
	$Rrbb$	1/2		4/72
	$rrbb$	1/4		4/144
(6) $Rrbb \times rrBB$	$RrBb$	1/2	4/36	4/72
	$rrBb$	1/2		4/72
(7) $Rrbb \times rrBb$	$RrBb$	1/4	8/36	8/144
	$rrBb$	1/4		8/144
	$Rrbb$	1/4		8/144
	$rrbb$	1/4		8/144
(8) $rrBB \times rrBB$	$rrBB$	100%	1/36	1/36
(9) $rrBb \times rrBB$	$rrBB$	1/2	4/36	4/72
	$rrBb$	1/2		4/72
(10) $rrBb \times rrBb$	$rrBB$	1/4	4/36	4/144
	$rrBb$	1/2		4/72
	$rrbb$	1/4		4/144

Summary of progeny genotypes:
1/9 $RRbb$
2/9 $RrBb$
2/9 $Rrbb$
1/9 $rrBB$
2/9 $rrBb$
1/9 $rrbb$

Summary of progeny phenotypes:
2/9 $R\text{-}B\text{-}$ = 2/9 red
1/3 $R\text{-}bb$ }
1/3 $rrB\text{-}$ } = 2/3 brown

1/9 $rrbb$ = 1/9 white

4 DUPLICATE DOMINANT GENES (15 : 1)

The 9 : 3 : 3 : 1 ratio is modified into a 15 : 1 ratio if the dominant alleles of both loci each produce the same phenotype without cumulative effect.

SOLVED PROBLEM 4.4

A plant of the genus *Capsella*, commonly called "shepherd's purse," produces a seed capsule, the shape of which is controlled by two independently assorting genes, represented by symbols *A* and *B*. When dihybrid plants were interpollinated, 6%

of the progeny were found to possess ovoid-shaped seed capsules. The other 94% of the progeny had triangular-shaped seed capsules. (*a*) What two-factor epistatic ratio is approximated by the progeny? (*b*) What is the genotype of the ovoid-shaped seed capsules?

Solution:

(*a*) Ninety-four percent of 16 is ~ 15 ($0.94 \times 16 = 15.04$, or $15/16 \times 100 = 93.75$) and 6% of 16 is ~ 1, thus, the ratio represented here is 15 triangular : 1 ovoid

(*b*) The genotype is the double recessive, *aabb*, which occurs 1/16 of the time. The dominant condition *A-* or *B-* is epistatic to either recessive condition, so *A-B-* (9/16), *A-bb* (3/16), *aaB-* (3/16) all exhibit the dominant phenotype ($9 + 3 + 3 = 15$).

5   DUPLICATE RECESSIVE GENES (9 : 7)

In the case where identical phenotypes are produced by both homozygous recessive genotypes, the F_2 ratio becomes 9 : 7. The genotypes *aaB-*, *A-bb*, and *aabb* produce one phenotype. Both dominant alleles, when present together, complement each other and produce a different phenotype.

SOLVED PROBLEM 4.5

Two white-flowered strains of the sweet pea (*Lathyrus odoratus*) were crossed, producing an F_1 with only purple flowers. Random crossing among the F_1 produced 96 progeny plants, 53 exhibiting purple flowers, and 43 with white flowers. (*a*) What phenotypic ratio is approximated by the F_2? (*b*) What type of interaction is involved? (*c*) What were the probable genotypes of the parental strains?

Solution:

(*a*) To determine the phenotypic ratio in terms of familiar sixteenths, the following proportion for white flowers may be made: $43/96 = x/16$, from which $x = 7.2$. That is, 7.2 white : 8.8 purple, or approximately a 7 : 9 ratio. We might just as well have arrived at the same conclusion by establishing the proportion for purple flowers: $53/96 = x/16$, from which $x = 8.8$ purple.

(*b*) A 7 : 9 ratio is characteristic of duplicate recessive genes where the recessive genotype at either or both of the loci produces the same phenotype.

(*c*) If *aa* or *bb* or both could produce white flowers, then only the genotype *A-B-* could produce purple. For two white parental strains (pure lines) to be able to produce an all-purple F1, they must be homozygous for different dominant-recessive combinations. Thus,

P:	*aaBB*	$\times$	*AAbb*
	white		white

F_1:	*AaBb*
	purple

$$F_2: \quad \begin{array}{l} 9/16\ A\text{-}B\text{-} \\ 3/16\ A\text{-}bb \\ 3/16\ aaB\text{-} \\ 1/16\ aabb \end{array} \left. \begin{array}{l} \\ \\ \\ \end{array} \right\} = \quad \begin{array}{l} = \quad 9/16\ \text{purple} \\ \\ 7/16\ \text{white} \\ \\ \end{array}$$

6 DOMINANT-AND-RECESSIVE INTERACTION (13 : 3)

Only two F_2 phenotypes result when a dominant genotype at one locus (e.g., A-) and the recessive genotype at the other (bb) produce the same phenotypic effect. Thus A-B-, A-bb, and aabb produce one phenotype and aaB- produces another in the ratio 13 : 3 (see Table 4-1).

Table 4-1. Summary of Epistatic Ratios

Genotypes	A-B-	A-bb	aaB-	aabb
Classical ratio	9	3	3	1
Dominant epistasis	12		3	1
Recessive epistasis	9	3	4	
Duplicate genes with cumulative effect	9	6		1
Duplicate dominant genes	15			1
Duplicate recessive genes	9	7		
Dominant and recessive interaction	13		3	

SOLVED PROBLEM 4.6

A dominant gene S in *Drosophila* produces a peculiar eye condition called "star." Its recessive allele S^+ produces the normal eye of wild type. The expression of S can be suppressed by the dominant allele of another locus, Su-S. The recessive allele of this locus, Su-S^+, has no effect on S^+. (a) What type of interaction is operative? (b) When a normal-eyed male of genotype Su-S/Su-S, S/S is crossed to a homozygous wild-type female of genotype Su-S^+/Su-S^+, S^+/S^+, what phenotypic ratio is expected in the F_1 and F_2?

Solution:

(a) Dominant and recessive interaction; both dominant and recessive alleles interact to result in an eye phenotype.

(b) P Su-S/Su-S, S/S × Su-S^+/Su-S^+, S^+/S^+
            wild-type male   wild-type female

   F_1:   Su-S/Su-S^+, S^+/S

All the F_1 offspring are wild type, due to the presence of the dominant Su-S allele. F_2: 3/16 s/-, Su-s^+/su-s^+ star eye: 13/16 wild type.

PEDIGREE ANALYSIS

SOLVED PROBLEM 4.7

The following pedigree involving two-factor epistasis shows the transmission of swine coat colors through three generations:

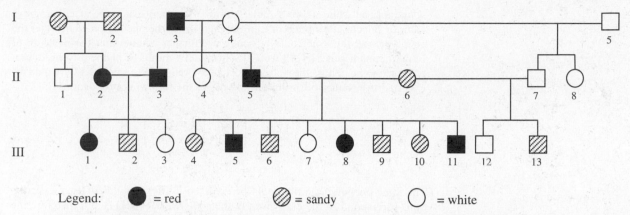

Legend: ● = red   ◑ = sandy   ○ = white

Assume the offspring of II5 × II6 shown in this pedigree occur in the ratio expected from the genotypes represented by their parents. How are these colors most likely inherited?

Solution:

Notice first that three phenotypes are expressed in this pedigree. This rules out epistatic combinations producing only two phenotypes such as those expressed in dominant and recessive interaction (13 : 3), duplicate dominant genes (15 : 1), and duplicate recessive genes (9 : 7). Epistatic gene interactions producing three phenotypes are those expressed in dominant epistasis (12 : 3 : 1), recessive epistasis (9 : 3 : 4), and dominant genes with cumulative action (9 : 6 : 1). Let us proceed to solve this problem by making an assumption and then applying it to the pedigree to see if the phenotypes shown there can be explained by our hypothesis.

Case 1. Assume dominant epistasis is operative. The genotypes responsible for the three phenotypes may be represented as follows: *A-B-* and *A-bb* = first phenotype, *aaB-* = second phenotype, and *aabb* = third phenotype. We must now determine which of the phenotypes represented in this pedigree corresponds to each of the genotypic classes. Obviously the only pure-line phenotype is the third one. Offspring of the mating *aabb* × *aabb* would all be phenotypically identical to the parents. The mating I4 × I5 appears to qualify in this respect and we shall tentatively assume that white coat color is represented by the genotype *aabb*. Certain matings between individuals with the dominant epistatic gene A could produce three phenotypically different types of offspring (e.g., *AaBb* × *AaBb*). Such a mating is observed between II2 and II3. Therefore, we might assume red color to be represented by the genotype *A-*. Sandy color must then be represented by genotype *aaB-*. Matings between sandy individuals could produce only sandy (*aaB-*) or white (*aabb*) progeny. However, sandy parents I1 × I2 produce white and red progeny (III1, II2). Therefore the assumption of dominant interaction must be wrong.

Case 2. Assume recessive epistasis to be operative. The genotypes responsible for the three phenotypes in this case may be represented as follows: *A-B-* as first phenotype, *A-bb* as second phenotype, and *aaB-* and *aabb* as third phenotype. As pointed out in case 1, matings between individuals of genotype *AaBb* are the only kind among identical phenotypes that are capable of producing all three phenotypes in the progeny. Thus *A-B-* should represent red (e.g., II2 × II3). The *aa* genotypes breed true, producing only white individuals (I4 × I5). Sandy is produced by genotype *A-bb*. Sandy × sandy (I1 × I2) could not produce the red offspring (II2). Therefore the assumption of recessive interaction must be wrong.

Case 3. Assume that duplicate genes with cumulative action are interacting. The genotypes responsible for the three phenotypes in this case may be represented as follows: *A-B-* as first phenotype, *A-bb* and *aaB-* as second phenotype, and *aabb* as third phenotype. As explained in the previous two cases, *A-B-* must be red and *aabb* must be white. If we assume that any dominant genotype at either the *A* locus or *B* locus contributes one unit of pigment to the phenotype, then either the genotype *aaB-* or *A-bb* could be sandy; we further assume that the presence of both dominant genes (*A-B-*) would contribute two units of pigment to produce a red phenotype. Thus, the mating II5 (*AaBb*) red × II6 (*aaBb*) sandy would be expected to produce offspring phenotypes in the following proportions:

Red	*A-B-*	$1/2 \cdot 3/4 = 3/8$
Sandy	*A-bb*	$\left. \begin{array}{l} 1/2 \cdot 1/4 = 1/8 \\ 1/2 \cdot 3/4 = 3/8 \end{array} \right\} 1/2$
	aaB-	
White	*aabb*	$1/2 \cdot 1/4 = 1/8$

The same phenotypic ratio would be expected if II6 were *Aabb*. These expectations correspond to the ones given in the pedigree (III4–III11) and therefore the hypothesis of duplicate genes with cumulative action is consistent with the data.

Nonepistatic Interactions

Genetic interaction may also occur without epistasis if the end products of different pathways each contribute to the same trait. This is often referred to as **complementary gene action**.

EXAMPLE 4.2
The dull-red eye color characteristic of wild-type flies is a mixture of two kinds of pigments (B and D) each produced from nonpigmented precursor compounds (A and C) by the action of different enzymes (e₁ and e₂) specified by different wild-type genes (g_1^+ and g_2^+).

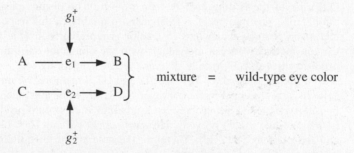

The recessive alleles at these two loci (g_1 and g_2) specify enzymatically inactive proteins. Thus, a genotype without either dominant allele would not produce any pigmented compounds and the eye color would be white.

Phenotypes	Genotypes	End Products
Wild type	$g_1^+/-, g_2^+/-$	B and D
Color B	$g_1^+/-, g_2/g_2$	B and precursor C
Color D	$g_1/g_1, g_2^+/-$	D and precursor A
White	$g_1/g_1, g_2/g_2$	None (A and C precursors unconverted)

In the above example, the genes for color B and color D produce phenotypes that are both dominant to white, but when they occur together they produce a novel phenotype (wild type) by interaction. If the two genes are assorting independently, the classical $9 : 3 : 3 : 1$ ratio will be seen in the progeny of the dihybrid parents.

EXAMPLE 4.3

A brown ommochrome pigment is produced in *Drosophila melanogaster* by a dominant gene st^+ on chromosome 3. A scarlet pterin pigment is produced by a dominant gene bw^+ on chromosome 2. The recessive alleles at these two loci produce no pigment. When pure scarlet flies are mated to pure brown flies, a different phenotype (wild type) appears in the progeny.

P:		brown	×	scarlet
		$st^+/st^+, bw/bw$		$st/st, bw^+/bw^+$
F_1:			wild type	
			$st^+/st, bw^+/bw$	

F_2:	9 $st^+/-, bw^+/-$	wild type
	3 $st^+/-, bw/bw$	brown
	3 $st/st, bw^+/-$	scarlet
	1 $st/st, bw/bw$	white

Interactions with Three or More Factors

Recall from Chapter 2 that the progeny from trihybrid parents are expected in the phenotypic ratio $27 : 9 : 9 : 9 : 3 : 3 : 1$. This classical ratio can also be modified whenever two or all three of the loci interact. Interactions involving four or more loci are also possible. Most genes probably depend to some extent upon other genes in the total genotype. The total phenotype depends upon interactions of the total genotype with the environment.

SOLVED PROBLEM 4.8

At least three loci are known to govern coat colors in mice. The genotype C- will allow pigment to be produced at the other two loci. The recessive genotype cc does not allow pigment production, resulting in "albino." The "agouti" pattern depends upon the genotype A-, and nonagouti upon the recessive aa. The color of the pigment may be black (B-) or chocolate (bb). Five coat colors may be produced by the action of alleles at these three loci:

Wild type (agouti, black)	A-B-C-
Black (nonagouti)	aaB-C-
Chocolate (nonagouti)	$aabbC$-
Cinnamon (agouti, chocolate)	A-bbC-
Albino	$----cc$

(*a*) What phenotypic frequencies are expected in the F_2 from crosses of pure black with albinos of type $AAbbcc$?

(b) A cinnamon male is mated to a group of albino females of identical genotype and among their progeny were observed 43 wild type, 40 cinnamon, 39 black, 41 chocolate, and 168 albino. What are the most probable genotypes of the parents?

Solution:

(a) P: $aaBBCC$ × $AAbbcc$ (b) P: $A\text{-}bbC\text{-}$ × $\text{-}\,\text{-}\,\text{-}\,\text{-}cc$
          pure black   albino   cinnamon male   albino females

      F$_1$: $AaBbCc$ F$_1$: 43 wild type $A\text{-}B\text{-}C\text{-}$
             wild type   40 cinnamon $A\text{-}bbC\text{-}$

      F$_2$: 27 $A\text{-}B\text{-}C\text{-}$ wild type   39 black $aaB\text{-}C\text{-}$
            9 $A\text{-}B\text{-}cc$ albino   41 chocolate $aabbC\text{-}$
            9 $A\text{-}bbC\text{-}$ cinnamon
            9 $aaB\text{-}C\text{-}$ black   <u>168</u> albino $\text{-}\,\text{-}\,\text{-}\,\text{-}cc$
            3 $A\text{-}bbcc$ albino   331 Total
            3 $aaB\text{-}cc$ albino
            3 $aabbC\text{-}$ chocolate
            1 $aabbcc$ albino

Summary of (a):   27/64 wild type
                               16/64 albino
                                9/64 cinnamon
                                9/64 black
                                3/64 chocolate

In part (b), the cinnamon progeny, $A\text{-}bbC\text{-}$, indicate b in the female parents. The black progeny, $aaB\text{-}C\text{-}$, indicate a in both parents, and B in the female parents. The chocolate progeny, $aabbC\text{-}$, indicate a in both parents, and b in the females. The albinos indicate c in the male. The genotype of the male is now known to be $AabbCc$. But the genotype of the albino females is known only to be $a\text{-}Bbcc$. They could be either $AaBbcc$ or $aaBbcc$.

Case 1. Assume the females to be $AaBbcc$.

       Parents: $AabbCc$ male   × $AaBbcc$ females

The expected phenotypic frequencies among the progency would be:

$A\text{-}BbCc$ wild type $3/4 \cdot 1/2 \cdot 1/2 = 3/16(331) =$ approx. 62
$A\text{-}bbCc$ cinnamon $3/4 \cdot 1/2 \cdot 1/2 = 3/16(331) =$ approx. 62
$aaBbCc$ black $1/4 \cdot 1/2 \cdot 1/2 = 1/16(331) =$ approx. 21
$aabbCc$ chocolate $1/4 \cdot 1/2 \cdot 1/2 = 1/16(331) =$ approx. 21
$\text{-}\,\text{-}\,\text{-}\,\text{-}cc$ albino $1 \cdot 1 \cdot 1/2 = 1/2(331) =$ approx. 166

Obviously, the expectations deviate considerably from the observations. Therefore, the females are probably not of genotype $AaBbcc$.

Case 2. Assume the females to be of genotype $aaBbcc$.

       Parents: $AabbCc$ male   × $aaBbcc$ females

The expected phenotypic frequencies among the progeny would be:

$AaBbCc$ wild type $1/2 \cdot 1/2 \cdot 1/2 = 1/8(331) =$ approx. 41
$AabbCc$ cinnamon $1/2 \cdot 1/2 \cdot 1/2 = 1/8(331) =$ approx. 41
$aaBbCc$ black $1/2 \cdot 1/2 \cdot 1/2 = 1/8(331) =$ approx. 41
$aabbCc$ chocolate $1/2 \cdot 1/2 \cdot 1/2 = 1/8(331) =$ approx. 41
$\text{-}\,\text{-}\,\text{-}\,\text{-}cc$ albino $1 \cdot 1 \cdot 1/2 = 1/2(331) =$ approx. 166

Now the expectations correspond very closely to the observations. Hence the genotype of the parental albino females is probably *aaBbcc*.

SOLVED PROBLEM 4.9

Lewis-a blood group substance appears on the human red blood cell when the dominant gene *Le* is present, but is absent if the dominant gene of the "secretor" locus *Se* is present. Suppose that from a number of families where both parents are Lewis-a negative of genotype *LeleSese*, we find that most of them have 3 Lewis-a positive : 13 Lewis-a negative children. In a few other families, suppose we find 2 Lewis-a negative : 1 Lewis-a positive. Furthermore, in families where both parents are secretors of genotype *Sese*, we find most of them exhibit a ratio of 3 secretor : 1 nonsecretor, but a few of them show 9 secretor : 7 nonsecretor. Propose a hypothesis to account for these results.

Solution: If only two loci are interacting, the dominant *Se* gene can suppress the expression of *Le*, resulting in Lewis-a negative blood type. When both parents are dihybrid, we expect a 13 : 3 ratio in the progeny, characteristic of dominant and recessive interaction.

P:	*LeleSese*	×	*LeleSese*
	Lewis-a negative		Lewis-a negative

F₁	9 *Le-Se-* ⎫		
	3 *leleSe* ⎬	=	13 Lewis-a negative
	1 *lelesese* ⎭		
	3 *Le-sese*	=	3 Lewis-a positive

The 9 : 7 ratio found in some families for the secretor trait indicates that two factors are again interacting. This is the ratio produced by duplicate recessive interaction; i.e., whenever the recessive alleles at either of two loci are present, a nonsecretor phenotype results. Let us symbolize the alleles of the second locus by *X* and *x*.

P:	*SeseXx*	×	*SeseXx*
	secretor		secretor
F1:	9 *Se-X-*	=	9 secretors
	3 *Se-xx* ⎫		
	3 *seseX-* ⎬	=	7 nonsecretors; the *xx* genotype
	1 *sesexx* ⎭		suppresses the expression of *Se*

If we assume the *x* gene to be relatively rare, then most families will have only the dominant gene *X*, but in a few families both parents will be heterozygous *Xx*. Let us assume that this is the case in those families that produce 2 Lewis-a negative : 1 Lewis-a positive.

P:	*LeleSeseXx*	×	*LeleSeseXx*
	Lewis-a negative		Lewis-a negative

F₁:	27 *Le-Se-X-*
	*9 *Le-Se-xx*
	*9 *Le-seseX-*
	9 *leleSe-X-*
	*3 *Le-sesexx*
	3 *leleSe-xx*
	3 *leleseseX-*
	1 *lelesesexx*

If *Se* suppresses *Le*, but *xx* suppresses *Se*, then only the genotypes marked with an asterisk (*) will be Lewis-a positive, giving a ratio of 21 Lewis-a positive : 43 Lewis-a negative. This is very close to a 1 : 2 ratio and indeed would appear to be such with limited data.

Pleiotropism

Many and perhaps most of the biochemical pathways in the living organism are interconnected and often interdependent. Products and intermediates of one pathway may be used in several other metabolic schemes. It is not surprising, therefore, that the phenotypic expression of a gene usually effects more than one trait. Sometimes one trait will be clearly evident (major effect) and other, perhaps seemingly unrelated ramifications (secondary effects) will be less evident to the casual observer. In other cases, a number of related changes may be considered together as a **syndrome**. All of the manifold phenotypic expressions of a single gene are spoken of as **pleiotropic** gene effects.

EXAMPLE 4.4
The syndrome called "sickle-cell anemia" in humans is due to an abnormal hemoglobin. This is the primary effect of the mutant gene. Subsidiary effects of the abnormal hemoglobin include the sickle shape of the cells and their tendency to clump together and clog blood vessels in various organs of the body. As a result, heart, kidney, spleen, and brain damage are common elements of the syndrome. Defective corpuscles are readily destroyed in the body, causing severe anemia.

Supplementary Problems

TWO-FACTOR INTERACTIONS

4.10. When homozygous yellow rats are crossed to homozygous black rats, the F_1 is all gray. Mating the F_1 among themselves produced an F_2 consisting of 10 yellow, 28 gray, 2 cream-colored, and 8 black. (*a*) How are these colors inherited? (*b*) Show, using appropriate genetic symbols, the genotypes for each color. (*c*) How many of the 48 F_2 rats were expected to be cream-colored? (*d*) How many of the 48 F_2 rats were expected to be homozygous?

4.11. Four comb shapes in poultry are known to be governed by two gene loci. The genotype *R-P-* produces walnut comb, characteristic of the Malay breed; *R-pp* produces rose comb, characteristic of the Wyandotte breed; *rrP-* produces pea comb, characteristic of the Brahma breed; *rrpp* produces single comb, characteristic of the Leghorn breed. (*a*) If pure Wyandottes are crossed with pure Brahmas, what phenotypic ratios are expected in the F_1 and F_2? (*b*) A Malay hen was crossed to a Leghorn cock and produced a dozen eggs, three of which grew into birds with rose combs and nine with walnut combs. What is the probable genotype of the hen? (*c*) Determine the proportion of comb types that would be expected in offspring from each of the following crosses: (1) *Rrpp* × *RrPP* (2) *rrPp* × *RrPp* (3) *rrPP* × *RRPp*

(4) $RrPp \times rrpp$ (5) $RrPp \times RRpp$ (6) $RRpp \times rrpp$ (7) $RRPP \times rrpp$ (8) $Rrpp \times Rrpp$ (9) $rrPp \times Rrpp$ (10) $rrPp \times rrpp$.

4.12. Listed below are 7 two-factor interaction ratios observed in progeny from various dihybrid parents. Suppose that in each case one of the dihybrid parents is testcrossed (instead of being mated to another dihybrid individual). What phenotypic ratio is expected in the progeny of each testcross? (*a*) $9:6:1$ (*b*) $9:3:4$ (*c*) $9:7$ (*d*) $15:1$ (*e*) $12:3:1$ (*f*) $9:3:3:1$ (*g*) $13:3$.

4.13. White fruit color in summer squash is governed by a dominant gene W and colored fruit by its recessive allele w. Yellow fruit is governed by an independently assorting hypostatic gene G and green by its recessive allele g. When dihybrid plants are crossed, the offspring appear in the ratio 12 white : 3 yellow : 1 green. What fruit color ratios are expected from the following crosses: (*a*) $Wwgg \times WwGG$ (*b*) $WwGg \times$ green   (*c*) $Wwgg \times wwGg$ (*d*) $WwGg \times Wwgg$?   (*e*) If two plants are crossed producing 1/2 yellow and 1/2 green progeny, what are the genotypes and phenotypes of the parents?

4.14. The Black Langshan breed of chickens has feathered shanks. When Langshans are crossed to the Buff Rock breed with unfeathered shanks, all the F_1 have feathered shanks. Out of 360 F_2 progeny, 24 were found to have nonfeathered shanks and 336 had feathered shanks. (*a*) What is the mode of interaction in this trait? (*b*) What proportion of the feathered F_2 would be expected to be heterozygous at one locus and homozygous at the other?

4.15. On chromosome 3 of corn there is a dominant gene A_1, which, together with the dominant gene (A_2) on chromosome 9, produces colored aleurone. All other genetic combinations produce colorless aleurone. Two pure colorless strains are crossed to produce an all-colored F_1. (*a*) What were the genotypes of the parental strains and the F_1? (*b*) What phenotypic proportions are expected among the F_2? (*c*) What genotypic ratio exists among the white F_2?

4.16. Two pairs of alleles govern the color of onion bulbs. A pure-red strain crossed to a pure-white strain produces an all-red F_1. The F_2 was found to consist of 47 white, 38 yellow, and 109 red bulbs. (*a*) What epistatic ratio is approximated by the data? (*b*) What is the name of this type of gene interaction? (*c*) If another F_2 is produced by the same kind of a cross, and eight bulbs of the F_2 are found to be of the double-recessive genotype, how many bulbs would be expected in each phenotypic class?

4.17. The color of corn aleurone is known to be controlled by several genes; A, C, and R are all necessary for color to be produced. The locus of a dominant inhibitor of aleurone color, I, is very closely linked to that of C. Thus, any one or more of the genotypes I-, aa-, cc-, or rr- produces colorless aleurone. (*a*) What would be the colored : colorless ratio among F_2 progeny from the cross $AAIICCRR \times aaiiCCRR$? (*b*) What proportion of the colorless F_2 is expected to be homozygous?

4.18. Suppose that crossing two homozygous lines of white clover, each with a low content of cyanide, produces only progeny with high levels of cyanide. When these F_1 progeny are backcrossed to either parental line, half the progeny has low cyanide content and the other half has high cyanide content. (*a*) What type of interaction may account for these results? (*b*) What phenotypic ratio is expected in the F_2? (*c*) If a $12:4$ ratio is observed among progeny from parents with high cyanide content, what are the parental genotypes? (*d*) If the low cyanide F_2, exclusive of the double recessives, are allowed to cross at random among themselves, what proportion of their progeny is expected to contain a high cyanide content?

4.19. In cultivated flowers called "stocks," the recessive genotype of one locus (aa) prevents the development of pigment in the flower, thus producing a white color. In the presence of the dominant allele A, alleles at another locus may be expressed as follows: C- = red, cc = cream. (*a*) When cream stocks of the genotype $Aacc$ are crossed to red stocks of the genotype $AaCc$, what phenotypic and genotypic proportions are expected in the progeny? (*b*) If cream stocks crossed to red stocks produce white progeny, what may be the genotypes of the parents? (*c*) When dihybrid red stocks are crossed together, what phenotypic ratio is expected among the progeny? (*d*) If red stocks crossed to white stocks produce progeny with red, cream, and white flowers, what are the genotypes of the parents?

4.20. An inhibitor of pigment production in onion bulbs (*I*-) exhibits dominant epistasis over another locus, the genotype *iiR*- producing red bulbs and *iirr* producing yellow bulbs. (*a*) A pure-white strain is crossed to a pure-red strain and produces an all-white F_1 and an F_2 with 12/16 white, 3/16 red, and 1/16 yellow. What were the genotypes of the parents? (*b*) If yellow onions are crossed to a pure-white strain of a genotype different from the parental type in part (*a*), what phenotypic ratio is expected in the F_1 and F_2? (*c*) Among the white F_2 of part (*a*), suppose that 32 were found to be of genotype *IiRR*. How many progeny are expected in each of the three F_2 phenotypic classes?

4.21. The color of the flower center in the common yellow daisy may be either purple-centered or yellow-centered. Two genes (*P* and *Y*) are known to interact in this trait. The results of two matings are given below:

1.   P: *PpYY*   ×   *PpYY*

                 purple-centered   purple-centered

       F_1:   3/4 *P-YY* purple-centered

                       1/4 *PpYY* yellow-centered

2.   P: *PpYY*   ×   *PpYY*

                 yellow-centered   yellow-centered

       F_1:   3/4 *ppY-* $\Big\}$

               1/4 *ppyy*   all yellow-centered

Determine the phenotypic ratios of progeny from the matings (*a*) *PpYy* × *PpYy*   (*b*) *PpYy* × *ppyy*   (*c*) *PPyy* × *ppYY*.

4.22. The aleurone of corn kernels may be either yellow, white, or purple. When pollen from a homozygous purple plant is used to fertilize a homozygous white plant, the aleurones of the resulting kernels are all purple. When homozygous yellow plants are crossed to homozygous white plants, only seeds with yellow aleurone are produced. When homozygous purple plants are crossed to homozygous yellow plants, only purple progeny appear. Some crosses between purple plants produce purple, yellow, and white progeny. Some crosses between yellow plants produce both yellow and white offspring. Crosses between yellow plants never produce purple progeny. Crosses among plants produced from seeds with white aleurone always produce only white progeny. (*a*) Can these results be explained on the basis of the action of a single-gene locus with multiple alleles? (*b*) What is the simplest explanation for the mode of gene action? (*c*) If plants with only dominant alleles at the two loci are crossed to plants grown from white seeds, what phenotypic proportions are expected among their F_2 progeny? (*d*) In part (*c*), how many generations of seeds must be planted in order to obtain an F_2 progeny phenotypically expressing the aleurone genes derived from the adult parent sporophytes? (*e*) What is the advantage of studying the genetics of seed traits rather than traits of the sporophyte?

4.23. Three fruit shapes are recognized in the summer squash (*Cucurbita pepo*): disk-shaped, elongated, and sphere-shaped. A pure disk-shaped variety was crossed to a pure elongated variety. The F_1 were all disk-shaped. Among 80 F_2, there were 30 sphere-shaped, 5 elongated, and 45 disk-shaped. (*a*) Reduce the F_2 numbers to their lowest ratio. (*b*) What types of interaction is operative? (*c*) If the sphere-shaped F_2 cross at random, what phenotypic proportions are expected in the progeny?

PEDIGREE ANALYSIS

4.24. The pedigree below illustrates a case of dominant epistasis. (*a*) What symbol represents the genotype *A-B-*? (*b*) What symbol represents the genotype *aaB-*? (*c*) What symbol represents the genotype *aabb*? (*d*) What type of epistasis would be represented if II2 × II3 produced, in addition to □ and ⊘, an offspring of type ●? (*e*) What type of interaction would be represented if III5 × III6 produced, in addition to ▨ and ○, an offspring of type ●?

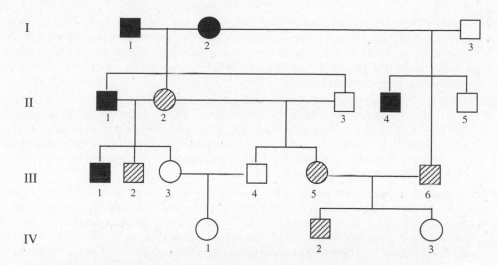

4.25. Given the following pedigree showing three generations of mink breeding, where open symbols represent wild type and solid symbols represent platinum, determine (*a*) the mode of inheritance of these coat colors, (*b*) the most probable genotypes of all individuals in the pedigree (use of familiar symbols such as *A, a* and *B, b* are suitable), (*c*) what phenotypic proportions are expected in progeny from III1 × III2.

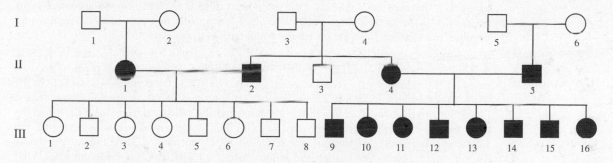

4.26. The pedigree below shows the genetic transmission of feather color in chickens. Open symbols represent white feathers, solid symbols represent colored feathers. Under the assumption of dominant and recessive interaction (given *A-* or *bb* or both = white, *aaB-* = color), assign genotypes to each individual in the pedigree. Indicate by (-) whatever genes cannot be determined.

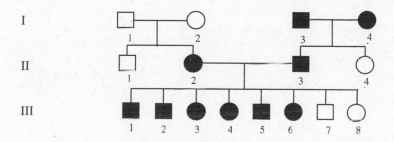

INTERACTIONS WITH THREE OR MORE FACTORS

4.27. A wheat variety with colored seeds is crossed to a colorless strain producing an all-colored F_1. In the F_2, 1/64 of the progeny has colorless seeds. (*a*) How many pairs of genes control seed color? (*b*) What were the genotypes of the parents and the F_1 (use your own symbols)?

4.28. In mice, spotted coat color is due to a recessive gene s and solid coat color to its dominant allele S. Colored mice possess a dominant allele C whereas albinos are homozygous recessive cc. Black is produced by a dominant allele B and brown by its recessive allele b. The cc genotype is epistatic to both the B and S loci. What phenotypic ratio is expected among the progeny of trihybrid parents?

4.29. A pure line of corn ($CCRR$) exhibiting colored aleurone is testcrossed to a colorless aleurone strain. Approximately 56% of the F_2 has colored aleurone, the other 44% being colorless. A pure line ($AARR$) with colored aleurone, when testcrossed, also produces the same phenotypic ratio in the F_2. (*a*) What phenotypic ratio is expected in the F_2 when a pure-colored line of genotype $AACCRR$ is testcrossed? (*b*) What proportion of the colorless F_2 is $aaccrr$? (*c*) What genotypic ratio exists among the colored F_2?

4.30. If a pure-white onion strain is crossed to a pure-yellow strain, the F_2 ratio is 12 white: 3 red: 1 yellow. If another pure-white onion is crossed to a pure-red onion, the F_2 ratio is 9 red : 3 yellow : 4 white. (*a*) What percentage of the white F_2 from the second mating would be homozygous for the yellow allele? (*b*) If the white F_2 (homozygous for the yellow allele) of part (*a*) is crossed to the pure-white parent of the first mating mentioned at the beginning of this problem, determine the F_1 and F_2 phenotypic expectations.

4.31. For any color to be developed in the aleurone layer of corn kernels, the dominant alleles at two loci plus the recessive condition at a third locus (*A-R-ii*) must be present. Any other genotypes produce colorless aleurone. (*a*) What phenotypic ratio of colored : colorless would be expected in progeny from matings between parental plants of genotype $AaRrIi$? (*b*) What proportion of the colorless progeny in part (*a*) would be expected to be heterozygous at one or more of the three loci? (*c*) What is the probability of picking from among the colored seeds in part (*a*) two seeds that, when grown into adult sporophytes and artificially crossed, would produce some colorless progeny with the triple-recessive genotype?

4.32. A dominant gene V is known in humans that causes certain areas of the skin to become depigmented, a condition called "vitiligo." Albinism is the complete lack of pigment production and is produced by the recessive genotype aa. The albino locus is epistatic to the vitiligo locus. Another gene locus, the action of which is independent of the previously mentioned loci, is known to be involved in a mildly anemic condition called "thalassemia." (*a*) When the adult progeny from parents both of whom exhibit vitiligo and a mild anemia is examined, the following phenotypic proportions are observed : 1/16 normal : 3/16 vitiligo : 1/8 mildly anemic : 1/12 albino : 3/8 vitiligo and mildly anemic : 1/6 albino and mildly anemic. What is the mode of genetic action of the gene for thalassemia? (*b*) What percentage of the viable albino offspring in part (*a*) would carry the gene for vitiligo? (*c*) What percentage of viable offspring with symptoms of mild anemia also show vitiligo?

4.33. When the White Leghorn breed of chickens is crossed to the White Wyandotte breed, all the F_1 birds have white feathers. The F_2 birds appear in the ratio 13 white: 3 colored. When the White Leghorn breed is crossed to the White Silkie breed, the F_1 is white and the F_2 is also 13 white: 3 colored. But when White Wyandottes are crossed to White Silkies, the F_1 is all colored and the F_2 appears in the ratio 9 colored: 7 white. (*a*) How are feather colors inherited in these breeds (use appropriate symbols in your explanation)? (*b*) Show, by use of your own symbols, the genotypes of each of the three breeds (assume the breed is homozygous for all loci under consideration). (*c*) What phenotypic ratio is expected among progeny from trihybrid parents? (*d*) What proportion of the white offspring of part (*c*) is expected to be dihybrid?

Review Questions

Matching Questions   Match the expected phenotypic ratios (right column) in progeny from dihybrid parents with the conditions specified in the left column.

1. Dominant epistasis
2. Recessive epistasis
3. Duplicate genes with cumulative action
4. Duplicate dominant interaction
5. Duplicate recessive interaction
6. Dominant and recessive interaction
7. No interaction

A. 9 : 6 : 1
B. 9 : 3 : 3 : 1
C. 9 : 3 : 4
D. 13 : 3
E. 12 : 3 : 1
F. 15 : 1
G. 9 : 7

Vocabulary   For each of the following definitions, give the appropriate term and spell it correctly.

1. Protein catalysts that speed the rate of chemical reactions.
2. The material(s) acted upon and changed by the catalyst in question 1.
3. The phenomenon in which a specific genotype(s) at one locus can prevent the phenotypic manifestation of a genotype(s) at one or more other loci.
4. Adjective describing a locus whose phenotypic manifestation is suppressed by the phenomenon in question 3.
5. A phenomenon wherein a single gene has more than one phenotypic effect.
6. A collection of phenotypic effects that collectively defines a disease, genetic or otherwise.

Multiple-Choice Questions   Questions 1–5 use the information in the diagram below.

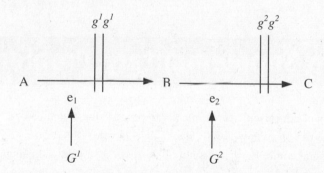

1. The genotype(s) capable of making C is/are   (a) $G^1g^1G^2g^2$ (b) $G^1G^1g^2g^2$ (c) $G^1g^1g^2g^2$ (d) $g^1g^1G^2G^2$ (e) more than one of the above
2. In a cross between genotypes $g^1g^1G^2G^2 \times G^1G^1g^2g^2$ what fraction of the F_2 is expected to be phenotypically B-positive and C-negative?   (a) 1/16   (b) 3/16   (c) 3/8   (d) 1/4   (e) none of the above
3. What fraction of the F_2 in the above cross (question 2) is expected to be C-negative?   (a) 3/4   (b) 1/2   (c) 9/16   (d) 7/16   (e) none of the above
4. What fraction of the F_2 (in question 2) is expected to be able to make enzyme 2, but not enzyme 1?   (a) 3/4   (b) 3/16   (c) 9/16   (d) 3/8   (e) none of the above
5. If a dihybrid is testcrossed, what fraction of the progeny is expected to be able to make substance C?   (a) 9/16   (b) 3/16   (c) 3/8   (d) 1/4   (e) none of the above

Questions 6–10 use the information in the following diagram. Enzyme e_3 becomes inactivated by binding to substance P.

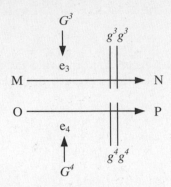

6. If the phenotype is determined by the presence or absence of substance N, what type of interaction exists? (*a*) dominant and recessive (*b*) duplicate recessive (*c*) duplicate dominant (*d*) duplicate genes with cumulative action (*e*) none of the above

7. Approximately (within rounding error) what percentage of the F_1 from $G^3g^3G^4g^4$ parents is expected to be N-positive? (*a*) 56 (*b*) 19 (*c*) 25 (*d*) 38 (*e*) none of the above

8. When dihybrids are testcrossed, the percentage of their progeny expected to be N-negative is (*a*) 25 (*b*) 81 (*c*) 75 (*d*) 38 (*e*) none of the above

9. Among the N-positive progeny from dihybrid parents, what fraction is expected to be P-positive? (*a*) 3/4 (*b*) 1/2 (*c*) 2/3 (*d*) 5/6 (*e*) none of the above

10. What fraction of the progeny from dihybrid parents is expected to be P-positive and N-negative? (*a*) 9/16 (*b*) 7/16 (*c*) 6/16 (*d*) 3/4 (*e*) none of the above

Answers to Supplementary Problems

4.10. (*a*) Two pairs of nonepistatic genes interact to produce these coat colors.

(*b*) *A-B-* (gray), *A-bb* (yellow), *aaB-* (black), *aabb* (cream) (*c*) 3 (*d*) 12

4.11. (*a*) F_1: all walnut comb; F_2: 9/16 walnut : 3/16 rose : 3/16 pea : 1/16 single (*b*) *RRPp*

(c)

	R-P-	R-pp	rrP-	rrpp
	Walnut	Rose	Pea	Single
1	3/4		1/4	
2	3/8	1/8	3/8	1/8
3	All			
4	1/4	1/4	1/4	1/4
5	1/2	1/2		
6		All		
7	All			
8		3/4		1/4
9	1/4	1/4	1/4	1/4
10			1/2	1/2

4.12. (*a*) $1 : 2 : 1$ (*b*) $1 : 1 : 2$ (*c*) $1 : 3$ (*d*) $3 : 1$ (*e*) $2 : 1 : 1$ (*f*) $1 : 1 : 1 : 1$ (*g*) $3 : 1$

4.13. (*a*) 3/4 white : 1/4 yellow   (*b*) and (*c*) 1/2 white : 1/4 yellow : 1/4 green   (*d*) 3/4 white : 1/8 yellow : 1/8 green   (*e*) yellow ($wwGg$) $\times$ green ($wwgg$)

4.14. (*a*) Duplicate dominant genes with only the double-recessive genotype producing nonfeathered shanks (*b*) 8/15

4.15. (*a*) P : $A_1A_1a_2a_2 \times a_1a_1A_2A_2$; F_1: $A_1a_1A_2a_2$;   (*b*) 9/16 colored : 7/16 colorless   (*c*) 1/7 $A_1A_1a_2a_2$: 2/7 $A_1a_1a_2a_2$: 1/7 $a_1a_1A_2A_2$: 2/7 $a_1a_1A_2a_2$: 1/7 $a_1a_1a_2a_2$

4.16. (*a*) $9 : 3 : 4$ (*b*) Recessive epistasis   (*c*) 32 white : 24 yellow : 72 red

4.17. (*a*) 13 colorless : 3 colored   (*b*) 3/13

4.18. (*a*) Duplicate recessive interaction   (*b*) 9 high cyanide : 7 low cyanide   (*c*) $A\text{-}Bb \times AABb$ or $AaB\text{-} \times AaBB$ (*d*) 2/9

4.19. (*a*) 3/8 red : 3/8 cream : 1/4 white : 1/8 $AACc$: 1/8 $AAcc$: 1/4 $AaCc$: 1/4 $Aacc$: 1/8 $aaCc$: 1/8 $aacc$ (*b*) $Aacc \times AaCc$ (or) $Aacc \times AaCC$ (*c*) 9 red : 3 cream : 4 white   (*d*) $AaCc \times aaCc$ (or) $AaCc \times aacc$

4.20. (*a*) $IIrr \times iiRR$ (*b*) F_1; all white : F_2: 12/16 white : 3/16 red : 1/16 yellow   (*c*) 16 yellow : 48 red : 192 white

4.21. (*a*) 9/16 purple-centered : 7/16 yellow-centered   (*b*) 1/4 purple-centered : 3/4 yellow-centered   (*c*) All purple-centered

4.22. (*a*) No   (*b*) Dominant epistasis where $Y\text{-}R\text{-}$ or $yyR\text{-}$ produce purple, $Y\text{-}rr =$ yellow, and $yyrr$ = white   (*c*) 3/4 purple : 3/16 yellow : 1/16 white   (*d*) One   (*e*) The appearance of seed traits requires one less generation of rearing than that for tissues found in the sporophyte.

4.23. (*a*) 9 disk : 6 sphere : 1 elongated   (*b*) Duplicate genes with cumulative effect   (*c*) 2/3 sphere : 2/9 disk : 1/9 elongate

4.24. (*a*) Solid symbol   (*b*) Diagonal lines   (*c*) Open symbol   (*d*) Recessive epistasis   (*e*) Duplicate genes with cumulative effect

4.25. (*a*) Duplicate recessive interaction   (*b*) $A\text{-}B\text{-}$ (wild type), $aa\text{-}\text{-}$ or $\text{-}\text{-}bb$ or $aabb$ (platinum); $AaB\text{-}$ (I1, 2), $A\text{-}Bb$ (I3, 4), $A\text{-}B\text{-}$ (I5, 6, II3), $aaBB$ (II1), $AAbb$ (II2), $AaBb$ (III1-III8), either aa or bb or both (II4, 5, III9-III16)   (*c*) 9 wild type : 7 platinum

4.26. The following set of genotypes is only one of several possible solutions: $aaB\text{-}$ (III1, 2, 3, 4, 5, 6), $A\text{-}$ or bb or both (II1), $a\text{-}\text{-}\text{-}$ (I1, 2), $aaBb$ (I3, 4, II2, 3), $aabb$ (II4, III7, 8)

4.27. (*a*) 3   (*b*) P : *AABBCC* × *aabbcc*; F₁ : *AaBbCc*

4.28. 27 solid black : 9 spotted black : 9 solid brown : 3 spotted brown : 16 albino

4.29. (*a*) 27 colored : 37 colorless   (*b*) 1/37   (*c*) 1/27 *AACCRR* : 2/27 *AACCRr* : 2/27 *AACcRR* : 4/27 *AACcRr* : 2/27 *AaCCRR*: 4/27 *AaCCRr* : 4/27 *AaCcRR* : 8/27 *AaCcRr*

4.30. (*a*) 25%   (*b*) F₁ : all white; F₂ : 52 white : 9 red : 3 yellow

4.31. (*a*) 9 colored : 55 colorless   (*b*) 48/55   (*c*) 16/81

4.32. (*a*) The gene for thalassemia is dominant to its normal allele, causing mild anemia when heterozygous, but is lethal when homozygous   (*b*) 75%   (*c*) 56.25%

4.33. (*a*) Three loci involved; one possesses a dominant inhibitor of color (*I*-) and the other two possess different recessive inhibitors of color (*cc* and *oo*). Only the genotype *iiC-O-* produces colored birds; all other genotypes produce white feathers.   (*b*) White Leghorn (*CCOOII*), White Wyandotte (*ccOOii*), White Silkie (*CCooii*)   (*c*) 55 white : 9 colored   (*d*) 20/55

Answers to Review Questions

Matching Questions
1. E   2. C   3. A   4. F   5. G   6. D   7. B

Vocabulary
1. enzymes   2. substrate(s)   3. epistasis   4. hypostatic   5. pleiotropism   6. syndrome

Multiple-Choice Questions
1. *a*   2. *b*   3. *d*   4. *b*   5. *d*   6. *a*   7. *b*   8. *c*   9. *a* (zero)   10. *d*

CHAPTER 5

The Genetics of Sex

The Importance of Sex

We are probably too accustomed to thinking of sex in terms of the males ♂ and females ♀ of our own or domestic species. Plants also have sexes; at least we know that there are male and female portions of a flower. Not all organisms, however, possess only two sexes. Some of the simplest forms of plant and animal life may have several sexes. For example, in one variety of the ciliated protozoan *Paramecium bursaria* there are eight sexes, or "mating types," all morphologically identical. Each mating type is physiologically incapable of conjugating with its own type, but may exchange genetic material with any of the seven other types within the same variety. In most complex organisms, the number of sexes is only two. These sexes may reside in different individuals or within the same individual. An animal possessing both male and female reproductive organs is usually referred to as a **hermaphrodite**. In plants where **staminate** (male) and **pistillate** (female) flowers occur on the same plant, the term of preference is **monoecious**. Moreover, most flowering plants have both male and female parts within the same flower (perfect flower). Relatively few angiosperms are **dioecious**, i.e., having the male and female elements in different individuals. Among the common cultivated crops known to be dioecious are asparagus, date palm, hemp, hops, and spinach.

Whether there are two or more sexes, or whether these sexes reside in the same or different individuals, is relatively unimportant. The importance of sex itself is that it is a mechanism that provides for the great amount of genetic variability characterizing most natural populations. The evolutionary process of natural selection depends upon this genetic variability to supply the raw material from which better adapted types usually survive to reproduce. Many subsidiary mechanisms have evolved to ensure cross-fertilization in most species as a means for generating new genetic combinations in each generation.

Sex-Determining Mechanisms

Most mechanisms for the determination of sex are under genetic control and may be classified into one of the following categories.

1   SEX CHROMOSOME MECHANISMS

    (a) Heterogametic Males. In humans, and most other mammals, there are two different, or **heteromorphic**, sex chromosomes, the X and the Y chromosomes. The presence of the Y chromosome determines maleness. Normal males have 22 pairs of autosomes and an X and a Y sex chromosome; females also have 22 pairs of autosomes, but have two X chromosomes. Since the male produces two kinds of gametes as far as the sex chromosomes are concerned, he is said to be the **hetero-gametic** sex. The female, producing only one kind of gamete, is the **homogametic** sex. Thus, assuming unbiased segregation and equal success of each type of gamete during fertilization, an equal number of each sex should be produced in each generation. The proportion of males to females is referred to as the **sex ratio**. This mode of sex determination is commonly referred to as the XY method.

EXAMPLE 5.1
XY Method of sex determination.

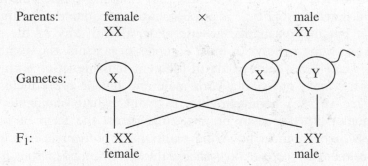

A gene called *SRY*, for **sex-determining region Y,** has been identified on the short arm of the Y chromosome. It encodes a gene product often referred to as the **testis-determining factor (TDF)**. *SRY* seems to be highly conserved in mammals. This gene, in combination with several other genes, such as the autosomal *DAX1, WT1, SF1,* and *SOX9* genes, encodes a DNA-binding protein that appears to control the expression of one or more other genes in a hierarchy or cascade of gene activation involved in testicular development and fertility (sperm production). In the absence of *SRY*, rudimentary gonadal tissue of the embryo would normally develop into an ovary. The location of *SRY* was aided by the discovery of rare exceptions to the rule that XX programs for femaleness and XY programs for maleness. It was found that normal-appearing but sterile XX human males have at least some of the *SRY* gene attached to one of their X chromosomes and normal-appearing human XY females have a Y chromosome that has lost a crucial part of the *SRY*.

In some insects, especially those of the orders Hemiptera (true insects) and Orthoptera (grasshoppers and roaches), males are also heterogametic, but produce either X-bearing sperm or gametes without a sex chromosome. In males of these species, the X chromosome has no homologous pairing partner because there is no Y chromosome present. Thus, males exhibit an odd number in their chromosome complement. The one-X and two-X condition determines maleness and femaleness, respectively. If the single X chromosome of the male is always included in one of the two types of gametes formed, then a 1 : 1 sex ratio is predicted in the progeny. This mode of sex determination is commonly referred to as the XO method where the O symbolizes the lack of a chromosome analogous to the Y of the XY system.

EXAMPLE 5.2
XO Method of sex determination.

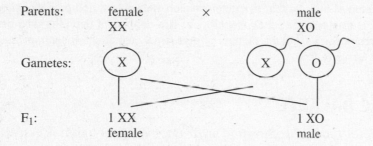

(b) Heterogametic Females. This method of sex determination is found in a comparatively large group of insects including the butterflies, moths, caddis flies, and silkworms, and in some birds and fishes. The 1-X and 2-X condition in these species determines femaleness and maleness, respectively. The females of some species (e.g., domestic chickens) have a chromosome similar to that of the Y in humans. In these cases, the chromosomes are sometimes labeled Z and W instead of X and Y, respectively, in order to call attention to the fact that the female (ZW) is the heterogametic sex and the male (ZZ) is the homogametic sex. The females of other species have no homologue to the single sex chromosome as in the case of the XO mechanism discussed previously. To point out this difference, the symbols ZZ and ZO may be used to designate males and females, respectively. A 1 : 1 sex ratio is expected in either case.

EXAMPLE 5.3
ZO Method of sex determination.

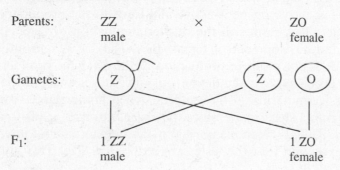

EXAMPLE 5.4
ZW Method of sex determination.

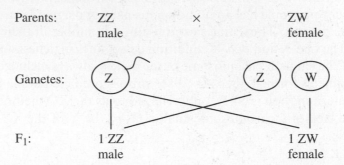

Parents: ZZ   ×   ZW
          male   female

Gametes: (Z)   (Z) (W)

F$_1$:   1 ZZ   1 ZW
          male   female

The W chromosome of the chicken is not a strong female-determining element.
Recent studies indicate that sex determination in chickens, and probably birds in
general, is similar to that of *Drosophila*, i.e., it is dependent upon the ratio between
the Z chromosomes and the number of autosomal sets of chromosomes (see next
section, Genic Balance).

2   GENIC BALANCE

The presence of the Y chromosome in *Drosophila*, although it is essential for male
fertility, apparently has nothing to do with the determination of sex. Instead, the
factors for maleness residing in all of the autosomes are "weighed" against the
factors for femaleness residing on the X chromosome(s). In fact, it is the ratio of X
chromosomes to haploid sets of autosomes that determines sex in *Drosophila*.
Using the letter A to represent a haploid set of chromosomes, a normal female
(2X : 2A) has an X : A ratio of 2 : 2, or 1.0, and therefore the balance is in favor
of femaleness. When only one X chromosome is present in a normal male
(XY : 2A), the ratio is 1 : 2 or 0.5. Several abnormal combinations of chromosomes have confirmed this hypothesis. For example, an individual with three sets
of autosomes and two X chromosomes (2X : 3A) has a ratio of 2 : 3 or 0.67, in
between the ratios for normal maleness and femaleness. This kind of fly, called
intersex, is sterile and has sexual characteristics intermediate between the male and
female. Ratios above 1.0 produce sterile **metafemales** (previously called **superfemales**) and ratios below 0.5 produce sterile **metamales**.

   The ratio determines sex by activating sex-specific gene expression of several
genes, such as **Sex-lethal (*Sxl*), transformer (*tra*), and doublesex (*dsx*)**. Simply
stated, in females, the *Sxl* gene is active and leads to the production of an active
tra gene product. This results in the further production of a female-specific *dsx*
gene product and development of female flies. The X : A chromosome ratio in
normal males does not result in production of the *Sxl* gene product; thus, the tra
protein is not produced and the default, or male developmental pathway, is followed. A recessive mutation in the *tra* gene, when homozygous, can transform a
diploid female into a sterile male, since the absense of *tra*$^+$ leads to maleness. The
X/X, *tra/tra* individuals resemble normal males in external and internal morphology with the exception that the testes are reduced in size. This mutation has no

effect in normal males. The presence of this mutation can considerably alter the sex ratio.

SOLVED PROBLEM 5.1

An autosomal recessive gene *tra*, when homozygous, transforms a *Drosophila* female (X/X) into a phenotypic male. All such "transformed" males are sterile. The gene is without effect in males (X/Y). A cross is made between a female heterozygous at the *tra* locus and a male homozygous recessive at the same locus (*tra/tra*). What is the expected sex ratio in the F_1 and F_2?

Solution: We will use a slash mark (/) to separate alleles or homologous chromosomes, and a comma (,) to separate one gene locus from another.

Parents: X/X, +/*tra* × X/Y, *tra/tra*
             normal female   normal male

Gametes: (X +) (X *tra*) (X *tra*) (Y *tra*)

F_1:

	X +	X *tra*
X *tra*	X/X, +/*tra* normal females	X/X, *tra/tra* "transformed" males
Y *tra*	X/Y, +/*tra* normal males	X/Y, *tra/tra* normal males

The F_1 phenotypic proportions thus appear as 3/4 males : 1/4 females.

F_2: The "transformed" F_1 males are sterile and hence do not contribute gametes to the F_2. Two kinds of matings must be considered. First mating = 1/2 of all possible matings:

First mating: X/X, +/*tra* × X/Y, +/*tra*
                  females   males

	X +	X *tra*
X +	X/X, +/+ female	X/X, +/*tra* female
X *tra*	X/X, +/*tra* female	X/X, *tra/tra* "transformed" male
Y +	X/Y, +/+ male	X/Y, +/*tra* male
Y *tra*	X/Y, +/*tra* male	X/Y, *tra/tra* male

Thus, F_2 offspring from this mating type appear in the proportions 3/8 female : 5/8 male. But this type of mating constitutes only half of all possible matings. Therefore, the contribution to the total F_2 from this mating is $= (1/2)(3/8) = 3/16$ female : $(1/2)(5/8) = 5/16$ male. Second mating $= 1/2$ of all possible matings:

Second mating: X/X, $+$/tra $\times$ X/Y, tra/tra
                              females   males

This is the same as the original parental mating and hence we expect 3/4 males : 1/4 females. Correcting these proportions for the frequency of this mating, we have $(1/2)(3/4) = 3/8$ males : $(1/2)(1/4) = 1/8$ females. Summary of the F_2 from both matings: males $= 5/16 + 3/8 = 11/16$; females $= 3/16 + 1/8 = 5/16$.

3 HAPLODIPLOIDY

Male bees are known to develop **parthenogenetically** (without union of gametes) from unfertilized eggs (**arrhenotoky**) and are therefore haploid. Females (both workers and queens) originate from fertilized (diploid) eggs. Sex chromosomes are not involved in this mechanism of sex determination, which is characteristic of the insect order Hymenoptera including the ants, bees, wasps, etc. The quantity and quality of food available to the diploid larva determines whether that female will become a sterile worker or a fertile queen. Thus, environment determines sterility or fertility but does not alter the genetically determined sex. The sex ratio of the offspring is under the control of the queen. Most of the eggs laid in the hive will be fertilized and develop into worker females. Those eggs that the queen chooses not to fertilize (from her store of sperm in the seminal receptacle) will develop into fertile haploid males. Queen bees usually mate only once during their lifetime.

4 SINGLE-GENE EFFECTS

(a) **Complementary Sex Determination (CSD) Factors.** In addition to haplodiploidy, members of the insect order Hymenoptera are known to produce males by homozygosity at a single-gene locus. This has been confirmed in the tiny parasitic wasp *Bracon hebetor* (often called *Habrobracon juglandis*), as well as in bees. At least nine sex alleles are known at this locus in *Bracon* and may be represented by $s^a, s^b, s^c, \ldots, s^i$. All females must be heterozygotes such as $s^a s^b$, $s^a s^c$, $s^d s^f$, etc. If an individual is homozygous for any of these alleles such as $s^a s^a$, $s^c s^c$, etc., it develops into a diploid male (usually sterile). Haploid males, of course, would carry only one of the alleles at this locus, e.g., s^a, s^c, or s^g, etc.

EXAMPLE 5.5

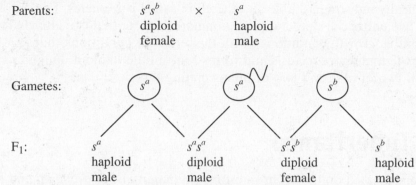

Parents: $s^a s^b$ × s^a
          diploid   haploid
          female   male

Gametes: s^a s^a s^b

F_1: s^a $s^a s^a$ $s^a s^b$ s^b
     haploid   diploid   diploid   haploid
     male   male   female   male

Among the diploid progeny we expect 1 $s^a s^a$ male : 1 $s^a s^b$ female.
Among the haploid progeny we expect 1 s^a male : 1 s^b male.

(b) "Mating Type" in Microorganisms. In microorganisms such as the alga *Chlamydomonas*, the fungi *Neurospora*, and yeast, sex is under the control of a single gene. Haploid individuals possessing the same allele of this "mating-type" locus usually cannot fuse with each other to form a zygote, but haploid cells containing opposite (complementary) alleles at this locus may fuse. Asexual reproduction in the single-celled motile alga *Chlamydomonas reinhardi* usually involves two mitotic divisions within the old cell wall (Fig. 5-1). Rupture of the sporangium releases the new generation of haploid **zoospores**. If nutritional requirements are satisfied, asexual reproduction may go on indefinitely. In unfavorable conditions where nitrogen balance is upset, daughter cells may be changed to gametes.

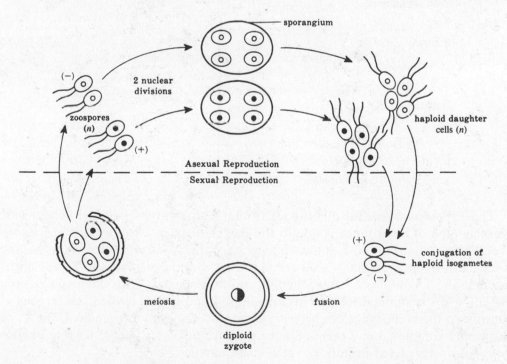

Fig. 5-1. Life cycle of *Chlamydomonas reinhardi*.

Genetically, there are two mating types, plus (+) and minus (−), which are morphologically indistinguishable and therefore called **isogametes**. Fusion of gametes unites two entire cells into a diploid nonmotile zygote that is relatively resistant to unfavorable growth conditions. With the return of conditions that favor growth, the zygote undergoes meiosis and forms four motile haploid daughter cells (zoospores), two of plus and two of minus mating type.

Sex-Linked Inheritance

Any gene located on the X chromosome (in mammals, *Drosophila*, and others) or on the analogous Z chromosome (in birds and other species) is said to be **sex-linked**, or **X-linked**. The first sex-linked gene found in *Drosophila* was the recessive white-eye mutation. Reciprocal crosses involving autosomal traits yield similar results. Reciprocal crosses are carried out by mating a male with one phenotype (e.g., black coat) and a female with another phenotype (e.g., white coat) and then repeating the cross with a male and female of phenotypes that are opposites of the first cross (i.e., a white-coated male and a black-coated female). Sex-linked traits do not show similar results in reciprocal crosses, as shown below. When white-eyed females are crossed with wild-type (red-eyed) males, all the male offspring have white eyes like their mother and all the female offspring have red eyes like their father.

EXAMPLE 5.6
Sex-linked inheritance.

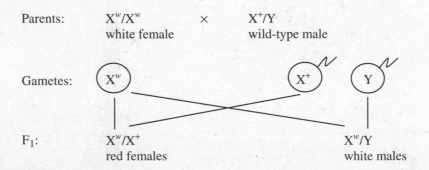

Parents: X^w/X^w × X^+/Y
white female   wild-type male

Gametes: X^w X^+ Y

F_1: X^w/X^+ X^w/Y
red females   white males

This crisscross mode of inheritance is characteristic of sex-linked genes. This peculiar type of inheritance is due to the fact that the Y chromosome carries no alleles homologous to those at the white locus on the X chromosome. Whereas the X chromosome carries 2000–3000 genes, in most organisms with the Y-type chromosome the Y contains no more than several dozen genes. Thus, males carry only one allele for sex-linked traits. This one-allelic condition is termed **hemizygous** in contrast to the homozygous or heterozygous possibilities in the female. If the F_1 of Example 5.6 mate among themselves to produce an F_2, a 1 red : 1 white phenotypic ratio is expected in both the males and females.

EXAMPLE 5.7

F_1: X^+/X^w × X^w/Y
red female   white male

F_2:

	X^+	X^w
X^w	X^+/X^w red female	X^w/X^w white female
Y	X^+/Y red male	X^w/Y white male

The reciprocal cross, where the sex-linked mutation appears in the male parent, results in the disappearance of the trait in the F_1 and its reappearance only in the males of the F_2. This type of skip generation inheritance also characterizes sex-linked genes.

EXAMPLE 5.8

Parents: X^+/X^+ × X^w/Y
red female   white male

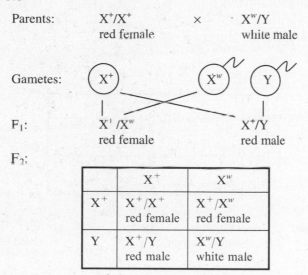

Gametes:

F_1: X^1/X^w X^+/Y
red female   red male

F_2:

	X^+	X^w
X^+	X^+/X^+ red female	X^+/X^w red female
Y	X^+/Y red male	X^w/Y white male

Thus, a 3 red : 1 white phenotypic ratio is expected in the total F_2 disregarding sex, but only the males show the mutant trait. The phenotypic ratio among the F_2 males is 1 red : 1 white. All F_2 females are phenotypically wild type.

Whenever working with problems involving sex linkage in this book, be sure to list the ratios for males and females separately unless specifically directed by the problem to do otherwise.

In normal diploid organisms with sex-determining mechanisms like those of humans or *Drosophila*, a trait governed by a sex-linked recessive gene usually manifests itself in the following manner: (1) it is usually found more frequently in the male than in the female of the species, (2) it fails to appear in females unless it also appeared in the paternal parent, (3) it seldom appears in both father and son, then only if the maternal parent is heterozygous. On the other hand, a trait governed by a sex-linked dominant gene usually manifests itself by (1) being found more frequently in the female than in the male of the species, (2) being found in all female offspring of a male that shows the trait, (3) failing to be transmitted to any son from a mother that did not exhibit the trait herself.

SOLVED PROBLEM 5.2

There is a dominant sex-linked gene B that places white bars on an adult black chicken as in the Barred Plymouth Rock breed. Newly hatched chicks, which will become barred later in life, exhibit a white spot on the top of the head. (*a*) Diagram the cross through the F_2 between a homozygous barred male and a nonbarred female. (*b*) Diagram the reciprocal cross through the F_2 between a homozygous nonbarred male and a barred female. (*c*) Will both of the above crosses be useful in sexing F_1 chicks at hatching?

Solution:

(*a*)

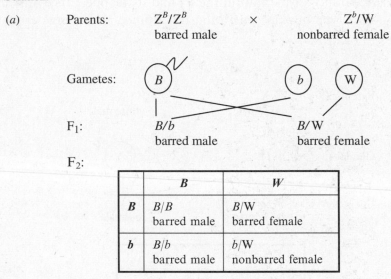

Parents: Z^B/Z^B × Z^b/W
           barred male   nonbarred female

Gametes: B b W

F_1: B/b B/W
        barred male   barred female

F_2:

	B	**W**
B	B/B barred male	B/W barred female
b	B/b barred male	b/W nonbarred female

(*b*) Parents: b/b × B/W
                 nonbarred male   barred female

Gametes: b   B W

F_1: b /B   b/W
        barred male   nonbarred female

F_2:

	b	**W**
B	B/b barred male	B/W barred female
b	B/b nonbarred male	b/W nonbarred female

(*c*) No. Only the cross shown in (*b*) would be diagnostic in sexing F_1 chicks at birth through the use of this genetic marker. Only male chicks will have a light spot on their heads.

SOLVED PROBLEM 5.3

A recessive sex-linked gene *h* prolongs the blood-clotting time, resulting in what is commonly called "bleeder's disease" (hemophilia). From the information in the pedigree, answer the following questions. (*a*) If II2 marries a normal man, what is the chance of her first child being a hemophilic boy? (*b*) Suppose her first child is actually hemophilic. What is the probability that her second child will be a hemophilic boy? (*c*) If II3 marries a hemophilic man, what is the probability that her first child will be normal? (*d*) If the mother of I1 was phenotypically normal, what phenotype was her father? (*e*) If the mother of I1 was hemophilic, what phenotype was her father?

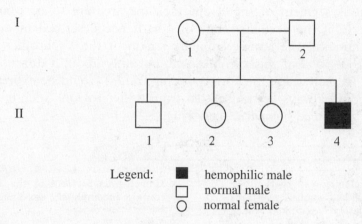

Legend: ■ hemophilic male
□ normal male
○ normal female

Solution:

(*a*) Since II4 is a hemophilic male (*h*Y), the hemophilic allele is on an X chromosome that he received from his mother (I1). But I1 is phenotypically normal and therefore must be heterozygous or a carrier of hemophilia of genotype *Hh*. Both I2 and II1 are normal males (*H*Y). Therefore, the chance of II2 being a carrier female (*Hh*) is 1/2. When a carrier woman marries a normal man (*H*Y), 25% of their children are expected to be hemophilic boys (*h*Y). The combined probability that she is a carrier and will produce a hemophilic boy is (1/2)(1/4) = 1/8.

(*b*) Because her first child was hemophilic, she must be a carrier. One-quarter of the children from carrier mothers (*Hh*) × normal fathers (*H*Y) are expected to be hemophilic boys (*h*Y).

(*c*) II3 (like II2) has a 50% chance of being a carrier of hemophilia (*Hh*). If she marries a hemophilic man (*h*Y), 1/2 of their children (both boys and girls) are expected to be hemophilic. The combined chance of II3 being a carrier and producing a hemophilic child is (1/2)(1/2) = 1/4. Therefore, the probability that her first child is normal is the complementary fraction, 3/4.

(*d*) It is impossible to deduce the phenotype of the father of I1 from the information given because the father could be either normal or hemophilic and still produce a daughter (I1) who is heterozygous normal (*Hh*), depending upon the genotype of the normal mother:

1.   *HH*   ×   *h*Y   =   *Hh*(I1)
   normal mother   hemophilic father   carrier daughter

2.   *Hh*   ×   *H*Y   =   *Hh*(I1)
   carrier mother   normal father   carrier daughter

(*e*) In order for a hemophilic mother (*hh*) to produce a normal daughter (*Hh*), her father must possess the dominant normal allele (*H*Y) and therefore would have normal blood-clotting time.

Variations of Sex Linkage

The sex chromosomes (X and Y) often are of unequal size, shape, and/or staining qualities. The fact that they pair during meiosis is an indication that they contain at least some homologous **pseudoautosomal** segments. Genes on the pseudoautosomal segments are said to be **incompletely sex-linked** or **partially sex-linked** and may recombine by crossing over in both sexes, just as do the gene loci on homologous autosomes. Special crosses are required to demonstrate the presence of such genes on the X chromosome, and few examples are known. Genes on the nonhomologous segment of the X chromosome are said to be **completely sex-linked** and exhibit the peculiar mode of inheritance described in the preceding sections. In humans, the nonrecombining region of the Y chromosome (NRY) makes up about 95% of this chromosome, but only about a dozen active genes reside there. In such cases, the corresponding trait(s) would be expressed only in males and would always be transmitted from father to son. Such completely Y-linked genes are called **holandric genes** (Fig. 5-2).

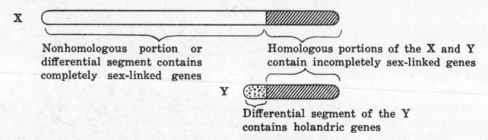

Fig. 5-2. Generalized diagram of X and Y chromosomes. The relative size of these chromosomes and the size of homologous and nonhomologous regions, as well as location of the centromeres (not shown), vary according to the species.

SOLVED PROBLEM 5.4
The recessive incompletely sex-linked gene called "bobbed" (*bb*) causes the bristles of *Drosophila* to be shorter and of smaller diameter than the normal bristles produced by its dominant wild-type allele (*bb*$^{+}$). Determine the phenotypic expectations of the F$_1$ and F$_2$ when bobbed females are crossed with each of the two possible heterozygous males.

Solution: Recall that an incompletely sex-linked gene has an allele on the homologous portion of the Y chromosome in a male. The wild-type allele in heterozygous males may be on either the X or the Y, thus making possible two types of crosses.

First Cross:

P: $X^{bb}X^{bb}$ × $X^{bb^+}Y^{bb}$
                   bobbed females   wild-type males

F_1: $X^{bb}X^{bb^+}$ and $X^{bb}Y^{bb}$
                all females wild type   all males bobbed

F_2:

	X^{bb}	X^{bb^+}
X^{bb}	$X^{bb}X^{bb}$ bobbed female	$X^{bb}X^{bb^+}$ wild-type female
Y^{bb}	$X^{bb}Y^{bb}$ bobbed male	$X^{bb^+}Y^{bb}$ wild-type male

Thus, 1/2 of the F_2 females are bobbed and 1/2 are wild type; 1/2 of the F_2 males are bobbed and 1/2 are wild type.

Second Cross:

P: $X^{bb}X^{bb}$ × $X^{bb}Y^{bb^+}$
                   bobbed female   wild-type male

F_1: $X^{bb}X^{bb}$ and $X^{bb}Y^{bb^+}$
               all females bobbed   all males wild type

F_2:

	X^{bb}	Y^{bb^+}
X^{bb}	$X^{bb}X^{bb}$ bobbed females	$X^{bb}Y^{bb^+}$ wild-type males

Thus, all F_2 females are bobbed and all males are wild type.

Sex-Influenced Traits

The genes governing sex-influenced traits may reside on any of the autosomes or on the homologous portions of the sex chromosomes. The expression of dominance or recessiveness by the alleles of sex-influenced loci is reversed in males and females due, in large part, to the difference in the internal environment provided by the sex hormones. Thus, examples of sex-influenced traits are most readily found in the higher animals with well-developed endocrine systems.

EXAMPLE 5.9

The gene for pattern baldness in humans exhibits dominance in men, but acts recessively in women.

Genotypes	Phenotypes	
	Men	Women
$b'b'$	Bald	Bald
$b'b$	Bald	Nonbald
bb	Nonbald	Nonbald

SOLVED PROBLEM 5.5

Let us consider two sex-influenced traits simultaneously, pattern baldness and short index finger, both of which are dominant in men and recessive in women. A heterozygous bald man with long index finger marries a heterozygous long-fingered, bald woman. Determine the phenotypic expectations for their children.

Solution: Let us first select appropriate symbols and define the phenotypic expression of the three genotypes in each sex.

Genotype	Males	Females	Genotype	Males	Females
B^1B^1	Bald	Bald	F^1F^1	Short-fingered	Short-fingered
$B'B^2$	Bald	Nonbald	F^1F^2	Short-fingered	Long-fingered
B^2B^2	Nonbald	Nonbald	F^2F^2	Long-fingered	Long-fingered

P: B^1B^2, F^2F^2 $\times$ B^1B^1, F^1F^2

   bald, long-fingered man   bald, long-fingered woman

F$_1$:

$1/2\ B^1B^1$

$1/2F^1F^2 = 1/4\ B^1B^1\ F^1F^2$ bald, short (men)/bald, long (women)

$1/2\ F^2F^2 = 1/4\ B^1B^1F^2F^2$ bald, long (men)/bald, long (women)

$1/2\ B^1B^2$

$1/2\ F^1F^2 = 1/4\ B^1B^2F^1F^2$ bald, short (men)/nonbald, long (women)

$1/2\ F^2F^2 = 1/4\ B^1B^2F^2F^2$ bald, long (men)/nonbald, long (women)

F$_1$ summary: **Men:**   1/2 bald, short-fingered : 1/2 bald, long-fingered

Women: 1/2 bald, long-fingered : 1/2 nonbald, long-fingered

Sex-Limited Traits

Some autosomal genes may only come to expression in one of the sexes, either because of differences in the internal hormonal environment or because of anatomical dissimilarities. For example, we know that bulls have many genes for milk production that they may transmit to their daughters, but they or their sons are unable to express this trait. The production of milk is therefore limited to variable expression in only the female sex. When the penetrance of a gene in one sex is zero, the trait will be **sex-limited**.

EXAMPLE 5.10
Chickens have a recessive gene for cock-feathering that is penetrant only in the male environment.

Genotypes	Phenotypes	
	Males	Females
HH	Hen-feathering	Hen-feathering
Hh	Hen-feathering	Hen-feathering
hh	Cock-feathering	Hen-feathering

SOLVED PROBLEM 5.6
Cock-feathering in chickens is a trait limited to expression only in males and is determined by the autosomal recessive genotype *hh*. The dominant allele *H* produces hen-feathered males. All females are hen-feathered regardless of genotype. A cock-feathered male is mated to three females, each of which produces a dozen chicks. Among the 36 progeny are 15 hen-feathered males, 18 hen-feathered females, and 3 cock-feathered males. What are the most probable genotypes of the three parental females?

Solution: In order for both hen-feathered (*H*-) and cock-feathered (*hh*) males to be produced, at least one of the females had to be heterozygous (*Hh*) or recessive (*hh*). The following female genotype possibilities must be explored:

(*a*) 2 *HH*, 1 *Hh* (*b*) 1 *HH*, 2 *Hh* (*c*) 1 *HH*, 1 *Hh*, 1 *hh* (*d*) 3 *Hh*
(*e*) 2 *Hh*, 1 *hh* (*f*) 1 *Hh*, 2 *hh* (*g*) 2 *HH*, 1 *hh* (*h*) 2 *hh*, 1 *HH*

Obviously, the more *hh* or *Hh* hen genotypes, proportionately the more cock-feathered males that are expected in the progeny. The ratio of 15 hen-feathered males : 3 cock-feathered males is much greater than the 1 : 1 ratio expected when all three females are heterozygous (*Hh*).

P: *hh*   ×   *Hh*
          cock-feathered male   hen-feathered females

F_1:   1/2 *Hh* hen-feathered males,   1/2 *hh* cock-feathered males

Possibility (*d*) is therefore excluded. Possibilities (*e*) and (*f*), which both contain one or more *hh* genotypes in addition to one or more *Hh* genotypes, must also be eliminated because these matings would produce even more cock-feathered males than possibility (*d*). In possibility (*g*), the 2 *HH* : 1 *hh* hens are expected to produce an equivalent ratio of 2 hen-feathered (*Hh*) : 1 cock-feathered (*hh*) males. This 2 : 1 ratio should be expressed in the 18 male offspring as 12 hen-feathered : 6 cock-feathered. These numbers compare fairly well with the observed 15 : 3, but possibility (*h*) would be even less favorable because even more cock-feathered males would be produced. Let us see if one of the remaining three possibilities will give us expected values closer to our observations.

Possibility (c):

$$P: \qquad hh \qquad \times \qquad \left.\begin{array}{l} 1/3\ HH \\ 1/3\ Hh \\ 1/3\ hh \end{array}\right\} \text{hen-feathered females}$$

cock-feathered male

F_1: $1/3(hh \times HH) = \quad 1/3 Hh$ hen-feathered males

$$1/3(hh \times Hh) = \begin{cases} (1/3)(1/2) = 1/6\ Hh \text{ hen-feathered males} \\ (1/3)(1/2) = 1/6\ hh \text{ cock-feathered males} \end{cases}$$

$1/3(hh \times hh) \quad = \quad 1/3\ hh$ cock-feathered males

Summary: Hen-feathered males $= 1/3 + 1/6 = 1/2$, cock-feathered males $= 1/3 + 1/6 = 1/2$. Again, this disagrees with the observations and must be excluded.

Possibility (b):

$$P: \qquad hh \qquad \times \qquad \left.\begin{array}{l} 1/3\ HH \\ \\ 2/3\ Hh \end{array}\right\} \text{hen-feathered females}$$

cock-feathered male

$$F_1: \qquad 1/3(hh \times HH) = \quad 1/3\ Hh \text{ hen-feathered males}$$

$$2/3(hh \times Hh) = \begin{cases} (2/3)(1/2) = 2/6\ Hh \text{ hen-feathered males} \\ (2/3)(1/2) = 2/6\ hh \text{ cock-feathered males} \end{cases}$$

Summary: Hen-feathered males $= 1/3 + 2/6 = 2/3$, cock-feathered males $= 2/6$ or $1/3$. These expectations are no closer to the observations than those of possibility (g).

Possibility (a):

$$P: \qquad hh \qquad \times \qquad \left.\begin{array}{l} 2/3\ HH \\ \\ 1/3\ Hh \end{array}\right\} \text{hen-feathered females}$$

cock-feathered male

$$F_1: \qquad 2/3(hh \times HH) = 2/3\ Hh \text{ hen-feathered males}$$

$$1/3(hh \times Hh) = \begin{cases} (1/3)(1/2) = 1/6\ Hh \text{ hen-feathered males} \\ (1/3)(1/2) = 1/6\ hh \text{ cock-feathered males} \end{cases}$$

Set the observation of 3 cock-feathered males equal to the 1/6, then $5 \times 3 = 15$ hen-feathered males should represent the 5/6. These expectations agree perfectly with the observations and therefore it is most probable that two of the females were HH and one was Hh.

Sex Reversal

Female chickens (ZW) that have laid eggs have been known to undergo not only a reversal of the secondary sexual characteristics such as development of cock-feathering, spurs, and crowing, but also the development of testes and even the production of sperm cells (primary sexual characteristics). This may occur when, for example, disease destroys the ovarian tissue, and in the absence of the female sex hormones the rudimentary testicular tissue present in the center of the ovary is allowed to proliferate. In solving problems involving sex reversals, it must be remembered that the functional male derived through sex reversal will still remain genetically female (ZW).

SOLVED PROBLEM 5.7

Suppose that a hen's ovaries are destroyed by disease, allowing its rudimentary testes to develop. Further suppose that this hen was carrying the dominant sex-linked gene B for barred feathers, and upon sex reversal was then crossed to a nonbarred female. What phenotypic proportions are expected in the F_1 and F_2?

Solution: Remember that sex determination in chickens is by the ZW method and that sex reversal does not change this chromosomal constitution. Furthermore, at least one sex chromosome (Z) is essential for life.

P: BW × bW
                     barred female   normal nonbarred female
                   sex reversed to a
                    functional male

F_1:

	B	**W**
b	Bb barred male	bW nonbarred female
W	BW barred female	WW lethal

The proportions are thus 3 males (all barred) : 2 females (half barred and half nonbarred).

F_2: Two equally frequent kinds of matings are possible among the F_1 birds. First mating = 1/2 of all matings.

                          Bb × bW
                     barred male   nonbarred female

Progeny Expectations	Correction for Frequency of Mating		Proportion of Total F_2
1/4 Bb	1/2	=	1/8 Bb barred males
1/4 bb	1/2	=	1/8 bb nonbarred males
1/4 BW	1/2	=	1/8 BW barred females
1/4 bW	1/2	=	1/8 bW nonbarred females

Second mating = 1/2 of all matings.

$$Bb \qquad \times \qquad BW$$
$$\text{barred male} \qquad\qquad \text{barred female}$$

Progeny Expectations	Correction for Frequency of Mating		Proportion of Total F_2
1/4 BB	1/2	=	1/8 BB barred males
1/4 Bb	1/2	=	1/8 Bb barred males
1/4 BW	1/2	=	1/8 BW barred females
1/4 bW	1/2	=	1/8 bW nonbarred females

$\left.\begin{array}{c} \\ \end{array}\right\} = 1/4$ (for the first two rows)

Summary of the F_2: Barred males $= 1/8 + 1/4 = 3/8$ Barred females $= 1/8 + 1/8 = 1/4$

Nonbarred males $= 1/8$ Nonbarred females $= 1/8 + 1/8 = 1/4$

Sexual Phenomena in Plants

Most flowering plants are monoecious and therefore do not have sex chromosomes. Indeed, the ability of mitotically produced cells with exactly the same genetic endowment to produce tissues with different sexual functions in a perfect flower speaks clearly for the bipotentiality of such plant cells. Well-known examples of dioecism usually are under the genetic control of a single-gene locus. However, at least one well-documented case of chromosomal sexuality is known in plants, i.e., in the genus *Melandrium* (a member of the pink family). Here, the Y chromosome determines a tendency to maleness just as it does in humans. Pistillate plants are XX and staminate plants are XY.

The ability of gametes produced by the same individual to unite and produce viable and fertile offspring is common among many families of flowering plants. Self-fertilization is also known to occur in a few of the lower animal groups. The perfect flowers of some monoecious plants fail to open (**cleistogamy**) until after the pollen has matured and accomplished self-fertilization. Self-fertilization is obligatory in barley, beans, oats, peas, soybeans, tobacco, tomato, wheat, and many other crops. In some species, self-fertilization as well as cross-fertilization may occur to varying degrees. For example, cotton and sorghum commonly experience more than 10% cross-fertilization. Still other monoecious species have developed genetic mechanisms that prevent self-fertilization or the development of zygotes produced by the union of identical gametes, making cross-fertilization obligatory. Self-incompatibility in monoecious species can become as efficient in enforcing cross-fertilization as the system exhibited under a dioecious mechanism of sex determination.

SOLVED PROBLEM 5.8
In monoecious corn, a recessive gene called "tassel-seed" (*ts*), when homozygous, produces only seeds where the staminate inflorescence (tassel) normally appears. No pollen is produced. Thus, individuals of genotype *ts/ts* are functionally reduced to a single sex, that of the female. On another chromosome, the recessive gene called "silkless" (*sk*), when homozygous, produces ears with no pistils (silks). Without silks,

none of these ears can produce seed and individuals of genotype sk/sk are reduced to performing only male functions (production of pollen in the tassel). The recessive gene for tassel-seed is epistatic to the silkless locus. (*a*) What sex ratio is expected in the F_1 and F_2 from the cross ts/ts, sk^+/sk^+ (female) × ts^+/ts^+, sk/sk (male)? (*b*) How could the genes for tassel-seed and silkless be used to establish male and female plants (dioecious) that would continue, generation after generation, to produce progeny in the ratio of 1 male : 1 female?

Solution:

(*a*)   P: ts/ts sk^+/sk^+ × ts^+/ts^+, sk/sk
                    female   male

F$_1$: ts^+/ts^+, sk^+/sk^+
                        monoecious
                    (both male and female flowers)

F$_2$:   9/16 $ts^+/-$, $sk^+/-$ = 9/16 monoecious
        3/16 $ts^+/-$, sk/sk = 3/16 male
        3/16 ts/ts, $sk^+/-$ ⎫
        1/16 ts/ts, sk/sk ⎭ 4/16 female

(*b*)   P: ts/ts, sk/sk × ts^+/ts^+, sk/sk
                    female   male

F$_1$:   1/2 ts^+/ts^+, sk/sk males
            1/2 ts/ts, sk/sk females

Subsequent generations would continue to exhibit a 1 : 1 sex ratio for these dioecious plants.

SOLVED PROBLEM 5.9

Pollen tubes containing the same self-incompatibility allele as that found in the diploid tissue of the style grow so slowly that fertilization cannot occur before the flower withers. Pollen produced by a plant of genotype S^1S^3 would be of two types, S^1 and S^3. If this pollen were to land on the stigma of the same plant (S^1S^3), none of the pollen tubes would grow. If these pollen grains (S^1 and S^3) were to alight on a stigma of genotype S^1S^2, then only the tubes containing the S^3 allele would be compatible with the alleles in the tissue of the style. If these pollen grains were to alight on a stigma of genotype S^2S^4, all of the pollen tubes would be functional. Four plant varieties (A, B, C, and D) are crossed, with the results listed in the table below.

		Male Parent			
		A	B	C	D
Female Parent	A	—	1/4 C, 1/4 D 1/4 E, 1/4 F	1/2 C 1/2 D	1/2 C 1/2 D
	B	1/4 C, 1/4 D 1/4 E, 1/4 F	—	1/2 C 1/2 E	1/2 D 1/2 F
	C	1/2 A 1/2 D	1/2 B 1/2 E	—	1/2 D 1/2 A
	D	1/2 A 1/2 C	1/2 B 1/2 F	1/2 A 1/2 C	—

Notice that two additional varieties (E and F) appear in the progeny. Determine the genotypes for all six varieties in terms of four self-sterility alleles (S^1, S^2, S^3, and S^4).

Solution: None of the genotypes are expected to be homozygous for the self-incompatibility alleles because pollen containing the same allele present in the maternal tissue is not functional and therefore homozygosity is prevented. Thus, six genotypes are possible with four self-incompatibility alleles: S^1S^2, S^1S^3, S^1S^4, S^2S^3, S^2S^4, S^3S^4. Crosses between genotypes with both alleles in common produce no progeny (A × A, B × B, etc.). Crosses between genotypes with only one allele in common produce offspring in the ratio of 1 : 1 (e.g., S^1S^2 female × S^1S^3 male = 1/2 S^1S^3 : 1/2 S^2S^3). Crosses between genotypes with none of their self-incompatibility alleles in common produce progeny in the ratio 1 : 1 : 1 : 1 (e.g., S^1S^2 × S^3S^4 = 1/4 S^1S^3: 1/4 S^1S^4: 1/4 S^2S^3: 1/4 S^2S^4. Turning now to the table of results, we find the cross B female × A male produces offspring in the ratio 1 : 1 : 1 : 1 and therefore neither B nor A have any alleles in common. If we assume that variety B has the genotype S^1S^4, then variety A must have the genotype S^2S^3 (the student's solution to this problem may differ from the one presented here in the alleles arbitrarily assigned as a starting point). The cross C male × A female produces offspring in the ratio 1 : 1, indicating one pair of alleles in common. Since we have already designated variety A to be of genotype S^2S^3, let us arbitrarily assign the genotype S^1S^2 to variety C. The cross D female × A male also indicates that one allele is held in common by these two varieties. Let us assign the genotype S^1S^3 to variety D. The genotype for variety E may now be determined from the cross C female × B male.

P: S^1S^2 (C)   × S^1S^4 (B)
F₁: 1/2 S^1S^4 = variety B   1/2 S^2S^4 = variety E

Likewise, the genotype for variety F may now be determined from the cross D female × B male.

P: S^1S^3 (D)   × S^1S^4 (B)
F₁: 1/2 S^1S^4 = variety B   1/2 S^3S^4 = variety F

Summary of genotypes for all six varieties:

$$A = S^2S^3 \quad B = S^1S^4 \quad C = S^1S^2 \quad D = S^1S^3 \quad E = S^2S^4 \quad F = S^3S^4$$

The student should confirm that the other results shown in the table are compatible with the genotypic assumptions shown above.

Supplementary Problems

SEX DETERMINATION AND SEX-LINKED INHERITANCE

Heterogametic Males (XY and XO Methods)

5.10.   A sex-linked recessive gene c produces red-green color blindness in humans. A normal woman whose father was color blind marries a color-blind man. (*a*) What genotypes are possible for the mother of the color-blind man? (*b*) What are the chances that the first child from this marriage will be a color-blind boy? (*c*) Of all the girls produced by these parents, what percentage is expected to be color blind? (*d*) Of all the children (sex unspecified) from these parents, what proportion is expected to be normal?

5.11. Sex determination in the grasshopper is by the XO method. The somatic cells of a grasshopper are analyzed and found to contain 23 chromosomes. (*a*) What sex is this individual? (*b*) Determine the frequency with which different types of gametes (number of autosomes and sex chromosomes) can be formed in this individual. (*c*) What is the diploid number of the opposite sex?

5.12. Male house cats may be black or yellow. Females may be black, tortoiseshell pattern, or yellow. (*a*) If these colors are governed by a sex-linked locus, how can these results be explained? (*b*) Using appropriate symbols, determine the phenotypes expected in the offspring from the cross yellow female × black male. (*c*) Do the same for the reciprocal cross of part (*b*). (*d*) A certain kind of mating produces females, half of which are tortoiseshell and half are black; half the males are yellow and half are black. What colors are the parental males and females in such crosses? (*e*) Another kind of mating produces offspring, 4 of which are yellow males, 4 yellow females, 4 black males, and 4 tortoiseshell females. What colors are the parental males and females in such crosses?

5.13. In the plant genus *Melandrium*, sex determination is similar to that in humans. A sex-linked gene *l* is known to be lethal when homozygous in females. When present in the hemizygous condition in males (*l*Y), it produces blotchy patches of yellow-green color. The homozygous or heterozygous condition of the wild-type allele (*LL* or *Ll*) in females, or the hemizygous condition in males (*L*Y) produces normal dark-green color. From a cross between heterozygous females and yellow-green males, predict the phenotypic ratio expected in the progeny.

5.14. The recessive gene for white eye color in *Drosophila* (*w*) is sex-linked. Another recessive sex-linked gene governing eye color is vermilion (*v*), which—when homozygous in females or hemizygous in males—together with the autosomal gene for brown eye (*bw/bw*), also produces white eye. White genotypes (*w*Y, *ww*) are epistatic to the other loci under consideration. (*a*) What phenotypic results are expected among progeny from mating a white-eyed male of genotype (*bw/bw*, *vw*+/Y) with a white-eyed female of genotype *bw*+/ *bw*, *vw/v* | *w*? (*b*) What phenotypic proportions are expected in the progeny from the mating of a vermilion female heterozygous at the brown locus but not carrying the white allele with a male that is white due to the *w* allele but heterozygous at the brown locus and hemizygous for the vermilion allele?

Heterogametic Females (ZW and ZO Methods)

5.15. Silver-colored plumage in poultry is due to a dominant sex-linked gene *S* and gold-colored plumage to its recessive allele *s*. List the phenotypic and genotypic expectations of the progeny from the matings   (*a*) *s*/W female × *S/S* male   (*b*) *s*/W female × *S/s* male   (*c*) *S*/W female × *S/s* male   (*d*) *S*/W female × *s/s* male.

5.16. In the Rosy Gier variety of carrier pigeon, a cross was made between gray-headed females and creamy-headed males. The F_1 ratio was 1 gray-headed female : 1 gray-headed male : 1 creamy-headed male. (*a*) How may these results be explained? (*b*) Diagram this cross using appropriate symbols.

5.17. Chickens have an autosomal dominant gene *C* that produces a short-legged phenotype called "creeper" in heterozygotes. Normal legs are produced by the recessive genotype *cc*. The homozygous dominant genotype *CC* is lethal. A dominant sex-linked gene *B* produces barred plumage, the recessive allele *b* produces nonbarred plumage. (*a*) Determine the phenotypic expectations among progeny (of both sexes) from the cross of a barred creeper female and a nonbarred creeper male. (*b*) Determine the phenotypic ratios within each sex for part (*a*). (*c*) Two chickens were mated and produced progeny in the following proportions: 1/12 nonbarred males, 1/6 nonbarred creeper females, 1/12 barred males, 1/12 nonbarred females, 1/6 nonbarred creeper males, 1/6 barred creeper males, 1/12 barred females, and 1/6 barred creeper females. What are the genotypes and phenotypes of the parents?

5.18. The presence of feathers on the shanks of the Black Langshan breed of chickens is due to the dominant alleles at either or both of two autosomal loci. Nonfeathered shanks are the result of the double-recessive genotype. A dominant sex-linked gene *B* places white bars on a black bird. Its recessive allele b produces nonbarred (black) birds. Trihybrid barred males with feathered shanks are mated to dihybrid nonbarred females with feathered shanks. Determine the F_1 phenotypic expectations.

Genic Balance

5.19. In *Drosophila*, the ratio between the number of X chromosomes and the number of sets of autosomes (A) is called the "sex index." Diploid females have a sex index (ratio X/A) $= 2/2 = 1.0$. Diploid males have a sex ratio of $1/2 = 0.5$. Sex index values between 0.5 and 1.0 give rise to intersexes. Values over 1.0 or under 0.5 produce weak and inviable flies called "superfemales" (metafemales) and "supermales" (metamales), respectively. Calculate the sex index and the sex phenotype in the following individuals: (*a*) AAX (*b*) AAXXY (*c*) AAAXX (*d*) AAXX (*e*) AAXXX (*f*) AAAXXX (*g*) AAY.

Haplodiploidy

5.20. If the diploid number of the honey bee is 16. (*a*) How many chromosomes will be found in the somatic cells of the drone (male)? (*b*) How many bivalents will be seen during the process of gametogenesis in the male? (*c*) How many bivalents will be seen during the process of gametogenesis in the female?

5.21. Seven eye colors are known in the honey bee, each produced by a recessive gene at a different locus: brick (*bk*), chartreuse (*ch*), ivory (*i*), cream (*cr*), snow (*s*), pearl (*pe*), and garnet (*g*). Suppose that a wild-type queen heterozygous at the brick locus (bk^+/bk) was to be artificially inseminated with a mixture of sperm from seven haploid drones each exhibiting a different one of the seven mutant eye colors. Further assume that the semen contribution of each male contains equal concentrations of sperm, that each sperm has an equal opportunity to enter fertilization, and that each zygote thus formed has an equal opportunity to survive. (*a*) What percentage of the drone offspring is expected to be brick-eyed? (*b*) What percentage of worker offspring is expected to be brick-eyed?

Single-Gene Effects

5.22. In the single-celled haploid plant *Chlamydomonas*, there are two mating types, (+) and (−). There is no morphological distinction between the (+) sex and the (−) sex in either the spore stage or the gamete stage (isogametes). The fusion of (+) and (−) gametes produces a $2n$ zygote that immediately undergoes meiosis to produce four haploid spores, two of which are (+) and two of which are (−). (*a*) Could a pair of genes for sex account for the 1 : 1 sex ratio? (*b*) Does the foregoing information preclude some other form of sex determination? Explain.

5.23. Sex determination in the wasp *Bracon* is either by sex alleles or haplodiploidy. A recessive gene called "veinless" (*v*) is known to assort independently of the sex alleles; the dominant allele v^+ results in wild type. For each of the eight crosses listed below, determine the relative frequencies of progeny phenotypes within each of three categories: (1) haploid males   (2) diploid males   (3) females. (*a*) v/v, $s^a/s^b \times v^+$, s^a (*b*) v/v, $s^a/s^b \times v^+$, s^c (*c*) v/v, $s^a/s^b \times v$, s^b (*d*) v/v, $s^a/s^b \times v$, s^c (*e*) v/v^+, $s^a/s^b \times v$, s^a (*f*) v/v^+, $s^a/s^b \times v$, s^c (*g*) v^+/v^+, $s^a/s^b \times v$, s^a, (*h*) v^+/v^+, $s^a/s^b \times v$, s^c.

VARIATIONS OF SEX LINKAGE

5.24. An Englishman by the name of Edward Lambert was born in 1717. His skin was like thick bark that had to be shed periodically. The hairs on his body were quill-like and he subsequently has been referred to as the "porcupine man." He had six sons, all of whom exhibited the same trait. The trait appeared to be transmitted from father to son through four generations. None of the daughters ever exhibited the trait. In fact, it has never been known to appear in females. (*a*) Could this be an autosomal sex-limited trait? (*b*) How is this trait probably inherited?

5.25. Could a recessive mutant gene in humans be located on the X chromosome if a woman exhibiting the recessive trait and a normal man had a normal son? Explain.

5.26. A holandric gene is known in humans that causes long hair to grow on the external ears. When men with hairy ears marry normal women, (a) what percentage of their sons would be expected to have hairy ears, (b) what proportion of the daughters is expected to show the trait, (c) what ratio of hairy-eared : normal children is expected?

SEX-INFLUENCED TRAITS

5.27. Suppose that a white forelock in humans is due to a sex-influenced gene w that is dominant in men and recessive in women. Using the allelic symbols w and w', indicate all possible genotypes and the phenotypes thereby produced in men and women.

5.28. The sex-influenced gene governing the presence of horns in sheep exhibits dominance in males but acts recessively in females. When the Dorset breed (both sexes horned) with genotype hh is crossed to the Suffolk breed (both sexes polled or hornless) with the genotype $h'h'$, what phenotypic ratios are expected in the F_1 and F_2?

5.29. The fourth (ring) finger of humans may be longer or shorter than the second (index) finger. The short index finger is thought to be produced by a gene that is dominant in men and recessive in women. What kinds of children and with what frequency would the following marriages be likely to produce: (a) heterozygous short-fingered man × short-fingered woman, (b) heterozygous long-fingered woman × homozygous short-fingered man, (c) heterozygous short-fingered man × heterozygous long-fingered woman, (d) long-fingered man × short-fingered woman?

5.30. In the Ayrshire breed of dairy cattle, mahogany-and-white color is dependent upon a gene C^M that is dominant in males and recessive in females. Its allele for red-and-white (CR) acts as a dominant in females but recessive in males. (a) If a red-and-white male is crossed to a mahogany-and-white female, what phenotypic and genotypic proportions are expected in the F_1 and F_2? (b) If a mahogany-and-white cow has a red-and-white calf, what sex is the calf? (c) What genotype is not possible for the sire of the calf in part (b)?

5.31. A sex-linked recessive gene in humans produces color-blind men when hemizygous and color-blind women when homozygous. A sex-influenced gene for pattern baldness is dominant in men and recessive in women. A heterozygous bald, color-blind man marries a nonbald woman with normal vision whose father was nonbald and color blind and whose mother was bald with normal vision. List the phenotypic expectations for their children.

SEX-LIMITED TRAITS

5.32. A dominant sex-limited gene is known to affect premature baldness in men but is without effect in women. (a) What proportion of the male offspring from parents, both of whom are heterozygous, is expected to be bald prematurely? (b) What proportion of all their children is expected to be prematurely bald?

5.33. In the clover butterfly, all males are yellow, but females may be yellow if they are of the homozygous recessive genotype yy or white if they possess the dominant allele Y-. What phenotypic proportions, exclusive of sex, are expected in the F_1 from the cross $Yy \times Yy$?

5.34. The barred plumage pattern in chickens is governed by a dominant sex-linked gene B. The gene for cock-feathering h is recessive in males, its dominant allele H produces hen-feathering. Normal females are hen-feathered regardless of genotype (sex-limited trait). Nonbarred females heterozygous at the hen-feathered locus are crossed to a barred, hen-feathered male whose father was cock-feathered and non-barred. What phenotypic proportions are expected among the progeny?

5.35. Cock-feathering is a sex-limited trait in chickens (see Example 5.10). In the Leghorn breed, all males are cock-feathered and all females are hen-feathered. In the Sebright bantam breed, both males and females are hen-feathered. In the Hamburg breed, males may be either cock-feathered or hen-feathered, but females are always hen-feathered. (a) How can these results be explained? (*b*) If the ovaries or testes are removed and the chickens are allowed to molt, they will become cock-feathered regardless of genotype. What kind of chemicals are involved in the expression of genotypes at this locus?

PEDIGREES

5.36. Could the trait represented by the solid symbols in the pedigree below be explained on the basis of (*a*) a dominant sex-linked gene, (*b*) a recessive sex-linked gene, (*c*) a holandric gene, (*d*) a sex-limited autosomal dominant, (*e*) a sex-limited autosomal recessive, (*f*) a sex-influenced autosomal gene dominant in males, (*g*) a sex-influenced autosomal gene recessive in males?

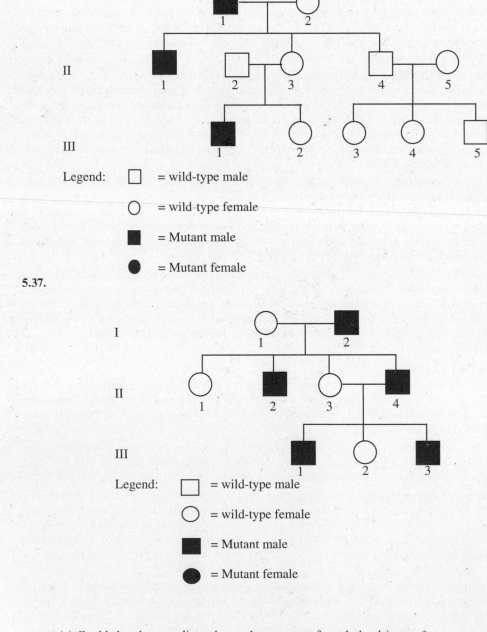

5.37.

Legend: ☐ = wild-type male

○ = wild-type female

■ = Mutant male

● = Mutant female

(*a*) Could the above pedigree be used as support for a holandric gene?

(b) Does the above pedigree contradict the assumption of a sex-linked recessive gene for the mutant trait?

(c) If a mating between III2 and III3 produced a mutant female offspring, which of the above two hypotheses would apply? List the genotype of each individual in the pedigree, using appropriate symbols.

5.38. Could the trait represented by the solid symbols in the pedigree shown below be produced by (a) an autosomal dominant, (b) an autosomal recessive, (c) a sex-linked dominant, (d) a sex-linked recessive, (e) a sex-limited gene, (f) a holandric gene, (g) a sex-influenced gene?

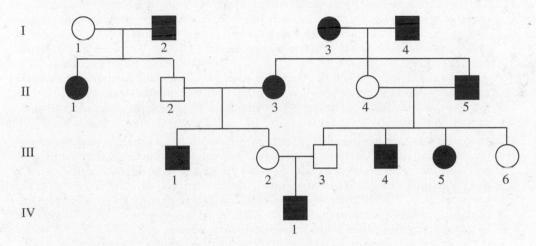

SEX REVERSAL

5.39. Suppose that a female undergoes sex reversal to become a functional male and is then mated to a normal female. Determine the expected F_1 sex ratios from such matings in species with (a) ZW method of sex determination, (b) XY method of sex determination.

5.40. The hemp plant is dioecious, probably resulting from an XY mechanism of sex determination. Early plantings (May–June) yield the normal 1 : 1 sex ratio. Late plantings in November, however, produce all female plants. If this difference is due to the length of daylight, it should be possible to rear both XY females and XY males under controlled conditions in the greenhouse. What sex ratio would be expected among seedlings grown early in the year from crosses between XY males and XY females?

SEXUAL PHENOMENA IN PLANTS

5.41. A completely pistillate inflorescence (female flower) is produced in the castor bean by the recessive genotype nn. Plants of genotype NN and Nn have mixed pistillate and staminate flowers in the inflorescence. Determine the types of flowers produced in the progeny from the following crosses: (a) NN female × Nn male (b) Nn female × Nn male (c) nn female × Nn male.

5.42. Asparagus is a dioecious plant in which maleness (staminate plants) is governed by a dominant gene P and femaleness (pistillate plants) by its recessive allele p. Sometimes pistillate flowers are found to have small nonfunctional anthers, and then again some staminate flowers may be found to possess immature pistils. Very rarely, a staminate plant may be found to produce seed, most likely by self-fertilization. (a) What sex ratio is expected among the F_1 from an exceptional staminate-seed plant of genotype Pp when selfed? (b) When the staminate F_1 plants from part (a) are crossed to normal pistillate plants (pp), what sex ratio is expected in the progeny? (c) What type of mating gives a 1 : 1 sex ratio?

5.43. Sex determination in the dioecious plant *Melandrium album* (*Lychnis dioica*) is by the XY method. A sex-linked gene governs leaf size; the dominant allele B produces broad leaves, and the recessive allele b produces narrow leaves. Pollen grains bearing the recessive allele are inviable. What phenotypic results are expected from the following crosses?

	Seed Parent		Pollen Parent
(a)	Homozygous broad-leaf	=	Narrow-leaf
(b)	Heterozygous broad-leaf	=	Narrow-leaf
(c)	Heterozygous broad-leaf	=	Broad-leaf

5.44. Two or more genes may cooperate to restrict selfing. An example is known in monoecious sorghum where the action of two complementary genes produces an essentially male plant by making the female structures sterile. Plants heterozygous at both loci (Fs_1/fs_1, Fs_2/fs_2) result in female sterile plants with no effect on their production of pollen. Whenever three dominant genes are present (Fs_1/Fs_1, Fs_2/fs_2, or Fs_1/fs_1, Fs_2/Fs_2), dwarf plants are produced that fail to develop a head. Although not yet observed, a genotype with all four dominant alleles would presumably also be dwarf and headless. All other genotypes produce normal plants. If these loci assort independently of one another, determine the F_1 phenotypic expectancies from the crosses   (a) Fs_1/fs_1, $Fs_2/fs_2 \times Fs_1/fs_1$, fs_2/fs_2 (b) Fs_1/fs_1, $fs_2/fs_2 \times Fs_1/fs_1$, fs_2/fs_2.

5.45. In some cases of self-incompatibility, pollen tube growth is so slow that the style withers and dies before fertilization can occur. Sometimes, if pollination is artificially accomplished in the bud stage, the pollen tube can reach the ovary before the style withers. In this case it is possible to produce a genotype homozygous at the self-sterility locus. (a) What would be the expected results from natural pollination of such a homozygote (S^1S^1) by a heterozygote containing one allele in common (S^1S^3)? (b) What would be the result of the reciprocal cross of part (a)? (c) What would be the result of natural pollination of S^1S^1 by S^2S^3. (d) Would the reciprocal cross of part (c) make any difference in the progeny expectations?

5.46. Two heteromorphic types of flowers are produced in many species of the plant genus *Primula*. One type, called "pin," has short anthers and a long style. The other type, called "thrum," has highly placed anthers and a short style. Thrum is produced by a dominant gene S and pin by the recessive allele s. The only pollinations that are compatible are those between styles and anthers of the same height, i.e., between thrum style and pin anther or between thrum anther and pin style. (a) What genotype do all thrum plants possess? (b) If both the pin and thrum are heterozygous for an independently segregating allelic pair Aa, what genotypic ratio is expected in the next generation?

5.47. The self-incompatibility mechanism of many plants probably involves a series of multiple alleles similar to that found in *Nicotiana*. In this species, pollen tubes grow very slowly or not at all down the style that contains the same allele at the self-incompatibility locus S. List the genotypic ratio of progeny sporophytes expected from the following crosses:

	Seed Parent		Pollen Parent
(a)	S^1S^2	$\times$	S^1S^2
(b)	S^1S^2	$\times$	S^1S^3
(c)	S^1S^2	$\times$	S^3S^4

(d)   How much of the pollen is compatible in each of the above three crosses?

5.48. A cross is made between two plants of the self-sterile genotype $S^1S^2 \times S^3S^4$. If all the F_1 progeny are pollinated only by plants of genotype S^2S^3, what genotypic proportions are expected in the F_2?

Review Questions

Vocabulary. For each of the following definitions, give the appropriate term and spell it correctly. Terms are single words unless indicated otherwise.

1. A form of reproduction involving the union of haploid gametes to form a diploid zygote.
2. An animal that has both male and female reproductive organs.
3. A flower having female but no male reproductive parts.
4. An adjective applicable to the sex that produces gametes bearing structurally different sex chromosomes (e.g., half X-bearing and half Y-bearing).
5. The two symbols that represent the sex chromosomes of female chickens, corresponding respectively to the X and Y chromosomes of mammals.
6. The mode of sex determination for *Drosophila*. (Two words.)
7. The mode of sex determination for bees.
8. An adjective applied to genes on the differential (nonhomologous) segment of the Y chromosome.
9. A class of traits governed by autosomal alleles whose dominance relationships are reversed in the two sexes as a consequence of sex hormone differences.
10. A class of autosomal traits having phenotypic variability in a population in only one sex; the other sex exhibits a single phenotype regardless of its genotype.

Multiple-Choice Questions   Choose the one best answer.

1. In an animal with the XO method of sex determination, which of the following could be the normal number of chromosomes in its somatic cells?   (*a*) 26 in males   (*b*) 17 in females   (*c*) 33 in females   (*d*) 13 in males   (*e*) more than one of the above
2. Suppose that in bees the dominant gene b^+ produces wild-type (brown) eyes, and its recessive allele b produces pink eye color. If a pink-eyed queen mates with a brown-eyed drone, their offspring would most likely be   (*a*) only wild-type progeny   (*b*) wild-type workers and pink-eyed drones   (*c*) only pink-eyed progeny   (*d*) workers = 2 wild : 2 pink-eye; all drones wild type   (*e*) insufficient information to allow a definitive answer
3. In the guinea pig pedigree shown below, supposedly involving sex-linked inheritance: (*a*) III could exhibit the dominant trait   (*b*) both I1 and II2 must be carriers of the gene responsible for the trait shown by II1   (*c*) the probability that the next offspring of II2 × II3 has the same phenotype as III1 is 0.5   (*d*) if II1 is crossed to a female genetically like his sister, 75% of his offspring is expected to be phenotypically like their aunt   (*e*) this pedigree is incompatible with a sex-linked explanation

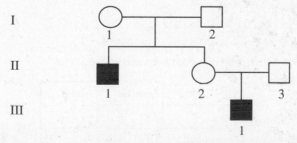

4. The presence of horns in the Dorset breed of sheep is due to a sex-influenced locus with horns dominant in males and recessive in females. Polled (hornless) males are mated to horned females. The fraction of the F_2 expected to be polled is   (*a*) 1/4   (*b*) 1/2   (*c*) 3/4   (*d*) 3/8   (*e*) none of the above

5. In the clover butterfly, all males are yellow, but females may be yellow if they are homozygous (cc) or white if they possess the dominant allele (C-). Matings between heterozygotes are expected to produce an F_1 generation containing (a) 1/2 yellow : 1/2 white (b) 3/4 white : 1/4 yellow (c) 3/8 white : 5/8 yellow (d) 9/16 yellow : 7/16 white (e) none of the above

6. If sex determination in a species ($2n = 14$) is determined by genic balance (as it is in *Drosophila*), then an intersex could have (a) 10 autosomes + 2X (b) 14 chromosomes + 2X (c) six pairs of autosomes + XY (d) 21 autosomes + 2X (e) none of the above

7. The presence of tusks is governed by a holandric gene in a certain mammalian species. When a tusked male is mated to nontusked females, among 100 of their F_2 progeny we would expect to find (a) 50 tusked males, 50 nontusked females (b) 25 tusked males, 25 tusked females, 25 nontusked males, 25 nontusked females (c) 50 nontusked females, 25 tusked males, 25 nontusked males (d) 50 nontusked males, 25 tusked females, 25 nontusked females (e) none of the above

8. In a bird species, blue beak is a sex-linked recessive trait; red beak is the alternative dominant trait. If a red-beaked male is mated to blue-beaked females, we would expect to find in the F_1 (a) all progeny red-beaked (b) all males red-beaked, all females blue-beaked (c) all males blue-beaked, all females red-beaked (d) all females red-beaked, 1/2 of males red-beaked, 1/2 of males blue-beaked (e) none of the above

9. Suppose that the testes of a male (in a species with an XO sex-determination mechanism) experiences a primary sex reversal and begins to produce only eggs. Long tail is a dominant sex-linked trait; short tail is its recessive alternative. If a long-tailed male undergoes a primary sex reversal (to function as a female) and is mated to a short-tailed male, which of the following is expected among the adult progeny? (a) 1/2 long-tailed progeny (b) 2/3 of males long-tailed (c) 1/4 of all progeny long-tailed females (d) 1/3 of all progeny long-tailed males (e) problem is ambiguous; insufficient information

10. A pair of codominant sex-linked alleles in a mammal produce red pigment when homozygous or hemizygous for A^1, colorless when homozygous or hemizygous for A^2, and pink when heterozygous. If a pink female is crossed to a white male, we expect among the progeny (a) 50% of females are white (b) 50% of all progeny are pink (c) 50% of males are pink (d) 25% of all progeny are white (e) none of the above

Answers to Supplementary Problems

5.10. (a) Cc or cc (b) 1/4 (c) 50% (d) 1/2

5.11. (a) Male (b) 1/2(11A + 1X): 1/2(11A) (c) 24

5.12. See Example 9.4. (a) A pair of codominant sex-linked alleles

	Females	Males
Black	$C^B C^B$	$C^B Y$
Tortoiseshell	$C^B C^Y$	—
Yellow	$C^Y C^Y$	$C^Y Y$

(b) All males yellow, all females tortoiseshell (c) All males black, all females tortoiseshell (d) Tortoise-shell female × black male (e) Tortoiseshell female × yellow male

5.13. 1/3 dark-green females : 1/3 males with yellow-green patches : 1/3 males dark green

5.14. (a) Males all white-eyed; females: 1/4 vermilion : 1/4 wild type : 1/4 white : 1/4 brown (b) Males and females: 3/4 vermilion : 1/4 white

5.15. (a) Silver females (S/W), silver males (S/s) (b) Males: 1/2 silver (S/s) : 1/2 gold (s/s); females: 1/2 silver (S/W) : 1/2 gold (s/W) (c) Males: all silver (1/2 S/S : 1/2 S/s); females: 1/2 silver (S/W) : 1/2 gold (s/W) (d) All males silver (S/s), all females gold (s/W)

5.16. (a) Sex-linked gene with one allele lethal when hemizygous in females or homozygous in males.

	Male	Female
Grey	HH	HW
Cream	HH^I	——
Lethal	$H^I H^I$	$H^I W$

(b) P: $HW \times HH^I$; F_1: 1/3 HH gray male : 1/3 HH^I cream male : 1/3 HW gray female

5.17. (a) 1/6 nonbar, normal-leg females : 1/6 bar, normal-leg male : 1/3 nonbar, creeper female : 1/3 bar, creeper male (b) Males: 2/3 bar, creeper : 1/3 bar, normal leg; females: 2/3 nonbar, creeper : 1/3 nonbar, normal leg (c) Bar, creeper male ($CcBb$) × nonbar, creeper female ($CcbW$)

5.18. Males and females: 15/32 bar, feathered : 15/32 nonbar, feathered : 1/32 bar, nonfeathered : 1/32 nonbar, nonfeathered

5.19. (a) 0.5 male (b) 1.0 female (c) 0.67 intersex (d) 1.0 female (e) 1.5 superfemale (g) lethal (f) 1.0 female (triploid)

5.20 (a) 8 (b) None; meiosis cannot occur in haploid males (c) 8

5.21. (a) 50% (b) 7.14%

5.22. (a) Yes (b) No. A sex chromosome mechanism could be operative without a morphological difference in the chromosomes, gametes, or spores.

5.23.

		Haploid Males		Diploid Males		Females	
		Wild Type	Veinless	Wild Type	Veinless	Wild Type	Veinless
(a)		0	All	All	0	All	0
(b)		0	All	0	0	All	0
(c)		0	All	0	All	0	All
(d)		0	All	0	0	0	All
(e)		1/2	1/2	1/2	1/2	1/2	1/2
(f)		1/2	1/2	0	0	1/2	1/2
(g)		All	0	All	0	All	0
(h)		All	0	0	0	All	0

5.24. (a) No. It is highly unlikely that a mutant autosomal sex-limited gene would be transmitted to all his sons through four generations without showing segregation. (b) Holandric gene (Y-linked)

5.25. Yes, if it was incompletely sex-linked and the father carried the dominant normal gene on the homologous portion of his Y chromosome.

5.26. (a) 100%   (b) None   (c) 1 hairy : 1 normal

5.27. Either allele w or w^l can be dominant in men and recessive in women; hence, two sets of genotypes are possible. Both are equally valid answers.

Genotypes 1	Men	Women	Genotypes 2
w dominant in men:			w' dominant in men:
ww	Forelock	Forelock	$w'w'$
ww'	Forelock	Normal	ww'
$w'w'$	Normal	Normal	ww

5.28. F_1: all males horned, all females polled; F_2 males: 3/4 horned : 1/4 polled; F_2 females: 3/4 polled : 1/4 horned

5.29. (a) All males short; females: 1/2 short : 1/2 long   (b) Same as (a)   (c) Males: 3/4 short : 1/4 long; females: 1/4 short : 3/4 long   (d) All males short, all females long

5.30. (a) F_1: $C^M C^R$ mahogany males, $C^M C^R$ red females; F_2 males and females : 1/4 $C^M C^M$: 1/2 $C^M C^R$: 1/4 $C^R C^R$; F_2 males: 3/4 mahogany : 1/4 red; F_2 females: 1/4 mahogany : 3/4 red   (b) Female   (c) $C^M C^M$

5.31.

Phenotype	Daughters	Sons
Bald, normal vision	1/8	3/8
Bald, color blind	1/8	3/8
Nonbald, normal vision	3/8	1/8
Nonbald, color blind	3/8	1/8

5.32. (a) 3/4   (b) 3/8

5.33. 5/8 yellow : 3/8 white

5.34. Males: 3/8 barred, hen-feathered : 1/8 barred, cock-feathered : 3/8 non-barred, hen-feathered : 1/8 nonbarred, cock-feathered; females: 1/2 barred, hen-feathered : 1/2 nonbarred, hen-feathered

5.35. (a) Leghorns are homozygous hh. Sebright bantams are homozygous HH. Hamburgs are segregating at this locus; one or the other allele has not been "fixed" in the breed. (b) Gonads are the source of steroid sex hormones as well as of reproductive cells. The action of these genes is dependent upon the presence or absence of these sex hormones.

5.36. (a) No   (b) Yes   (c) No   (d) Yes   (e) Yes   (f) Yes   (g) No

5.37. (a) Yes   (b) No   (c) Sex-linked recessive gene; Aa (I1, II1, II3, III2), aY (I2, II2, II4, III1, III3)

5.38. (a)-(f) No   (g) Yes (if black is dominant in males and recessive in females)

5.39. (a) 2 females : 1 male   (b) All females

5.40. 2 males : 1 female

5.41. (a) All mixed   (b) 3/4 mixed : 1/4 pistillate   (c) 1/2 mixed : 1/2 pistillate

5.42. (a) 3/4 staminate : 1/4 pistillate   (b) 2/3 staminate : 1/3 pistillate   (c) $Pp \times pp$

5.43. (*a*) Only broad-leaved males   (*b*) 1/2 broad-leaved males : 1/2 narrow-leaved males   (*c*) All females broad-leaved; 1/2 males broad-leaved : 1/2 males narrow-leaved

5.44. (*a*) 1/8 dwarf : 1/4 female sterile : 5/8 normal (monoecious)   (*b*) 1/2 normal : 1/4 female sterile : 1/4 dwarf

5.45. (*a*) All S^1S^3 (*b*) No progeny   (*c*) $1/2\ S^1S^2 : 1/2\ S^1S^3$ (*d*) No

5.46. (*a*) *Ss*   (*b*) 1/8 *AASs* : 1/4 *AaSs* : 1/8 *aaSs* : 1/8 *AAss* : 1/4 *Aass* : 1/8 *aass*

5.47. (*a*) None   (*b*) $1/2\ S^1S^3 : 1/2\ S^2S^3$ (*c*) $1/4\ S^1S^3 : 1/4\ S^1S^4 : 1/4\ S^2S^3 : 1/4\ S^2S^4$ (*d*) a = none, b = 2, c = all

5.48. $1/4\ S^1S^2 : 1/3\ S^2S^3 : 1/12\ S^1S^3 : 1/12\ S^2S^4 : 1/4\ S^3S^4$

Answers to Review Questions

Vocabulary

1. sexual
2. hermaphrodite
3. pistillate
4. heterogametic
5. Z, W
6. genic balance
7. haplodiploidy
8. holandric
9. sex-influenced
10. sex-limited

Multiple-Choice Questions

1. *d* 2. *b* 3. *b* 4. *b* 5. *c* 6. *d* 7. *a* 8. *c* 9. *d* 10. *a*

CHAPTER 6

Linkage and Chromosome Mapping

Recombination Among Linked Genes

1 LINKAGE

When two or more genes reside on the same chromosome, they are said to be **linked**. They may be linked together on one of the autosomes or connected together on the sex chromosome (Chapter 5). Genes on different chromosomes are distributed into gametes independently of one another (Mendel's Law of Independent Assortment). Genes on the same chromosome, however, tend to stay together during the formation of gametes. Thus, the results of testcrossing dihybrid individuals will yield different results, depending upon whether the genes are linked or on different chromosomes.

EXAMPLE 6.1
Genes on different chromosomes assort independently during meiosis, giving a
1 : 1 : 1 : 1 testcross ratio.

Parents: $AaBb$ × $aabb$

Gametes: AB Ab aB ab ab

F_1:   1/4 $AaBb$: 1/4 $Aabb$: 1/4 $aaBb$: 1/4 $aabb$

EXAMPLE 6.2
Linked genes do not assort independently, but tend to stay together in the same combinations as they were in the parents. Genes to the left of the slash line (/) are on one chromosome and those to the right are on the homologous chromosome. Very closely linked genes may not recombine in the formation of gametes.

Parents:	*AB/ab*	×	*ab/ab*

Gametes: (*AB*) (*ab*)   (*ab*)

F₁: 1/2 *AB/ab* : 1/2 *ab/ab*

Large deviations from a 1 : 1 : 1 : 1 ratio on the testcross progeny of a dihybrid could be used as evidence for linkage. Linked genes do not always stay together, however, because homologous nonsister chromatids may exchange segments of varying length with one another during meiotic prophase. Recall from Chapter 1 that homologous chromosomes pair with one another in a process called "synapsis" and that the points of genetic exchange, called "chiasmata," produce recombinant gametes through crossing over.

2   CROSSING OVER

In preparation for meiosis, the DNA of each chromosome replicates, producing two genetically identical (barring mutation) sister chromatids. During prophase I, homologous chromosomes form pairs called synapses [Fig. 6-1(*a*)] with the aid of proteins in the synaptonemal complex. Very large protein complexes, called **recombination modules** [about 90 nanometers (nm) in diameter], occur at intervals along the synaptonemal complex; each of these recombination modules is thought to function as a multienzyme "recombination machine" that affects synapsis and recombination. A **nick** is the removal of a phosphodiester bond between adjacent nucleotides in a DNA strand. Endonucleases in the recombination modules nick a single strand of each chromatid, allowing nonsister strands to be exchanged [Fig. 6-1(*b*)], and thus affecting the recombination of linked genes. A DNA polymerase may extend the exchanged strands, and an enzyme called DNA ligase repairs the nicks [Fig. 6-1(*c*)]. If the top chromatid strand is rotated by 180°, a cross-shaped structure called a **chi (χ) form** can be seen under the microscope. This structure is also referred to as the **Holliday model** after R. Holliday who proposed it in 1964 [Fig. 6-1(*d*)]. An endonuclease nicks the two previously uncut strands at tetranucleotide sequences 5′-(A/T)TT(G/C)-3′. Gaps and nicks are then repaired, creating four recombinant chromatids [Fig. 6-1(*e*)] which will segregate during the second meiotic division to be incorporated into different gametes. Note that if only the *A* and *B* loci are being studied in the progeny of dihybrid parents (*AB/ab*), two of the four possible gametes will retain the linkage relationships of the dihybrid parents (*AB* and *ab*) and are thus referred to as **parental** or **noncrossover types**; the two other gametes will be **recombinant** or **crossover types** (*Ab* and *aB*). Thus, each crossover or chiasma event is expected to produce four gametes (*AB, Ab, aB, ab*) with equal frequencies. However, if a crossover between the two genes under study does not occur in every meiosis, then among all of the gametes (both those with and without crossovers in this region) produced by a dihybrid individual, the frequency of noncrossover-type gametes will exceed that of crossover-type gametes.

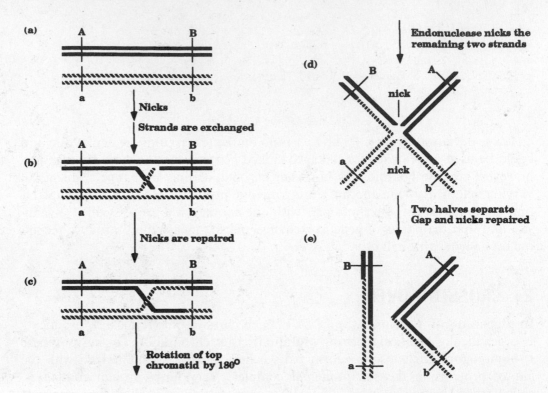

Fig. 6-1. General recombination and the formation of a Holliday intermediate. (*a*) Two homologous chromatids synapse. (*b*) A strand of each homologous chromatid is nicked and exchanged. (*c*) DNA synthesis may extend the exchanged strands subsequent to DNA nick repair. (*d*) Branch migration and formation of the Holliday intermediate. (*e*) The nicking of the uncut strands of the chromatids creates separate recombined chromatids. Nicks in the recombined chromatids are repaired by a ligase enzyme. (From W. D. Stansfield, J. S. Colomé, R. J. Cano, *Schaum's Molecular and Cellular Biology*, McGraw-Hill, New York, 1996.)

The alleles of double heterozygotes (dihybrids) at two linked loci may appear in either of two positions relative to one another. If the two dominant (or wild-type) alleles are on one chromosome and the two recessives (or mutants) on the other (*AB/ab*), the linkage relationship is called **coupling phase**. When the dominant allele of one locus and the recessive allele of the other occupy the same chromosome (*Ab/aB*), the relationship is termed **repulsion phase**. Parental and recombinant gametes will be of different types, depending upon how these genes are linked in the parent.

EXAMPLE 6.3   Coupling Parent: *AB/ab*

Parental: (*AB*)   (*ab*)

Gametes:

Recombinant: (*Ab*)   (*aB*)

EXAMPLE 6.4   Repulsion Parent: *Ab/aB*

Noncrossover: (*Ab*)   (*aB*)

Gametes:

Crossover: (*AB*)   (*ab*)

SOLVED PROBLEM 6.1

In the human pedigree below where the male parent does not appear, it is assumed that he is phenotypically normal. Both hemophilia (h) and color blindness (c) are sex-linked recessives. Insofar as possible, determine the genotypes for each individual in the pedigree.

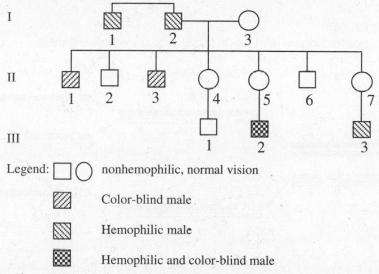

Legend: □○   nonhemophilic, normal vision

        ▨   Color-blind male

        ▨   Hemophilic male

        ▨   Hemophilic and color-blind male

Solution: The linkage relationship of the males' genes on their single X chromosome is obvious from their phenotype. Thus, I1, I2, and III3 are all hemophilic with normal color vision and therefore must be hC/Y. Nonhemophilic, color-blind males II1 and II3 must be Hc/Y. Normal males II2, II6, and III1 must possess both dominant alleles HC/Y. III2 is both hemophilic and color blind and therefore must possess both recessives hc/Y. Now let us determine the female genotypes. I3 is normal but produces sons, half of which are color blind and half normal. The X chromosome contributed by I3 to her color-blind sons II1 and II3 must have been Hc; the X chromosome she contributed to her normal sons II2 and II6 must have been HC. Therefore, the genotype for I3 is HC/Hc.

Normal females II4, II5, and II7 each receive hC from their father (I2), but could have received either Hc or HC on the X chromosome they received from their mother (I3). II4 has a normal son (III1) to which she gives HC; therefore, II4 is probably hC/HC, although it is possible for II4 to be hC/Hc and produce an HC gamete by crossing over. II5, however, could not be hC/HC and produce a son with both hemophilia and color blindness (III2); therefore, II5 must be hC/Hc in order to give the crossover gamete hc to her son.

3   CHIASMA FREQUENCY

A pair of synapsed chromosomes (bivalent) consists of four chromatids called a tetrad. Every tetrad usually experiences at least one chiasma somewhere along its length. Generally speaking, the longer the chromosome, the greater the number of chiasmata. Each type of chromosome within a species has a characteristic (or average) number of chiasmata. The frequency with which a chiasma occurs between any two genetic loci also has a characteristic or average probability. The further apart two genes are located on a chromosome, the greater the opportunity for a chiasma to occur between them. The closer two genes are linked, the smaller the chance for a chiasma occurring between them. These chiasmata probabilities are useful in predicting the proportions of parental and recombinant gametes expected to be formed from a given genotype. The percentage of cross-over (recombinant) gametes formed by a given genotype is a direct reflection of the frequency with which a chiasma forms between the genes in question. Only when a crossover forms between the gene loci under consideration will recombination be detected.

EXAMPLE 6.5
Crossing over outside the *A–B* region fails to recombine these markers.

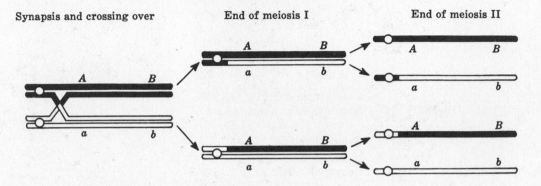

Synapsis and crossing over   End of meiosis I   End of meiosis II

When a chiasma forms between two gene loci, only half of the meiotic products will be of crossover type. Therefore, chiasma frequency is twice the frequency of cross-over products.

$$\text{Chiasma \%} = 2(\text{crossover \%}) \quad \text{or} \quad \text{Crossover \%} = 1/2(\text{chiasma \%})$$

EXAMPLE 6.6
If a chiasma forms between the loci of genes *A* and *B* in 30% of the tetrads of an individual of genotype *AB/ab*, then 15% of the gametes will be recombinant (*Ab* or *aB*) and 85% will be parental (*AB* or *ab*).

EXAMPLE 6.7
Suppose progeny from the testcross *Ab/ab* × *ab/ab* were found in the proportions 40% *Ab/ab*, 40% *Ab/ab*, 10% *Ab/ab*, and 10% *ab/ab*. The genotypes *Ab/ab* and *ab/ab* were produced from crossover gametes. Thus, 20% of all gametes formed by the dihybrid parent were crossover types. This means that a chiasma occurs between these two loci in 40% of all tetrads.

4 MULTIPLE CROSSOVERS

When two-strand double crossovers occur between two genetic markers, the products, as detected through the progeny phenotypes, are only parental types [Fig. 6-2(*a*)]. However, a third gene locus *c* between the outside markers allows detection of double crossovers [Fig. 6-2(*b*)].

(a) Two-loci model

Synapsis and crossing over   End of meiosis I   End of meiosis II

(b) Three-loci model

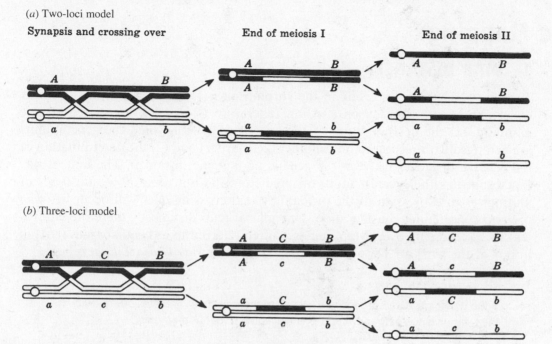

Fig. 6-2.   Multiple crossovers with two loci segregating (a) and three loci segregating (b).

If there is a certain probability that a crossover will form between the A and C loci and another independent probability of a crossover forming between the C and B loci, then the probability of a double crossover is the product of the two independent probabilities.

> **EXAMPLE 6.8**
> If a crossover between the A and C loci occurs in 20% of the tetrads and between C and B loci in 10% of the tetrads in an individual of genotype ACB/acb, then 2% (0.2 × 0.1) of the gametes are expected to be of double-crossover types AcB and aCb.

Odd numbers of two-strand crossovers (one, three, five, etc.) between two gene loci produce detectable recombinations between the outer markers, but even numbers of two-strand crossovers (two, four, six, etc.) do not.

5   LIMITS OF RECOMBINATION

If two gene loci are so far apart in the chromosome that the probability of a chiasma forming between them is 100%, then 50% of the gametes will be parental type (noncrossover) and 50% recombinant (crossover) type. When such dihybrid individuals are testcrossed, they are expected to produce progeny in a 1 : 1 : 1 : 1 ratio as would be expected for genes on different chromosomes. Recombination between two linked genes cannot exceed 50% even when multiple crossovers occur between them.

Genetic Mapping

1 MAP DISTANCE

The places where genes reside in the chromosome (loci) are positioned in linear order analogous to beads on a string. There are two major aspects to genetic mapping: (1) the determination of the linear order with which the genetic units are arranged with respect to one another (gene order) and (2) the determination of the relative distances between the genetic units (gene distance). The unit of distance that has the greatest utility in predicting the outcome of certain types of matings is an expression of the probability that crossing over will occur between the two genes under consideration. One unit of map distance is therefore equivalent to 1% crossing over. Map units are often referred to as **centimorgans (cM)** in honor of the work of Thomas Hunt Morgan, a famous *Drosophila* geneticist.

> **EXAMPLE 6.9**
> If the genotype *Ab/aB* produces 8% each of the crossover gametes *AB* and *ab*, then the distance between *A* and *B* is estimated to be 16 map units.

> **EXAMPLE 6.10**
> If the map distance between the loci *B* and *C* is 12 units, then 12% of the gametes of genotype *BC/bc* should be crossover types; i.e., 6% *Bc* and 6% *bC*.

Each chiasma produces 50% crossover products. Fifty percent crossing over is equivalent to 50 map units. If the average (mean) number of chiasmata is known for a chromosome pair, the total length of the map for that linkage group may be predicted:

$$\text{Total length} = \text{mean number of chiasmata} \times 50$$

2 TWO-POINT TESTCROSS

The easiest way to detect crossover gametes in a dihybrid is through the testcross progeny. Suppose we testcross dihybrid individuals in coupling phase (*AC/ac*) and find in the progeny phenotypes 37% dominant at both loci, 37% recessive at both loci, 13% dominant at the first locus and recessive at the second, and 13% dominant at the second locus and recessive at the first. Obviously, the last two groups (genotypically *Ac/ac* and *aC/ac*) were produced by crossover gametes from the dihybrid parent. Thus, 26% of all gametes (13 + 13) were of crossover types and the distance between the loci *A* and *C* is estimated to be 26 map units, or 26 cM.

3 THREE-POINT TESTCROSS

Double crossovers usually do not occur between genes less than 5 map units apart. For genes further apart, it is advisable to use a third marker between the other two in order to detect any double crossovers. Suppose that we testcross trihybrid individuals of genotype ABC/abc and find in the progeny the following:

36% ABC/abc	9% Abc/abc	4% ABc/abc	1% AbC/abc
36% abc/abc	9% aBC/abc	4% abC/abc	1% aBc/abc
72% Parental type :	18% Single crossovers : between A and B (region I)	8% Single crossovers : between B and C (region II)	2% Double crossovers

To find the distance A–B we must count all crossovers (both singles and doubles) that occurred in region I $= 18\% + 2\% = 20\%$ or 20 map units between the loci A and B. To find the distance B–C we must count all crossovers (both singles and doubles) that occurred in region II $= 8\% + 2\% = 10\%$ or 10 map units between the loci B and C. The A–C distance is therefore 30 map units when double crossovers are detected in a three-point linkage experiment and 26 map units when double crossovers are undetected in the two-point linkage experiment above.

Without the middle marker B, double crossovers would appear as parental types and hence we underestimate the true map distance (crossover percentage). In this case the 2% double crossovers would appear with the 72% parental types, making a total of 74% parental types and 26% recombinant types. Therefore, for any three linked genes whose distances are known, the amount of detectable crossovers (recombinants) between the two outer markers A and C when the middle marker B is missing is (A-B crossover percentage) plus (B-C crossover percentage) minus ($2 \times$ double-crossover percentage). This procedure is appropriate only if a crossover in the A–B region occurs independently of that in the B–C region (see Interference and Coincidence, page 184).

EXAMPLE 6.11
Given distances A–$B = 20$, B–$C = 10$, A–$C = 30$ map units, the percentage of detectable crossovers from the dihybrid testcross $AC/ac \times ac/ac = 0.20 + 0.10 - 2(0.20)(0.10) = 0.30 - 2(0.02) = 0.30 - 0.04 = 0.26$ or 26% (13% Ac/ac and 13% aC/ac).

SOLVED PROBLEM 6.2
A kidney-bean-shaped eye is produced by a recessive gene k on the third chromosome of *Drosophilia*. Orange eye color, called "cardinal," is produced by the recessive gene cd on the same chromosome. Between these two loci is a third locus with a recessive allele e that produces ebony body color. Homozygous "kidney," cardinal females are mated to homozygous ebony males. The trihybrid F_1 females are then testcrossed to produce the F_2. Among 4000 F_2 progeny are the following:

1761 kidney, cardinal	97 kidney
1773 ebony	89 ebony, cardinal
128 kidney, ebony	6 kidney, ebony, cardinal
138 cardinal	8 wild type

(a) Determine the linkage relationships in the parents and F_1 trihybrids.
(b) Estimate the map distances.

Solution:

(*a*) The parents are homozygous lines:

$$k\ e^+\ cd/k\ e^+\ cd \qquad \times \qquad k^+\ e\ cd^+/k^+\ e\ cd^+$$
kidney, cardinal females ebony males

The F$_1$ is then trihybrid

$$k\ e^+\ cd/k^+\ e\ cd^+$$
wild type

The linkage relationships in the trihybrid F$_1$ can also be determined directly from the F$_2$. By far the most frequent F$_2$ phenotypes are kidney, cardinal (1761) and ebony (1773), indicating that kidney and cardinal were on one chromosome in the F$_1$ and ebony on the other.

(*b*) Crossing over between the loci *k* and *e* produces the kidney, ebony (128) and cardinal (138) offspring. Double crossovers are the triple mutants (6) and wild type (8). Altogether there are $128 + 138 + 6 + 8 = 280$ crossovers between *k* and *e*:

$$280/4000 = 0.07 \text{ or } 7\% \text{ crossing over } = 7 \text{ map units}$$

Crossovers between *a* and *cd* produced the single-crossover types kidney (97) and ebony, cardinal (89). Double crossovers again must be counted in this region.

$$97 + 89 + 6 + 8 = 200 \text{ crossovers between } a \text{ and } cd$$

$$200/4000 = 0.05 \text{ or } 5\% \text{ crossing over } = 5 \text{ map units}$$

4 GENE ORDER

The additivity of map distances allows us to place genes in their proper linear order. Three linked genes may be in any one of three different orders, depending upon which gene is in the middle. We will ignore left and right end alternatives for the present. If double crossovers do not occur, map distances may be treated as completely additive units. When we are given the distances $A–B = 12$, $B–C = 7$, $A–C = 5$, we should be able to determine the correct order.

Case 1. Let us assume that *A* is in the middle.

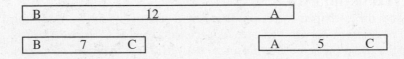

The *C* loci do not align. Therefore, *A* cannot be in the middle.

Case 2. Let us assume that *B* is in the middle.

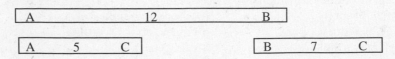

Again, the *C* loci do not align. Therefore, *B* cannot be in the middle.

Case 3. Let us assume that C is in the middle.

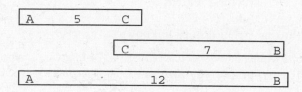

Now, all three loci align. Therefore, C must be in the middle.

(a) Linkage Relationships from a Two-Point Testcross. Parental combinations will tend to stay together in the majority of the progeny and the crossover types will always be the least frequent classes. From this information, the mode of linkage (coupling or repulsion) may be determined for the dihybrid parent.

EXAMPLE 6.12

P: Dihybrid parent   ×   Testcross parent
                    Aa, Bb   *ab/ab*
            (linkage relationships unknown)

F₁: 42% *AaBb* ⎫   8% *Aabb* ⎫
                        ⎬ Parental types   ⎬ Recombinant types
          42% *aabb* ⎭   8% *aaBb* ⎭

The testcross parent contributes *ab* to each progeny. The remaining genes come from the dihybrid parent. Thus *A* and *B* must have been on one chromosome of the dihybrid parent and *a* and *b* on the other, i.e., in coupling phase (*AB/ab*), because these were the combinations that appeared with greatest frequency in the progeny.

EXAMPLE 6.13

P: Dihybrid parent   ×   Testcross parent
                    Aa,Bb   *ab/ab*
            (linkage relationships unknown)

F₁: 42% *Aabb* ⎫   8% *AaBb* ⎫
                        ⎬ Parental types   ⎬ Recombinant types
          42% *aaBb* ⎭   8% *aabb* ⎭

By reasoning similar to that in Example 6.12, *A* and *b* must have been on one chromosome of the dihybrid parent and *a* and *B* on the other, i.e., in repulsion phase (*Ab/aB*).

(b) Linkage Relationships from a Three-Point Testcross. In a testcross involving three linked genes, the parental types are expected to be most frequent and the double crossovers to be the least frequent. The gene order is determined by manipulating the parental combinations into the proper order for the production of double-crossover types.

EXAMPLE 6.14

P: Trihybrid Parent × Testcross Parent
  Aa, Bb, Cc *abc/abc*
  (linkage relationships unknown)

F₁: 36% *Aabbcc* 9% *aabbCc* 4% *AabbCc* 1% *AaBbCc*
  36% *aaBbCc* 9% *AaBbcc* 4% *aaBbcc* 1% *aabbcc*
  72% 18% 8% 2%

The 72% group is composed of parental types because noncrossover gametes are always produced in the highest frequency. Obviously, the only contribution the testcross parent makes to all the progeny is *abc*. Thus, the trihybrid parent must have had *A, b,* and *c* on one chromosome and *a, B,* and *C* on the other. But which locus is in the middle? Again, three cases can be considered.

Case 1. Can we produce the least frequent double-crossover types (2% of the F₁) if the *B* locus is in the middle?

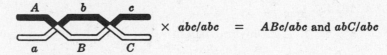

A b c
a B C × *abc/abc* = *ABc/abc* and *abC/abc*

These are not double-crossover types and therefore the B locus is not in the middle.

Case 2. Can we produce the double-crossover types if the *C* locus is in the middle? Remember to keep *A, b,* and *c* on one chromosome and *a, B,* and *C* on the other when switching different loci to the middle position.

A c b
a C B × *acb/acb* = *ACb/acb* and *acB/acb*

These are not double-crossover types and therefore the *C* locus is not in the middle.

Case 3. Can we produce the double-crossover types if the *A* locus is in the middle?

b A c
B a C × *bac/bac* = *bac/bac* and *BAC/bac*

These are the double-crossover types and we conclude that the *A* locus is in the middle. Now that we know the gene order and the parental linkage relationships, we can deduce the single crossovers. Let us designate the distance *B–A* as region I, and the *A–C* distance as region II. Single crossovers in region I:

b A c
B a C × *bac/bac* = *baC/bac* and *BAc/bac*

Single crossovers in region II:

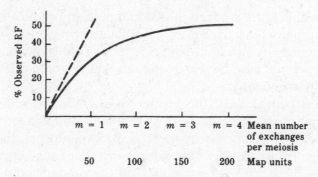

$$\times \ bac/bac \ = \ bAC/bac \text{ and } Bac/bac$$

5   RECOMBINATION PERCENTAGE VS. MAP DISTANCE

In two-point linkage experiments, the chance of double (and other even-numbered) crossovers occurring undetected increases with the unmarked distance (i.e., without segregating loci) between genes. Hence, closely linked genes give the best estimate of crossing over. Double crossovers do not occur within 10–12 map units in *Drosophila*. Minimum double-crossover distance varies by species. Within this minimum distance, recombination percentage is equivalent to map distance. Outside it, the relationship becomes nonlinear (Fig. 6-3). True map distance will thus be underestimated by the recombination fraction, with the two becoming virtually independent at large distances.

Fig. 6-3.   Relationship between observed recombination frequency (RF) and real map units (solid line). Dashed line represents the relationship for very small mean numbers of exchanges per meiosis (m). (From D. T. Suzuki, A. J. F. Griffiths, and R. C. Lewontin, *An Introduction to Genetic Analysis*, 2nd ed., W. H. Freeman and Co., San Francisco, 1976.)

6   GENETIC VS. PHYSICAL MAPS

The frequency of crossing over usually varies in different segments of the chromosome, but is a highly predictable event between any two gene loci. Therefore, the actual physical distances (in nucleotide base pairs) between linked genes bears no direct relationship to the map distances calculated on the basis of crossover percentages. For human chromosomes 1 cM is approximately equivalent to $1-3 \times 10^6$ bp of DNA depending on the particular region of the chromosome. The linear order, however, is identical in both cases. A **genetic map** is drawn in centimorgan (CM) units, while a **physical map** is drawn in base pairs (bp) of DNA.

7 COMBINING MAP SEGMENTS

Segments of map determined from three-point linkage experiments may be combined whenever two of the three genes are held in common.

EXAMPLE 6.15

Consider three map segments.

1. a (8) b (10) c
2. c (10) b (22) d
3. c (30) $e(2)d$

Superimpose each of these segments by aligning the genes shared in common.

1. a (8) b (10) c
2. d (22) b (10) c
3. $d(2)e$ (30) c

Then combine the three segments into one map.

The a to d distance $= (d$ to $b) - (a$ to $b) = 22 - 8 = 14$.
The a to e distance $= (a$ to $d) - (d$ to $e) = 14 - 2 = 12$.

$d(2)e$ (12) a (8) b (10) c

Additional segments of map added in this manner can produce a total linkage map over 100 map units long. However, as explained previously, the maximum recombination between any two linked genes is 50%. That is, genes very far apart on the same chromosome may behave as though they were on different chromosomes (assorting independently).

All other factors being equal, the greater the number of individuals in an experiment, the more accurate the linkage estimates should be. Therefore, in averaging the distances from two or more replicate experiments, the linkage estimates may be weighted according to the sample size. For each experiment, multiply the sample size by the linkage estimate. Add the products and divide by the total number of individuals from all experiments.

EXAMPLE 6.16

Let $n =$ number of individuals, $d =$ map distance.

Experiment	n	d	nd
1	239	12.3	2940
2	652	11.1	7237
3	966	12.9	12,461
Total	1857		22,638

$22{,}638/1857 = 12.2$ map units (weighted average)

SOLVED PROBLEM 6.3

The map distances for six genes in the second linkage group of the silkworm *Bombyx mori* are shown in the table below. Construct a genetic map that includes all of these genes.

	Gr	*Rc*	*S*	*Y*	*P*	*oa*
Gr	—	25	1	19	7	20
Rc	25	—	26	6	32	5
S	1	26	—	20	6	21
Y	19	6	20	—	26	1
P	7	32	6	26	—	27
oa	20	5	21	1	27	—

Solution:

Step 1. It makes little difference where one begins to solve this kind of problem, so we shall begin at the top. The *Gr–Rc* distance is 25 map units, and the *Gr–S* distance is 1 unit. Therefore, the relationship of these three genes may be either

(*a*) S(1)Gr _____ (25) _____ Rc

or

(*b*) Gr(1)S _____ (24) _____ Rc

The table, however, tells us that the distance *S–Rc* is 26 units. Therefore, alternative (*a*) must be correct, i.e., *Gr* is between *S* and *Rc*.

Step 2. The *Gr–Y* distance is 19 units. Again, two alternatives are possible:

(*c*) S(1)Gr _____ (19) _____ Y (6) Rc

or

(*d*) Y _____ (18) _____ S(1)Gr _____ (25) _____ Rc

In the table we find that the distance *Y–Rc* = 6. Hence, possibility (*c*) must be correct, i.e., *Y* lies between the loci of *Gr* and *Rc*.

Step 3. The distance *Gr–P* is 7 map units. Two alternatives for these loci are

(*e*) S(1)Gr (7) P _____ (12) _____ Y (6) Rc

or

(*f*) P (6) S(1)Gr _____ (19) _____ Y (6) Rc

The distance *P–S* is read from the table, and thus alternative (*f*) must be correct.

Step 4. There are 20 units between *Gr* and *oa*. These two genes may be in one of two possible relationships:

(*g*) P (6) S(1)Gr _____ (19) _____ Y(1)oa (5) Rc

or

(*h*) oa _____ (13) _____ P (6) S(1)Gr _____ (19) _____ Y (6) Rc

The table indicates that *Y* and *oa* are 1 map unit apart. Therefore, (*g*) is the completed map.

8   INTERFERENCE AND COINCIDENCE

In most of the higher organisms, the formation of one chiasma actually reduces the probability of another chiasma forming in an immediately adjacent region of the chromosome. This reduction in chiasma formation may be simplistically thought of as being due to a physical inability of the chromatids to bend back upon themselves within certain minimum distances. The net result of this **interference** is the observation of fewer double-crossover types than would be expected according to map distances. The strength of interference varies in different segments of the chromosome and is usually expressed in terms of a **coefficient of coincidence**, or the ratio between the observed and the expected double crossovers.

$$\text{Coefficient of coincidence} = \frac{\%\ \text{observed double crossovers}}{\%\ \text{expected double crossovers}}$$

Coincidence is the complement of interference.

$$\text{Coincidence} + \text{interference} = 1.0$$

When interference is complete (1.0), no double crossovers will be observed and coincidence becomes zero. When we observe all the double crossovers expected, coincidence is unity and interference becomes zero. When interference is 30% operative, coincidence becomes 70%, etc.

EXAMPLE 6.17
Given the map distances $A–B = 10$ and $B–C = 20$, then $0.1 \times 0.2 = 0.02$ or 2% double crossovers are expected if there is no interference. Suppose we observe 1.6% double crossovers in a testcross experiment.

$$\text{Coincidence} = 1.6/2.0 = 0.8$$

This simply means that we observed only 80% of the double crossovers that were expected on the basis of combining independent probabilities (map distances).

$$\text{Interference} = 1.0 - 0.8 = 0.2$$

Thus, 20% of the expected double crossovers did not form due to interference.

The percentage of double crossovers that will probably be observed can be predicted by multiplying the expected double crossovers by the coefficient of coincidence.

EXAMPLE 6.18.
Given a segment of map (below) with 40% interference, we expect $0.1 \times 0.2 = 0.02$ or 2% double crossover on the basis of combining independent probabilities. However, we will observe only 60% of those expected because of the interference. Therefore, we should observe $0.02 \times 0.6 = 0.012$ or 1.2% double-crossover types.

a _____ (10) _____ b _____ (20) _____ c

Linkage Estimates from F$_2$ Data

1 SEX-LINKED TRAITS

In organisms where the male is XY or XO, the male receives only the Y chromosome from the paternal parent (or no chromosome, homologous with the X in the case of XO sex determination). The Y contains, on its differential segment, no alleles homologous to those on the X chromosome received from the maternal parent. Thus, for completely sex-linked traits the parental and recombinant gametes formed by the female can be observed directly in the F$_2$ males, regardless of the genotype of the F$_1$ males.

EXAMPLE 6.19

Consider in *Drosophila* the recessive sex-linked bristle mutant scute (*sc*), and on the same chromosome the gene for vermilion eye color (*v*).

P: $\dfrac{++}{++}$ × $\dfrac{sc\ v}{\longrightarrow}$ ($\longrightarrow$ = Y chromosome)

  wild-type females   scute, vermilion males

F$_1$: $\dfrac{++}{sc\ v}$ and $\dfrac{++}{\longrightarrow}$

  wild-type females   wild-type males

F$_2$:

	♀ \ ♂	+ +	Y
Parental Gametes	+ +	+ +/+ + wild type	+ +/Y wild type
	sc v	+ +/*sc v* wild type	*sc v*/Y scute, vermilion
Crossover Gametes	+ *v*	+ +/+ *v* wild type	+ *v*/Y vermilion
	sc +	+ +/*sc* + wild type	*sc* +/Y scute
		Females	Males

If the original parental cross involves scute, vermilion females and wild-type males then both male and female progeny of the F$_2$ can be used to estimate the percentage of crossing over. The student should verify that this expectation is valid.

SOLVED PROBLEM 6.4

White eyes (*w/w* females; *w/Y* males) in *Drosophila* can be produced by the action of a sex-linked recessive gene. White eyes can also be produced through the interaction of two other genes: the recessive sex-linked gene *v* for vermilion eye color, and the autosomal recessive gene *bw* for brown eye color. Consider the parental cross: *bw/ bw, w$^+$v$^+$/w v* (brown-eyed females) × *bw/bw, w v/Y* (white-eyed males), where the

F_1 progeny consists of 70 brown-eyed and 130 white-eyed individuals. Estimate the distance between the sex-linked genes w and v.

Solution:

F_1:

♀ \ ♂	$bw/, w\ v/$	$bw/, Y/$
Parental Types $bw/, w^+ v^+ /$	$bw/bw,\ w^+ v^+ /w\ v$ brown	$bw/bw,\ w^+ v^+ /Y$ brown
$bw/, w\ v/$	$bw/bw,\ w\ v/w\ v$ white	$bw/bw,\ w\ v/Y$ white
Recombinant Types $bw/, w^+ v/$	$bw/bw,\ w^+\ v/w\ v$ white	$bw/bw,\ w^+\ v/Y$ white
$bw/, w\ v^+ /$	$bw/bw,\ w\ v^+ /w\ v$ white	$bw/bw,\ w\ v^+ /Y$ white

Only the genotypes of the brown offspring are known for certain. The 70 brown offspring constitute only one-half of the offspring produced by noncrossover maternal gametes. Therefore, we estimate that 70 of the white individuals were also produced by noncrossover maternal gametes. Thus, 140 out of 200 F_1 flies are estimated to be parental-type offspring = 70%. The other 30% must be crossover types. The best estimate of linkage between the white and vermilion loci would be 30 map units.

2 AUTOSOMAL TRAITS

An alternative to the testcross method for determining linkage and estimating distances is by allowing dihybrid F_1 progeny to produce an F_2 either by random mating among the F_1 or, in the case of plants, by selfing the F_1. However, this is a poor substitution. Such an F_2 that obviously does not conform to the 9 : 3 : 3 : 1 ratio expected for genes assorting independently may be considered evidence for linkage. Two methods for estimating the degree of linkage from F_2 data are presented below.

(a) **Square-Root Method.** The frequency of double-recessive phenotypes in the F_2 may be used as an estimator of the frequency of noncrossover gametes when the F_1 is in coupling phase, and as an estimator of the frequency of crossover gametes when the F_1 is in repulsion phase.

EXAMPLE 6.20

F_1 in coupling phase. *AB/ab*

F_2: The frequency of *ab* gametes $= \frac{1}{2}$ of the frequency of all noncrossover gametes. If the crossover percentage is 20%, we would expect 80% noncrossover gametes (40% *AB* and 40% *ab*). The probability of two *ab* gametes uniting to form the double-recessive *ab/ab* $= (0.4)^2 = 0.16$ or 16%. Now, if we do not know the crossover percentage, but the F_2 data tell us that 16% are double recessive, then the percentage of noncrossover gametes = $2\sqrt{\text{freq. of double recessives}} = 2\sqrt{0.16} = 2(0.4) = 0.8$ or 80%. If 80% are noncrossovers, the other 20% must be crossover types. Therefore the map distance between *A* and *B* is estimated at 20 units.

EXAMPLE 6.21

F_1 in repulsion phase. *Ab/aB*

F_2: The reasoning is similar to that in Example 6.20. With 20% crossing over we expect 10% of the gametes to be *ab*. The probability of 2 of these gametes uniting to form the double recessive $(ab/ab) = (0.1)^2 = 0.01$ or 1%. Now, if we do not know the crossover percentage, but the F_2 data tell us that 1% are double recessives, then the percentage of crossover gametes = $2\sqrt{\text{req. of double recessives}} = 2\sqrt{0.01} = 2(0.1) = 0.2$ or 20%.

(b) Product-Ratio Method. An estimate of the frequency of recombination from double-heterozygous (dihybrid) F_1 parents can be ascertained from F_2 phenotypes *R-S*, *R-ss*, *rrS-*, and *rrss* appearing in the frequencies a, b, c and d, respectively. The ratio of crossover to parental types, called the **product ratio**, is a function of recombination.

$$\text{For coupling data: } x = bc/ad$$
$$\text{For repulsion data: } x = ad/bc$$

The recombination fraction represented by the value of x may be read directly from a product-ratio table (Table 6-1). The product-ratio method utilizes all of the F_2 data available and not just the double-recessive class as in the square-root method. The product-ratio method should therefore yield more accurate estimates of recombination than the square-root method.

EXAMPLE 6.22

Coupling data:

P: $RS/RS \times rs/rs$

F_1: RS/rs (coupling phase)

F_2:

	Phenotypes	Numbers
(a)	R-S-	1221
(b)	R-ss	219
(c)	rrS-	246
(d)	rrss	243

$$x \text{ (for coupling data)} = \frac{bc}{ad} = \frac{(219)(246)}{(1221)(243)} = \frac{53,874}{296,703} = 0.1816$$

Locating the value of x in the body of the coupling column (Table 6-1), we find that 0.1816 lies between the values 0.1777 and 0.1948, which corresponds to recombination fractions of 0.28 and 0.29, respectively. Therefore, without interpolation, recombination is approximately 28%.

Table 6-1. Recombination Fraction Estimated by the Product-Ratio Method

Recombination Fraction	Ratio of Products		Recombination Fraction	Ratio of Products	
	ad/bc (Repulsion)	*bc/ad* (Coupling)		*ad/bc* (Repulsion)	*bc/ad* (Coupling)
0.00	0.000000	0.000000	0.26	0.1608	0.1467
0.01	0.000200	0.000136	0.27	0.1758	0.1616
0.02	0.000801	0.000552	0.28	0.1919	0.1777
0.03	0.001804	0.001262	0.29	0.2089	0.1948
0.04	0.003213	0.002283	0.30	0.2271	0.2132
0.05	0.005031	0.003629			
0.06	0.007265	0.005318	0.31	0.2465	0.2328
0.07	0.009921	0.007366	0.32	0.2672	0.2538
0.08	0.01301	0.009793	0.33	0.2892	0.2763
0.09	0.01653	0.01262	0.34	0.3127	0.3003
0.10	0.02051	0.01586	0.35	0.3377	0.3259
0.11	0.02495	0.01954	0.36	0.3643	0.3532
0.12	0.02986	0.02369	0.37	0.3927	0.3823
0.13	0.03527	0.02832	0.38	0.4230	0.4135
0.14	0.04118	0.03347	0.39	0.4553	0.4467
0.15	0.04763	0.03915	0.40	0.4898	0.4821
0.16	0.05462	0.04540	0.41	0.5266	0.5199
0.17	0.06218	0.05225	0.42	0.5660	0.5603
0.18	0.07033	0.05973	0.43	0.6081	0.6034
0.19	0.07911	0.06787	0.44	0.6531	0.6494
0.20	0.08854	0.07671	0.45	0.7013	0.6985
0.21	0.09865	0.08628	0.46	0.7529	0.7510
0.22	0.1095	0.09663	0.47	0.8082	0.8071
0.23	0.1211	0.1078	0.48	0.8676	0.8671
0.24	0.1334	0.1198	0.49	0.9314	0.9313
0.25	0.1467	0.1328	0.50	1.0000	1.0000

Source: F. R. Immer and M. T. Henderson, "Linkage studies in barley," *Genetics*, 28: 419–440, 1943.

EXAMPLE 6.23

Repulsion data:

P: $Ve/Ve \times vE/vE$
F$_1$: Ve/vE (repulsion phase)

	Phenotypes	Numbers
(a)	V-E-	36
(b)	V-ee	12
(c)	vvE-	16
(d)	vvee	2

$$x \text{ (for repulsion data)} = \frac{ad}{bc} = \frac{(36)(2)}{(12)(16)} = \frac{72}{192} = 0.3750$$

Locating the value of x in the body of the repulsion column, we find that 0.3750 lies between the values 0.3643 and 0.3927, which corresponds to recombination fractions of 0.36 and 0.37, respectively. Therefore recombination is approximately 36%.

Use of Genetic Maps

1   PREDICTING RESULTS OF A DIHYBRID CROSS

If the map distance between any two linked genes is known, the expectations from any type of mating may be predicted by use of a Punnett square-type analysis.

EXAMPLE 6.24

Given genes A and B that are 10 map units apart and parents AB/AB males $\times$ ab/ab females, the F$_1$ will all be heterozygous in coupling phase AB/ab. Ten percent of the F$_1$ gametes are expected to be of crossover types (5% Ab and 5% aB). Ninety percent of the F$_1$ gametes are expected to be parental types (45% AB and 45% ab). The F$_2$ can be derived by use of a Punnett square, combining independent probabilities by multiplication.

		Parental Types		Crossover Types	
		0.45 **AB**	**0.45** **ab**	**0.05** **Ab**	**0.05** **aB**
Parental Types	**0.45** **AB**	0.2025 AB/AB	0.2025 AB/ab	0.0225 AB/Ab	0.0225 AB/aB
	0.45 **ab**	0.2025 ab/AB	0.2025 ab/ab	0.0225 ab/Ab	0.0225 ab/aB
Crossover Types	**0.05** **Ab**	0.0225 Ab/AB	0.0225 Ab/ab	0.0025 Ab/Ab	0.0025 Ab/aB
	0.05 **aB**	0.0225 aB/AB	0.0225 aB/ab	0.0025 aB/Ab	0.0025 aB/aB

Summary of Phenotypes: 0.7025 or 70.25% A-B-
0.0475 or 4.75% A-bb
0.0475 or 4.75% aaB-
0.2025 or 20.25% $aabb$

SOLVED PROBLEM 6.5
Elongate tomato fruit is produced by plants homozygous for a recessive gene o, round fruit shape is produced by the dominant allele at this locus (O). A compound inflorescence is the result of another recessive gene s, simple inflorescence is produced by the dominant allele at this locus (S). A Yellow Pear variety (with elongate fruit and simple inflorescence) is crossed to a Grape Cluster variety (with round fruit and compound inflorescence). The F_1 plants are randomly crossed to produce the F_2. Among 259 F_2, 126 round, simple : 63 round, compound : 66 long, simple : 4 long, compound are found. Estimate the amount of recombination by the "square-root method."

Solution:

P: oS/oS × Os/Os
    Yellow Pear variety   Grape Cluster variety
    (long, simple)   (round, compound)

F_1: oS/Os
        (round, simple)

F_2:

		Parental Gametes		Crossover Gametes	
		oS	*Os*	*os*	*OS*
Parental Gametes	*oS*	*oS/oS* long, simple	*oS/Os* round, simple	*oS/os* long, simple	*oS/OS* round, simple
	Os	*os/oS* round, simple	*Os/Os* round, compound	*Os/os* round, compound	*Os/OS* round, simple
Crossover Gametes	*os*	*os/oS* long, simple	*os/Os* round, compound	*os/os* long, compound	*os/OS* round, simple
	OS	*oS/oS* round, simple	*OS/Os* round, simple	*OS/os* round, simple	*OS/OS* round, simple

Notice that the double-recessive phenotype (long, compound) occupies only 1 of the 16 frames in the gametic checkerboard. This genotype is produced by the union of two identical double-recessive gametes (o, s). If we let x = the frequency of formation of os gametes, then x^2 = frequency of occurrence of the os/os genotype (long, compound phenotype) = 4/259 = 0.0154. Thus, $x = \sqrt{0.0154} = 0.124$. But x estimates only half of the crossover gametes. Therefore $2x$ estimates all of the crossover gametes = 2(0.124) = 0.248 or 24.8% recombination.

2 PREDICTING RESULTS OF A TRIHYBRID TESTCROSS

Map distances or crossover percentages may be treated as any other probability estimates. Given a particular kind of mating, the map distances involved, and either the coincidence or interference for this region of the chromosome, we should be able to predict the results in the offspring generation.

EXAMPLE 6.25

Parents: $AbC/aBc \quad \times \quad abc/abc$

Map: $\quad a \quad (10) \quad b \quad\quad (20) \quad\quad c$

Interference: $\quad\quad 40\%$

Given the above information, the expected kinds and frequencies of progeny genotypes and phenotypes can be determined as follows.

Step 1. For gametes produced by the trihybrid parent, determine the parental
types, single crossovers in each of the two regions, and the double-crossover types. Let interval A–B = region I and interval B–C = region II.

F_1:	<u>Step 1</u>	<u>Steps 2–5</u>	
Parental types	AbC	35.6%	
	aBc	35.6%	71.2%
Singles in region I	ABc	4.4%	
	abC	4.4%	8.8%
Singles in region II	Abc	9.4%	
	aBC	9.4%	18.8%
Double crossovers	ABC	0.6%	
	abc	0.6%	1.2%
		100.0%	

Step 2. The frequency of double crossovers expected to be observed is calculated
by multiplying the two decimal equivalents of the map distances by the
coefficient of coincidence.

$$0.1 \times 0.2 \times 0.6 = 0.012 \text{ or } 1.2\%$$

This percentage is expected to be equally divided (0.6% each) between the
two double-crossover types.

Step 3. Calculate the single crossovers in region II (between b and c) and correct it
for the double crossovers that also occurred in this region:

$$20\% - 1.2\% = 18.8\%$$

equally divided into two classes = 9.4% each.

Step 4. The single crossovers in region I (between a and b) are calculated in the
same manner as step 3:

$$10\% - 1.2\% = 8.8\%$$

divided equally among the two classes = 4.4% each.

Step 5. Total all the single crossovers and all the double crossovers and subtract
from 100% to obtain the percentage of parental types:

$$100 - (8.8 + 18.8 + 1.2) = 71.2\%$$

to be equally divided among the two parental classes = 35.6% each.

For convenience, we need not write out the entire genotype or phenotype of the progeny because, for example, when the gamete *AbC* from the trihybrid parent unites with the gamete produced by the testcross parent (*abc*), obviously the genotype is *AbC/abc*. Phenotypically, it will exhibit the dominant trait at the *A* locus, the recessive trait at the *B* locus, and the dominant trait at the *C* locus. All this could be predicted directly from the gamete *AbC*.

An alternative method for predicting F_1 progeny types is by combining the probabilities of crossovers and/or noncrossovers in appropriate combinations. This method can be used only when there is no interference.

EXAMPLE 6.26

Parents: ABC/abc × abc/abc

Coincidence: 1.0

	I		II	
Map:	a (10) b		(20)	c

No. of progeny: 2000

Step 1. Determine the parental, single-crossover and double-crossover progeny types expected.

F_1:	Step 1	Steps 2–5
Parental types	*ABC*	720
	abc	720
Singles in region I	*Abc*	80
	aBC	80
Singles in region II	*ABc*	180
	abC	180
Double crossovers	*AbC*	20
	aBc	20
		2000

Step 2. The number of double crossovers expected to appear in the progeny is $0.1 \times 0.2 \times 2000 = 40$, equally divided between the two double-crossover types (20 each).

Step 3. The probability of a single crossover occurring in region I is 10%. Hence, there is a 90% chance that a crossover will not occur in that region. The combined probability that a crossover will not occur in region I and will occur in region II is $(0.9)(0.2) = 0.18$ and the number of region II single-crossover progeny expected is $0.18(2000) = 360$, equally divided between the two classes (180 each).

Step 4. Likewise the probability of a crossover occurring in region I and not in region II is $0.1(0.8) = 0.08$ and the number of region I single-crossover progeny expected is $0.08(2000) = 160$, equally divided among the two classes (80 each).

Step 5. The probability that a crossover will not occur in region I and region II is 0.9(0.8) = 0.72 and the number of parental-type progeny expected is 0.72(2000) = 1440, equally divided among the two parental types (720 each).

SOLVED PROBLEM 6.6

Several three-point testcrosses were made in maize utilizing the genes booster (B, a dominant plant color intensifier), liguleless leaf (lg_1), virescent seedling (v_4, yellowish-green), silkless (sk, abortive pistils), glossy seedling (gl_2), and tassel-seed (ts_1, pistillate terminal inflorescence). Using the information from the following testcrosses, map this region of the chromosome.

Testcross 1. Trihybrid parent is heterozygous for booster, liguleless, tassel-seed.

Testcross Progeny

71 booster, liguleless, tassel-seed	17 tassel-seed
111 wild type	24 booster, liguleless
48 liguleless	6 booster
35 booster, tassel-seed	3 liguleless, tassel-seed

Testcross 2. Trihybrid parent is heterozygous for booster, liguleless, tassel-seed.

Testcross Progeny

57 tassel-seed	21 liguleless, tassel-seed
57 booster, liguleless	21 booster, liguleless, tassel-seed
20 wild type	8 booster, tassel-seed
31 booster	7 liguleless

Testcross 3. Trihybrid parent is heterozygous for booster, liguleless, silkless.

Testcross Progeny

52 silkless	56 booster, liguleless
8 booster, silkless	13 liguleless
2 booster, liguleless, silkless	131 liguleless, silkless
148 booster	

Testcross 4. Trihybrid parent is heterozygous for booster, liguleless, silkless.

Testcross Progeny

6 booster	3 liguleless, silkless
137 booster, silkless	30 silkless
291 booster, liguleless, silkless	34 booster, liguleless
142 liguleless	339 wild type

Testcross 5. Trihybrid parent is heterozygous for liguleless, virescent, glossy.

Testcross Progeny

431 wild type	128 virescent, glossy
399 liguleless, virescent, glossy	153 liguleless
256 virescent	44 glossy
310 liguleless, glossy	51 virescent, liguleless

Testcross 6. Trihybrid parent is heterozygous for booster, liguleless, virescent

Testcross Progeny

60 wild type	18 virescent
37 liguleless, booster, virescent	23 liguleless, booster
32 virescent, booster	11 booster
34 liguleless	12 virescent, liguleless

Testcross 7. Trihybrid parent is heterozygous for virescent, liguleless, booster.

Testcross Progeny

25 booster	8 booster, virescent
11 booster, liguleless	2 liguleless, booster, virescent

Solution: Following the procedures established in this chapter, we determine from each of the testcrosses the gene order (which gene is in the middle) and the percent crossing over in each region. Note that the results of testcrosses 1 and 2 may be combined, recognizing that the linkage relationships are different in the trihybrid parents. Likewise, the results of 3 and 4 may be combined, as well as those of testcrosses 6 and 7. The analyses of these seven testcrosses are summarized below in tabular form.

Testcross No.	Trihybrid Parent	Parental Type Progeny		Recombinant Progeny						Total
				Region I		Region II		DCO		
1	$+++/lg_1\ B\ ts_1$	111	71	35	48	17	24	6	3	315
2	$++ts_1/lg_1\ B\ +$	57	57	31	21	20	21	8	7	222
		296		135		82		24		
										537
				25.1%		15.3%		4.5%		
3*	$+\ B\ +\ /lg_1\ +\ sk$	148	131	52	56	8	13	0	2	410
4	$+++/lg_1\ B\ sk$	339	291	137	142	30	34	6	3	982
		909		387		85		11		
				27.8%		6.1%		0.8%		1392
5	$+++/lg_1\ gl_2\ v_4$	431	399	128	153	256	310	44	51	
		830		281		566		95		1772
				15.9%		31.9%		5.4%		
6	$+++/lg_1\ B\ v_4$	60	37	32	34	18	23	11	12	227
7†	$+\ B\ +\ /lg_1\ +\ v_4$	25	0	0	11	8	0	0	2	46
		122		77		49		25		273
				28.2%		17.9%		9.2%		

* Note that in testcross 3, only seven phenotypes appeared, whereas we expected eight. We suspect that the missing phenotype (wild type) is a double-crossover (DCO) type because DCO types are expected to be less frequent than the others. The two phenotypes with the highest numbers should derive from the parental (non-crossover gametes). Thus, the booster phenotype indicates that the three genes B, lg_1^+, and sk were on one parental chromosome; likewise, the other high-frequency progeny phenotype (liguleless and silkless) indicates that the three genes lg_1, sk, and B^+ were on the homologous chromosome, but we do not know which of these loci is in the middle. Assuming that the least frequent phenotype (booster, liguleless, silkless) is one of the double-crossover types, we can infer that the booster locus is in the middle; i.e., if parents were lg_1^+, B, sk^+/lg_1, B^+, sk the double crossovers would be $+++$ (wild type) and lg_1, B, sk (liguleless, booster, silkless). This inference is confirmed by testcross 4 where all eight progeny phenotypes are present.

[†] In testcross 7, only four phenotypes appeared in the progeny (no explanation given for the missing phenotypes). One might be tempted to eliminate such bizarre results from a report, but it would be scientifically unethical to do so. Data selection or alteration would be considered fraudulent. A scientist must report all the data or give reasons for failing to do so. However, it is possible to establish the gene order from testcross 6, so that the type of progeny (noncrossovers, single crossovers, and double crossovers) can be identified unambiguously.

To find the map distances between lg_1 and B in the first two testcrosses, we add the double crossovers (4.5%) to the region I single crossovers (25.1%) = 29.6% or 29.6 map units. Likewise, to find the map distance between B and ts_1, we add 4.5% to the region II single crossovers (15.3%) = 19.8% or 19.8 map units. Thus, this segment of map becomes

$$lg_1 \qquad (29.6) \qquad B \qquad (19.8) \qquad ts_1$$

Three other map segments are similarly derived from testcross data:

Testcrosses 3 and 4 $\quad lg_1 \qquad (28.6) \qquad B \quad (6.9) \quad sk$

Testcross 5 $\qquad lg_1 \qquad (21.3) \qquad gl_2 \qquad (37.3) \qquad v_4$

Testcrosses 6 and 7 $\quad lg_1 \qquad (37.4) \qquad B \qquad (27.1) \qquad v_4$

Now let us combine all four maps into one:

$$lg_1 \quad (21.3) \quad gl_2 \quad B \ (6.9) \ sk \qquad ts_1 \qquad v_4$$

29.6
28.6
37.4

19.8

27.1

37.3

The weighted-average distance lg_1–B is determined next:

Expt.	No. in Expt. x	Distance y	xy
1, 2	537	29.6	15,895.2
3, 4	1392	28.6	39,811.2
6, 7	273	37.4	10,210.2
	2202		65,916.6

65,916.6/2202 = 29.9 map units (weighted average)

The sk–ts_1 distance = $(B$–$ts_1)$–$(B$–$sk)$ = 19.8 − 6.9 = 12.9 map units. The gl_2–B distance has two estimators:

1. $(lg_1$–$B) - (lg_1$–$gl_2)$ = 29.9 − 21.3 = 8.6
2. $(gl_2$–$v_4) - (B$–$v_4)$ = 37.3 − 27.1 = 10.2

All other factors being equal, the second estimate is likely to be less accurate because of the greater distances involved. There is no easy way to accurately average these two values. We will arbitrarily use the estimate of 8.6 map units until more definitive experimental results are obtained.

Likewise, the ts_1–v_4 distance has two estimators:

1. $(B-v_4) - (B-ts_1) = 27.1 - 19.8 = 7.3$
2. $(gl_2-v_4) - [(gl_2-B) + (B-ts_1)] = 37.3 - (8.6 + 19.8) = 8.9$

Again, the second estimate is likely to be less accurate because of the distances involved, and we will use 7.3 map units as our estimate of the ts_1–v_4 distance. The unified map now appears as follows:

$$lg_1 \qquad (21.3) \qquad gl_2 \quad (8.6) \quad B \quad (6.9) \quad sk \qquad (12.9) \qquad ts_1 \quad (7.3) \quad v_4$$

Additional experimental data may considerably modify certain portions of this genetic map. It should always be remembered that these maps are only estimates, and as such are continually subject to refinements.

Crossover Suppression

Many extrinsic and intrinsic factors are known to contribute to the crossover rate. Among these are the effects of sex, age, temperature, proximity to the centromere or heterochromatic regions (darkly staining regions presumed to carry little genetic information), chromosomal aberrations such as inversions, and many more. Two specific cases of crossover suppression are presented in this section: (1) complete absence of crossing over in male *Drosophila* and (2) the maintenance of balanced lethal systems as permanent trans heterozygotes through the prevention of crossing over.

1 ABSENCE OF CROSSING OVER IN MALE *DROSOPHILA*

One of the unusual characteristics of *Drosophila* is the apparent absence of crossing over in males. This fact is shown clearly by the nonequivalent results of reciprocal crosses.

EXAMPLE 6.27
Testcross of heterozygous females.
    Consider two genes on the third chromosome of *Drosophila*, hairy (*h*) and scarlet (*st*), approximately 20 map units apart.

P: $h\ +/+\ st$ × $h\ st/h\ st$
          wild-type females   hairy, scarlet males

F_1:

		♂
	♀	**h st**
80% Parental Types	40% $h\ +$	$h\ +/\ h\ st = 40\%$ hairy
	40% $+\ st$	$+\ st/h\ st = 40\%$ scarlet
20% Recombinant Types	10% $h\ st$	$h\ st/h\ st = 10\%$ hairy and scarlet
	10% $+\ +$	$+\ +/h\ st = 10\%$ wild type

EXAMPLE 6.28

Test cross of heterozygous males (reciprocal cross of Example 6.27).

P: $h\ st/h\ st$ × $h+/+st$
          hairy, scarlet females   wild-type males

F_1:

	♀	
♂		**h st**
Only Parental Types	$h\ +$	$h\ +/\ h\ st = 50\%$ hairy
	$+\ st$	$+\ st/h\ st = 50\%$ scarlet

When dihybrid males are crossed to dihybrid females (both in repulsion phase) the progeny will always appear in the ratio $2:1:1$ regardless of the degree of linkage between the genes. The double-recessive class never appears. The student should verify that these expectations are valid.

Drosophila is not unique in this respect. For example, crossing over is completely suppressed in female silkworms. Other examples of complete and partial suppression of crossing over are common in the genetic literature.

2   BALANCED LETHAL SYSTEMS

A gene that is lethal when homozygous and linked to another lethal with the same mode of action can be maintained in permanent dihybrid condition in repulsion phase when associated with a genetic condition that prevents crossing over (see "inversions" in Chapter 7). Balanced lethals breed true and their behavior simulates that of a homozygous genotype. These systems are commonly used to maintain laboratory cultures of lethal, semilethal, or sterile mutants.

EXAMPLE 6.29

Two dominant genetic conditions, curly wings (*Cy*) and plum eye color (*Pm*), are linked on chromosome 2 of *Drosophila* and associated with a chromosomal inversion that prevents crossing over. Both *Cy* and *Pm* are lethal when homozygous.

Half the progeny from repulsion heterozygotes die, and the viable half are repulsion heterozygotes just like the parents.

P: $Cy\ Pm^+/Cy^+\ Pm$ × $Cy\ Pm^+/Cy^+\ Pm$
       curly, plum females   curly, plum males

F_1:

	$Cy\ Pm^+$	$Cy^+\ Pm$
$Cy\ Pm^+$	$Cy\ Pm^+/Cy\ Pm^+$ dies	$Cy\ Pm^+/Cy^+\ Pm$ curly, plum
$Cy^+\ Pm$	$Cy^+\ Pm/Cy\ Pm^+$ curly, plum	$Cy^+\ Pm/Cy^+\ Pm$ dies

Balanced lethals may be used to determine on which chromosome an unknown genetic unit resides (see Solved Problem 6.7). Sex-linked genes make themselves known through the nonequivalence of progeny from reciprocal matings (Chapter 5). Without the aid of a balanced lethal system, the assignment of an autosomal gene to a particular linkage group may be made through observation of the peculiar genetic ratios obtained from abnormal individuals possessing an extra chromosome (trisomic) bearing the gene under study (Chapter 7).

SOLVED PROBLEM 6.7

Suppose you are given a strain of *Drosophila* exhibiting an unknown abnormal genetic trait (mutation). We mate the mutant females to males from a balanced lethal strain ($Cy\ Pm^+/Cy^+\ Pm$, $D\ Sb^+/D^+\ Sb$) where curly wings (Cy) and plum eye (Pm) are on chromosome 2 and dichaete wing (D) and stubble bristles (Sb) are on chromosome 3. Homozygosity for either curly, plum, dichaete, or stubble is lethal. The trait does not appear in the F_1. The F_1 males with curly wings and stubby bristles are then backcrossed to the original mutant females. In the progeny the mutation appears in equal association with curly and stubble. *Drosophila melanogaster* has a haploid number of 4 including an X, 2, 3, and 4 chromosome. (*a*) Determine whether the mutation is a dominant or a recessive. (*b*) To which linkage group (on which chromosome) does the mutation belong?

Solution:

(*a*) If the mutation were a dominant (let us designate it *M*), then each member of the strain (pure line) would be of genotype *MM*. Since the trait does not appear in our balanced lethal stock, they must be homozygous recessive (M^+M^+). Crosses between these two lines would be expected to produce only heterozygous genotypes (M^+M) and would be phenotypically of the mutant type. But since the mutant type did not appear in the F_1, the mutation must be a recessive (now properly redesignated *m*). The dominant wild-type allele may now be designated m^+.

(*b*) Let us assume that this is a sex-linked recessive mutation. The F_1 males receive their single X chromosome from the mutant female (*mm*). Therefore, all males of the F_1 should exhibit the mutant trait because males would be hemizygous for all sex-linked genes (*m*Y). Since the mutant type did not appear in the F_1, our recessive mutation could not be sex-linked.

Let us assume that our recessive mutation is on the second chromosome. The curly, stubble F_1 males carry the recessive in the heterozygous condition ($Cy\ m^+/Cy^+\ m$, Sb/Sb^+). Notice that we omit the designation of loci with which we are not concerned. When these carrier males are then backcrossed to the original mutant females (Cy^+ m/Cy^+ m, Sb^+/Sb^+), the F_2 expectations are as follows:

Cy m^+/Cy^+ m, Sb/Sb^+ curly, stubble Cy m^+/Cy^+ m, Sb/Sb^+ mutant, stubble
Cy m^+/Cy^+ m, Sb^+/Sb^+ curly Cy m^+/Cy^+ m, Sb^+/Sb^+ mutant

Note that the mutant cannot appear with curly. Therefore, our recessive mutation is not on chromosome 2.

Let us then assume that our mutant gene is on the third chromosome. When F_1 carrier males (Cy/Cy^+, Sb m^+/Sb^+ m) are backcrossed to the original mutant females (Cy^+/Cy^+, Sb^+ m/Sb^+ m), the F_2 expectations are as follows:

Cy/Cy^+, Sb m^+/Sb^+ m curly, stubble Cy^+/Cy^+, Sb m^+/Sb^+ m stubble
Cy/Cy^+, Sb^+ m/Sb^+ m curly, mutant Cy^+/Cy^+, Sb^+ m/Sb^+ m mutant

Note that the mutant cannot appear with stubble. Hence, our recessive mutation is not on chromosome 3.

If the mutant is not sex-linked, not on 2 nor on 3, then obviously it must be on the fourth chromosome. Let us prove this. When F_1 carrier males (Cy/Cy^+, Sb/Sb^+, m^+/m) are backcrossed to the original mutant females (Cy^+/Cy^+, Sb^+/Sb^+, m/m), the F_2 expectations are as follows:

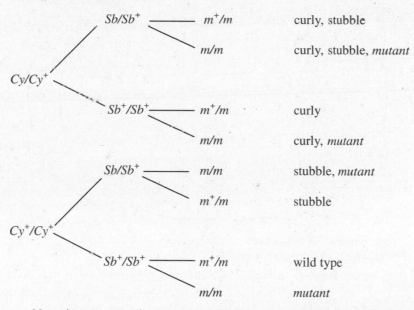

Note that our recessive mutant occurs in equal association with curly and stubble, which satisfies the conditions of the problem. We conclude that this mutation is on the fourth chromosome.

Tetrad Analysis in Fungi

Fungi that produce sexual spores (**ascospores**) housed in a common sac (**ascus**; **asci**, plural) are called **ascomycetes**. One of the simplest ascomycetes is the unicellular baker's yeast *Saccharomyces cerevisiae* (Fig. 6-4). Asexual reproduction is by budding, a mitotic process usually with unequal cytokinesis. The sexual cycle involves the union of entire cells of opposite mating type, forming a diploid zygote. The diploid cell may reproduce diploid progeny asexually by budding or haploid progeny by meiosis. The four haploid nuclei form ascospores enclosed by the ascus. Rupture of the ascus releases the haploid spores, which then germinate into new yeast cells.

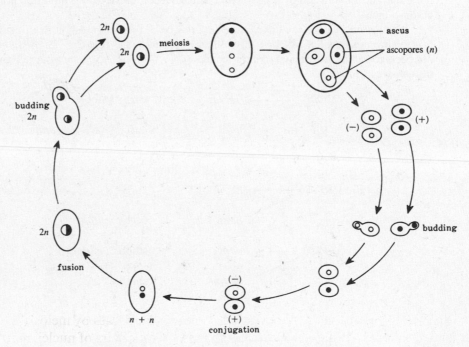

Fig. 6-4. Life cycle of *Saccharomyces cerevisiae*.

Another ascomycete of interest to geneticists is the bread mold *Neurospora crassa* (Fig. 6-5). The fungal mat or **mycelium** (**mycelia**, plural) is composed of intertwined filaments called hyphae. The tips of **hyphae** (**hypha**, singular) may pinch off asexual spores called **conidia** (**conidium**, singular), which germinate into more hyphae. The vegetative hyphae are segmented, with several haploid nuclei in each segment. Hyphae from one mycelium may anastomose with hyphae of another mycelium to form a mixture of nuclei in a common cytoplasm called a **heterokaryon**. A pair of alleles, *A* and *a*, governs the two mating types. Sexual reproduction occurs only when cells of opposite mating type unite. Specialized regions of the mycelium produce immature female fruiting bodies (**protoperithecia**) from which extrude receptive filaments called **trichogynes**. A conidium or hypha

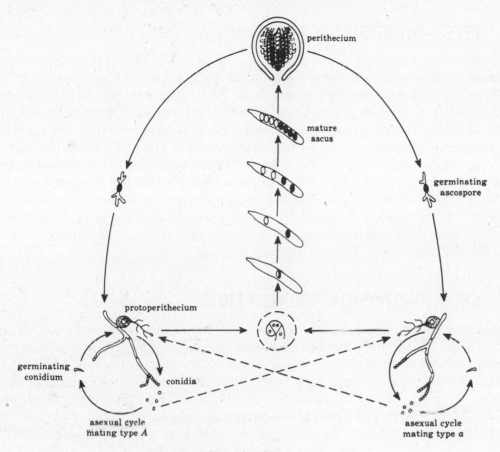

Fig. 6-5. Life cycle of *Neurospora crassa.*

from the opposite mating type fuses with the trichogyne, undergoes several kar-yokineses, and fertilizes many female nuclei. Each of the resulting diploid zygotes lies within an elongated sac called the ascus. The zygote divides by meiosis to form four nuclei, followed by a mitotic division that yields four pairs of nuclei, maturing into eight ascospores. A mature fruiting body (**perithecium**) may contain over 100 asci, each containing eight ascospores. The confines of the ascus force the polar organization of division to orient lengthwise in the ascus and also prevent the meiotic or mitotic products from slipping past each other. Each of the four chro-matids of first meiotic prophase are now represented by a pair of ascospores in tandem order within the ascus.

In the case of yeast the ascospores representing the four chromatids of meiosis are in no special order, but in the bread mold *Neurospora* the ascospores are linearly ordered in the ascus in the same sequence as the chromatids were on the meiotic metaphase plate. The recovery and investigation of all of the products from a single meiotic event is called **tetrad analysis**.

Each ascus of *Neurospora*, when analyzed for a segregating pair of alleles, reveals one of two linear ratios: (1) 4 : 4 ratio, attributed to **first-division segrega-tion** or (2) 2 : 2 : 2 : 2 ratio, resulting from **second-division** segregation.

1 FIRST-DIVISION SEGREGATION

A cross between a culture with a wild-type (c^+) spreading form of mycelial growth and one with a restricted form of growth called "colonial" (c) is diagrammed in Fig. 6-6(a). If the ascospores are removed one by one from the ascus in linear order and each is grown as a separate culture, a linear ratio of 4 colonial : 4 wild type indicates that a first-division segregation has occurred. That is, during first meiotic anaphase both of the c^+ chromatids moved to one pole and both of the c chromatids moved to the other pole. The 4 : 4 ratio indicates that *no crossing over* has occurred between the gene and its centromere. The further the gene locus is from the centromere, the greater is the opportunity for crossing over to occur in this region. Therefore, if the meiotic products of a number of asci are analyzed and most of them are found to exhibit a 4 : 4 pattern, then the locus of c must be close to the centromere.

2 SECOND-DIVISION SEGREGATION

Let us now investigate the results of a crossover between the centromere and the c locus [Fig. 6-6(b)]. Note that crossing over in meiotic prophase results in a c^+ chromatid and a c chromatid being attached to the same centromere. Hence, c^+ and c fail to separate from each other during first anaphase. During second anaphase, sister chromatids move to opposite poles, thus affecting segregation of c^+ from c. The 2 : 2 : 2 : 2 linear pattern is indicative of a second-division segregation ascus produced by crossing over between the gene and its centromere.

Recombination Mapping with Tetrads

1 ORDERED TETRADS

The frequency of crossing over between the centromere and the gene in question is a reflection of its map distance from the centromere. Thus, the percentage of asci showing second-division segregation is a measure of linkage intensity. It must be remembered, however, that one crossover event gives one second-division ascus, but that only half of the ascospores in that ascus are recombinant type. Therefore, to convert second-division asci frequency to crossover frequency, we divide the former by 2.

2 UNORDERED TETRADS

The meiotic products of most ascomycetes seldom are in a linear order as in the ascus of *Neurospora*. Let us analyze unordered tetrads involving two linked genes from the cross $+ + \times ab$. The fusion nucleus is diploid ($+ +/ab$) and immediately undergoes meiosis. If a crossover does not occur between these two loci or if a two-strand double crossover occurs between them, the resulting meiotic products

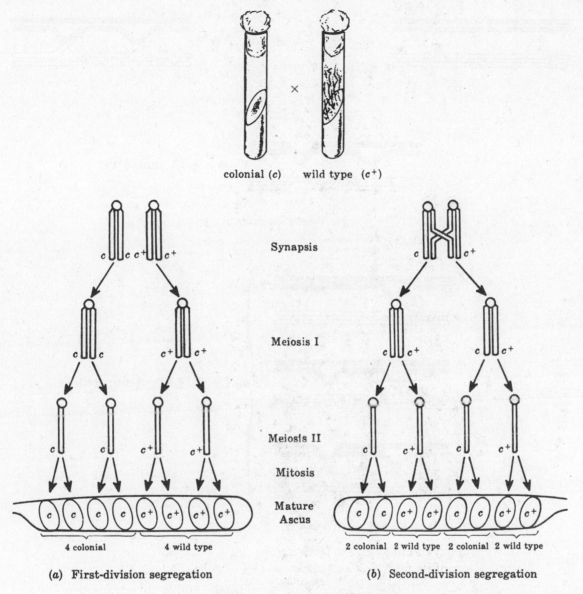

Fig. 6-6. *Neurospora* spore patterns.

will be of two kinds, equally frequent, resembling the parental combinations. Such a tetrad (Fig. 6-7) is referred to as a **parental ditype (PD)**.

A four-strand double crossover between the two genes results in two kinds of products, neither of which are parental combinations. This tetrad is called a **nonparental ditype (NPD)**, and is the rarest of the tetrad double crossovers (Fig. 6-8).

A **tetratype (TT)** is produced by either a single crossover or a three-strand double crossover (of two types) between the two genes (Fig. 6-9).

Whenever the numbers of parental ditypes and nonparental ditypes are statistically nonequivalent, this may be considered evidence for linkage between the two genes. To estimate the amount of recombination between the two markers, we use the formula:

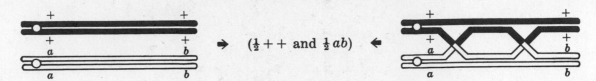

$$(\tfrac{1}{2}++ \text{ and } \tfrac{1}{2}ab)$$

Fig. 6-7. Parental ditypes.

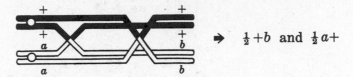

$$\tfrac{1}{2}+b \text{ and } \tfrac{1}{2}a+$$

Fig. 6-8. Non-parental ditypes.

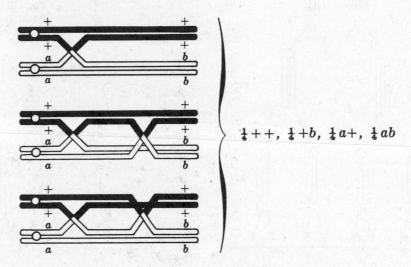

$$\tfrac{1}{4}++, \ \tfrac{1}{4}+b, \ \tfrac{1}{4}a+, \ \tfrac{1}{4}ab$$

Fig. 6-9. Tetratypes.

$$\text{Recombination frequency} = \frac{\text{NPD} + 1/2\ \text{TT}}{\text{total number of tetrads}}$$

The derivation of the above formula becomes clear when we analyze these diagrams and see that all of the products from an NPD tetrad are recombinant, but only half of the products from a TT tetrad are recombinant. Recombination frequency is not always equivalent to crossover frequency (map distance). If a third genetic marker was present midway between the loci of a and b, the three-strand double crossovers could be distinguished from the single crossovers and crossover frequency could thus be determined. Recombination frequency analysis of two widely spaced genes thereby can establish only minimum map distances between the two genes.

SOLVED PROBLEM 6.8

A strain of *Neurospora* requiring methionine (*m*) was crossed to a wild-type (m^+) strain with the results shown below. How far is this gene from its centromere?

No. Asci observed	Spores			
	1 + 2	3 + 4	5 + 6	7 + 8
6	+	*m*	+	*m*
5	*m*	+	+	*m*
6	*m*	+	*m*	+
7	+	*m*	*m*	+
40	*m*	*m*	+	+
36	+	+	*m*	*m*
100 total				

Solution: Noncrossover asci are those that appear with the greatest frequency — 40 + 36 = 76 out of 100 total. The other 24/100 are crossover types. While 24% of the asci are crossover types, only half of the spores in these asci are recombinant. Therefore, the distance from the gene to the centromere is 12 map units. The origin of the crossover-type asci is as follows:

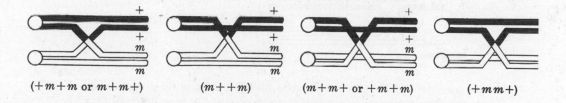

(+*m*+*m* or *m*+*m*+)   (*m*++*m*)   (*m*+*m*+ or +*m*+*m*)   (+*mm*+)

SOLVED PROBLEM 6.9

Two strains of *Neurospora*, one mutant for gene *a*, the other mutant for gene *b*, are crossed. Results are shown below. Determine the linkage relationships between these two genes.

Pattern	Percent of Asci	Spores			
		1 + 2	3 +4	5 + 6	7+ 8
1	79	*a* +	*a* +	+ *b*	+ *b*
2	14	*a* +	+ +	*a b*	+ *b*
3	6	*a* +	*a b*	+ +	+ *b*
4	1	*a* +	+ *b*	*a* +	+ *b*

Solution: Pattern 1 represents the noncrossover types showing first-division segregation (4 : 4) for both *a* and *b*. Pattern 2 shows second-division segregation (2 : 2 : 2 : 2) for *a*, but first-division segregation for *b*. Genes that show high frequency of second-division segregation are usually further from the centromere than genes with low frequency of second-division segregation. Judging by the relatively high frequency of pattern 2 these are probably single crossovers, and we suspect that *a* is more distal from its centromere than *b*.

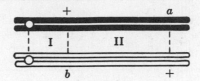

Pattern 3, indicating first-division segregation for *a* and second-division segregation for *b*, cannot be generated by a single crossover if *a* and *b* are linked as shown above, but requires a double crossover.

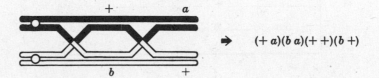

$$(+ a)(b\, a)(+ +)(b +)$$

Furthermore, pattern 4 could be produced from the linkage relationships as assumed above by a single crossover in region I.

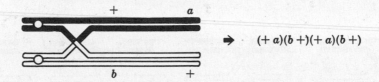

$$(+ a)(b +)(+ a)(b +)$$

Double crossovers (DCO) are expected to be much less frequent than single crossovers (SCO). Under the above assumptions, double-crossover pattern 3 is more frequent than one of the single-crossover patterns 4. This does not make sense, and thus our assumption must be wrong. The locus of *a* must be further from the centromere than *b*, but it need not be on the same side of the centromere with *b*. Let us place *a* on the other side of the centromere.

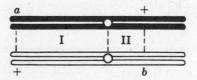

Now a single crossover in region I produces pattern 2, a single crossover in region II produces pattern 3, and a two-strand double crossover (I, II) produces pattern 4. The percentage of asci are numerically acceptable under this assumption.

The distance *a*–centromere = 1/2(SCO I + DCO) = 2(14 + 1) = 7.5 map units
The distance centromere–*b* = 1/2(SCO II + DCO) = 2(6 + 1) = 3.5 map units

SOLVED PROBLEM 6.10

The cross $(abc) \times (+ + +)$ is made in an ascomycete with unordered tetrads. From the analysis of 100 asci, determine the linkage relationships between these three loci as completely as the data allow.

1. 40 $(abc)(abc)(+ + +)(+ + +)$ 3. 10 $(a + c)(+ + c)(ab +)(+ b +)$
2. 42 $(ab +)(ab +)(+ + c)(+ + c)$ 4. 8 $(a + +)(+ + +)(abc)(+ bc)$

Solution: Pattern 1 is parental ditype (PD) for ab, ac, and bc. Pattern 2 is PD for ab, non-parental ditype (NPD) for ac and bc. Pattern 3 is tetratype (TT) for ab and ac, NPD for bc. Pattern 4, is TT for ab and ac, PD for bc. For each pair of markers the relative frequencies of each type of tetrad are as follows:

	PD	NPD	TT
ab	40 42 82/100 = 0.82	0	10 8 18/100 = 0.18
ac	40/100 = 0.40	42/100 = 0.42	10 8 18/100 = 0.18
bc	40 8 48/100 = 0.48	42 10 52/100 = 0.52	

For the ab pair, PDs are not equivalent with NPDs. Thus, a and b must be linked. For pairs ac and bc, PDs are roughly equivalent with NPDs. Thus, c must be assorting independently on another chromosome.

The recombination frequency between a and b is

$$\frac{\text{NPD} + \frac{1}{2}TT}{\text{total \# tetrads}} = \frac{0 + \frac{1}{2}(18)}{100} = 9\%$$

The recombination frequency in percent is equal to 9 map units. Single crossovers between either b and its centromere or c and its centromere, or both, would produce TTs for bc. Since none occurred we can assume that the loci of b and c are both very near their respective centromeres.

Supplementary Problems

RECOMBINATION AMONG LINKED GENES

6.11. There is 21% crossing over between the locus of *p* and that of *c* in the rat. Suppose that 150 primary oocytes could be scored for chiasmata within this region of the chromosome. How many of these oocytes would be expected to have a chiasma between these two genes?

6.12. The longest chromosome in the sweet pea has a minimum uncorrected map length (based on known genetic markers) of 118 units. Cytological observations of the longest chromosome in meiotic cells revealed an average chiasmata frequency of 2.96 per tetrad. Calculate the maximum number of crossover units remaining in this chromosome for the mapping of new genes outside the range already known.

GENETIC MAPPING

6.13. The distances between eight loci in the second chromosome of *Drosophila* are presented in the following table. Construct a genetic map to include these eight loci. The table is symmetrical above and below the diagonal.

	d	dp	net	J	ed	ft	cl	ho
d	—	18	31	10	20	19	14.5	27
dp		—	13	28	2	1	3.5	9
net			—	41	11	12	16.5	4
J				—	30	29	24.5	37
ed					—	1	5.5	7
ft						—	4.5	8
cl							—	12.5
ho								—

6.14. The recessive gene *sh* produces shrunken endosperm in corn kernels and its dominant allele sh^+ produces full, plump kernels. The recessive gene *c* produces colorless endosperm and its dominant allele c^+ produces colored endosperm. Two homozygous plants are crossed, producing an F_1 all phenotypically plump and colored. The F_1 plants are testcrossed and produce 149 shrunken, colored : 4035 shrunken, colorless : 152 plump, colorless : 4032 plump, colored. (*a*) What were the phenotypes and genotypes of the original parents? (*b*) How are the genes linked in the F_1? (*c*) Estimate the map distance between *sh* and *c*.

6.15. The presence of one of the Rh antigens on the surface of the red blood cells (Rh-positive) in humans is produced by a dominant gene *R*; Rh-negative cells are produced by the recessive genotype *rr*. Oval-shaped erythrocytes (elliptocytosis or ovalocytosis) are caused by a dominant gene *E*; its recessive allele *e* produces normal red blood cells. Both of these genes are linked approximately 20 map units apart on one of the autosomes. A man with elliptocytosis, whose mother had normally shaped erythrocytes and a homozygous Rh-positive genotype and whose father was Rh-negative and heterozygous for elliptocytosis, marries a normal Rh-negative woman. (*a*) What is the probability of their first child being Rh-negative and elliptocytotic? (*b*) If their first child is Rh-positive, what is the chance that it will also be elliptocytotic?

6.16. The Rh genotypes, as discussed in Problem 6.15, are given for each individual in the pedigree shown below. Solid symbols represent elliptocytotic individuals. (*a*) List the *E* locus genotypes for each individual in the pedigree. (*b*) List the gametic contribution (for both loci) of the elliptocytotic individuals (of genotype *Rr*) beside each of their offspring in which it can be detected. (*c*) How often in part (*b*) did *R* segregate with *E*, and *r* with *e*? (*d*) On the basis of random assortment, in how many of the offspring in part (*b*) would we expect to find *R* segregating with *e*, or *r* with *E*? (*e*) If these genes assort independently, calculate the probability of *R* segregating with *E*, and *r* with *e* in all 10 cases. (*f*) Is the solution to part (*c*) suggestive of linkage between these two loci? (*g*) Calculate part (*e*) if the siblings III1 and II2 were identical twins (developed from a single egg). (*h*) How are these genes probably linked in I1?

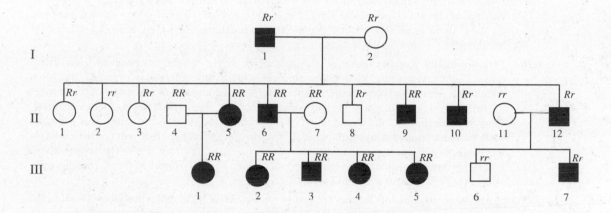

6.17. Two recessive genes in *Drosophila* (*b* and *vg*) produce black body and vestigial wings, respectively. When wild-type flies are testcrossed, the F_1 are all dihybrid in coupling phase. Testcrossing the female F_1 produced 1930 wild type : 1888 black and vestigial : 412 black: 370 vestigial. (*a*) Calculate the distance between *b* and *vg*. (*b*) Another recessive gene *cn* lies between the loci of *b* and *vg*, producing cinnabar eye color. When wild-type flies are testcrossed, the F_1 are all trihybrid. Testcrossing the F_1 females produced 664 wild type : 652 black, cinnabar, vestigial : 72 black, cinnabar : 68 vestigial : 70 black : 61 cinnabar, vestigial : 4 black, vestigial : 8 cinnabar. Calculate the map distances. (*c*) Do the *b*–*vg* distances calculated in parts (*a*) and (*b*) coincide? Explain. (*d*) What is the coefficient of coincidence?

6.18. In corn, a dominant gene *C* produces colored aleurone; its recessive allele *c* produces colorless. Another dominant gene *Sh* produces full, plump kernels; its recessive allele *sh* produces shrunken kernels due to collapsing of the endosperm. A third dominant gene *Wx* produces normal starchy endosperm and its recessive allele *wx* produces waxy starch. A homozygous plant from a seed with colorless, plump, and waxy endosperm is crossed to a homozygous plant from a seed with colored, shrunken, and starchy endosperm. The F_1 is testcrossed to a colorless, shrunken, waxy strain. The progeny seed exhibit the following phenotypes: 113 colorless, shrunken, starchy : 4 colored, plump, starchy : 2708 colorless, plump, waxy : 626 colorless, plump, starchy : 2 colorless, shrunken, waxy : 116 colored, plump, waxy : 2538 colored, shrunken, starchy : 601 colored, shrunken, waxy. (*a*) Construct a genetic map for this region of the chromosome. Round all calculations to the nearest tenth of a percent. (*b*) Calculate the interference in this region.

6.19. A gene called "forked" (*f*) produces shortened, bent, or split bristles and hairs in *Drosophila*. Another gene called "outstretched" (*od*) results in wings being carried at right angles to the body. A third gene called "garnet" (*g*) produces pinkish eye color in young flies. Wild-type females heterozygous at all three loci were crossed to wild-type males. The F_1 data appear below.

F_1:	Females:	all wild type
	Males:	57 garnet, outstretched
		419 garnet, forked
		60 forked
		1 outstretched, forked
		2 garnet
		439 outstretched
		13 wild type
		9 outstretched, garnet, forked
		1000

(*a*) Which gene is in the middle? (*b*) What was the linkage relationship between alleles at the forked and outstretched loci in the maternal parent? (*c*) What was the linkage relationship between alleles at the forked and garnet loci in the maternal parent? (*d*) On what chromosome do these three genes reside?·(*e*) Calculate the map distances. (*f*) How much interference is operative?

6.20. Five sex-linked recessive genes of *Drosophila* (*ec, sc, v, cv,* and *ct*) produce traits called echinus, scute, vermilion, crossveinless, and cut, respectively. Echinus is a mutant producing rough eyes with large facets. Scute manifests itself by the absence or reduction in the number of bristles on certain parts of the body. Vermilion is a bright orange-red eye color. Crossveinless prevents the development of supporting structures in the wings. Cut produces scalloped and pointed wings with manifold (pleiotropic) effects in other parts of the body. At the beginning of our experiments we do not know the gene order. From the results of the following three experiments, construct a genetic map for this region of the X chromosome. Whenever possible use weighted averages.

Experiment 1. Echinus females crossed to scute, crossveinless males produced all wild-type females and all echinus males in the F_1. When the F_1 females were testcrossed, the results (including both male and female progeny) were as follows:

810	echinus	89	scute
828	scute, crossveinless	62	echinus, scute
88	crossveinless	103	echinus, crossveinless

Experiment 2. Crossveinless females crossed to echinus, cut males produced all wild-type females and all crossveinless males in the F_1. When the F_1 females were testcrossed, the results (including both male and female progeny) were as follows:

2207	crossveinless	223	crossveinless, cut
2125	echinus, cut	217	echinus
273	echinus, crossveinless	5	wild type
265	cut	3	echinus, crossveinless, cut

Experiment 3. Cut females crossed to vermilion, crossveinless males produced all wild-type females and cut males in the F_1. When the F_1 females were testcrossed, the results (including both male and female progeny) were as follows:

766	vermilion, crossveinless	73	vermilion
759	cut	85	crossveinless, cut
140	vermilion, cut	2	wild type
158	crossveinless	2	vermilion, crossveinless, cut

LINKAGE ESTIMATES FROM F_2 DATA

6.21 Two recessive sex-linked genes are known in chickens (ZW method of sex determination; Chapter 5): rapid feathering (*sl*) and gold plumage (*s*). The dominant alleles produce slow feathering (*Sl*) and silver plumage (*S*), respectively. Females of the Silver Penciled Rock breed, with slow feathering and silver plumage, are crossed to males of the Brown Leghorn breed, with rapid feathering and gold plumage. The F_2 progeny data appear below:

	Slow, silver	Rapid, silver	Slow, gold	Rapid, gold
Males	94	40	7	127
Females	117	28	7	156

(*a*) Determine the F_1 genotypes and phenotypes. (*b*) In what linkage phase are the F_1 males? (*c*) Calculate the amount of recombination expected to occur between these two loci in males.

6.22. Assume the genotype AB/AB is testcrossed and produces an F_2 consisting of 37 *A-B-*, 11 *A-bb*, 12 *aaB-*, and 4 *aabb*. Estimate the percentage recombination between *A* and *B* by the square-root method and by the product ratio method.

6.23 Two recessive genes in the third linkage group of corn produce crinkly leaves and dwarf plants, respectively. A pure crinkly plant is pollinated by a pure dwarf plant. The F_2 progeny consist of 104 normal : 43 dwarf : 51 crinkly : 2 dwarf, crinkly. Using the square-root method, estimate the amount of recombination between these two loci.

6.24. The duplicate recessive genes r_1 and r_2 produce a short, velvetlike fur called "rex." Two rex rabbits of different homozygous genotypes were mated and produced an F_1 that was then testcrossed to produce 64 rex and 6 normal testcross progeny. (*a*) Assuming independent assortment, how many normal and rex phenotypes would be expected among 70 progeny? (*b*) Do the data indicate linkage? (*c*) What is the genotype and the phenotype of the F_1? (*d*) What is the genotype and phenotype of the testcross individuals? (*e*) Calculate the map distance.

USE OF GENETIC MAPS

6.25. Two loci are known to be in linkage group IV of the rat. Kinky hairs in the coat and vibrissae (long nose "whiskers") are produced in response to the recessive genotype *kk* and a short, stubby tail is produced by the recessive genotype *st/st*. The dominant alleles at these loci produce normal hairs and tails, respectively. Given 30 map units between the loci of *k* and *st*, determine the expected F_1 phenotypic proportions from heterozygous parents that are (*a*) both in coupling phase, (*b*) both in repulsion phase, (*c*) one in coupling and the other in repulsion phase.

6.26. Deep-yellow hemolymph (blood) in silkworm larvae is the result of a dominant gene *Y* at locus 25.6 (i.e., 25.6 crossover units from the end of the chromosome). Another dominant mutation *Rc*, 6.2 map units from the *Y* locus, produces a yellowish-brown cocoon (rusty). Between these two loci is a recessive mutant *oa* governing mottled translucency in the larval skin, and mapping at locus 26.7. The loci *Rc* and *oa* are separated by 5.1 crossover units. An individual that is homozygous for yellow blood, mottled translucent larval skin, and wild-type cocoon color is crossed to an individual of genotype $Y^+ oa^+ Rc/Y^+ oa^+ RC$ that spins a rusty cocoon. The F_1 males are then testcrossed to produce 3000 F_2 progeny. Coincidence is assumed to be 10%. (*a*) Predict the numbers within each phenotypic class that will appear in the F_2 (to the nearest whole numbers). (*b*) On the basis of probabilities, how many more F_2 progeny would need to be produced in order to recover one each of the DCO phenotypes?

6.27. The eyes of certain mutant *Drosophila* have a rough texture due to abnormal facet structure. Three of the mutants that produce approximately the same phenotype (mimics) are sex-linked recessives: roughest (*rst*), rugose (*rg*), and roughex (*rux*). The loci of these genes in terms of their distances from the end of the X chromosome are 2, 11, and 15 map units, respectively. (*a*) From testcrossing wild-type females of genotype *rst + rux/+ rg +* predict the number of wild-type and rough-eyed flies expected among 20,000 progeny. Assume no interference. (*b*) Approximately how many rough-eyed progeny flies are expected for every wild-type individual? (*c*) If the females of part (*a*) were of genotype *rst rq rux/+ + +*, what would be the approximate ratio of wild-type : rough-eyed progeny?

6.28. In Asiatic cotton, a pair of factors (*R* and *r*) controls the presence or absence, respectively, of anthocyanin pigmentation. Another gene, about 10 map units away from the *R* locus, controls chlorophyll production. The homozygous recessive genotype at this locus (*yy*) produces a yellow (chlorophyll-deficient) plant that dies early in the seedling stage. The heterozygote *Yy* is phenotypically green and indistinguishable from the dominant homozygote *YY*. Obviously, testcrosses are not possible for the *Y* locus. When dihybrids are crossed together, calculate the expected phenotypic proportions among the seedlings and among the mature F_1 when parents are (*a*) both in coupling phase, (*b*) both in repulsion phase, (*c*) one in coupling and one in repulsion phase. (*d*) Which method [in parts (*a*), (*b*), or (*c*)] is expected to produce the greatest mortality?

CROSSOVER SUPPRESSION

6.29. A black-bodied *Drosophila* is produced by a recessive gene *b* and vestigial wings by another recessive gene *vg* on the same chromosome. These two loci are approximately 20 map units apart. Predict the progeny phenotypic expectations from (*a*) the mating of repulsion phase females × coupling-phase males, (*b*) the reciprocal cross of part (*a*), (*c*) the mating where both parents are in repulsion phase.

6.30. Poorly developed mucous glands in the female silkworm *Bombyx mori* cause eggs to be easily separated from the papers on which they are laid. This is a dominant genetic condition; its wild-type recessive allele Ng^+ produces normally "glued" eggs. Another dominant gene *C*, 14 map units from *Ng*, produces a golden-yellow color on the outside of the cocoon and nearly white inside. Its recessive wild-type allele C^+ produces normally pigmented or wild-type cocoon color. A pure "glueless" strain is crossed to a pure golden strain. The F_1 females are then mated to their brothers to produce the F_2. Predict the number of individuals of different phenotypes expected to be observed in a total of 500 F_2 offspring. (*Hint*: Crossing over does not occur in female silkworms.)

6.31. Two autosomal recessive genes, "dumpy" (*dp*, a reduction in wing size) and "net" (*net*, extra veins in the wing), are linked on chromosome 2 of *Drosophila*. Homozygous wild-type females are crossed to net, dumpy males. Among 800 F_2 offspring were found: 574 wild type : 174 net, dumpy : 25 dumpy : 27 net. Estimate the map distance.

6.32. Suppose that an abnormal genetic trait (mutation) appeared suddenly in a female of a pure culture of *Drosophila melanogaster*. We mate the mutant female to a male from a balanced lethal strain [*Cy/Pm, D/Sb*, where curly (*Cy*) and plum (*Pm*) are on chromosome 2 and dichaete (*D*) and stubble (*Sb*) are on chromosome 3]. About half of the F_1 progeny (both males and females) exhibit the mutant phenotype. The F_1 mutant males with curly wings and stubble bristles are then mated to unrelated virgin wild-type females. In the F_2 the mutant trait never appears with stubble. Recall that this species of *Drosophila* has chromosomes X, 2, 3, and 4. Could the mutation be (*a*) an autosomal recessive, (*b*) a sex-linked recessive, (*c*) an autosomal dominant, (*d*) a sex-linked dominant? (*e*) In which chromosome does the mutant gene reside? (*f*) Suppose the mutant trait in the F_2 appeared in equal association with curly and stubble. In which chromosome would the mutant gene reside? (*g*) Suppose the mutant trait in the F_2 appeared only in females. In which chromosome would the mutant gene reside? (*h*) Suppose the mutant trait in the F_2 never appeared with curly. In which chromosome would the mutant gene reside?

RECOMBINATION MAPPING WITH TETRADS

6.33. Given the adjoining meiotic metaphase orientation in *Neurospora*, determine the simplest explanation to account for the following spore patterns.

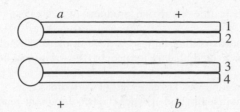

(a) $(+b)(a+)(a+)(+b)$ (e) $(ab)(++)(a+)(+b)$ (i) $(ab)(a+)(++)(+b)$
(b) $(a+)(+b)(+b)(a+)$ (f) $(a+)(++)(ab)(+b)$ (j) $(a+)(ab)(++)(+b)$
(c) $(++)(a+)(ab)(+b)$ (g) $(ab)(+b)(++)(a+)$ (k) $(a+)(ab)(+b)(++)$
(d) $(++)(ab)(a+)(+b)$ (h) $(a+)(+b)(++)(ab)$ (l) $(ab)(a+)(+b)(++)$

6.34. A certain strain of *Neurospora* cannot grow unless adenine is in the culture medium. Adenineless is a recessive mutation (*ad*). Another strain produces yellow conidia (*ylo*). Below are shown the results from crossing these two strains. Calculate the map distance between these two genes.

No. of Asci	Meiotic Products
106	$(ad+)(ad+)(+ylo)(+ylo)$
14	$(ad+)(ad\ ylo)(++)(+ylo)$

6.35. A riboflavineless strain (*r*) of *Neurospora* is crossed with a tryptophaneless strain (*t*) to give

No. of Asci	Meiotic Products	No. of Asci	Meiotic Products
129	$(r+)(r+)(+t)(+t)$	13	$(rt)(r+)(+t)(++)$
1	$(+t)(r+)(r+)(+t)$	17	$(r+)(rt)(++)(+t)$
2	$(r+)(+t)(r+)(+t)$	17	$(r+)(rt)(+t)(++)$
1	$(r+)(+t)(+t)(r+)$	2	$(r+)(++)(rt)(+t)$
15	$(rt)(r+)(++)(+t)$	1	$(rt)(++)(r+)(+t)$

Construct a map that includes these two genes.

6.36. Two of the genes *s*, *t*, and *u* are linked; the third assorts independently and is very tightly linked to its centromere. Analyze the unordered tetrads produced by the cross $(stu) \times (+++)$. (*Hint*: See Problem 6.10.)

No. of Tetrads	Tetrads
59	$(stu)(stu)(+++)(+++)$
53	$(s+u)(s+u)(+t+)(+t+)$
26	$(st+)(+t+)(s+u)(++u)$
30	$(s++)(+++)(stu)(+tu)$
32	$(stu)(+t+)(s+u)(+++)$

Review Questions

Vocabulary   For each of the following definitions, give the appropriate term and spell it correctly. Terms are single words unless indicated otherwise.

1. The recovery and investigation of all the products of a single meiotic event (as occurs in ascomycete fungi).
2. A cross-shaped cytological structure that is observed in paired homologous chromosomes during first meiotic prophase, and that is formed by the reciprocal breaking and rejoining of nonsister chromatids.
3. The genetic event that recombines linked genes. (Two words.)
4. The linkage arrangement in a dihybrid in which the two dominant genes are on one chromosome and their corresponding recessive alleles are on the homologous chromosome. (Two words.)
5. The linkage arrangement in a dihybrid in which, on each chromosome of a homologous pair, there is one dominant and one recessive gene. (Two words.)
6. A unit of map distance equivalent to 1% crossing over.
7. The ratio of the double-crossover progeny observed in a three-point testcross relative to the double-crossover progeny expected on the basis of treating map units as if they were independent variables. (One to three words.)
8. The complement of the coefficient of coincidence.
9. A genetic system that can maintain dihybrid genotypes in repulsion-phase generation after generation. (Two words.)
10. Investigation of the products of a given meiotic event in ascomycete fungi. (Two words.)

Multiple-Choice Questions   Choose the one best answer.

1. If a chiasma forms between the loci of genes A and B in 20% of the tetrads of an individual of genotype AB/ab, the percentage of gametes expected to be Ab is (a) 40 (b) 20 (c) 10 (d) 5 (e) none of the above
2. The average number of chiasmata that forms in one pair of homologous chromosomes is 1.2. The total length (in map units) for this linkage group is expected to be (a) 120 (b) 60 (c) 50 (d) 30 (e) none of the above

For questions 3–7, use the following information. Distances $A-B = 15$ map units, $B-C = 8$ map units, and $A-C = 23$ map units.

3. In an individual of genotype AbC/aBc, the percentage of gametes expected to be ABC is (a) 23 (b) 11.5 (c) 5.75 (d) 0.6 (e) none of the above
4. In the cross $Abc/abC \times abc/abc$, the percentage of Abc/abc progeny expected is (a) 39.7 (b) 23.4 (c) 77.1 (d) 48.6 (e) none of the above
5. If from the cross $ABC/abc \times abc/abc$ we observe among 1000 progeny one ABC/abc and one aBc/abc, then the coefficient of coincidence is estimated to be approximately (within rounding error) (a) 0.054 (b) 0.167 (c) 0.296 (d) 0.333 (e) none of the above
6. The percentage of progeny expected to be ABC/abc from the cross $ABC/abc \times abc/abc$ is (a) 15.0 (b) 7.5 (c) 6.9 (d) 23.0 (e) 18.7
7. If the $B-C$ distance of 8 map units is a weighted average of two experiments (experiment 1 = 7.3 map units based on a sample size of 100; experiment 2 = 8.4 map units), the sample size of experiment 2 was approximately (a) 116 (b) 144 (c) 167 (d) 175 (e) 190
8. From the cross $AB/ab \times AB/ab$ we observe 9% of the progeny to be ab/ab. The estimate of the recombination percentage between these two loci is (a) 50 (b) 40 (c) 35 (d) 20 (e) none of the above

The following information applies to questions 9 and 10. A trihybrid testcross (order of loci unknown) produced 35 AbC/abc, 37 aBc/abc, 8 ABc/abc, 10 abC/abc, 3 ABC/abc, 5 abc/abc, 1 Abc/abc, 1 aBC/abc

9. The gene order (a) is CBA (b) is BAC (c) is BCA (d) cannot be determined from the information provided (e) is none of the above
10. The distance (in map units) between the A and C loci is estimated to be (a) 30 (b) 25 (c) 20 (d) 15 (e) none of the above

Answers to Supplementary Problems

6.11. 63

6.12. 30 units

6.13. *net*   (4)   *ho*   (7)   *ed*(1) *ft*(1) *dp* (3.5) *cl*   (14.5)   *d*   (10)   *J*

6.14. (*a*) *sh c/sh c* (shrunken, colorless) × *sh⁺ c⁺/sh⁺ c⁺* (plump, colored)   (*b*) *sh⁺ c⁺/sh c* (coupling phase)   (*c*) 3.6 map units

6.15. (*a*) 2/5   (*b*) 1/5

6.16. (*a*) All open symbols = *ee*, all solid symbols = *Ee*   (*b*) *re* (II1, II2, III6), *RE* (II5, II9, III2-5, (7)   (*c*) All 10 cases   (*d*) 5 each   (*e*) 1/1024   (*f*) Yes   (*g*) 1/512   (*h*) *RE/re* (coupling phase)

6.17. (*a*) 17 map units   (*b*) *b-cn* = 8.9 map units; *cn-vg* = 9.5 map units   (*c*) No. In part (*a*), two-point testcross cannot detect double crossovers that then appear as parental types, thus underestimating the true map distance.   (*d*) 0.89

6.18. (*a*) c   (3.5)   *sh*   (18.4)   *wx*   (*b*) 86.1%

6.19. (*a*) *f*   (*b*) Repulsion phase   (*c*) Coupling phase   (*d*) X chromosome (sex-linked)   (*e*) *g–f* = 12.0, *f–od* = 2.5 (*f*) None

6.20.

Experiment 1:   *sc*   (7.6)   *ec*   (9.7)   *cv*

Experiment 2:   *ec*   (10.3)   *cv*   (8.4)   *ct*

Experiment 3:   *cv*   (8.2)   *ct*   (15.2)   *v*

Combined Map: *sc*   (7.6)   *ec*   (10.1)   *cv*   (8.3)   *ct*   (15.2)   *v*

6.21. (*a*) *Sl S/sl s* (slow, silver males) × *sl s/W* (rapid, gold females)   (*b*) Coupling phase   (*c*) 14.2%

6.22. 50%

6.23. 20%

6.24. (*a*) 52.5 rex : 17.5 normal   (*b*) Yes   (*c*) R_1r_2/r_1R_2, normal   (*d*) r_1r_2/r_1r_2, rex   (*e*) 17.14 map units

6.25.

Phenotype	(*a*)	(*b*)	(*c*)
Normal	0.6225	0.5225	0.5525
Kinky hair	0.1275	0.2275	0.1975
Stub tail	0.1275	0.2275	0.1975
Kinky hair, stub tail	0.1225	0.0225	0.0525

6.26. (*a*) 1408 yellow, mottled : 1408 rusty : 16 yellow, rusty : 16 mottled : 76 yellow, mottled, rusty : 76 wild type. *Note*: Rounding errors may allow one whole individual difference in each of these phenotypic classes. (*b*) 32,651

6.27. (*a*) 19,964 rough-eyed : 36 wild type   (*b*) Approximately 1 : 555   (*c*) Approximately 1 : 1.289

6.28. (*a*)(*b*)(*c*)

	Seedlings			Adult		
	(*a*)	(*b*)	(*c*)	(*a*)	(*b*)	(*c*)
Normal chlorophyll, anthocyanin present	0.7025	0.5025	0.5225	0.9367	0.6700	0.6967
Normal chlorophyll, anthocyanin absent	0.0475	0.2475	0.2275	0.0633	0.3300	0.3033
Yellow, anthocyanin present	0.0475	0.2475	0.2275			
Yellow, anthocyanin absent	0.2025	0.0025	0.0225			

(*d*) All three crosses = 25% mortality

6.29. (*a*) 55% wild type : 20% black : 20% vestigial : 5% black, vestigial

(*b*), (*c*) 50% wild type : 25% black : 25% vestigial

6.30. 250 glueless, golden : 125 glueless : 125 golden

6.31. 13 map units

6.32. A dominant trait may appear suddenly in a population by mutation of a recessive wild-type gene to a dominant allelic form. Such a mutant individual would be heterozygous. (*a*) No (*b*) No (*c*) Yes (*d*) No (*e*) Chromosome 3 (*f*) Chromosome 4 (*g*) X chromosome (sex-linked) (*h*) Chromosome 2

6.33. (*a*) (1, 3) (*b*) (2,4) (*c*) (1, 3)(1, 3) or (1, 3)(2, 4) (*d*) (1, 3)(2, 3) or (1, 3)(1, 4) (*e*) (2, 3)(1, 3) or (1, 4)(1, 3) (*f*) (2, 3)(2, 3) or (1, 4)(2, 3) (*g*) (2, 4)(1, 3) or (2, 4)(2, 4) (*h*) (2, 4)(2, 3) or (2, 4)(1, 4) (*i*) (1, 3) (*j*) (2, 3) (*k*) (2, 4) (*l*) (1, 4)

6.34. 5.83 map units

6.35. •——(1.8) *r*——————(16.4)——————— *t*

6.36. Genes *u* and *s* are linked on the same side of the centromere with *u* closest to the centromere. Centromere to *u* = 16 units, *u* to *s* = 14 units. Gene *t* is on another chromosome and very closely linked to its centromere.

Answers to Review Questions

Vocabulary
1. tetrad analysis
2. chiasma, or χ (chi)
3. crossing over
4. coupling phase
5. repulsion phase
6. centimorgan
7. coincidence, coincidence coefficient, or coefficient of coincidence
8. interference
9. balanced lethals
10. tetrad analysis

Multiple-Choice Questions
1. *d*   2. *b*   3. *d*   4. *a*   5. *b*   6. *c*   7. *d*   8. *b*   9. *c*   10. *c*

CHAPTER 7

Cytogenetics

The Union of Cytology with Genetics

Perhaps one reason that Mendel's discoveries (1865) were not appreciated by the scientific community of his day was that the mechanics of mitosis and meiosis had not yet been discovered. During the years 1870–1900 rapid advances were made in the study of cells (**cytology**). At the turn of the 20th century, when Mendel's laws were rediscovered, the cytological basis was available to render the statistical laws of genetics intelligible in terms of physical units. **Cytogenetics** is the hybrid science that attempts to correlate cellular events, especially those of the chromosomes, with genetic phenomena.

Chromosome Structure

Eukaryotic DNA is primarily associated with a basic class of proteins known as **histones**. Together, the DNA, histones, and nonhistone chromosomal proteins form **nucleoprotein**, or **chromatin**. In the first level of packaging (Fig. 7-1), about 150 bp of the double-stranded DNA molecule are wrapped around an octomer of four pairs of different histone proteins to form a **nucleosome** core particle. A fifth kind of histone protein occupies the **linker DNA** that connects one core particle with another (analogous to beads on a string). This "string" first coils up into a solenoid of 30-nm diameter and then into a filament of 300-nm diameter. Even higher levels of compaction occur during cell division when, viewed in the light microscope, the chromatin material appears to condense from an amorphous chromatin mass into distinctive chromosomes. Nonhistone proteins (including various DNA and RNA polymerases, regulatory proteins, etc.) can also be found associated with chromatin, but they are not responsible for the basic structure of chromatin.

The highly compacted chromatin that can be seen in the light microscope during interphase is called **heterochromatin**; the much more open form of chro-

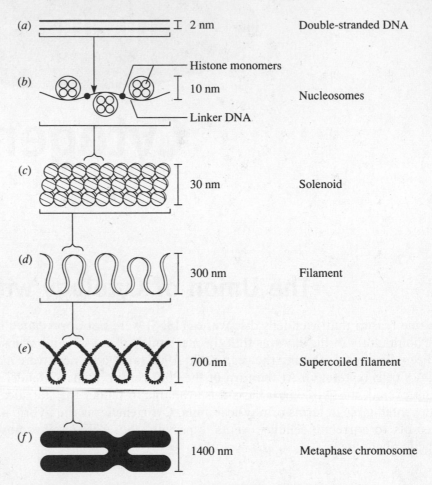

Fig. 7-1. Structural levels of organization hypothesized to occur during the condensation of a chromosome. (After J. D. Watson et al., *Molecular Biology of the Gene*, 4th ed., Benjamin/Cummings Publishing Company, Inc, 1987).

matin that is difficult to see is called **euchromatin**. Eukaryotic genes cannot be expressed (i.e., cannot serve as templates for RNA sythesis) if they are tightly bound to histones. So the first step in gene activation requires a dissociation of the DNA from the histones. Thus, euchromatic regions are thought to contain active genes whereas heterochromatic regions are thought to contain silenced genes or highly repetitive DNA sequences. DNA sequences surrounding centromeres are one example of a chromosomal region that is generally heterochromatic. In *Drosophila*, up to 30% of the genome is heterochromatic. Various mechanisms exist to regulate which regions become heterochromatic.

EXAMPLE 7.1
In female mammals, one of the two X chromosomes in each somatic cell appears highly condensed (heterochromatic), indicating that it is genetically inactive. In some cells one of the X chromosomes is inactive, and in other cells the other X chromosome is inactive. Thus, the female is a mosaic consisting of a mixture of two cell types with regard to the activity of sex-linked genes.

EXAMPLE 7.2
The stained giant polytene chromosomes of *Drosophila* (see Fig. 7-2, page 227) consist of alternating dark and light bands. The dark bands are heterochromatin; the light bands are euchromatin. At various stages during larval development, specific regions appear to decondense and form fluffy puffs of open chromatin material. At least some of the genes in these puffs are actively synthesizing RNA. Furthermore, the pattern of these puffed regions changes during larval development, indicating that different groups of genes are being activated and silenced as cellular differentiation proceeds.

The DNA in a metaphase chromosome appears to be attached to a protein **scaffold** constructed from two high-molecular-weight proteins. This association persists in metaphase chromosomes that have had their histones and most of their nonhistones removed. How this elegent scaffold becomes assembled and its role in the induction of superhelicity of the DNA is not known.

SOLVED PROBLEM 7.1
There are approximately 6.4×10^9 nucleotide pairs per diploid human cell. Each nucleotide pair occupies 3.4 angstroms (Å) of the DNA double helix. If the average length of a human chromosome at metaphase is about 6 micrometers (μm), what is the average packing ratio (i.e., the ratio of extended DNA to condensed DNA lengths)?

Solution: The total extended length of DNA per cell is

$$(3.4 \text{ Å per nucleotide pair}) \times (6.4 \times 10^9 \text{ nucleotide pairs}) = 2.2 \times 10^{10} \text{ Å}$$

Since $1 \text{ Å} = 10^{-10}$ meter (m), and 100 cm = 1 m,

$$(2.2 \times 10^{10} \text{ Å}) \times (10^{-10} \text{ m/Å}) = 2.2 \text{ m or 220 cm}$$

Because there are 23 chromosome pairs in a human diploid cell, the average extended length of DNA per chromosome is 220 cm per 46 chromosomes = 4.8 cm per chromosome. A micrometer is one-millionth of a meter (10^{-6} m) or 10^{-4} cm. Thus, the packing ratio of an average human chromosome is

$$\text{(Extended DNA length)/(condensed DNA length)} = 4.8 \text{ cm}/(6 \times 10^{-4} \text{ cm})$$
$$= 8.0 \times 10^3$$

or 8000 times longer when extended than when condensed in metaphase.

Variation in Chromosome Number

Each species has a characteristic number of chromosomes. Most higher organisms are diploid, with two sets of homologous chromosomes: one set donated by the father, the other set by the mother. Variation in the number of sets of chromosomes (**ploidy**) is commonly encountered in nature. It is estimated that one-third of the angiosperms (flowering plants) have more than two sets of chromosomes (**polyploid**). The term **euploidy** is applied to organisms with chromosomes that are multiples of some basic number (*n*), while **aneuploidy** refers to chromosome numbers that are not exact multiples of *n*.

1 EUPLOIDY

(a) **Monoploid**. One set of chromosomes (*n*) is characteristically found in the nuclei of some less complex organisms such as fungi. Monoploids in complex multicellular organisms are usually smaller and less vigorous than the normal diploids. A notable exception exists in male bees and wasps. Monoploid plants are known but are usually sterile.

(b) **Diploid**. Two sets of chromosomes (*2n*) is typical of most animals and complex multicellular organisms. A diploid state arises from the union of two haploid gametes.

(c) **Triploid**. Three sets of chromosomes (*3n*) can originate from the union of a monoploid gamete (*n*) with a diploid gamete (*2n*). The extra set of chromosomes of the triploid is distributed in various combinations to the germ cells, resulting in genetically unbalanced gametes. Because of the sterility that characterizes triploids, they are not commonly found in natural populations.

(d) **Tetraploid.** Four sets of chromosomes (*4n*) can arise in body cells by the somatic doubling of the chromosome number. Doubling is accomplished either spontaneously or it can be induced in high frequency by exposure to chemicals such as the alkaloid colchicine. Tetraploids are also produced by the union of unreduced diploid (*2n*) gametes.

  (i) **Autotetraploid**. The prefix "auto" indicates that the ploidy involves only homologous chromosome sets. Somatic doubling of a diploid produces four sets of homologous chromosomes (autotetraploid). Union of unreduced diploid gametes from the same species would accomplish the same result. Meiotic chromosome pairing usually produces quadrivalents (four synapsing chromosomes) that can produce genetically balanced gametes if disjunction is by twos, i.e., two chromosomes of the quadrivalent going to one pole and the other two to the opposite pole. If **disjunction** is not stabilized in this fashion for all quadrivalents, the gametes will be genetically unbalanced. Sterility will be expressed in proportion to the production of unbalanced gametes.

  (ii) **Allotetraploid**. The prefix "allo" indicates that nonhomologous sets of chromosomes are involved. The union of unreduced (*2n*) gametes from different diploid species could produce, in one step, an allotetraploid that appears and behaves like a new species. Alternatively, two diploid plant species may hybridize to produce a sterile diploid F_1. The sterility results from the failure of each set of chromosomes to provide sufficient genetic homology to effect pairing. The sterile diploid can become fertile if it undergoes doubling of the chromosome number. The allotetraploid thus produced has two matched sets of chromosomes that can pair just as effectively as in the diploid. Double diploids of this kind, found only in plants, are called **amphidiploids**.

EXAMPLE 7.3

Let the diploid set of chromosomes of one species be *AA* and that of the other species be *BB*.

P:	*AA*	×	*BB*
F_1:		*AB*	(sterile hybrid)
		↓	(chromosome doubling)
Amphidiploid:		*AABB*	(fertile)

(e) Polyploid. This term can be applied to any cell with more than $2n$ chromosomes. Ploidy levels higher than tetraploid are not commonly encountered in natural populations, but some of our most important crops are polyploid. For example, common bread wheat is hexaploid ($6n$), some strawberries are octaploid ($8n$), etc. Some triploids as well as tetraploids exhibit a more robust phenotype than their diploid counterparts, often having larger leaves, flowers, and fruits (**gigantism**). Many commercial fruits and ornamentals are polyploids. Sometimes, a specialized tissue within a diploid organism will be polyploid. For example, some liver cells of humans are polyploid. A common polyploid with which the reader should already be familiar is the triploid endosperm tissue of corn and other grains. Polyploids offer an opportunity for studying **dosage effects**; i.e., how two or more alleles of one locus behave in the presence of a single dose of an alternative allele. Because of double fertilization in angiosperms (Fig. 1-10), the genotype of sperm nuclei in pollen grains can influence not only the traits of the plant embryo, but also those of the seed endosperm. This phenomenon is known as **xenia**.

EXAMPLE 7.4

In corn, starchy endosperm is governed by a gene *S* that shows xenia with respect to its allele for sugary endosperm (*s*). Four genotypes are possible for these triploid cells: starchy = (S)(S)(S), (S)(S)(s), (s)(s)(S); sugary = (s)(s)(s). If pollen contains (S) sperm, the resulting endosperm will be starchy regardless of the genetic contribution of the seed parent (xenia).

The term *haploid*, strictly applied, refers to the gametic chromosome number. For diploids ($2n$), the haploid number is n; for an allotetraploid ($4n$), the haploid (reduced) number is $2n$; for an allohexaploid ($6n$) the haploid number is $3n$; etc. Lower organisms such as bacteria and viruses are called haploids because they have a single set of genetic elements. However, since they do not form gametes comparable to those of higher organisms, the term "monoploid" would seem to be more appropriate.

2 ANEUPLOIDY

Variations in chromosome number may occur that do not involve whole sets of chromosomes, but only parts of a set. The term *aneuploidy* is given to variations of this nature, and the suffix "-somic" generally refers to a particular organism and its chromosome number (which may be an abnormal situation).

(a) Monosomic. Diploid organisms that are missing one chromosome of a single pair are monosomics with the genomic formula $2n - 1$. The single chromosome without a pairing partner may go to either pole during meiosis, but more frequently will lag at anaphase and fails to be included in either nucleus. Monosomics can thus form two kinds of gametes, (n) and $(n - 1)$. In plants, the $n - 1$ gametes seldom function. In animals, loss of one whole chromosome often results in genetic unbalance, which is manifested by high mortality or reduced fertility.

(b) Trisomic. Diploids that have one extra chromosome are represented by the chromosomal formula $2n + 1$. One of the pairs of chromosomes has an extra member, so that a trivalent structure may be formed during meiotic prophase. If two chromosomes of the trivalent go to one pole and the third goes to the opposite pole, then gametes will be $(n + 1)$ and (n), respectively. Trisomy can produce different phenotypes, depending upon which chromosome of the complement is present in triplicate. In humans, the presence of one small extra chromosome (autosome 21) has a very deleterious effect resulting in **Down syndrome**, formerly called "mongolism."

(c) Tetrasomic. When one chromosome of an otherwise diploid organism is present in quadruplicate, this is expressed as $2n + 2$. A quadrivalent may form for this particular chromosome during meiosis that then has the same problem as that discussed for autotetraploids.

(d) Double Trisomic. If two different chromosomes are each represented in triplicate, the double trisomic can be symbolized as $2n + 1 + 1$.

(e) Nullosomic. An organism that has lost a chromosome pair is a nullosomic. The result is usually lethal to diploids $(2n - 2)$. Some polyploids, however, can lose two homologues of a set and still survive. For example, several nullosomics of hexaploid wheat $(6n - 2)$ exhibit reduced vigor and fertility but can survive to maturity because of the genetic redundancy in polyploids.

SOLVED PROBLEM 7.2
Suppose that an autotetraploid of genotype $AAaa$ forms only diploid gametes by random assortment from the quadrivalents (formed by synapsis of four chromosomes) during meiosis. Recall that chromosomes separate during the first meiotic division; sister chromatids separate during the second meiotic division. The A locus is so close to the centromere that crossing over in this region is negligible. (a) Determine the expected frequencies of zygotic genotypes produced by selfing the autotetraploid. (b) Calculate the expected reduction in the frequency of progeny with a recessive phenotype in comparison with that of a selfed diploid of genotype Aa.

Solution:
(a) Let us identify each of the four genes as follows: A^1, A^2, a^1, a^2 (A^1 and A^2 represent identical dominant alleles; a^1 and a^2 represent identical recessive alleles at the A locus).

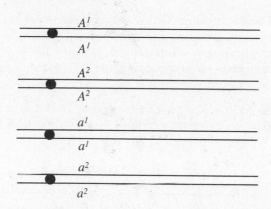

For genes that are tightly linked to their centromeres, the distribution of alleles into gametes follows the same pattern as chromosomal assortment. Let us first use a checkerboard to determine the kinds and frequencies of different combinations of alleles in pairs expected in the diploid gametes of the autotetraploid.

	A^1	A^2	a^1	a^2
A^1	—	A^1A^2	A^1a^1	A^1a^2
A^2		—	A^2a^1	A^2a^2
a^1			—	a^1a^2
a^2				—

Because sister chromotids separate at meiosis II, the diagonal of the above table represents the nonexistent possibility of a given chromosome (or identical allele) with itself in a gamete. The table is symmetrical on either side of the diagonal. Ignoring the superscript identification of alleles, the expected ratio of possible diploid gametes is 1 AA : 4 Aa : 1 aa = 1/6 AA : 2/3 Aa : 1/6 aa. Using these diploid gametic expectations, let us now construct a zygotic checkerboard for the prediction of tetraploid progeny genotypes.

	1/6 AA	2/3 Aa	1/6 aa
1/6 AA	1/36 $AAAA$	2/18 $AAAa$	1/36 $AAaa$
2/3 Aa	2/18 $AAAa$	4/9 $AAaa$	2/18 $Aaaa$
1/6 aa	1/36 $AAaa$	2/18 $Aaaa$	1/36 $aaaa$

Ratio of offspring genotypes: 1/36 $AAAA$ (quadruplex) : 8/36 $AAAa$ (triplex) : 18/36 $AAaa$ (duplex) : 8/36 $Aaaa$ (simplex) : 1/36 $aaaa$ (nulliplex).

(b) If one dose of the dominant allele is sufficient to phenotypically mask one or more doses of the recessive allele, then the phenotypic ratio is expected to be 35A : 1a. One-quarter of the offspring of a selfed diploid heterozygote (Aa) is expected to be of the recessive phenotype. The reduction in the frequency of the recessive trait is from 1/4 to 1/36, or ninefold. When homozygous genotypes produce a less desirable phenotype than heterozygotes, polyploidy can act as a buffer to reduce the incidence of homozygotes.

SOLVED PROBLEM 7.3

Assume that an autotetraploid of genotype $AAaa$ has the A locus 50 or more map units from the centromere, so that the equivalent of a single crossover always occurs between the centromere and the A locus. In this case, the chromatids will assort independently. Further assume that random assortment of chromatids to the gametes occurs by twos. Determine (*a*) the expected genotypic ratio of the progeny that results from selfing this autotetraploid, and (*b*) the expected increase in the incidence of heterozygous genotypes compared with selfed diploids of genotype *Aa*.

Solution:

(*a*) Let each of the genes of the duplex tetraploid be labeled as shown in the illustration below. All capital letters represent identical dominant genes; all lowercase letters represent identical recessive alleles.

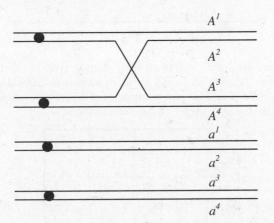

As in Problem 7.2, let us first use a Punnett square to determine the kinds and frequencies of different combinations of alleles (in pairs) expected in the gametes of the autotetraploid. Note in the illustration that (for example) alleles A^1 and A^2, originally on sister chromatids, can enter the same gamete if crossing over occurs between the centromere and the *A* locus. Likewise, any other pair of alleles could enter a gamete by the same mechanism (chromatid assortment). Random assortment (by twos) of eight chromatids of the autotetraploid during meiosis is shown in the following table. Note that the diagonal of the table represents the nonsense union of any given allele with itself (e.g., A^1 with A^1). The table is symmetrical on either side of the diagonal.

	A^1	A^2	A^3	A^4	a^1	a^2	a^3	a^4
A^1	—	A^1A^2	A^1A^3	A^1A^4	A^1a^1	A^1a^2	A^1a^3	A^1a^4
A^2		—	A^2A^3	A^2A^4	A^2a^1	A^2a^2	A^2a^3	A^2a^4
A^3			—	A^3A^4	A^3a^1	A^3a^2	A^3a^3	A^3a^4
A^4				—	A^4a^1	A^4a^2	A^4a^3	A^4a^4
a^1					—	a^1a^2	a^1a^3	a^1a^4
a^2						—	a^2a^3	a^2a^4
a^3							—	a^3a^4
a^4								—

Ignoring the superscript identification of alleles, the expected ratio of possible diploid gametes is 6 *AA* : 16 *Aa* : 6 *aa* or 3 : 8 : 3. Using these gametic expectations, we can now construct a zygotic checkerboard to generate the expected progeny.

	3 *AA*	8 *Aa*	3 *aa*
3 *AA*	9 *AAAA*	24 *AAAa*	9 *AAaa*
8 *Aa*	24 *AAAa*	64 *AAaa*	24 *Aaaa*
3 *aa*	9 *AAaa*	24 *Aaaa*	9 *aaaa*

Summary of progeny genotypes: 9 *AAAA* (quadruplex) : 48 *AAAa* (triplex) : 82 *AAaa* (duplex) : 48 *Aaaa* (simplex) : 9 *aaaa* (nulliplex).

(b) The total of the numbers (3 + 8 + 3) along the top or side of the preceding table is 14. Thus, the total of all crossproducts in the table is $14 \times 14 = 196$. Nine genotypes are homozygous *AAAA*, and another nine are homozygous *aaaa*. Thus, all other genotypes (196 − 18 = 178) are heterozygotes. Selfing the autotetraploid produces 178/196 = 91% heterozygous progeny. Selfing a diploid of genotype *Aa* produces 50% heterozygous progeny. The increase from 50 to 91% is 41/50 = 82%.

SOLVED PROBLEM 7.4

Pericarp is the outermost layer of the corn kernel and is maternal in origin. A dominant gene *B* produces brown pericarp, and its recessive allele *b* produces colorless pericarp. Tissue adjacent to the pericarp is aleurone (triploid). Purple pigment is deposited in the aleurone when the dominant gene *C* is present; its recessive allele *c* results in colorless aleurone. Aleurone is actually a single specialized layer of cells of the endosperm. The color of endosperm itself is modified by a pair of alleles. Yellow is governed by the dominant allele *Y* and white by the recessive allele *y*. Both *C* and *Y* show xenia to their respective alleles. A plant that is *bbCcYy* is pollinated by a plant of genotype *BbCcYy*. (a) What phenotypic ratio is expected among the progeny kernels? (b) If the F_1 is pollinated by plants of genotype *bbccyy*, in what color ratio will the resulting F_2 kernels be expected to appear?

Solution:

(a) If pericarp is colorless, then the color of the aleurone shows through. If aleurone is also colorless, then the color of the endosperm becomes visible. Since the maternal parent is *bb*, the pericarp on all F_1 seeds will be colorless. Any seeds with *C* will have purple aleurone. Only if the aleurone is colorless (*ccc*) can the color of the endosperm be seen.

Parents: *bbCcYy* female × *BbCcYy* male

F_1: 3/4 *C--* = 3/4 purple

  1/4 *ccc* — 3/4 *Y--* = 3/16 yellow

   — 1/4 *yyy* = 1/16 white (colorless)

(b) Half of the F_1 embryos is expected to be *Bb* and will thus lay down a brown pericarp around their seeds in the F_2; the other half is expected to be *bb* and will envelop its seeds with a colorless pericarp. Thus, half of the seeds of the F_1 plants will be brown. Of the remaining half that has colorless pericarp, we need show only as much of the genotype as is necessary to establish the phenotype.

$\underline{bb}$	F_1		Diploid Fusion Nucleus			Sperm Nucleus		Triploid Tissue
$1/2 \times 1/4\ CC$	=		$1/8\ CC$	+		c	=	$1/8\ CCc$ purple
$1/2 \times 1/2\ Cc$	=	$1/4$	$\begin{cases} 1/8\ CC \\ 1/8\ cc \end{cases}$ $\begin{cases} 1/16\ YY \\ 1/16\ yy \end{cases}$	+ + +		c c cy cy	= = = =	$1/8\ CCc$ purple $1/16\ cccYYy$ yellow $1/16\ cccyyy$ white
$1/2 \times 1/4\ cc$	=	$1/8$	$\begin{cases} 1/16\ ccYY \\ 1/16\ ccyy \end{cases}$	+ +		cy cy	= =	$1/16\ cccYYy$ yellow $1/16\ cccyyy$ white

Summary of F_2 seed colors: 1/2 brown : 1/4 purple : 1/8 yellow : 1/8 white.

Variation in Chromosome Size

In general, chromosomes of most organisms are too small and too numerous to be considered as good subjects for cytological investigation. *Drosophila* was considered to be a favorable organism for genetic studies because it produces large numbers of progeny within the confines of a small bottle in a short interval of time. Many distinctive phenotypes can be recognized in laboratory strains. It was soon discovered that crossing over does not occur in male fruit flies, thereby making it especially useful for genetic analyses. Later, its unusual sex mechanism was found to be a balance between male determiners on the autosomes and female determiners on the sex chromosomes. Although it had been known for over 30 years that some species of dipterans had extra large chromosomes in certain organs of the body, their utility in cytogenetic studies of *Drosophila* was not recognized until about 1934. There are only four pairs of chromosomes in the diploid complement of *D. melanogaster*, but their size in reproductive cells and most body cells is quite small. Unusually large chromosomes, 100 times as large as those in other parts of the body, are found in the larval salivary gland cells. Each giant **polytene chromosome** (Fig. 7-2) is composed of hundreds of chromatids paired along their identical DNA sequences throughout their length. Furthermore, each pair of homologous polytene chromosomes is also constantly synapsed in these somatic cells. Distinctive crossbandings (appearing when chromosomes are stained) represent regions (called **chromomeres**) of the chromatid bundle containing highly coiled or condensed DNA that is interspersed between regions of less condensation. The crossbanding pattern of each chromosome is characteristic of each species, but the pattern may change in a precise sequence at various stages of development. Chromosomal aberrations (translocations, inversions, duplications, deletions, etc.) can often be easily recognized in these polytene chromosomes under the light microscope.

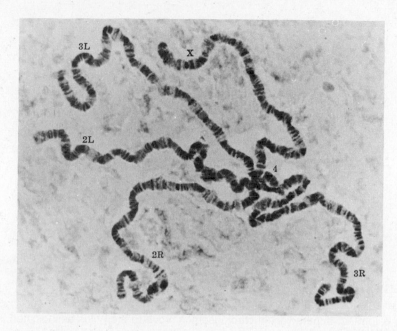

Fig. 7-2. Salivary gland chromosomes of *Drosophila melanogaster*. (Courtesy of B. P. Kaufmann.)

Variation in the Arrangement of Chromosome Segments

1 TRANSLOCATIONS

Chromosomes occasionally undergo spontaneous rupture, or can be induced to rupture in high frequency by ionizing radiation. The broken ends of such chromosomes behave as though they were "sticky" and may rejoin into nonhomologous combinations (**translocations**). A reciprocal translocation involves the exchange of segments between two nonhomologous chromosomes. During meiosis, an individual that is structurally heterozygous for a reciprocal translocation (i.e., two structually normal chromosomes and two chromosomes that are attached to nonhomologous pieces, as shown in Example 7.5) must form a cross-shaped configuration in order to effect pairing or synapsis of all homologous segments. A structural heterozygote may or may not be genetically heterozygous at one or more loci, but this is of no concern for the present purpose. In many of the following diagrams, only chromosomes (not chromatids) are shown and centromeres are omitted for the sake of simplicity.

EXAMPLE 7.5
Assume that a reciprocal translocation occurs between chromosomes 1 and 2 and between chromosomes 3 and 4.

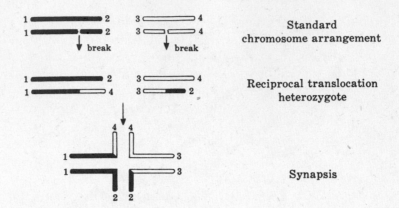

Standard
chromosome arrangement

Reciprocal translocation
heterozygote

Synapsis

The only way that functional gametes can be formed from a translocation heterozygote is by the alternate disjunction of chromosomes.

EXAMPLE 7.6
At the end of the meiotic prophase begun in Example 7.5, a ring of four chromosomes is formed. If the adjacent chromosomes move to the poles as indicated in
the diagram below, all of the gametes will contain some extra segments (duplications) and some pieces will be missing (deficiencies).

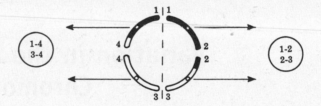

EXAMPLE 7.7
By forming a "figure-eight," alternate disjunction produces functional gametes.

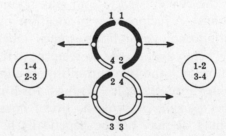

Translocation heterozygotes have several distinctive manifestations. (1) If an
organism produces gametes with equal facility by either segregation of adjacent
chromosomes (Example 7.6) or by alternate chromosomes (Example 7.7), semisterility occurs because only the latter mechanism produces functional gametes. (2)
Some genes that formerly were on nonhomologous chromosomes will no longer
appear to be assorting independently. (3) The phenotypic expression of a gene may
be modified when it is translocated to a new position in the genome. **Position**

effects are particularly evident when genes in euchromatin (lightly staining areas usually containing genetic elements) are shifted near heterochromatic regions (darker staining areas presumably devoid of active genes).

In the evening primrose of the genus *Oenothera*, an unusual series of reciprocal translocations has occurred involving all seven of its chromosome pairs. If we label each chromosome end with a different number, the normal set of seven chromosomes would be 1-2, 3-4, 5-6, 7-8, 9-10, 11-12, and 13-14; a translocation set would be 2-3, 4-5, 6-7, 8-9, 10-11, 12-13, and 14-1. A multiple translocation heterozygote like this would form a ring of 14 chromosomes during meiosis. The presence of different lethals in each of the two haploid sets of seven chromosomes enforces structural heterozygosity. Since only alternate segregation from the ring can form viable gametes, each group of seven chromosomes behaves as though it were a single large linkage group with recombination confined to the pairing ends of each chromosome. Each set of seven chromosomes that is inherited as a single unit is called a **Renner complex**. This is an example of a **translocation complex.**

EXAMPLE 7.8

In *Oenothera lamarckiana*, one of the Renner complexes is called *gaudens* and the other is called *velans*. This species is largely self-pollinated. The lethals become effective in the zygotic stage so that only the *gaudens-velans* (G-V) zygotes are viable. *Gaudens-gaudens* (G-G) or *velans-velans* (V-V) zygotes are lethal.

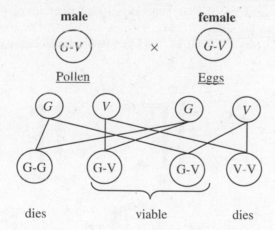

EXAMPLE 7.9

The two complexes in *Oenothera muricata* are called *rigens* (R) and *curvans* (C). Gametic lethals in each complex act differentially in the gametophytes. Pollen with the *rigens* complex are inactive; eggs with the *curvans* complex are inhibited. Only the *curvans* pollen and the *rigens* eggs are functional to give the *rigens-curvans* complex in the zygote.

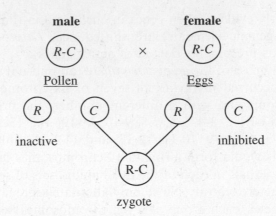

SOLVED PROBLEM 7.5

In 1931, Curt Stern found two different translocations in *Drosophila* from which he developed females possessing heteromorphic X chromosomes. One X chromosome had a piece of the Y chromosome attached to it; the other X was shorter and had a piece of chromosome IV attached to it. Two sex-linked genes were used as markers for detecting crossovers; the recessive trait carnation eye color (*car*) and the dominant trait bar eye (*B*). Dihybrid bar females with heteromorphic chromosomes (both mutant alleles on the X portion of the X-IV chromosome) were crossed to hemizygous carnation males with normal chromosomes. The results of this experiment provided cytological proof that genetic crossing over involves an actual physical exchange between homologous chromosome segments. Diagram the expected cytogenetic results of this cross showing all genotypes and phenotypes.

Solution:

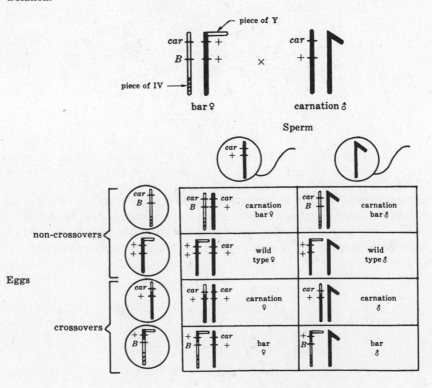

The existence of a morphologically normal X chromosome in recombinant male progeny with carnation eyes provides cytological proof that genetic crossing over is correlated with physical exchange between homologous chromosomes in the parents. Similarly, all other phenotypes correlate with the cytological picture. Chromosomal patterns other than the ones shown above may be produced by crossing over outside the inverted region.

SOLVED PROBLEM 7.6

The centromere of chromosome V in corn is about 7 map units from the end. The gene for light yellow (virescent) seedling (*v*) is 10 map units from this end, and a gene called brevis (*bv*) that shortens internode length is 12 map units from this end. The break point of a translocation (*T*) is 20 map units from this end. A translocation heterozygote involving chromosomes V and VIII of genotype $+ \ bv \ t/v \ + \ T$ is pollinated by a normal (nontranslocated, *t*) plant of genotype *v bv t/v bv t*. If gametes are formed exclusively by alternate segregation from the ring of chromosomes formed by the translocation heterozygote, predict the ratio of progeny genotypes and phenotypes from this cross (considering multiple crossovers to be negligible).

Solution: First let us diagram the effect that crossing over will have between the centromere and the point of translocation. We will label the ends of chromosome V with 1-2, and of chromosome VIII with 3-4. A cross-shaped pairing figure is formed during meiosis.

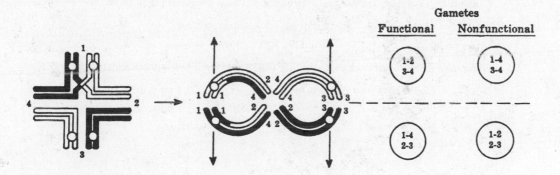

Alternate segregation produces half functional and half nonfunctional (duplication-deficiency) gametes. Note that the nonfunctional gametes derive only from the crossover chromatids. Thus, recovery of chromatids that experience a crossover between the centromere and the point of translocation is prevented. The combination of genes in this region of the chromosome is prevented from being broken up by crossing over and are thus transmitted as a unit. This situation is analogous to the block of genes within an inversion heterozygote that are similarly held together as a genetic unit. Noncrossover chromatids will form two types of functional gametes with equal frequency: $+ \ bv \ t$ and $v \ + \ T$. Expected zygotes are: $1/2 \ + \ bv$ $t/v \ bv \ \ t =$ brevis, homozygous for the normal chromosome order and $1/2 \ v \ + \ T/v \ bv \ t =$ virescent, heterozygous for the translocation.

SOLVED PROBLEM 7.7

Shrunken endosperm of corn is governed by a recessive gene *sh* and waxy endosperm by another recessive *wx*. Both of these loci are linked on chromosome 9. A plant that is heterozygous for a translocation involving chromosomes 8 and 9 and that developed from a plump, starchy kernel is pollinated by a plant from a shrunken, waxy kernel with normal chromosomes. The progeny are

171 shrunken, starchy, normal ear
205 plump, waxy, semisterile ear
82 plump, starchy, normal ear
49 shrunken, waxy, semisterile ear
17 shrunken, starchy, semisterile ear
40 plump, waxy, normal ear
6 plump, starchy, semisterile ear
3 shrunken, waxy, normal ear

(*a*) How far is each locus from the point of translocation? (*b*) Diagram and label the pairing figure in the plump, starchy parent.

Solution:

(*a*) The point of translocation may be considered as a gene locus because it produces a phenotypic effect, namely, semisterility. The conventional symbol for translocation is T, and t is used for the normal chromosome without a translocation. Gene order in the parents must be

$$\frac{+\ wx\ T}{sh\ +\ t} \times \frac{sh\ wx\ t}{sh\ wx\ t}$$

in order for double crossovers to produce the least frequent phenotypes:

$$+\ +\ T = \text{plump, starchy, semisterile ear}$$
$$sh\ wx\ t = \text{shrunken, waxy, normal ear}$$

The map distances are calculated in the usual way for a three-point testcross.

Distance $sh-wx = (82 + 49 + 6 + 3)/573 = 24.4$ map units
Distance $wx-T = (17 + 40 + 6 + 3)/573 = 11.5$ map units
Distance $sh-T = 24.4 + 11.5 = 35.9$ map units

(*b*)

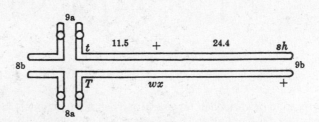

2 INVERSIONS

Assume that the normal order of segments within a chromosome is (1-2-3-4-5-6) and that breaks occur in regions 2-3 and 5-6, and that the broken piece is reinserted in reverse order. The inverted chromosome now has segments (1-2-5-4-3-6). One way in which inversions might arise is shown in Fig. 7-3. An inversion heterozygote has one chromosome in the inverted order and its homologue in the normal order. During meiosis the synaptic configuration attempts to maximize the pairing between homologous regions in the two chromosomes. This is usually accomplished by a loop in one of the chromosomes. Crossing over within the inverted segment gives rise to crossover gametes that are inviable because of duplications and deficiencies. Chromatids that are not involved in crossing over will be viable. Thus, as we have seen with translocations, inversions produce semisterility and altered linkage relationships. Inversions are sometimes called "crossover suppressors." Actually they do not prevent crossovers from occurring but they do prevent the crossover products from functioning. Genes within the inverted segment are thus held together and transmitted as one large linked group. **Balanced lethal** systems (Chapter 6) involve either a translocation or an inversion to prevent the recovery of crossover products and thus maintain heterozygosity generation after generation. In some organisms, these "inversions" have a selective advantage under certain environmental conditions and become more prevalent in the population than the standard chromosome order. Two types of inversion heterozygotes will be considered in which crossing over occurs within the inverted segment.

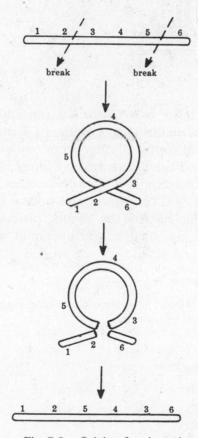

Fig. 7-3.   Origin of an inversion.

(a) Pericentric Inversion. The centromere lies within the inverted region. First meiotic anaphase figures appear normal unless crossing over occurs within the inversion. If a single two-strand crossover occurs within the inversion, the two chromatids of each chromosome will usually have arms of unequal length (unless there are chromosome segments of equal length on opposite sides of the inversion). Half of the meiotic products in this case (resulting from crossing over) are expected to contain duplications and deficiencies and would most likely be nonfunctional. The other half of the gametes (noncrossovers) are functional; one-quarter have the normal segmental order, one-quarter have the inverted arrangement.

EXAMPLE 7.10

Diagram of a pericentric inversion heterozygote with crossing over in region 3-4.

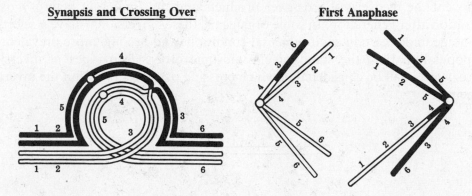

(b) Paracentric Inversion. The centromere lies outside the inverted segment. Crossing over within the inverted segment produces a **dicentric** chromosome (possessing two centromeres) that forms a bridge from one pole to the other during first anaphase. The bridge will rupture somewhere along its length and the resulting fragments will contain duplications and/or deficiencies. Also, an acentric fragment (without a centromere) will be formed; and since it usually fails to move to either pole, it will not be included in the meiotic products. Again, half of the products are nonfunctional, one-quarter are functional with a normal chromosome, and one-quarter are functional with an inverted chromosome.

EXAMPLE 7.11

Diagram of a paracentric inversion heterozygote with crossing over in region 4-5.

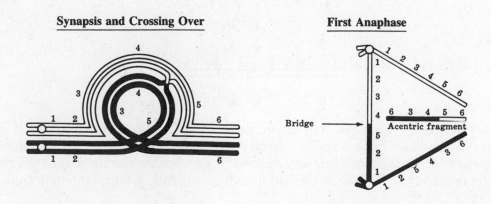

SOLVED PROBLEM 7.8

An inversion heterozygote possesses one chromosome in the normal order:

a b c d e f g h

and one in the inverted order:

a b f e d c g h

A four-strand double crossover occurs in the areas *f-e* and *d-c*. Diagram and label the first anaphase figures.

Solution: A somewhat easier way to diagram the synapsing chromosomes when crossing over is only within the inversion as shown below. This is obviously not representative of the actual pairing figure. Let the crossover in the *c-d* region involve strands 2 and 3, and the crossover in the *e-f* region involve strands 1 and 4.

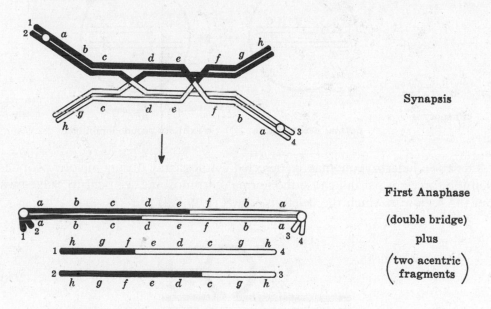

Synapsis

First Anaphase

(double bridge)

plus

$\left(\begin{array}{c}\text{two acentric}\\\text{fragments}\end{array}\right)$

Variation in the Number of Chromosome Segments

1 DELETIONS (DEFICIENCIES)

Loss of a chromosomal segment may be so small that it includes only a single gene or part of a gene. In this case, the phenotypic effects may resemble those of a mutant allele at that locus. For example, the "notch" phenotype of *Drosophila* is a sex-linked deletion that acts like a dominant mutation; a deletion at another sex-linked locus behaves as a recessive mutation, producing yellow body color when homozygous. Deletions never backmutate to the normal condition, because a lost piece of chromosome cannot be replaced. In this way, as well as others to be explained in subsequent chapters, a deletion can be distinguished from a point mutation. A loss of any considerable portion of a chromosome is usually lethal to a diploid organism because of genetic unbalance. When an organism heterozygous for a pair of alleles, *A* and *a*, loses a small portion of the chromosome bearing the dominant allele, the recessive allele on the other chromosome will become expressed phenotypically. This is called **pseudodominance**, but it is a misnomer because the condition is **hemizygous** rather than **dizygous** at this locus.

EXAMPLE 7.12
A deficiency in the segment of chromosome bearing the dominant gene *A* allows the recessive allele *a* to become phenotypically expressed.

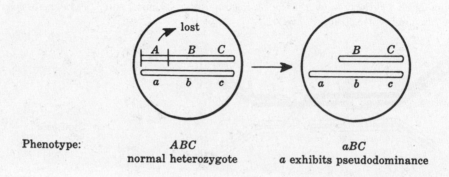

Phenotype:	*ABC*	*aBC*
	normal heterozygote	*a* exhibits pseudodominance

A deletion heterozygote may be detected cytologically during meiotic prophase when the forces of pairing cause the normal chromosome segment to bulge away from the region in which the deletion occurs (Fig. 7-4).

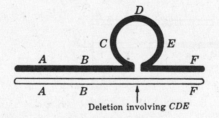

Deletion involving *CDE*

Fig. 7-4. Synapsis in a deletion heterozygote.

Overlapping deletions have been extensively used to locate the physical position of genes in the chromosome (cytological mapping).

EXAMPLE 7.13
A laboratory stock of *Drosophila* females is heterozygous in coupling phase for two linked genes at the tip of the X chromosome, *ac* (achaete) and *sc* (scute). A deletion in one chromosome shows pseudodominance for both achaete and scute. In other individuals, another deletion displays pseudodominance only for achaete. Obviously, these two deletions overlap. In the giant chromosomes of *Drosophila*, the absence of these segments of chromosome is easily seen. The actual location of the scute gene resides in the band or bands that differentiate the two overlapping deletions.

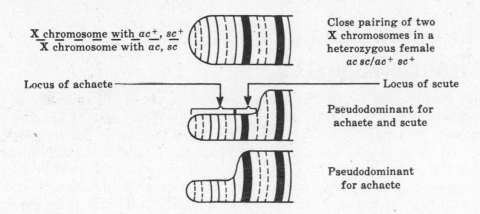

X chromosome with ac^+, sc^+
X chromosome with ac, sc

Close pairing of two
X chromosomes in a
heterozygous female
ac sc/ac^+ sc^+

Locus of achaete ——————————————— Locus of scute

Pseudodominant for
achaete and scute

Pseudodominant
for achaete

2   DUPLICATIONS (ADDITIONS)

Extra segments in a chromosome may arise in a variety of ways. Generally speaking, their presence is not as deleterious to the organism as a deficiency. It is assumed that some duplications are useful in the evolution of new genetic material. Because the old genes can continue to provide for the present requirements of the organism, the superfluous genes may be free to mutate to new forms without a loss in immediate adaptability. Genetic redundancy, of which this is one type, may protect the organism from the effects of a deleterious recessive gene or from an otherwise lethal deletion. During meiotic pairing, the chromosome bearing the duplicated segment bulges away from its normal homologue to maximize the juxtaposition of homologous regions. In some cases, extra genetic material is known to cause a distinct phenotypic effect. Relocation of chromosomal material without altering its quantity may result in an altered phenotype (**position effect**).

EXAMPLE 7.14
A reduced eye size in *Drosophila*, called "bar eye," is known to be associated with a duplicated region on the X chromosome. Genetically, the duplication behaves as a dominant factor. Wild-type flies arise in homozygous bar-eye cultures with a frequency of about 1 in 1600. With approximately the same frequency, a very small eye called "double-bar" is also produced. These unusual phenotypes apparently arise in a pure bar culture by improper synapsis and unequal crossing over as shown below, where the region (*a-b-c-d*) is a duplication.

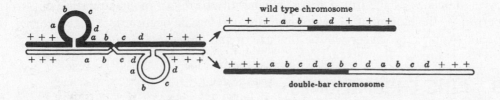

wild type chromosome

double-bar chromosome

Variation in Chromosome Morphology

1 ISOCHROMOSOMES

It has already been shown that a translocation can change the structure of the chromosome both genetically and morphologically. The length of the chromosome may be longer or shorter, depending upon the size of the translocated piece. An inversion does not normally change the length of the chromosome, but, if the inversion includes the centromere (pericentric), the position of the centromere may be changed considerably. Deletions or duplications, if viable, may sometimes be detected cytologically by a change in the size of the chromosome (or banding pattern in the case of the giant chromosomes of *Drosophila*), or by the presence of "bulges" in the pairing figure. Chromosomes with unequal arm lengths may be changed to **isochromosomes**, having arms of equal length and genetically homologous with each other, by an abnormal transverse division of the centromere. The telocentric X chromosome of *Drosophila* may be changed to an "attached-X" form by a misdivision of the centromere (Fig. 7-5).

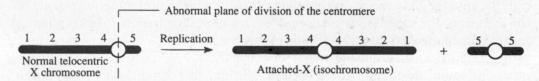

Fig. 7-5. Origin of attached-X chromosome. Segment 5 is nonessential heterochromatin.

2 BRIDGE-BREAKAGE-FUSION-BRIDGE CYCLES

The shape of a chromosome may change at each division once it has broken. Following replication of a broken chromosome, the broken ends of the sister chromatids may be fused by DNA repair mechanisms. Such broken ends are said to be "sticky." When the chromatids move to opposite poles, a bridge is formed. The bridge will break somewhere along its length and the cycle repeats at the next division. This sequence of events is called the **bridge-breakage-fusion-bridge cycle**. Mosaic tissue appearing as irregular patches of an unexpected phenotype on a background of normal tissue (**variegation**) can be produced by such a cycle. The size of the unusual tissue generally bears an inverse relationship to the period of development at which the original break occurred; i.e., the earlier the break occurs, the larger will be the size of the abnormal tissue.

3 RING CHROMOSOMES

Chromosomes are not always rod-shaped. Occasionally, ring chromosomes are encountered in plants or animals. If breaks occur at each end of a chromosome, the broken ends may become joined to form a ring chromosome (Fig. 7-6). If an

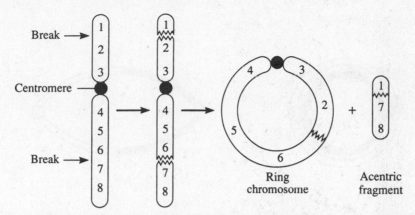

Fig. 7-6. Formation of a ring chromosome.

acentric fragment is formed by union of the end pieces, it will soon be lost. The phenotypic consequences of these deletions vary, depending on the specific genes involved. Crossing over between ring chromosomes can lead to bizarre anaphase figures.

EXAMPLE 7.15

A single exchange in a ring homozygote produces a double bridge at first anaphase.

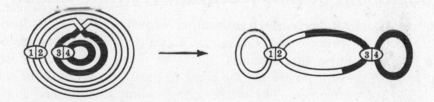

SOLVED PROBLEM 7.9

Data from *Drosophila* studies indicate that noncrossover (NCO) rings are recovered in equal frequencies with NCO rods from ring-rod heterozygotes. What light does this information shed on the occurrence of sister-strand crossing over?

Solution: Let us diagram the results of a sister-strand crossover in a rod and in a ring chromosome.

(*a*) Rod chromosome:

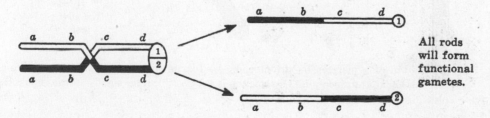

All rods will form functional gametes.

(*b*) Ring chromosome:

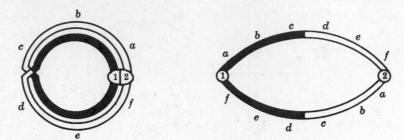

The double bridge at anaphase will rupture and produce nonfunctional gametes with duplications or deficiences. These would fail to be recovered in viable offspring. The fact that both rings and rods are recovered with equal frequency argues against the occurrence of sister-strand crossing over.

Modern techniques (involving autoradiography with labeled thymidine or fluorescence microscopy of cultured cells that have incorporated 5-bromodeoxyuridine in place of thymine) reveal that some sister-strand exchanges occur by a repair mechanism when DNA is damaged. One of the initiating steps that transforms a normal cell to a cancer cell is DNA damage. Hence, screening chemicals for their ability to induce sister-strand exchanges is one method for detecting potential cancer-inducing agents (carcinogens).

4   ROBERTSONIAN TRANSLOCATION (CENTRIC FUSION)

A whole arm fusion (**Robertsonian translocation**, named after W. R. B. Robertson who described the process in 1911) is an eucentric, reciprocal translocation between two acrocentric chromosomes where the break in one chromosome is near the front of the centromere and the break in the other chromosome is immediately behind its centromere. The smaller chromosome thus formed consists of largely inert heterochromatic material near the centromeres; it usually carries no essential genes and tends to become lost. A Robertsonian translocation thus results in a reduction of the chromosome number (Fig. 7-7).

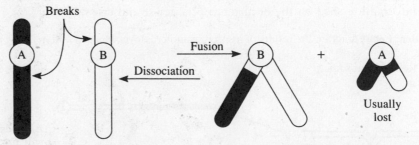

Fig. 7-7.   Formation of a metacentric chromosome by fusion of two acrocentric chromosomes (Robertsonian translocation). Dissociation is not possible if the small chromosome is lost.

EXAMPLE 7.16

Humans have 46 chromosomes whereas the great apes (chimpanzees, gorillas, and orangutans) have 48. It seems likely that humans evolved from a common human/ape ancestor by (among other structural changes) centric fusion of two acrocentrics to produce a single large chromosome (2) containing the combined genetic content of the two acrocentrics. Structural rearrangements of chromosomes may lead to reproductive isolation and the formation of new species. The mule is a hybrid from crossing the horse ($2n = 64$) and the ass or donkey ($2n = 62$). The mule is sterile because there is insufficient homology between the two sets of chromosomes to pair successfully at meiosis.

SOLVED PROBLEM 7.10

Yellow body color in *Drosophila* is produced by a recessive gene y at the end of the X chromosome. A yellow male is mated to an attached-X female (XX) heterozygous for the y allele. Progeny are of two types: yellow females and wild-type females. What insight does this experiment offer concerning the stage (two strand or four strand) at which crossing over occurs? (Reminder: the Y chromosome does not determine maleness in *Drosophila*; see Genic Balance, Chapter 5).

Solution: Let us assume that crossing over occurs in the two-strand stage, i.e., before the chromosome replicates into two chromatids.

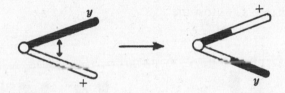

The yellow male produces gametes with either a y-bearing X chromosome or one with the Y chromosome that is devoid of genetic markers. Trisomic X (XXX) flies seldom survive (superfemales). Those with XXY will be viable heterozygous wild-type attached-X females. Crossing over fails to produce yellow progeny when it occurs in the two-strand stage.

Let us assume that crossing over occurs after replication of the chromosome, i.e., in the four-strand stage:

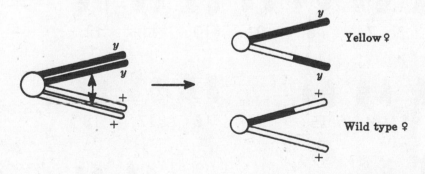

The appearance of yellow females in the progeny is proof that crossing over occurs in the four-strand stage.

Human Cytogenetics

The diploid human chromosome number of 46 (23 pairs) was established by Tjio and Levan in 1956. When grouped as homologous pairs, the somatic chromosome complement (**karyotype**) of a cell becomes an **idiogram**. Formerly, a chromosome could be distinguished only by its length and the position of its centromere at the time of maximum condensation (late prophase). No single autosome could be easily identified, but a chromosome could be assigned to one of seven groups (A–G) according to the "Denver system" of classification (Fig. 7-8). Group A consists of large, metacentric chromosomes (1–3); group B contains submedian chromosomes (4, 5); group C has medium-sized chromosomes with submedian centromeres (pairs 6–12); group D consists of medium-sized chromosomes (pairs 13–15) with one very short arm (acrocentric); group E chromosomes (16–18) are a little shorter than those in group D and have median or submedian centromeres; group F (19, 20) contains short, metacentric chromosomes; and group G has the smallest acrocentric chromosomes (21, 22). The X and Y sex chromosomes are not members of the autosome groups, and are usually placed together in one part of the idiogram. The Y chromosome may vary in size from one individual to another but usually has the appearance of a group G autosome. The X chromosome has the appearance of a group C autosome.

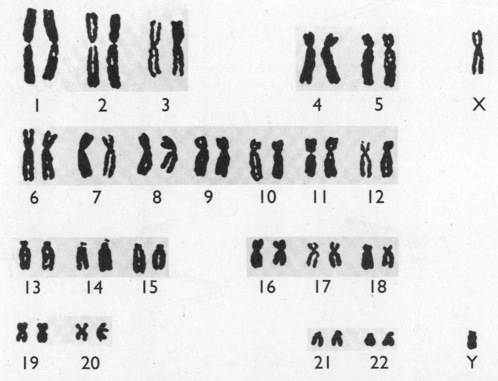

Fig. 7-8. Idiogram of the chromosomes of a normal human male. [From L. P. Wisniewski and K. Hirschhorn (eds.), *A Guide to Human Chromosome Defects*, 2nd ed., The March of Dimes Birth Defects Foundation, White Plains, N.Y.; BD:OAS XVI(6), 1980, with permission.]

Special staining techniques (e.g., Giemsa, quinacrine) allow specific banding patterns (G bands, Q bands, etc.) for each chromosome to be revealed. This allows for the individual identification of each chromosome in the karyotype (Fig. 7-9). Now, fluorescent probes are being used to "paint" chromosomes with different colors that correspond to a different gene or region sequence.

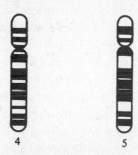

4   5

Fig. 7-9. Diagrams of the banding patterns that distinguish human chromosomes 4 and 5.

The X chromosome can be identified in many nondividing (interphase) cells of females as a dark-staining mass called **sex chromatin** or **Barr body** (after Dr. Murray L. Barr) that is attached to the nuclear membrane. The analogue of sex chromatin in certain white blood cells is a "drumstick" appendage attached to the multilobed nucleus of neutrophilic leukocytes. Dr. Mary Lyon theorizes that sex chromatin results from condensation [**heterochromatinization** (darkly stained)] and inactivation of any X chromosomes in excess of one per cell. Sex-linked traits are not expressed more intensely in females with two doses of X-linked genes than in males with only one X chromosome. At a particular stage early in development of females, one of the two X chromosomes in a cell becomes inactivated as a **dosage compensation mechanism**. Different cells inactivate one of the two chromosomes in an apparently random manner, but subsequently all derived cells retain the same functional chromosome. Females are thus a mixture of two kinds of cells; in some cells one X chromosome is active, and in different cells the other X chromosome is active. The same principle applies to mammals other than humans.

Prenatal screening of babies for gross chromosomal aberrations (polyploidy, aneuploidy, deletions, translocations, etc.), as well as sex prediction, is now possible. A fluid sample can be taken from the amniotic fluid that surrounds the fetus *in utero*, a process termed **amniocentesis**. The cells found in this fluid are of fetal origin. Such cells can be cultured *in vitro* in a highly nutritive solution and then treated with colchicine to stop division at metaphase. The cells are then placed into a hypotonic salt solution to cause the cells to swell and burst, resulting in release of their chromosomes. This preparation is placed on a slide, stained, and photographed under a microscope. Individual chromosomes are then cut from the resulting photograph and matched as homologous pairs to form an idiogram.

Mosaicism is the presence in an individual of two or more cell lineages of different chromosomal constitution, each cell lineage being derived from the same zygote. For example, a **bilateral gynandromorph** is a unique type of fruit fly that has had half its body develop as a male and the other half as a female; the different cell lineages are maleness and femaleness. In contrast, fusion of cell lines from different zygotes produces a **chimera**. Mosaicism results from abnormal postzygotic (mitotic) divisions of three kinds: (1) nondisjunction during the first cleavage division of the zygote, (2) nondisjunction during later mitotic divisions, and (3) anaphase lag, in which one member of a chromosome pair fails to segregate chomatids from the metaphase plate, and that chromatid fails to be included in the daughter cell nuclei (the entire chromosome is thus lost). Assuming that nondisjunction of chromatids affects only one member of a pair of chromosomes of the diploid set, the expected mosaic karyotype from nondisjunction during the first cleavage division of a zygote would be that half of the individual's cells are trisomic ($2n + 1 = 47$) and the other half are monosomic ($2n - 1 = 45$). If the first cleavage division is normal, but the second cleavage division involves a nondisjunctional event, three cell lineages would be established (45/46/47). Each line should "breed true," barring further mitotic abnormalities. In addition, XX/XO, sex-chromatin-positive Turner syndrome (Example 7.17) mosaics may result from anaphase lag of the sex chromosomes in females. A condition that may resemble Turner syndrome or be a hermaphrodite with physical characteristics of both sexes (XY/XO) may result from anaphase lag of the sex chromosomes in males.

EXAMPLE 7.17
Aneuploid females with only one X chromosome (XO) have a karyotype with $2n - 1 = 45$. They are called **Turner females** (after Henry Turner, who first described them), and they exhibit a group of characteristics that together define Turner syndrome: short stature, webbing of neck skin, underdeveloped gonads, shieldlike chest, and impaired intelligence. Non-mosaic Turner females are sex-chromatin-negative.

EXAMPLE 7.18
Abnormal males possessing an extra X chromosome (XXY) have a karyotype with $2n + 1 = 47$. They are called **Klinefelter males** (after Harry Klinefelter, who first described them), and they exhibit **Klinefelter syndrome**: sterility, long limbs, feminine breast development (**gynecomastia**), sparse body hair, and mental deficiency. Klinefelter males are sex-chromatin-positive. If some portion of the extra X chromosome is not inactivated, this could account for the phenotypic differences not only between XXY Klinefelter males and XY normal males, but also between XO Turner females and XX normal females.

EXAMPLE 7.19
XXX "**superfemales**" (**metafemales**) are karyotyped as $2n + 1 = 47$ trisomics and exhibit two Barr bodies. These individuals may range phenotypically from normal fertile females to nearly like those with Turner syndrome. They have a high incidence of mental retardation.

EXAMPLE 7.20
Tall, trisomic XYY males were first discovered in relatively high frequencies in penal and mental institutions. The presence of an extra Y chromosome was thought to predispose such a male to antisocial behavior, hence the name "tall-aggressive syndrome." Subsequently, more XYY males have been found among the non-institutionalized population, casting doubt upon the validity of the above

hypothesis. XYY males do tend to have subnormal IQs, however, and this may contribute to impulsive behavior.

EXAMPLE 7.21

Down syndrome (named after the physician Langdon Down and formerly called mongolism or mongolian idiocy) is usually associated with a trisomic condition for one of the smallest human autosomes (21). It is the most common chromosomal abnormality in live births (1/900 births). These individuals are mentally retarded, short, possess eye folds resembling those of Mongolian races, have stubby fingers and a swollen tongue. Women over age 35 have a greater chance of conceiving a child with Down syndrome: at 35, a woman has a 1 in 400 chance; while by age 40, this chance rises to 1 in 110; and at age 45, 1 in 35. Nondisjunction of chromosome pair 21 during spermatogenesis can also produce a child with Down syndrome, but paternal age does not seem to be associated with its incidence. In about 2–5% of the cases, the normal chromosome number is present ($2n = 46$), but the extra chromosome 21 is attached (translocated) to one of the larger autosomes.

EXAMPLE 7.22

Human autosomal monosomics are rarer than trisomics, possibly because harmful recessive mutants on the remaining homologue are hemizygous and can be expressed. Most cases of autosomal monosomy are mosaics of normal diploid ($2n$) and monosomic ($2n - 1$) cells resulting from mitotic nondisjunctions. Mosaics involving sex chromosomes are also known: e.g., XO : XX, XO : XY, XXY : XX, as well as autosomal mosaics such as 21-21 : 21-21-21, etc.

EXAMPLE 7.23

Deletion of part of the long arm of chromosome 22 produces an abnormality known as a **Philadelphia chromosome** (so named because it was discovered in that city). It is found only in the bone marrow (along with chromosomally normal cells) in approximately 90% of patients with chronic myelogenous leukemia (a kind of cancer). Usually, the missing piece of chromosome 22 can be found translocated to one of the larger autosomes (most frequently chromosome 9). See Example 11.11.

Supplementary Problems

VARIATION IN CHROMOSOME NUMBER

7.11. Abyssinian oat (*Avena abyssinica*) appears to be a tetraploid with 28 chromosomes. The common cultivated oat (*Avena sativa*) appears to be a hexaploid in this same series. How many chromosomes does the common oat possess?

7.12. The European raspberry (*Rubus idaeus*) has 14 chromosomes. The dewberry (*Rubus caesius*) is a tetraploid with 28 chromosomes. Hybrids between these two species are sterile F_1 individuals. Some unreduced gametes of the F_1 are functional in backcrosses. Determine the chromosome number and level of ploidy for each of the following: (a) F_1 (b) F_1 backcrossed to *R. idaeus*   (c) F_1 backcrossed to *R. caesius*   (d) chromosome doubling of F_1 (*R. maximus*).

7.13. There are 13 pairs of chromosomes in Asiatic cotton (*Gossypium arboreum*) and also 13 pairs in an American species *G. thurberi*. Interspecific crosses between *arboreum* and *thurberi* are sterile because of highly irregular chromosome pairing during meiosis. The American cultivated cotton (*G. hirsutum*) has 26 pairs of chromosomes. Crosses of *arboreum* × *hirsutum* or *thurberi* × *hirsutum* produce triploids with 13 bivalents (pairs of chromosomes) and 13 univalents (single unpaired chromosomes). How can this cytological information be used to interpret the evolution of *G. hirsutum*?

7.14. If two alleles, *A* and *a*, exist at a locus, five genotypic combinations can be formed in an autotetraploid: quadruplex (*AAAA*), triplex (*AAAa*), duplex (*AAaa*), simplex (*Aaaa*), and nulliplex (*aaaa*). Assume that *A* exhibits xenia over *a*. For each of these five genotypes, determine the expected phenotypic ratio of *A* : *a* when, (a) the locus is tightly linked to its centromere (chromosomal assortment) and the genotype is selfed, (b) the locus is assorting chromosomally and the genotype is testcrossed, (c) the locus is far from its centromere so that chromatids assort independently and the genotype is selfed, (d) the locus assorts by chromatids and the genotype is testcrossed.

7.15. The flinty endosperm character in maize is produced whenever two or all three of the alleles in this triploid tissue are *F*. In the presence of its alternative allele *F'* in double or triple dose, a floury endosperm is produced. White endosperm color is produced by a triple dose of a recessive allele *y*, its dominant allele *Y* exhibiting xenia and producing yellow endosperm. The loci of *F* and *Y* assort independently. (a) In crosses between parents of genotype *FF'Yy*, what phenotypic ratio is expected in the progeny seed? (b) Pollen from a plant of genotype *FF'Yy* is crossed onto a plant of genotype *FFyy*. Compare the phenotypic ratios produced by this cross with its reciprocal cross.

7.16. The diploid number of an organism is 12. How many chromosomes would be expected in the following: (a) a monosomic   (b) a trisomic   (c) a tetrasomic   (d) a double trisomic   (e) a nullisomic   (f) a monoploid   (g) a triploid   (h) an autotetraploid?

7.17. Sugary endosperm of corn is regulated by a recessive gene *s* on chromosome IV and starchy endosperm by its dominant alllele *S*. Assuming *n* + 1 pollen grains are nonfunctional, predict the genotypic and phenotypic ratios of endosperm expected in the progeny from the cross of (a) diploid *ss* pollinated by trisomic-IV of genotype *SSs*, (b) diploid *Ss* pollinated by trisomic-IV of genotype *SSs*.

7.18. A dominant gene w^+ produces yellow flowers in a certain plant species and its recessive allele *w* produces white flowers. Plants trisomic for the chromosome bearing the color locus will produce *n* and *n* + 1 functional female gametes, but viable pollen has only the *n* number. Find the phenotypic ratio expected from each of the following crosses:

	Seed Parent		Pollen Parent
(a)	+ + w	×	+ + w
(b)	+ w w	×	+ + w
(c)	+ + w	×	+ w
(d)	+ w w	×	+ w

7.19. Normal women possess two sex chromosomes (XX) and normal men have a single X chromosome plus a Y chromosome that carries male determiners. Rarely, a woman is found with marked abnormalities of primary and secondary sexual characteristics, and she has only one X chromosome (XO). The phenotypic expressions of this monosomic-X state is called Turner syndrome. Likewise, men are occasionally discovered with an XXY constitution that exhibits corresponding abnormalities and is called Klinefelter syndrome. Color blindness is a sex-linked recessive trait. (*a*) A husband and wife both have normal vision, but one of their children is a color-blind Turner girl. Diagram this cross, including the gametes that produced this child. (*b*) In another family the mother is color blind and the father has normal vision. Their child is a Klinefelter with normal vision. What gametes produced this child? (*c*) Suppose the same parents in part (*b*) produced a color-blind Klinefelter. What gametes produced this child? (*d*) The normal diploid number for humans is 46. A trisomic condition for chromosome 21 results in Down syndrome. At least one case of Down-Klinefelter has been recorded. How many chromosomes would this individual be expected to possess?

VARIATION IN ARRANGEMENT OF CHROMOSOME SEGMENTS

7.20. Colorless aleurone of corn kernels is a trait governed by a recessive gene *c* and is in the same linkage group (IX) with another recessive gene *wx* that governs waxy endosperm. In 1931, Creighton and McClintock found a plant with one normal IX chromosome, but its homologue had a knob on one end and a translocated piece from another chromosome on the other end. A dihybrid colored, starchy plant with the heteromorphic IX chromosome shown below was testcrossed to a colorless, waxy plant with normal chromosomes. The results of this experiment provided cytological proof that genetic crossing over involves an actual physical exchange between homologous chromosome segments. Diagram the results of this cross, showing all genotypes and phenotypes.

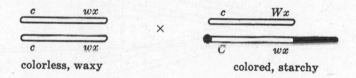

colorless, waxy × colored, starchy

7.21. Nipple-shaped tips on tomato fruit is the phenotypic expression of a recessive gene *nt* on chromosome V. A heterozygous plant (*Nt/nt*) that is also heterozygous for a reciprocal translocation involving chromosomes V and VIII is testcrossed to a plant with normal chromosomes. The progeny were 48 normal fruit, fertile : 19 nipple fruit, fertile : 11 normal fruit, semisterile : 37 nipple fruit, semisterile. What is the genetic position of the locus of gene *Nt* with respect to the point of translocation?

7.22. Given a pericentric inversion heterozygote with one chromosome in normal order (1 2 3 4.5 6 7 8) and the other in the inverted order (1 5.4 3 2 6 7 8), diagram the first anaphase figure after a four-strand double crossover occurs: one crossover involves the regions between 4 and the centromere (.); the other crossover occurs between the centromere and 5.

7.23. A four-strand double crossover occurs in an inversion heterozygote. The normal chromosome order is (.1 2 3 4 5 6 7 8); the inverted chromosome order is (.1 2 7 6 5 4 3 8). One crossover is between 1 and 2 and the other is between 5 and 6. Diagram and label the first anaphase figures.

7.24. A chromosome with segments in the normal order is (.*a b c d e f g h*). An inversion heterozygote has the abnormal order (.*a b f e d c g h*). A three-strand double crossover occurs involving the regions between *a* and *b* and between *d* and *e*. Describe the first and second anaphase figures.

7.25. A species of the fruit fly is differentiated into five races on the basis of differences in the banding patterns of one of its giant chromosomes. Eight regions of the chromosome are designated *a–h*. If each of these races is separated by a single overlapping inversion, devise a scheme to account for the evolution of the five races: (1) *adghfcbe* (2) *fhgdacbe* (3) *fhcadgbe* (4) *fhgbcade* (5) *fadghcbe*

VARIATION IN THE NUMBER OF CHROMOSOME SEGMENTS

7.26. In higher animals, even very small deficiencies, when homozygous, are usually lethal. A recessive gene w in mice results in an abnormal gait called "waltzing." A waltzing male was crossed to several homozygous normal females. Among several hundred offspring, one was found to be a waltzer female. Presumably, a deficiency in the chromosome carrying the w^+ allele caused the waltzing trait to appear as pseudodominant. The pseudodominant waltzer female was then crossed to a homozygous normal male and produced only normal offspring. (*a*) List two possible genotypes for the normal progeny from the above cross. (*b*) Suppose that two males, one of each genotype produced in part (*a*), were backcrossed to their pseudodominant mother and each produced 12 zygotes. Assuming that homozygosity for the deletion is lethal, calculate the expected combined number of waltzer and normal progeny.

VARIATION IN CHROMOSOME MORPHOLOGY

7.27. Vermilion eye color in *Drosophila* is a sex-linked recessive condition; bar eye is a sex-linked dominant condition. An attached-X female with vermilion eyes, also having a Y chromosome ($\hat{X}XY$), is mated to a bar-eyed male. (*a*) Predict the phenotypic ratio that is expected in the F_1 flies. (*b*) How much death loss is anticipated in the F_1 generation? (*Hint:* see Problem 5.19) (*c*) What phenotypic ratio is expected in the F_2?

7.28. Two recessive sex-linked traits in *Drosophila* are garnet eye (*g*) and forked bristle (*f*). The attached-X chromosomes of females heterozygous for these genes are diagrammed below.

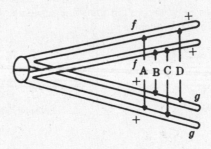

A crossover between two chromatids attached to the same centromere is called a reciprocal exchange; a crossover between two chromatids attached to different centromeres is a non-reciprocal exchange. Approximately 7% of the daughters from these attached-X females were $+ +/f g$, 7% were $f +/+ +$, 7% were $f g/+ g$ and the remainder were $f +/+ g$. (*a*) Which of the single exchanges (A, B, C, or D in the diagram) could produce the daughters (1) $+ +/ f g$ and $f +/+ g$, (2) $f +/+ +$ and $f g/+ g$? (*b*) Are chromatids attached to the same centromere more likely to be involved in an exchange than chromatids attached to different centromeres? (*c*) Does the fact that neither homozygous wild-type nor garnet-forked progeny were found shed any light on the number of chromatids that undergo exchange at any one locus?

7.29. Given the ring homozygote at the left (below), diagram the first anaphase figure when crossovers occur at position (*a*) A and B, (*b*) A and C, (*c*) A and D.

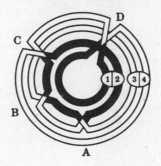

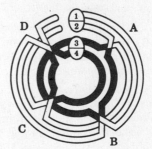

7.30. Given the ring-rod heterozygote at the right (above), diagram the first anaphase figure when crossovers occur at positions (*a*) A and B, (*b*) A and C, (*c*) A and D.

HUMAN CYTOGENETICS

7.31. Meiotic nondisjunction of the sex chromosomes in either parent can produce a child with Klinefelter syndrome (XXY) or Turner syndrome (XO). Color blindness is due to a sex-linked recessive gene. (*a*) If a color-blind woman and man with normal vision produce a color-blind Klinefelter child, in which parent did the nondisjunctional event occur? (*b*) If a heterozygous woman with normal vision and a man with normal vision produce a color-blind Klinefelter child, how can this be explained?

7.32. Explain what type of abnormal sperm unites with a normal egg to produce an XYY offspring. Specifically, how does such an abnormal gamete arise?

7.33. In mosaics of XX and XO cell lines, the phenotype may vary from complete Turner syndrome to a completely normal-appearing female. Likewise, in XO/XY mosaics, the phenotypic variation ranges from complete Turner syndrome to a normal-appearing (but infertile) male. How can these variations be explained?

7.34. Suppose that part of the short arm of one chromosome 5 becomes nonreciprocally attached to the long-arm end of one chromosome 13 in the diploid set. This is considered to be a "balanced translocation" because essentially all of the genetic material is present and the phenotype is normal. One copy of the short arm of chromosome 5 produces *cri du chat* syndrome (mental retardation, facial abnormalities); three copies lead to early postnatal death. If such a translocation individual has children by a chromosomally normal partner, predict the (*a*) chromosomal and (*b*) phenotypic expectations.

7.35. About 2% of patients with Down syndrome have a normal chromosome number of 46. The extra chromosome 21 has been nonreciprocally translocated onto another autosome of the D or G group. These individuals are referred to as translocation mongols, and, because this condition tends to be hereditary, it is also called familial mongolism. (*a*) Suppose that one phenotypically normal parent has 45 chromosomes, one of which is a translocation of the centromere and long arm of a D-group chromosome (either 14 or 15) and the long arm minus the centromere of a G-group chromosome (21). The short arms of each chromosome (presumably carrying no vital genes) are lost in previous cell divisions. If gametes from this translocated parent unite with those from a normal diploid individual, predict the chromosomal and phenotypic expectations in their progeny. (*b*) Assuming that in one parent the translocation is between chromosomes 21 and 22, that the centromere of the translocation is that of chromosome 22 (like centromeres go to opposite poles), and that the other parent is a normal diploid, predict the chromosomal and phenotypic expectations in their children. (*c*) Make the same analysis as in part (*b*), assuming that the centromere of the 21/22 translocation chromosome is that of chromosome 21. (*d*) Assuming that in one parent the translocation involves 21/21 and the other parent is a normal diploid, predict the chromosomal and phenotypic expectations in their children. (*e*) Among the live offspring of parts (*c*) and (*d*), what are the risks of having a Down child?

7.36. The photograph accompanying this problem is at the back of the book. It shows the chromosomes from a human cell. Cut out the chromosomes and construct an idiogram. Do not look at the answer until you have solutions to the following questions. (*a*) Is the specimen from a male or a female? (*b*) What possible kinds of chromosomal abnormalities may be present in this patient?

Review Questions

Vocabulary For each of the following definitions, give the appropriate term and spell it correctly. Terms are single words unless indicated otherwise.

1. A cell or organism containing three sets of chromosomes.
2. A cell or organism produced by doubling the chromosome number of an interspecific hybrid.
3. Any variation in chromosome number that does not involve whole sets of chromosomes.
4. A cell or organism having a genomic formula $2n - 1$.
5. An adjective applicable to a giant chromosome consisting of hundreds of chromatid strands.
6. Exchange of pieces between two nonhomologous chromosomes. (Two words.)
7. Altered phenotypic expression of a gene as a consequence of movement from its normal location. (Two words.)
8. A chromosomal aberration that, with the help of crossing over within the aberration, can lead to "bridge and fragment" formation. (Two words.)
9. Phenotypic expression of a recessive gene as a consequence of loss of a chromosomal segment bearing the corresponding dominant allele.
10. The arrangement of the somatic chromosome complement (karyotype) of a cell in groups of homologous pairs.

Multiple-Choice Questions Choose the one best answer.

1. A treatment often used to induce polyploidy experimentally in plants is (*a*) X-rays (*b*) gibberellic acid (*c*) colchicine (*d*) acridine dyes (*e*) azothioprene
2. A mechanism that can cause a gene to move from one linkage group to another is (*a*) translocation (*b*) inversion (*c*) crossing over (*d*) duplication (*e*) dosage compensation
3. If during synapsis a certain kind of abnormal chromosome is always forced to bulge away from its normal homologue, the abnormality is classified as (*a*) an inversion (*b*) a duplication (*c*) an isochromosome (*d*) a deficiency (*e*) none of the above
4. If four chromosomes synapse into a cross-shaped configuration during meiotic prophase, the organism is heterozygous for a (*a*) pericentric inversion (*b*) deletion (*c*) translocation (*d*) paracentric inversion (*e*) none of the above
5. A segment of chromosome may be protected from recombination by (*a*) an inversion (*b*) a translocation (*c*) balanced lethals (*d*) more than one of the above (*e*) all of the above
6. A person with Klinefelter syndrome is considered a (*a*) monosomic (*b*) triploid (*c*) trisomic (*d*) deletion heterozygote (*e*) none of the above
7. Given a normal chromosome with segments labeled C123456 (C = centromere), a homologue containing an inversion including regions 3–5, and a single two-strand crossover between regions 4 and 5; then the acentric fragment present during first meiotic anaphase is (*a*) 63456 (*b*) 12344321 (*c*) 65521 (*d*) 654321 (*e*) none of the above
8. Pseudodominance may be observed in heterozygotes for (*a*) a deletion (*b*) a duplication (*c*) a paracentric inversion (*d*) a reciprocal translocation (*e*) more than one of the above
9. The most easily recognized characteristic of an inversion heterozygote in plants is (*a*) gigantism (*b*) semisterility (*c*) a cross-shaped chromosome configuration during meiosis (*d*) pseudodominance (*e*) none of the above
10. If the garden pea has 14 chromosomes in its diploid complement, how many double trisomics could theoretically exist? (*a*) 6 (*b*) 9 (*c*) 16 (*d*) 21 (*e*) none of the above. (*Hint*: See formula (2.1).]

Answers to Supplementary Problems

7.11. 42

7.12. (*a*) 21, triploid   (*b*) 28, tetraploid   (*c*) 35, pentaploid   (*d*) 42, hexaploid

7.13. Half of the chromosomes of *hirsutum* have homology with *arboreum*, and the other half with *thurberi*. Doubling the chromosome number of the sterile hybrid (*thurberi* × *arboreum*) could produce an amphidiploid with the cytological characteristics of *hirsutum*.

7.14.

Genotype	(*a*)	(*b*)	(*c*)	(*d*)
Quadruplex	All *A*	All *A*	All *A*	All *A*
Triplex	All *A*	All *A*	783 *A* : 1 *a*	27 *A* : 1 *a*
Duplex	35 *A* : 1 *a*	5 *A* : 1 *a*	20.8 *A* : 1 *a*	3.7 *A* : 1 *a*
Simplex	3 *A* : 1 *a*	1 *A* : 1 *a*	2.48 *A* : 1 *a*	0.87 *A* : 1 *a*
Nulliplex	All *a*	All *a*	All *a*	All *a*

7.15. (*a*) 3/8 flinty, yellow : 1/8 flinty, white : 3/8 floury, yellow : 1/8 floury, white

(*b*)

	Original Cross	Reciprocal Cross
Flinty, white	1/2	1/4
Flinty, yellow	1/2	1/4
Floury, white		1/4
Floury, yellow		1/4

7.16. (*a*) 11   (*b*) 13   (*c*) 14   (*d*) 14   (*e*) 10   (*f*) 6   (*g*) 18   (*h*) 24

7.17. (*a*) 2 *Sss* (starchy) : 1 *sss* (sugary)   (*b*) 1/3 *SSS* : 1/6 *SSs* : 1/3 *Sss* : 1/6 *sss* : 5/6 starchy : 1/6 sugary

7.18. (*a*) 17 yellow : 1 white   (*b*) 5 yellow : 1 white   (*c*) 11 yellow : 1 white   (*d*) 3 yellow : 1 white

7.19. (*a*) P: $X^C X^c$ × $X^C Y$; gametes : X^c and O ; F_1: $X^C O$ (*b*) X^c and $X^C Y$ (*c*) $X^c X^c$ and Y (*d*) 48

7.20.

	Noncrossovers		Crossovers	
	c ———— *Wx*	*C* ●———— *wx*	*c* ———— *wx*	*C* ●———— *Wx*
c ———— *wx*	*c* ———— *Wx*	*C* ●———— *wx*	*c* ———— *wx*	*C* ●———— *Wx*
	c ———— *wx*	*c* ———— *wx*	*c* ———— *wx*	*c* ———— *wx*
	colorless, starchy	colored, waxy	colorless, waxy	colored, starchy

7.21. 26.1 map units from the point of translocation.

7.22.

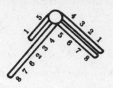

7.23.

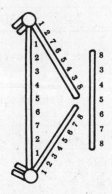

7.24. First anaphase: a diad, a loop chromatid, and an acentric fragment; second anaphase: the diad splits into two monads and the loop forms a bridge. The acentric fragment formed during meiosis I would not be expected to be present at meiosis II.

7.25.

7.26. (*a*) + /*w* and + /(−) (heterozygous deficiency)   (*b*) 9 waltzers : 12 normals

7.27. (*a*) All daughters have vermilion eyes (X̂XY); all sons have bar eyes (XY).   (*b*) 50% death loss; nullo-X is lethal (YY); superfemales (X̂XX) usually die.   (*c*) Same as part (*a*)

7.28. (*a*) (1) D or C   (2) B or A   (*b*) No. Reciprocal vs. nonreciprocal exchanges are occurring in a 1 : 1 ratio, indicating that chromatids attached to the same centromere are involved in an exchange with the same frequency as chromatids attached to different centromeres. Daughters with genotype *f* +/+ *g* resulting from single crossovers of type C cannot be distinguished from nonexchange chromatids. (*c*) Two exchanges in the garnet-forked region involving all four strands, as well as one nonreciprocal exchange between *f* and the centromere, are required to give homozygous wild-type and garnet-forked daughters. Their absence is support for the assumption that only two of the four chromatids undergo exchange at any one locus.

7.29.

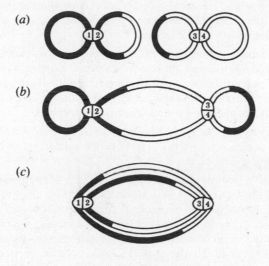

7.30.

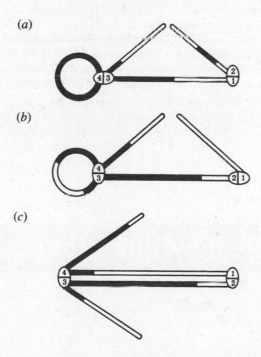

7.31. (a) Either nondisjunction of the two X chromosomes occurred in the mother in the first meiotic division or nondisjunction of the two sister chromatids occurred in the second meiotic division. (b) Nondisjunction during the second meiotic division of the sister chromatids of the X chromosome bearing the recessive color-blind gene would produce an egg with two X chromosomes bearing only the color-blind alleles. Alternatively, if crossing over occurs between the centromere and the color-blind locus and is followed by nondisjunction of the X chromosomes at the first meiotic division, one of the four meiotic products would be expected to contain two recessive color-blind alleles.

7.32. A sperm bearing two Y chromosomes is produced by nondisjunction of the Y sister chromatids during the second meiotic division. The other product of that same nondisjunctional second meiotic division would contain no sex chromosome; when united with a normal egg, an XO Turner female would be expected.

7.33. If mitotic nondisjunction occurs early in embryogenesis, mosaicism is likely to be widespread throughout the body. If it occurs late in embryogenesis, mosaicism may be limited to only one organ or to one patch of tissue. If chromosomally abnormal cells are extensive in reproductive tissue or in endocrine tissues responsible for gamete and/or sex hormone production, the effects on sterility are likely to be more intensively expressed.

7.34. (a) 1 normal karyotype : 1 balanced translocation : 1 deficient for short arm of chromosome 5 : 1 with three copies of the short arm of chromosome 5   (b) 2 normal : 1 *cri du chat* syndrome : 1 early childhood death

7.35. (a) 1 chromosomally and phenotypically normal ($2n = 46$) : 1 translocation carrier, phenotypically normal ($2n - 1 = 45$) : 1 monosomic ($2n = 45$) for a G-group chromosome (incompatible with life; aborted early in pregnancy) : 1 translocation Down trisomic for the long arm of chromosome 21 ($2n = 46$). Among the live-born offspring we expect 1/3 chromosomally normal : 1/3 translocation carriers : 1/3 Down syndrome. (b) 1 chromosomally and phenotypically normal ($2n = 46$) : 1 that is a 21/22 translocation carrier, phenotypically normal ($2n - 1 = 45$) : 1 monosomic for chromosome 21 and aborted early in pregnancy ($2n - 1 = 45$) : 1 with a 21/22 translocation chromosome who is essentially trisomic for the long arms of 21 ($2n = 46$) and phenotypically Down syndrome. (c) 1 chromosomally and phenotypically normal ($2n = 46$) : 1 that is a 21/22 translocation carrier, phenotypically normal ($2n - 1 = 45$) : 1 monosomic for 22 and aborted early in pregnancy ($2n - 1 = 45$) : 1 with a 21/22 translocation chromosome who is essentially trisomic for the long arms of 22 ($2n = 46$), phenotype unspecified. (d) 1 monosomic ($2n - 1 = 45$) for 21 and aborted early in pregnancy : 1 with a 21/21 translocation chromosome who is essentially trisomic for the long arms of 21 and phenotypically Down. (e) 1 in 3 for part (c); 100% for part (d).

7.36. (a) Male   (b) The idiogram contains an extra G-group chromosome ($2n + 1 = 47$). It cannot be determined whether the extra chromosome is 21, 22, or Y. If the patient has the physical characteristics of Down syndrome, the extra chromosome is 21.

Answers to Review Questions

Vocabulary

1. triploid
2. allotetraploid (amphidiploid)
3. aneuploidy
4. monosomic
5. polytene
6. reciprocal translocation
7. position effect
8. paracentric inversion
9. pseudodominance
10. idiogram

Multiple-Choice Questions

1. *c*   2. *a*   3. *b*   4. *c*   5. *e*   6. *c*   7. *a*   8. *a*   9. *b*   10. *d*

CHAPTER 8

Quantitative Genetics

Qualitative vs. Quantitative Traits

The classical Mendelian traits encountered in the previous chapters have been qualitative in nature; i.e., traits that are easily classified into distinct phenotypic categories. These discrete phenotypes are under the genetic control of only one or a very few genes with little or no environmental modification to obscure the gene effects. In contrast to this, the variability exhibited by many traits fails to fit into separate phenotypic classes (discontinuous variability), but instead forms a spectrum of phenotypes that blend imperceptively from one type to another (continuous variability). Economically important traits such as body weight gains, mature plant heights, egg or milk production records, and yield of grain per acre are **quantitative** or **complex traits** (historically referred to as **metric traits**) with continuous variability. The basic differences between qualitative and quantitative traits involve the number of genes contributing to the phenotypic variability and the degree to which the phenotype can be modified by environmental factors. Quantitative traits may be governed by many genes (perhaps 10–100 or more), each contributing such a small amount to the phenotype that their individual effects cannot be detected by Mendelian methods. All genes act in concert with other genes. Thus, more than one gene may contribute to a given trait. For this reason, quantitative traits are also referred to as **polygenic traits**. Genes responsible for phenotypes associated with quantitative traits are called **quantitative trait loci** or **QTLs**.

Furthermore, each gene usually has effects on more than one trait (pleiotropy; Chapter 4). The idea that each character is controlled by a single gene (the one-gene-one-trait hypothesis) has often been falsely attributed to Mendel. But even he recognized that a single factor (or gene) might have manifold effects on more than one trait. For example, he observed that purple flowers are correlated with brown seeds and a dark spot on the axils of leaves; similarly, white flowers are correlated with light-colored seeds and no axillary spots on the leaves. In *Drosophila*, many loci (e.g., genes named *dumpy*, *cut*, *vestigial*, *apterous*) are known to contribute to a complex character such as wing development. Each of these genes also has pleiotropic effects on other traits. For example, the gene for vestigial wings also effects the halteres (balancers), bristles, egg production in females, and longevity.

Protein-coding genes encode products such as enzymes that participate in multistep biochemical pathways (Chapter 4) or proteins that regulate the activity of one or more other genes in metabolic, regulatory, or developmental pathways. Because of the complex interactions within these pathways, a gene product acting at any one step might have phenotypic effects on two or more characters. For a given gene, some of its pleiotropic effects may be relatively strong for certain traits, whereas its effects on other traits may be so weak that they are difficult or impossible to identify by Mendelian techniques. It is the totality of these pleiotropic effects of numerous loci that constitutes the genetic base of a quantitative character. In addition to this genetic component, the phenotypic variability of a quantitative trait in a population usually has an environmental component. It is the task of the geneticist to determine the magnitude of the genetic and environmental components of the total phenotypic variability of each quantitative trait in a population. In order to accomplish this task, statistics is used. Only some of the more easily understood rudiments of this branch of genetics will be presented in this chapter. Table 8-1 summarizes some of the major differences between quantitative and qualitative genetics.

Table 8-1.   Differences Between Qualitative and Quantitative Genetics

Qualitative Genetics	Quantitative Genetics
1. Characters of kind	1. Characters of degree
2. Discontinuous variation; discrete phenotypic classes	2. Continuous variation; phenotypic measurements form a spectrum
3. Single-gene effects discernible	3. Polygenic control; effects of single genes too slight to be detected
4. Concerned with *individual matings* and their progeny	4. Concerned with a population of organisms consisting of all possible kinds of matings
5. Analyzed by making *counts* and *ratios*	5. Statistical analyses give estimates of population parameters such as the mean and standard deviation

Polygenic Traits

In the early days of Mendelian genetics it was thought that there was a fundamental difference in the essence of qualitative and quantitative traits. One of the classical examples that helped to bridge the gap between these two kinds of traits is the multiple-gene model developed around 1910 by the Swedish geneticist Nilsson-Ehle to explain kernel color in wheat. When he crossed a certain red strain to a white strain he observed that the F_1 was all light red and that approximately 1/16 of the F_2 was as extreme as the parents; i.e., 1/16 was white and 1/16 was red. He interpreted these results in terms of two genes, each with a pair of alleles exhibiting cumulative effects. In the following explanation, the use of capital and lowercase letters does not imply dominant and recessive allelic interactions, but rather **additive gene action** (i.e., a polygenic trait with additive genes) in which each R gene makes an equal contribution to redness and each r allele contributes nothing to color of the wheat kernel. Note that the phenotype of the F_1 is intermediate between the two parental types and that the average phenotype of the F_2 is the same as that of the F_1 but is a much more variable population; i.e., the F_2 contains many more phenotypes (and genotypes) than in the F_1. The student should recognize the F_2 distribution as an expansion of the binomial $(a + b)^4$, where $a =$ "active alleles" R_1 or R_2, b = "inactive alleles" r_1 or r_2, and $a = b = 1/2$ in the F_1.

P: $R_1R_1R_2R_2$ × $r_1r_1r_2r_2$
             red   white

F_1: $R_1r_1R_2r_2$
                   light red

F_2:

$$1/16\ R_1R_1R_2R_2 : 4/16 \begin{cases} R_1R_1R_2r_2 \\ R_1r_1R_2R_2 \end{cases} : 6/16 \begin{cases} R_1R_1r_2r_2 \\ R_1r_1R_2r_2 \\ r_1r_1R_2R_2 \end{cases} : 4/16 \begin{cases} R_1r_1r_2r_2 \\ r_1r_1R_2r_2 \end{cases} : 1/16\ r_1r_1r_2r_2$$

      red   medium   light red   very   white
             red   light red

In this model situation, each of the "active" alleles R_1 or R_2 adds an equal amount of red to the phenotype, so that the genotype of white contains neither of these alleles and a fully red genotype contains only R_1 and R_2 alleles. These results are plotted as histograms in Fig. 8-1.

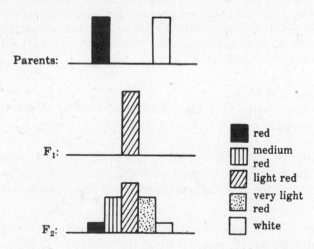

Fig. 8-1. Wheat color as an example of a polygenic trait.

Certain other strains of wheat with dark-red kernels when crossed to whites exhibit an F_1 phenotype intermediate between the two parental types, but only 1/64 of the F_2 is white. In this case, the F_1 is probably segregating for three pairs of genes and only the genotype $r_1r_1r_2r_2r_3r_3$ produces white. Of course, there would be more shades of red exhibited in the F_2 here than in the previous case where only two genes are segregating. Even if the environment does not modify these color phenotypes (which it probably does to some extent), the ability of the eye to measure subtle differences in shading would probably cause difficulties with this many phenotypes and colour discernment would become impossible if four or five genes were each contributing to kernel color.

Thus, these multiple-gene models, which are adequate to explain certain examples wherein discontinuous variation is still evident, may (by conceptual extension to include more genes plus environmental modifications) be useful in understanding the origin of continuous variation characterizing truly quantitative traits.

A rough estimate of the number of gene loci contributing to a polygenic trait can be obtained by determining the fraction of the F_2 (resulting from selfing the F_1 hybrid between two pure varieties) that is as extreme in its phenotype as that of one of the pure parental strains.

Number of gene loci	1	2	3	. . .	n
Fraction of F_2 as extreme as one parent	1/4	1/16	1/64	. . . .	$(1/4)^n$

SOLVED PROBLEM 8.1

Two homozygous varieties of *Nicotiana longiflora* have mean corolla lengths of 40.5 and 93.3 mm. The average of the F_1 hybrids from these two varieties was of intermediate length. Among 444 F_2 plants, none was found to have flowers either as long as or as short as the average of the parental varieties. Estimate the minimal number of pairs of alleles segregating from the F_1.

Solution: If four pairs of alleles were segregating from the F_1, we expect $(1/4)^4 = 1/256$ of the F_2 to be as extreme as one or the other parental average. Likewise, if five pairs of alleles were segregating, we expect $(1/4)^5 = 1/1024$ of the F_2 to be as extreme as one parent or the other. Since none of the 444 F_2 plants had flowers this extreme, more than four loci (minimum of five loci) are probably segregating from the F_1.

SOLVED PROBLEM 8.2

The mean internode length of the Abed Binder variety of barley was found to be 3.20 mm. The mean length in the Asplund variety was 2.10 mm. Crossing these two varieties produced an F_1 and F_2 with average internode lengths of 2.65 mm. About 6% of the F_2 had an internode length of 3.2 mm, and another 6% had a length of 2.1 mm. Determine the most probable number of gene pairs involved in internode length and the approximate contribution each gene makes to the phenotype.

Solution: With one pair of genes we expect about 1/4 or 25% of the F_2 to be as extreme as one of the parents. With two pairs of genes we expect approximately 1/16 or 6.25% as extreme as one parent. Thus, we may postulate two pairs of genes. Let A and B represent growth factors and a and b represent null genes

P: $AA\ BB$ × $aa\ bb$
Abed Binder   Asplund
4 growth genes = 3.2 mm   0 growth genes = 2.1 mm

F_1: $Aa\ Bb$
2 growth genes = 2.65 mm

The difference $2.65 - 2.10 = 0.55$ mm is the result of two growth genes. Therefore, each growth gene contributes 0.275 mm to the phenotype.

F_2:

No. of Growth Genes	Mean Internode Length (mm)	Frequency	Genotypes
4	3.200	1/16	$AABB$
3	2.925	1/4	$AaBB$, $AABb$
2	2.650	3/8	$AAbb$, $AaBb$, $aaBB$
1	2.375	1/4	$aaBb$, $Aabb$
0	2.100	1/16	$aabb$ (physiological minimum due to residual genotype)

SOLVED PROBLEM 8.3

A large breed of chicken, the Golden Hamburg, was crossed to small Sebright Bantams. The F_1 was intermediate in size. The mean size of the F_2 was about the same as that of the F_1, but the variability of the F_2 was so great that a few individuals were found to exceed the size of either parental type (**transgressive variation**). If all of the alleles contributing to size act with equal and cumulative effects, and if the parents are considered to be homozygous, how can these results be explained?

Solution: Let capital letters stand for growth genes (active alleles) and small letters stand for alleles that do not contribute to growth (null alleles). For simplicity, we will consider only four loci.

P: $aa\ BB\ CC\ DD$ × $AA\ bb\ cc\ dd$
           large Golden Hamburg   small Sebright Bantam
               6 active alleles   2 active alleles

F_1: $Aa\ Bb\ Cc\ Dd$
                        intermediate-sized hybrid
                            4 active alleles

F_2:   Some genotypes could segregate out in the F_2 with phenotypic values that exceed that of the parents. For example:

$AA\ BB\ CC\ DD$ 8 active alleles $\}$ larger than
$Aa\ BB\ CC\ DD$ 7 active alleles $\}$ Golden Hamburg

$Aa\ bb\ cc\ dd$ 1 active allele $\}$ smaller than
$aa\ bb\ cc\ dd$ 0 active alleles $\}$ Sebright Bantams
                     (physiological
                     minimum)

The Normal Distribution

The study of a quantitative trait in a large population usually reveals that very few individuals possess the extreme phenotypes and that progressively more individuals are found nearer the average value for that population. This type of symmetrical distribution is characteristically bell-shaped, as shown in Fig. 8-2, and is called a **normal distribution**. It is approximated by the **binomial distribution** $(p + q)^n$ when the power of the binomial is very large and p and q are both $1/n$ or greater; p and q represent the probabilities of alternative independent events, $p + q = 1$.

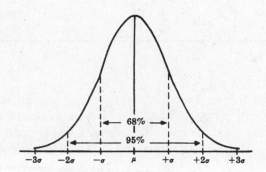

Fig. 8-2. A normal distribution.

1 AVERAGE MEASUREMENTS

The average phenotypic value for a normally distributed trait is expressed as the **arithmetic mean** $\overline{X}$ (read "X bar"). The arithmetic mean is the sum of the individual measurements ($\sum X$) divided by the number of individuals measured (N). The Greek letter "sigma" ($\sum$) directs the statistician to sum what follows.

$$\overline{X} = \frac{\sum_{i=1}^{N} X_i}{N} = \frac{X_1 + X_2 + X_3 + \cdots + X_N}{N} \tag{8.1}$$

It is usually not feasible to measure every individual in a population; therefore, measurements are usually made on a sample from that population in order to estimate the population value (parameter). If the sample is truly representative of the larger population of which it is a part, then the arithmetic mean will be an accurate estimate of the mean of the entire population (μ). Note that letters from the English alphabet are used to represent **statistics**, i.e., measurements derived from a sample, whereas Greek letters are used to represent **parameters**, i.e., attributes of the population from which the sample was drawn. Parameters are seldom known and must be estimated from results gained by sampling. Obviously, the larger the sample size, the more accurately the statistic estimates the parameter.

2   MEASUREMENTS OF VARIABILITY

(a) Standard Deviation. Consider the three normally distributed populations shown in Fig. 8-3. Populations A and C have the same mean, but C is much more variable than A. Populations A and B have different means, but otherwise appear to have the same shape (dispersion). Therefore, in order to adequately define a normal distribution, we must know not only its mean but also how much variability exists. One of the most useful measures of variability in a population for genetic purposes is the **standard deviation**, symbolized by the lowercase Greek letter "sigma" (σ). A sample drawn from this population at random will have a sample standard deviation (s). To calculate s, the sample mean is subtracted $(X_i - \overline{X})$ from each individual measurement (X_i) and the deviation is squared $(X_i - \overline{X})^2$, summed over all individuals in the sample

$$\left[\sum_{i=1}^{n} (X_i - \overline{X})^2 \right]$$

and divided by $n - 1$, where n is the sample size. The calculation is completed by taking the square root of this value.

$$s = \sqrt{\frac{\sum_{i=1}^{n} (X_i - \overline{X})^2}{n - 1}} \tag{8.2}$$

To calculate σ, we substitute the total population size (N) for n in the above formula. For samples less than about 30, the appropriate correction factor for the denominator should be $n - 1$; for sample sizes greater than this, it makes little difference in the value of s whether n or $n - 1$ is used in the denominator.

All other things being equal, the larger the sample size, the more accurately the statistic s should estimate the parameter. Calculators can be used to accumulate

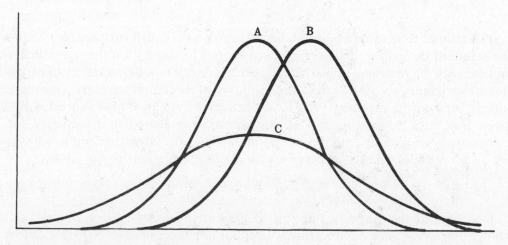

Fig. 8-3.   Comparison of three populations (A, B, C) with respect to means and variances (see text).

squared numbers. This usually makes it easier to calculate s by the equivalent formula

$$s = \sqrt{\frac{\sum X^2 - [(\sum X)^2/n]}{n-1}} \tag{8.3}$$

It is the property of every normal distribution that approximately 2/3 of the measurements (68%) will lie within plus or minus one standard deviation from the mean ($\mu \pm \sigma$). Approximately 19/20 of the measurements (95%) will lie within two standard deviations of the mean ($\mu \pm 2\sigma$). More than 99% of the measurements will be found within plus or minus three standard deviations of the mean ($\mu \pm 3\sigma$).

> **EXAMPLE 8.1**
>
> The mean height of a sample from a plant population is 56 in; the sample standard deviation is 6 in. This indicates that approximately 2/3 of the sample will be found between the values $56 + 6 = 50$ in to 62 in. Approximately 25% of all plants in this sample will measure smaller than $56 - (2 \times 6) = 56 - 12 = 44$ in and 25% will measure larger than $56 + (2 \times 6) = 68$ in.

The standard deviation can be plotted on a normal distribution by locating the point of inflection of the curve (point of maximum slope). A perpendicular constructed from the baseline that intersects the curve at this point is one standard deviation from the mean (Fig. 8-2).

(b) Coefficient of Variation. Traits with relatively large average metric values generally are expected to have correspondingly larger standard deviations than traits with relatively small average metric values. Furthermore, since different traits may be measured in different units, the **coefficients of variation** are useful for comparing their relative variabilities. Dividing the standard deviation by the mean renders the coefficient of variation independent of the units of measurement.

$$\text{Coefficient of variation} = \sigma/\mu \text{ for a population}$$
$$= s/\overline{X} \text{ for a sample} \tag{8.4}$$

(c) Variance. The square of the standard deviation is called **variance** (σ^2). Unlike the standard deviation, however, variance cannot be plotted on the normal curve and can only be represented mathematically. Variance is widely used as an expression of variability because of the additive nature of its components. By a technique called "analysis of variance," the total phenotypic variance (σ_P^2) expressed by a given trait in a population can be statistically fragmented or partitioned into components of genetic variance (σ_G^2), nongenetic (or environmental) variance (σ_E^2), and variance due to genotype-environment interactions (σ_{GE}^2). Thus,

$$\sigma_P^2 = \sigma_G^2 + \sigma_E^2 + \sigma_{GE}^2 \tag{8.5}$$

It is beyond the scope of this text to present the analysis of variance, but a knowledge of variance components is essential to a discussion of breeding theory. Both the genetic variance and environmental variance can be further partitioned by this technique, so that the relative contributions of a number of factors influencing a quantitative trait can be ascertained. In order to simplify discussion, we shall ignore the genotype-environment interaction component.

EXAMPLE 8.2

An analysis of variance performed on the birth weights of humans produced the following results:

Variance Component	Percent of Total Phenotypic Variance
Offspring genotype	16
Sex	2
Maternal genotype	20
Maternal environment	24
Chronological order of child	7
Maternal age	1
Unaccountable variations (error)	30
Total	**100**

(d) Variance Method of Estimating the Number of Genes. A population such as a line, a breed, a variety, a strain, a subspecies, etc., is composed of individuals that are more nearly alike in their genetic composition than those in the species as a whole. Phenotypic variability will usually be expressed, even in a group of organisms that are genetically identical. All such variability within pure lines is obviously environmental in origin. Crosses between two pure lines produce a genetically uniform hybrid F_1. Phenotypic variability in the F_1 is likewise nongenetic in origin. In the formation of the F_2 generation, gene combinations are reshuffled and dealt out in new combinations to the F_2 individuals. It is a common observation that the F_2 generation is much more variable than the F_1 from which it was derived (Fig. 8-4).

In a normally distributed trait, the means of the F_1 and F_2 populations tend to be intermediate between the means of the two parental lines. If there is no change in the environment from one generation to the next, then the environmental variation of the F_2 should be approximately the same as that of the F_1. An increase in phenotypic variance of the F_2 over that of the F_1 may then be attributed to genetic causes. Thus, the genotypic variance of the F_2 (σ^2_{GF2}) is equal to the phenotypic variance of the F_2 (σ^2_{PF2}) minus the phenotypic variance of the F_1 (σ^2_{PF1}):

$$\sigma^2_{GF2} = \sigma^2_{PF2} - \sigma^2_{PF1}$$

The genetic variance of the F_2 is expressed by the formula $\sigma^2_{GF2} = (a^2 N)/2$, where a is the contribution of each active allele and N is the number of pairs of genes involved in the quantitative trait. An estimate of a is obtained from the formula

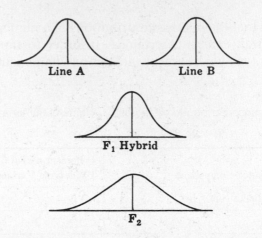

Fig. 8-4. Crosses between inbred lines A and B produce an intermediate F_1 hybrid population, which in turn produces a more variable F_2 population.

$a = D/2N$, where D is the numerical difference between the two parental means. Making substitutions and solving for N,

$$\sigma_{PF2}^2 - \sigma_{PF1}^2 = \sigma_{GF2}^2 = a^2 N/2 = D^2/8N$$

from which

$$N = \frac{D^2}{8(\sigma_{PF2}^2 - \sigma_{PF1}^2)} \tag{8.6}$$

This formula is an obvious oversimplification since it assumes that all genes are contributing cumulatively the same amount to the phenotype, and that there is no dominance, no linkage, and no interaction. Much more sophisticated formulas have been developed to take such factors into consideration, but these are beyond the scope of a first genetics course.

SOLVED PROBLEM 8.4
A representative sample of lamb weaning weights is shown below. Determine the weight limits within which 95% of all lambs from this population are expected to be found at weaning time.

81	81	83	101	86
65	68	77	66	92
94	85	105	60	90
94	90	81	63	58

Solution: The standard deviation is calculated as follows:

$$\overline{X} = \frac{\sum X}{N} = \frac{1620}{20} = 81.0$$

X	X − $\overline{X}$	$(X − \overline{X})^2$
81	0	0
65	−16	256
94	+13	169
94	+13	169
81	0	0
68	−13	169
85	+4	16
90	+9	81
83	+2	4
77	−4	16
105	+24	576
81	0	0
101	+20	400
66	−15	225
60	−21	441
63	−18	324
86	+5	25
92	+11	121
90	+9	81
58	−23	529
$\sum X = 1620$	$\sum(X − \overline{X}) = 0$	$\sum(X − \overline{X})^2 = 3602$

$$s = \sqrt{\frac{\sum(X − \overline{X})^2}{N − 1}} = \sqrt{\frac{3602}{19}} = 13.77$$

95% of all weaning weights are expected to lie within ± 2s of the mean. Thus, $\overline{X} \pm 2s$ = 81.0 ± 2(13.77) = 81.0 ± 27.54. The upper limit is 108.54 and the lower limit is 53.46.

SOLVED PROBLEM 8.5

The Flemish breed of rabbits has an average body weight of 3600 grams (g). The Himalayan breed has a mean of 1875 g. Matings between these two breeds produce an intermediate F_1 with a standard deviation of 162 g. The variability of the F_2 is greater, as indicated by a standard deviation of 230 g. (*a*) Estimate the number of pairs of factors contributing to mature body weight in rabbits. (*b*) Estimate the average quantitative contribution of each active allele.

Solution:

(a) From formula (8.6),

$$N = \frac{D^2}{8(\sigma^2_{PF2} - \sigma^2_{PF1})} = \frac{(3600 - 1875)^2}{8(230^2 - 162^2)} = 13.95 \text{ or approx. 14 pairs}$$

(b) The difference $3600 - 1875 = 1725$ g is attributed to 14 *pairs* of factors or 28 active alleles. The average contribution of each active allele is

$$1725/28 = 61.61 \text{ g.}$$

SOLVED PROBLEM 8.6

In a normally distributed population that has a phenotypic mean of 55 units, a total genetic variance for the trait of 35 units², and an environmental variance of 14 units², between what two phenotypic values will approximately 68% of the population be found?

Solution:

$$\sigma^2_P = \sigma^2_G + \sigma^2_E = 35 + 14 = 49, \qquad \sigma_P = 7$$

68% of a normally distributed population is expected to be found within the limits $\mu \pm \sigma = 55 \pm 7$, or between 48 units and 62 units.

Types of Gene Action

Alleles may interact with one another in a number of ways to produce variability in their phenotypic expression. The following models may help us understand various modes of gene action.

1. With dominance lacking, i.e., **additive genes**, or a heterozygote that is intermediate between the two homozygotes, each A^1 allele is assumed to contribute nothing to the phenotype (null allele), whereas each A^2 allele contributes one unit to the phenotype (active allele).

   Scale of phenotypic value:

   |0| |1| |2|

   Genotype:

   A^1A^1 A^1A^2 A^2A^2

2. With **partial or incomplete dominance**, the heterozygote contributes almost as much as the A^2A^2 homozygote.

   Scale of phenotypic value:

   |0| |1| |2|

   Genotype:

   A^1A^1 A^1A^2 A^2A^2

3. In **complete dominance**, identical phenotypes are produced by the heterozygote and A^2A^2 homozygote.

   Scale of phenotypic value:

   |0| |2|

   Genotype:

   A^1A^1 A^1A^2
                                   A^2A^2

4. In **overdominance**, the phenotype of the heterozygote is beyond the range of either homozygote.

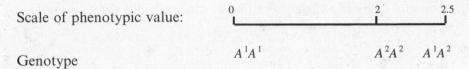

Scale of phenotypic value:

Genotype

If allelic interaction is completely additive, a linear phenotypic effect is produced. In Fig. 8-5, a constant increment (i) is added to the phenotype for each A^2 allele in the genotype.

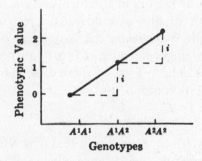

Fig. 8-5.   Additive gene action.

Even if complete dominance is operative, an underlying component of additivity (linearity) is still present (solid line in Fig. 8-6). The deviations from the additive scheme (dotted lines) due to many such genes with partial or complete dominance can be statistically estimated from appropriately designed experiments. The genetic contributions from such effects appear in the dominance component of variance (σ_D^2).

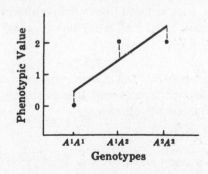

Fig. 8-6.   Dominant gene action.

In a much more complicated way, deviations from an underlying additive scheme could be shown to exist for the interactions between genes at different loci (**epistatic relationships**). The contribution to the total genetic variance (σ_G^2) made by these genetic elements can be partitioned into a component called the epistatic or interaction variance (σ_I^2).

The sum of the additive gene effects produced by additive genes and by the additive contribution of genes with dominance or epistatic effects appears in the additive component of genetic variance (σ_A^2).

Thus, the total genetic variance can be partitioned into three fractions:

$$\sigma_G^2 = \sigma_A^2 + \sigma_D^2 + \sigma_I^2 \tag{8.7}$$

Heritability

One of the most important factors in the formulation of effective breeding plans for improving the genetic quality of crops and livestock is a knowledge of the relative contribution made by genes to the variability of a trait under consideration. The variability of phenotypic values for a quantitative trait (as expressed in the phenotypic variance) can, at least in theory, be partitioned into genetic and nongenetic (environmental) components.

$$\sigma_P^2 = \sigma_G^2 + \sigma_E^2$$

Heritability (symbolized H^2 or H in some texts) is the proportion of the total phenotypic variance due to all types of gene effects; H represents the corresponding ratio of standard deviations and is not used in this book.

$$H^2 = \frac{\sigma_G^2}{\sigma_P^2} \tag{8.8}$$

The heritability of a given trait may be any number from 0 to 1.

EXAMPLE 8.3
If all of the phenotypic variability of a trait is genetic in nature (as is true for most classical Mendelian traits, such as blood types), then environmental effects are absent and heritability equals 1; i.e., if $\sigma_G^2 = \sigma_P^2$, then $H^2 = 1$.

EXAMPLE 8.4
If all of the phenotypic variability of a trait is environmental in nature (as is true for any trait within a genetically homozygous line), then heritability of the trait is zero; i.e., if $\sigma_E^2 = \sigma_P^2$, then $\sigma_G^2 = 0$ and $H^2 = 0/\sigma_P^2 = 0$.

EXAMPLE 8.5
If half of the phenotypic variability is due to gene effects, then heritability is 50%, i.e., if $\sigma_G^2 = (1/2)\sigma_P^2$, then $2\sigma_G^2 = \sigma_P^2$ and $H^2 = 1/2 = 50\%$. *Note*: We cannot say that because heritability of a quantitative trait (such as seed production in a given plant population) is 0.5 that 50% of any one individual's seed productivity is due to gene effects, as explained in the following scenario. One plant may have a superior genotype for seed production, but was grown in a poor environment (e.g., shade, disease, insect damage, etc.) and thus produced fewer seeds than the average number for the population. Another plant in that same population, having only an average genotype, might have grown in an enriched environment (e.g., accidentally fertilized by animal dung) and was thus able to produce a seed number near its

genetic potential that actually exceeded the number of seeds produced by the aforementioned plant. Likewise, it is illogical to say that a population with a high heritability for a given trait is genetically superior to another population (in the same or different species) with a lower heritability for that same trait.

EXAMPLE 8.6

If the environmental component of variance is three times as large as the genetic component, heritability is 25%, i.e., if $\sigma_E^2 = 3\sigma_G^2$, then

$$H^2 = \frac{\sigma_G^2}{\sigma_G^2 + \sigma_E^2} = \frac{\sigma_G^2}{\sigma_G^2 + 3\sigma_G^2} = \frac{1}{4} = 25\%$$

The parameter of heritability involves all types of gene action and thus forms a broad estimate of heritability, or referred to as **broad sense heritability**, or **broad heritability** (H^2) or the degree of genetic determination. In the case of complete dominance, when a gamete bearing the active dominant allele A^2 unites with a gamete bearing the null allele A^1, the resulting phenotype might be two units. When two A^2 gametes unite, the phenotypic result would still be two units. On the other hand, if genes lacking dominance (additive genes) are involved, then the A^2 gamete will add one unit to the phenotype of the resulting zygote regardless of the allelic contribution of the gamete with which it unites, Thus, only the additive genetic component of variance (i.e., the variance of breeding values) has the quality of predictability necessary in the formulation of breeding plans. Nonallelic (epistatic) interactions may exist, but they usually make a minor contribution to the total genetic variance, their effects are difficult to access, and they will be ignored in the simple models of this book. If they do occur, all nonadditive genetic variance will be included in the dominance variance. The concept of additive variance does not necessarily involve the assumption of additive gene action. Additive variance can be produced from genes with additive gene action and/or any degree of dominance and/or epistasis. Heritability expressed as the ratio of the additive genetic variance to the phenotypic variance is termed **narrow sense heritability**, or **narrow heritability**:

$$h^2 = \frac{\sigma_A^2}{\sigma_P^2} \tag{8.9}$$

Unless otherwise specified in the problems of this book, heritability in the narrow sense is to be employed. It must be emphasized that the heritability of a trait applies only to a given population living in a particular environment. A genetically different population (perhaps a different variety, breed, race, or subspecies of the same species) living in an identical environment is likely to have a different heritability for the same trait. Likewise, the same population is likely to exhibit different heritabilities for the same trait when measured in different environments because a given genotype does not always respond to different environments in the same way. There is no one genotype that is adaptively superior in all possible environments. That is why natural selection tends to create genetically different populations within a species, each population being potentially adapted to local conditions rather than generally adapted to all environments in which the species is found.

Several methods can be used to estimate heritabilities of quantitative traits.

1   VARIANCE COMPONENTS

Consider the simple, single-locus model (below) with alleles b^1 and b^2.

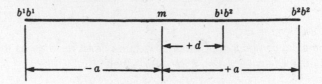

The midparent value $m = (1/2)(b^1b^1 + b^2b^2)$. If the heterozygote does not have a phenotypic value equal to m, some degree of dominance (d) exists. If no dominance exists, then the alleles are completely additive. However, quantitative traits are governed by many loci and it might be possible that genotype b^1b^2 is dominant in a positive direction whereas genotype c^1c^2 is dominant in a negative direction, so that they cancel each other, giving the illusion of additivity. Dominance of all types can be estimated from the variances of F_2 and backcross generations. All of the phenotypic variance within pure lines b^1b^1 and b^2b^2, as well as in their genetically uniform F_1 (b^1b^2), is environmental. Hence, the phenotypic variances of each pure parental line (V_{P1} and V_{P2}) as well as that of the F_1 (V_{F1}) serve to estimate the environmental variance (V_E). The F_2 segregates $(1/4)b^1b^1 : (1/2)b^1b^2 : (1/4)b^2b^2$. If each genotype departs from the midparent value as shown in the above model, then the average phenotypic value of F_2 should be $(1/4)(-a) + 1/2(+d) + 1/4(+a) = (1/2)d$. The contribution that each genotype makes to the total is its squared deviation from the mean (m) multiplied by its frequency $[f(x - \bar{x})^2]$. Therefore, the total F_2 variance (all genetic in this model) is the mean of squared deviations from the mean:

$$= \tfrac{1}{4}\left(-a - \tfrac{1}{2}d\right)^2 + \tfrac{1}{2}\left(d - \tfrac{1}{2}d\right)^2 + \tfrac{1}{4}\left(a - \tfrac{1}{2}d\right)^2$$

$$= \tfrac{1}{4}\left(a^2 + ad + \tfrac{1}{4}d^2\right) + \tfrac{1}{2}\left(\tfrac{1}{4}d^2\right) + \tfrac{1}{4}\left(a^2 - ad + \tfrac{1}{4}d^2\right) = \tfrac{1}{2}a^2 + \tfrac{1}{4}d^2$$

If we let $a^2 = A$, $d^2 = D$, and $E =$ environmental component, then the total F_2 phenotypic variance (V_{F2}) $= (1/2)A + (1/4)D + E$, representing the additive genetic variance (V_A) + the dominance genetic variance (V_D) + the environmental variance (V_E), respectively. Likewise it can be shown that V_{B1} (the variance of backcross progeny $F_1 \times P_1$) or V_{B2} (the variance of backcross progeny $F_1 \times P_2$), $= (1/4)A + (1/4)D + E$, and $V_{B1} + V_{B2} = (1/2)A + (1/2)D + 2E$. The degree of dominance is expressed as

$$\frac{d}{a} = \sqrt{\frac{D}{A}} \tag{8.10}$$

Heritability can be easily calculated from these variance components. The same is true of variance components derived from studies of identical (monozygotic) vs. nonidentical (fraternal, dizygotic) twins. If twins reared together tend to be treated more alike than unrelated individuals, the heritabilities will be overestimated. This problem, and the fact that the environmental variance of fraternal twins tends to be greater than for identical twins, can be largely circumvented by studying twins that have been reared apart.

2 GENETIC SIMILARITY OF RELATIVES

If offspring phenotypes were always exactly intermediate between the parental values regardless of the environment, then such traits would have a narrow heritability of 1.0. On the other hand, if parental phenotypes (or phenotypes of other close relatives) could not be used to predict (with any degree of accuracy) the phenotypes of offspring (or other relatives), then such traits must have very low (or zero) heritabilities.

(a) Regression Analysis. The **regression coefficient** (b) is an expression of how much (on the average) one variable (Y) may be expected to change per unit change in some other variable (X).

$$b = \frac{\sum_{i=1}^{n}(X_i - \overline{X})(Y_i - \overline{Y})}{\sum_{i=1}^{n}(X_i - \overline{X})^2} = \frac{\sum XY - (\sum X \sum Y)/n}{\sum X^2 - (\sum X)^2/n} \tag{8.11}$$

EXAMPLE 8.7

If for every egg laid by a group of hens (X) the average production by their respective female progeny (Y) is 0.2, then the regression line of Y on X would have a slope (b) of 0.2 (symbol Δ = increment of change).

$$b = \frac{\Delta Y}{\Delta X} = \frac{0.2}{1.0} = 0.2$$

The regression line of Y on X has the formula

$$a = \overline{Y} - b\overline{X} \tag{8.12}$$

where a is the "Y intercept" (the point where the regression line intersects the Y axis), and $\overline{X}$ and $\overline{Y}$ are the respective mean values. The regression line also goes through the point $(\overline{X}, \overline{Y})$ and establishing these two points allows the regression line to be drawn. Any X value can then be used to predict the corresponding Y value. Let $\hat{Y}$ = estimate of Y from X; then

$$\hat{Y} = a + bX \qquad \text{(formula for a straight line)}$$

Since daughters receive only a sample half of their genes from each parent, the daughter-dam regression estimates only one-half of the narrow heritability of a trait (e.g., egg production in chickens). If the variances in the two populations are equal ($s_x = s_y$), then

$$h^2 = 2b_{(\text{daughter}-\text{dam})} \tag{8.13}$$

Similarly, the regression of offspring on the average of their parents (midparent) is also an estimate of heritability

$$h^2 = b_{(\text{offspring}-\text{midparent})} \tag{8.14}$$

Full sibs (having the same parents) are expected to share 50% of their genes in common; half-sibs share 25% of their genes. Therefore,

$$h^2 = 2b_{(\text{full sibs})} \tag{8.15}$$

$$h^2 = 4b_{(\text{half-sibs})} \tag{8.16}$$

If the variances of the two populations are unequal, the data can be converted to standardized variables (as discussed later in this chapter) and the resulting regression coefficients equated to heritabilities as described above.

(b) Correlation Analysis. The statistical **correlation coefficient** (r) measures how closely two sets of data are associated, is dimensionless, and has the limits ± 1. If all of the data points fall on the regression line, there is complete correlation. The regression coefficient (b) and the correlation coefficient (r) always have the same sign.

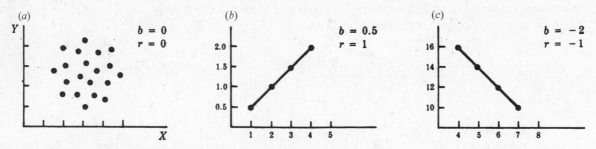

Fig. 8-7.   (a) X and Y are uncorrelated events. (b) Perfect positive correlation. For each unit gain in X there is one-half unit gain in Y. (c) Perfect negative correlation. For each unit gain in X, there is a two unit decrease in Y.

The correlation coefficient (r) of Y on X is defined as the linear change of Y, in standard deviations, for each increase of one standard deviation in X. The covariance (cov) of X and Y can be calculated from the following formula:

$$\text{cov}(X, Y) = \frac{\sum_{i=1}^{n}(X_i - \overline{X})(Y_i - \overline{Y})}{n - 1} \tag{8.17}$$

The covariance becomes the numerator in the formula for the correlation coefficient.

$$r = \text{covariance/geometric mean of variances}$$

$$= \text{cov}(X, Y)/s_x s_y$$

$$= \frac{\sum[(X - \overline{X})(Y - \overline{Y})]/(n-1)}{\sqrt{[\sum(X - \overline{X})^2/(n-1)][\sum(Y - \overline{Y})^2/(n-1)]}}$$

$$= \frac{\sum XY - [(\sum X \sum Y)/n]}{\sqrt{(\sum X^2 - [(\sum X)^2/n])(\sum Y^2 - [(\sum Y)^2/n])}} \tag{8.18}$$

Notice that the numerators in the formulas for r and b are equivalent. Regression and correlation coefficients are related by

$$b = r\left(\frac{s_y}{s_x}\right) \tag{8.19}$$

so that if the variances of X and Y are identical, then $b = r$. If the data are first converted to standardized variables, then the sample has a mean of 0 and a standard deviation of ± 1. Using standardized variables, regression and correlation coefficients become identical. Heritabilities can be estimated from r just as they can from b.

EXAMPLE 8.8
The correlation coefficient of Y offspring and midparent (X) is equivalent to narrow heritability; $h^2 = r$.

EXAMPLE 8.9
If all the variation between offspring and one parent (e.g., their sires) is genetic, then r should equal 0.5; if $r = 0.2$, then $h^2 = 2(0.2) = 0.4$.

EXAMPLE 8.10
If litter mates were phenotypically correlated for a trait by r = 0.15, then $h^2 = 2(0.15) = 0.3$.

EXAMPLE 8.11
If the correlation coefficient for half-sibs is 0.08, then $h^2 = 4(0.08) = 0.32$.

All unbiased estimates of heritability based on correlations between relatives depend upon the assumption that there are no environmental correlations between relatives. Experimentally, this can be fostered by randomly assigning all individuals in the study to their respective environments (field plots, pens, etc.), but this obviously is not possible for humans. Relatives such as full sibs usually share the same maternal and family environment and are likely to show a greater correlation among themselves in phenotype than should rightly be attributed to common heredity. For this reason, the phenotypic correlation between sire and offspring is more useful for calculating heritabilities because sires often do not stay in the same environments with their offspring while mothers or siblings are prone to do so.

(c) Response to Selection. Let us assume we wished to increase the birth weight of beef cattle by selecting parents who themselves were relatively heavy at birth. Assume our initial population ($\bar{P}_1$) has a mean birth weight of 80 lb with a 10 lb standard deviation [Fig. 8-8(a)]. Further suppose that we will save all animals for breeding purposes that weigh over 95 lb at birth. The mean of these animals that have been selected to be parents of the next generation ($\bar{P}_p$) is 100 lb.

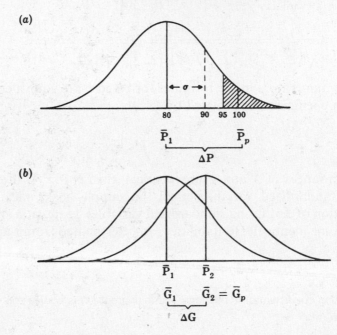

Fig. 8-8. Selection for birth weight in beef cattle. (a) Parental generation. Shaded area represents individuals selected to produce the next generation. (b) Progeny generation (right) compared with parental generation.

The difference $\bar{P}_p - P_1$, is called the **selection differential**, symbolized ΔP (read "delta P"), and is sometimes referred to as "reach." Some individuals with an inferior genotype are expected to have high birth weights largely because of a favorable intrauterine environment. Others with a superior genotype may possess a low birth weight because of an unfavorable environment. In a large, normally distributed population, however, the plus and minus effects produced by good and poor environments are assumed to cancel each other so that the average phenotype ($\bar{P}_1$) reflects the effects of the average genotype ($\bar{G}_1$). Random mating among the selected group produces an offspring generation [Fig. 8-8(b)], with its phenotypic mean ($\bar{P}_2$) also reflecting its average genotypic mean ($\bar{G}_2$). Furthermore, the mean genotype of the parents ($\bar{G}_p$) will be indicated in the mean phenotype of their offspring ($\bar{P}_2$) because only genes are transmitted from one generation to the next. Assuming the environmental effects remain constant from one generation to the next, we can attribute the difference $\bar{G}_2 - \bar{G}_1$ to the selection of genes for high birth weight in the individuals that we chose to use as parents for the next generation. This difference ($\bar{G}_2 - \bar{G}_1$) is called **genetic gain** or **genetic response**, symbolized ΔG. If all of the variability in birth weight exhibited by a population was due solely to additive gene effects, and the environment was contributing nothing at all, then by selecting individuals on the basis of their birth weight records we would actually be selecting the genes that are responsible for high birth weight. That is, we will not be confused by the effects that a favorable environment can produce with a mediocre genotype or by the favorable interaction ("nick") of a certain combination of genes that will be broken up in subsequent generations. Realized heritability is defined as the ratio of the genetic gain to the selection differential:

$$h^2 = \frac{\Delta G}{\Delta P} \qquad\qquad (8.20)$$

EXAMPLE 8.12
If we gained in the offspring all that we "reached" for in the parents, then heritability is unity; i.e., if $\bar{P}_2 - \bar{P}_1 = 100 - 80 = 20$, and $\Delta P = \bar{P}_p - \bar{P}_1 = 100 - 80 = 20$, then $h^2 = \Delta G/\Delta P = 20/20 = 1$.

EXAMPLE 8.13
If selection of parents with high birth weights fails to increase the mean birth weight of their offspring over that of the mean in the previous generation, then heritability is zero; i.e., if $\bar{P}_2$ and $\bar{P}_1 = 80$, then $\Delta G = \bar{P}_2 - \bar{P}_1 = 0$ and $h^2 = \Delta G/\Delta P = 0/20 = 0$.

EXAMPLE 8.14
If the mean weight of the offspring is increased by half the selection differential, then heritability of birth weight is 50%; i.e., if $\Delta G = 1/2\Delta P$, $\Delta P = 2\Delta G$, $h^2 = 0.5 = 50\%$. This is approximately the heritability estimate actually found for birth weight in one population of beef cattle [Fig. 8-8(b)].

Most quantitative traits are not highly heritable. What is meant by high or low heritability is not rigidly defined, but the following values are generally accepted.

High heritability > 0.5
Medium heritability = 0.2 − 0.5
Low heritability < 0.2

SOLVED PROBLEM 8.7
Identical twins are derived from a single fertilized egg (monozygotic). Fraternal twins develop from different fertilized eggs (dizygotic) and are expected to have about half of their genes in common. Left-hand middle-finger-length measurements were taken on the fifth birthday in samples (all of the same sex) of identical twins, fraternal twins, and unrelated individuals from a population of California Caucasians. Using only variances between twins and between unrelated members of the total population, devise a formula for estimating the heritability of left-hand middle-finger length at 5 years of age in this population.

Solution: Let V_i = phenotypic variance between identical twins, V_f = phenotypic variance between fraternal twins, and V_t = phenotypic variance between randomly chosen pairs from the total population of which these twins are a part. All of the phenotypic variance between identical twins is nongenetic (environmental); thus, $V_i = V_e$. The phenotypic variance between fraternal twins is partly genetic and partly environmental. Since fraternal twins are 50% related, their genetic variance is expected to be only half that of unrelated individuals; thus, $V_f = (1/2)V_g + V_e$. The difference $(V_f - V_i)$ estimates half of the genetic variance.

$$V_f - V_i = \underbrace{(1/2)V_g + V_e}_{V_f} - V_e = (1/2)V_g$$

Therefore, heritability is twice that difference divided by the total phenotypic variance.

$$h^2 = \frac{2(V_f - V_i)}{V_i}$$

SOLVED PROBLEM 8.8

Two homozygous varieties of *Nicotiana longiflora* were crossed to produce F_1 hybrids. The average variance of corolla length for all three populations was 8.76. The variance of the F_2 was 40.96. Estimate the heritability of flower length in the F_2 population.

Solution: Since the two parental varieties and the F_1 are all genetically uniform, their average phenotypic variance is an estimate of the environmental variance (V_e). The phenotypic variance of the F_2 (V_t) is partly genetic and partly environmental. The difference ($V_t - V_e$) is the genetic variance (V_g).

$$h^2 = \frac{V_g}{V_t} - \frac{(V_t - V_e)}{V_t} = \frac{40.96 - 8.76}{40.96} = 0.79$$

SOLVED PROBLEM 8.9

The heading rate data are recorded in the table below for two pure varieties of wheat, their F_1 and F_2 progenies, and the first backcross generations.

Generation	Mean	Phenotypic Variance
P_1 (Ramona)	13.0	11.04
P_2 (Baart)	27.6	10.32
F_1	18.5	5.24
F_2	21.2	40.35
B_1	15.6	17.35
B_2	23.4	34.29

Find: (a) the best estimate of the environmental variance (V_E), (b) the broad heritability (H^2) for this trait in this population, (c) the additive genetic variance for this trait, (d) the narrow heritability (h^2) estimate, (e) the dominance variance (V_D), and (f) the degree of dominance.

Solution:
(a) Since all of the phenotypic variance within pure lines and their genetically uniform F_1 progeny is environmental, the mean of these variances produces the best estimate of the

environmental variance (V_E), assuming that there has been no change in the environment from one generation to the next.

$$V_E = (V_{F1} + V_{P1} + V_{P2})/3 = (5.24 + 11.04 + 10.32)/3 = 8.89$$

(b) The phenotypic variance of the F_2 has a genetic and an environmental component.

$$V_{F2} = V_G + V_E$$
$$V_G = V_{F2} - V_E = 40.35 - 8.89 = 31.46$$
$$H^2 = V_G/V_{F2} = 31.46/40.35 = 0.78$$

(c) The total genetic variance has both an additive and a dominance component. From page 270 we have

$$V_{F2} = (1/2)A + (1/4)D + E = 40.35 \tag{1}$$

$$V_{B1} + V_{B2} = (1/2)A + (1/2)D + 2E = 17.35 + 34.29 = 51.64 \tag{2}$$

where A = sum of all squared additive deviations from the mean, $1/2A = V_A$ and $1/4D = V_D$. Multiplying formula (1) by a factor of 2 and subtracting formula (2), we have

$$2V_{F2} = \quad A + (1/2)D + 2E = 80.70$$
$$\underline{V_{B1} + V_{B2} = (1/2)A + (1/2)D + 2E = 51.64}$$
$$(1/2)A \qquad\qquad = 29.06 = V_A$$

(d) The narrow heritability estimate (h^2) is

$$h^2 = V_A/V_{F2} = 29.06/40.35 = 0.72$$

Notice that h^2 is smaller than H^2, calculated in part (b). This indicates that most of the genetic variance is additive and relatively less is due to dominance variance.

(e) The dominance variance is calculated thus:

$$V_G = V_A + V_D$$
$$V_D = V_G - V_A = 31.46 - 29.06 = 2.40$$

(f) In step (c) the following formula was given

$$V_{F2} = (1/2)A + (1/4)D + E$$

Thus, if $(1/2)A = V_A$ and $(1/2)D = V_D$, then $A = 2V_A$ and $D = 4V_D$, and the degree of dominance (formula (8.10)) is

$$\sqrt{\frac{D}{A}} = \frac{\sqrt{4(2.40)}}{\sqrt{2(29.06)}} = \frac{\sqrt{9.60}}{\sqrt{58.12}} = \frac{3.10}{7.62} = 0.41$$

SOLVED PROBLEM 8.10

The pounds of grease fleece weight was measured in a sample from a sheep population. The data listed below is for the average of both parents (X, midparent) and their offspring (Y).

X	11.8	8.4	9.5	10.0	10.9	7.6	10.8	8.5	11.8	10.5
Y	7.7	5.7	5.8	7.2	7.3	5.4	7.2	5.6	8.4	7.0

(a) Calculate the regression coefficient of offspring on midparent and estimate the heritability of grease fleece weight in this population. (b) Plot the data and draw the regression line. (*c*) Calculate the correlation coefficient and from that estimate the heritability.

Solution:

(*a*)

X	Y	X^2	Y^2	XY
11.8	7.7	139.24	59.29	90.86
8.4	5.7	70.56	32.49	47.88
9.5	5.8	90.25	33.64	55.10
10.0	7.2	100.00	51.84	72.00
10.9	7.3	118.81	53.29	79.57
7.6	5.4	57.76	29.16	41.04
10.8	7.2	116.64	51.84	77.76
8.5	5.6	72.25	31.36	47.60
11.8	8.4	139.24	70.56	99.12
10.5	7.0	110.25	49.00	73.50
$\sum X = 99.8$	$\sum Y = 67.3$	$\sum X^2 = 1015.00$	$\sum Y^2 = 462.47$	$\sum XY = 684.43$

$$n = 10$$

$$\frac{(\sum X)^2}{n} = \frac{(99.8)^2}{10} = \frac{9960.04}{10} = 996.0$$

$$\frac{\sum X \sum Y}{n} = \frac{(99.8)(67.3)}{10} = \frac{6716.54}{10} = 671.65$$

$$b = \frac{\sum XY - [(\sum X \sum Y)/n]}{\sum X^2 - [(\sum X)^2/n]} = \frac{684.43 - 671.65}{1015 - 996} = \frac{12.78}{19} = 0.6726$$

The regression of offspring on midparent is an estimate of heritability:

$$h^2 = 0.67$$

(*b*) Data plot and regression line:

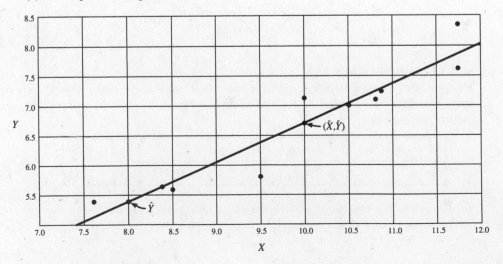

The regression line goes through the intersection of the two means $(\overline{X}, \overline{Y})$; $\overline{X} = 99.8/10 = 9.98$; $\overline{Y} = 67.3/10 = 6.73$. The regression line intersects the Y axis at the Y intercept (a; formula 8.12).

$$a = \overline{Y} - b\overline{X} = 6.73 - 0.673(9.98) = 0.01$$

Now let us choose a value of X that is distant from $(\overline{X}, \overline{Y})$ but easily plotted on the graph (e.g., $X = 8.0$). The corresponding value of Y is estimated to be

$$\hat{Y} = a + bX = 0.01 + 0.673(8.0) = 5.39$$

These two points $[(\overline{X}, \overline{Y})$ and $\hat{Y}]$ establish the regression line with slope $b = 0.67$. For every 1 lb increase in midparent values, offspring tend to produce 0.67 lb.

(c) The correlation coefficient (r) is calculated using formula (8.18):

$$r = \frac{\overbrace{\sum XY - [(\sum X \sum Y)/n]}^{\text{Same numerator as } b}}{\underbrace{\sqrt{\left\{\sum X^2 - \left[(\sum X)^2/n\right]\right\}\left\{\sum Y^2 - \left[(\sum Y)^2/n\right]\right\}}}_{\text{Same denominator as } b}}$$

$$\sum Y^2 = 462.47$$
$$\frac{(\sum Y)^2}{n} = \frac{4529.29}{10} = 452.93$$

$$r = \frac{12.78}{\sqrt{(19)(462.47 - 452.93)}} = \frac{12.78}{\sqrt{181.26}} = \frac{12.78}{13.46} = 0.95$$

Therefore, the X and Y values are very highly positively correlated. Note that two variables can be highly correlated without also being nearly equal. Two variables are perfectly correlated if for one unit change in one variable there is a constant change (either plus or minus) in the other. Negative correlations for heritability estimates are biologically meaningless. Different traits, however, may have negative genetic correlations (e.g., total milk production vs. butterfat percentage in dairy cattle); many of the same genes that contribute positively to milk yield also contribute negatively to butterfat content.

$$h^2 = b = r\left(\frac{s_y}{s_x}\right)$$

$$s_y = \sqrt{\frac{\sum Y^2 - \left[(\sum Y)^2/n\right]}{n-1}} = \sqrt{\frac{462.47 - 452.93}{9}} = 1.03$$

$$s_y = \sqrt{\frac{\sum X^2 - \left[(\sum X)^2/n\right]}{n-1}} = \sqrt{\frac{1015 - 996}{9}} = 1.45$$

$$h^2 = b = 0.95\left(\frac{1.03}{1.45}\right) = 0.95(0.71) = 0.67$$

SOLVED PROBLEM 8.11

The total genetic variance of 180-day body weight in a population of swine is 250 lb^2. The variance due to dominance effects is 50 lb^2. The variance due to epistatic effects is 20 lb^2. The environmental variance is 350 lb^2. What is the heritability estimate (narrow sense) of this trait?

Solution:

$$\sigma_P^2 = \sigma_G^2 + \sigma_E^2 = 250 + 350 = 600$$

$$\sigma_G^2 = \sigma_A^2 + \sigma_D^2 + \sigma_I^2, \qquad 250 = \sigma_A^2 + 50 + 20, \qquad \sigma_A^2 = 180$$

$$h^2 = \sigma_A^2/\sigma_P^2 = 180/600 = 0.3$$

SOLVED PROBLEM 8.12

The heritability of feedlot rate of gain in beef cattle is 0.6. The average rate of gain in the population is 1.7 lb/day. The average rate of gain of the individuals selected from this population to be the parents of the next generation is 2.8 lb/day. What is the expected average daily gain of the progeny in the next generation?

Solution:

$$\Delta P = \overline{P}_p - \overline{P} = 2.8 - 1.7 - 1.1$$

$$\Delta G = h^2(\Delta P) = 0.6(1.1) = 0.66$$

$$\overline{P}_2 = \overline{P}_1 + \Delta G = 2.36 \text{ lb/day}$$

Selection Methods

Artificial selection is operative when humans determine which individuals will be allowed to leave offspring (and/or the number of such offspring). Likewise, **natural selection** allows only those individuals to reproduce that possess traits adaptive to the environments in which they live. There are several methods by which artificial selection can be practiced.

1 MASS SELECTION

If heritability of a trait is high, most of the phenotypic variability, is due to genetic variation. Thus, a breeder should be able to make good progress by selecting from the masses those that excel phenotypically because the offspring-parent correlation should be high. This is called **mass selection**, but it is actually based on the individual's own performance record or phenotype. As the heritability of a trait declines, so does the prospect of making progress in improving the genetic quality of the selected line. In practice, selection is seldom made on the basis of one characteristic alone. Breeders usually desire to practice selection on several criteria simultaneously. However, the more traits selected for, the less selection "pressure" can be exerted on each trait. Selection should thus be limited to the two or three traits that the breeder considers to be the most important economically. It is probable that individuals scoring high in trait A will be mediocre or even poor in trait B (unless the two traits have a positive genetic correlation, i.e., some of the genes increasing trait A are also contributing positively to trait B). The breeder therefore must make compromises, selecting on a "total merit" basis some individuals that would probably not be saved for breeding if selection was being practiced on the basis of only a single trait.

The model used to illustrate the concept of genetic gain [Fig. 8-9(*a*)], wherein only individuals that score above a certain minimum value for a single trait would be saved for breeding, must now be modified to represent the more probable situation in which selection is based on the total merit of two or more traits [Fig. 8.9(*b*)].

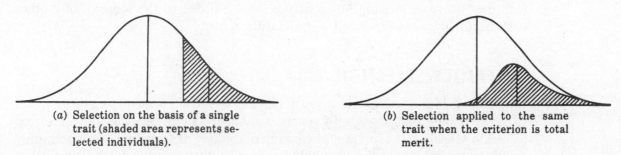

(*a*) Selection on the basis of a single trait (shaded area represents selected individuals).

(*b*) Selection applied to the same trait when the criterion is total merit.

Fig. 8-9.   Selection on single trait vs. total merit. The mean of selected individuals (those allowed to reproduce the next generation) is indicated by the vertical line in the shaded area.

In selecting breeding animals on a "total merit" basis, it is desirable to reduce the records of performance on the important traits to a single score called the **selection index**. The index number has no meaning by itself, but is valuable in comparing several individuals on a relative basis. The methods used in constructing an index may be quite diverse, but they usually take into consideration the heritability and the relative economic importance of each trait in addition to the genetic and phenotypic correlations between the traits.

An index (I) for three traits may have the general form

$$I = a\text{A}' + b\text{B}' + c\text{C}'$$

where *a*, *b*, and *c* are coefficients correcting for the relative heritability and the relative economic importance for traits A, B, and C, respectively, and where A', B', and C' are the numerical values of traits A, B, and C expressed in "standardized form." A **standardized variable** (X') is computed in a sample by the formula

$$X' = \frac{X - \overline{X}}{s} \tag{8.21}$$

where X is the record of performance made by an individual, $\overline{X}$ is the average performance of the population, and s is the standard deviation of the trait. In comparing different traits, one is confronted by the fact that the mean and the variability of each trait is different and often the traits are not even expressed in the same units.

EXAMPLE 8.15
An index for poultry might use egg production (expressed in numbers of eggs per laying season), egg quality (expressed in terms of grades such as AA, A, B, etc.), and egg size (expressed in ounces per dozen).

The standardized variable, however, is a pure number (i.e., independent of the units used) based on the mean and standard deviation. Therefore, any production record or score of a quantitative nature can be added to the score of any other such trait if they are expressed in standardized form.

2 FAMILY SELECTION

When both broad and narrow heritabilities of a trait are low, environmental variance is high compared with genetic variance. **Family selection** (also referred to as **kin selection**) is most useful when heritabilities of traits are low and family members resemble one another only because of their genetic relationship. It is usually more practical to first reduce environmental variance before initiating selective breeding programs. Another way to minimize the effects of an inflated environmental variance is to save for breeding purposes all members of families that have the highest average performance, even though some members of such families have relatively poor phenotypes. In practice, it is not uncommon to jointly use more than one selection method; e.g., choosing only the top 50% of individuals in only the families with the highest averages.

Family selection is most beneficial when members of a family have a high average genetic relationship to one another but the observed resemblance is low. If inbreeding increases the average genetic relationship within a family more than the increases in phenotypic resemblance, the gain from giving at least some weight to family averages may become relatively large.

3 PEDIGREE SELECTION

In this method, consideration is given to the merits of ancestors. Rarely should pedigree selection be given as much weight as the individual's own merit unless the selected traits have low inheritabilities and the merits of the parents and grandparents are much better known than those of the individual in question. It may be useful for characteristics that can only be seen in the opposite sex or for traits that will not be manifested until later in life, perhaps even after slaughter or harvest. The value of pedigree selection depends upon how closely related the ancestor is to the individual in the pedigree, upon how many ancestors' or colateral ancestors' records exist, upon how completely the merits of such ancestors are known, and upon the degree of heritability of the selected traits.

4 PROGENY TEST

A **progeny test** is a method of estimating the breeding value of an animal by the performance or phenotype of its offspring. It has its greatest utility for those traits that (1) can be expressed only in one sex (e.g., estimating the genes for milk production possessed by a bull), (2) cannot be measured until after slaughter (e.g., carcass characteristics), or (3) have low heritabilities so that individual selection is apt to be highly inaccurate.

Progeny testing cannot be practiced until after the animal reaches sexual maturity. In order to progeny-test a male, he must be mated to several females. If the sex ratio is 1 : 1, then obviously every male in a flock or herd cannot be tested. Therefore, males that have been saved for a progeny test have already been selected by some other criteria earlier in life. The more progeny each male is allowed to produce the more accurate the estimate of his "transmitting ability" (**breeding value**), but, in so doing, fewer males can be progeny-tested. If more animals could be tested, the breeder would be able to save only the very best for widespread use in the herd or flock. Thus, a compromise must be made, in that the breeder fails to test as many animals as desired because of the increased accuracy that can be gained by allotting more females to each male under test.

The information from a progeny test can be used in the calculation of the "equal-parent index" (sometimes referred to as the "midparent index"). If the progeny receive a sample half of each of their parents' genotypes and the plus and minus effects of Mendelian errors and errors of appraisal tend to cancel each other in averages of the progeny and dams, then average of progeny = sire/2 + (average of dams)/2 or

$$\text{Sire} = 2(\text{average of progeny}) - (\text{average of dams}) \qquad (8.22)$$

SOLVED PROBLEM 8.13

Fifty gilts (female pigs) born each year in a given herd can be used for proving sires. Average litter size at birth is 10 with 10% mortality to maturity. Only the 5 boars (males) with the highest sire index will be saved for further use in the herd. If each test requires 18 mature progeny, how much culling can be practiced among the progeny-tested boars; i.e., what proportion of those tested will not be saved?

Solution: Each gilt will produce an average of $10 - (0.1)(10) = 9$ progeny raised to maturity. If 18 mature progeny are required to prove a sire, then each boar should be mated to 2 gilts. (50 gilts)/(2 gilts per boar) = 25 boars can be proved. $20/25 = 4/5 = 80\%$ of these boars will be culled.

SOLVED PROBLEM 8.14

Given the following pedigree with butterfat records on the cows and equal-parent indices on the bulls, estimate the index for young bull X (*a*) using information from A and B, (*b*) when the record made by B is only one lactation, and that made in another herd.

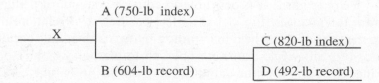

Solution:

(a) The midparent index (estimate of transmitting ability) for X is $(750 + 604)/2 = 677$.

(b) Since we cannot rely on B's record, we should use information from C and D, recalling that X is separated by two Mendelian segregations from the grandparents. Then $X = 750/2 + 820/4 + 492/4 = 703$.

Mating Methods

Once the selected individuals have been chosen, they may be mated in various ways. The process known as "breeding" includes the judicious selection and mating of individuals for particular purposes.

1   RANDOM MATING (PANMIXIS)

If the breeder places no mating restraints upon the selected individuals, their gametes are likely to randomly unite by chance alone. This is commonly the case with **outcrossing** (non-self-fertilizing) plants. Wind or insects carry pollen from one plant to another in essentially a random manner. Even livestock such as sheep and range cattle are usually bred panmicticly. The males locate females as they come into heat and inseminate them without any artificial restrictions as they forage for food over large tracts of grazing land. Most of the food that reaches our table is produced by random mating because it is the most economical mating method; relatively little manual labor is expended by the shepherd or herdsman other than keeping the flock or herd together, warding off predators, etc. This mating method is most likely to generate the greatest genetic diversity among the progeny.

2   POSITIVE ASSORTATIVE MATING

This method involves mating individuals that are more alike, either phenotypically or genotypically, than the average of the selected group.

(a) Based on Genetic Relatedness. Inbreeding is the mating of individuals that are more closely related than the average of the population to which they belong. Figure 8-10(a) shows a pedigree in which no inbreeding is evident because there is no common ancestral pathway from B to C (D, E, F, and G all being unrelated). In the inbred pedigree of Fig. 8-10(b), B and C have the same parents and thus are full sibs (brothers/sisters). In the standard pedigree form shown in Fig. 8-10(b), sires appear on the upper lines and dams on the lower lines. Thus, B and D are males; C and E are females. It is desirable to convert a standard pedigree into an arrow diagram for analysis [Fig. 8-10(c)]. The **coefficient of relationship** (R) estimates the percentage of genes held in common by two individuals because of their common ancestry. Since one transmits only a sample half of one's genotype to one's offspring, each arrow in the diagram represents a probability of 2. The sum

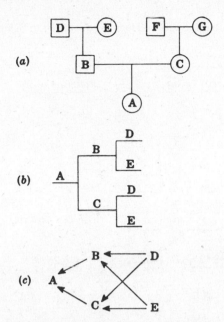

Fig. 8-10. Pedigree diagrams. (*a*) No inbreeding. (*b*) An inbred pedigree. (*c*) Arrow diagram for pedigree (b).

(Σ) of all pathways between two individuals through common ancestors is the coefficient of relationship.

EXAMPLE 8.17

In the arrow diagram of Fig. 8-10(*c*), there are two pathways connecting B and C. The coefficient of relationship between individuals B and C (R_{BC}) = $\sum(1/2)^s$, where *s* is the number of steps (arrows) from B to the common ancestor and back to C.

B and C probably contain $(1/2)(1/2) = 1/4$ of their genes in common through ancestor D.

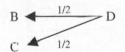

Similarly, B and C probably contain 1/4 of their genes in common through ancestor E.

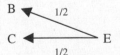

The sum of these two pathways is the coefficient of relationship between the full sibs B and C: $R_{BC} = 1/4 + 1/4 = 1/2$ or 50%.

When matings occur only between closely related individuals (inbreeding), the genetic effect is an increase in homozygosity. The most intense form of inbreeding is self-fertilization. If we start with a population containing 100 heterozygous individuals (*Aa*) as shown in Table 8.2, the expected number of homozygous genotypes is increased by 50% due to selfing in each generation.

Table 8.2. Expected Increase in Homozygosity Due to Selfing

Generation	Genotypes			Percent Heterozygosity	Percent Homozygosity
	AA	*Aa*	*aa*		
0		100		100	0
1	25	50	25	50	50
2	25 12.5	25	12.5 25	25	75
3	37.5 6.25	12.5	6.25 37.5	12.5	87.5
4	43.75 3.125	6.25	3.125 43.75	6.25	93.75

Other less intense forms of inbreeding produce a less rapid approach to homozygosity, as shown graphically in Fig. 8-11. As homozygosity increases in a population, due to either inbreeding or selection, the genetic variability of the population decreases. Since heritability depends upon the relative amount of genetic variability, it also decreases, so that in the limiting case (pure line) heritability becomes zero, meaning there is no genetic variation for selection to act upon.

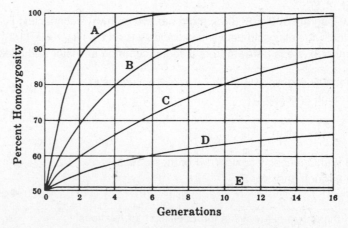

Fig. 8-11. Increase in percentage of homozygosity under various systems of inbreeding. (A) self-fertilization, (B) full sibs, (C) double first cousins, (D) single first cousins, (E) second cousins.

When population size is reduced to a small isolated unit containing less than about 50 individuals, inbreeding very likely will result in a detectable increase in genetic uniformity. The **coefficient of inbreeding** (symbolized by F) is a useful indicator of inbreeding at two levels.

1. On an *individual basis*, the coefficient of inbreeding indicates the probability that the two alleles at any locus are identical by descent, i.e., they are both replication products of a gene present in a common ancestor.

2. On a *population basis*, the coefficient of inbreeding indicates the percentage of all loci that were heterozygous in the base population that now have probably become homozygous due to the effects of inbreeding. The base population is that point in the history of the population from which we desire to begin a calculation of the effects of inbreeding. Many loci are probably homozygous at the time we establish our base population. The inbreeding coefficient then measures the *additional* increase in homozygosity due to matings between closely related individuals.

The coefficient of inbreeding (F) can be determined for an individual in a pedigree by several similar methods.

1. If the common ancestor is not inbred, the inbreeding coefficient of an individual (F_x) is half the coefficient of relationship between the sire and dam (R_{SD}):

$$F_x = (1/2)R_{SD} \qquad (8.23)$$

2. If the common ancestors are not inbred, the inbreeding coefficient is given by

$$F_x = \sum (1/2)^{P_1 + P_2 + 1} \qquad (8.24)$$

where p_1 is the number of generations (arrows) from one parent back to the common ancestor and p_2 is the number of generations from the other parent back to the same ancestor.

3. If the common ancestors are inbred (F_A), the inbreeding coefficient of the individual must be corrected for this factor:

$$F_x = \sum [(1/2)^{P_1 + P_2 + 1}(1 + F_A)] \qquad (8.25)$$

4. The coefficient of inbreeding of an individual may be calculated by counting the number of arrows (n) that connect the individual through one parent back to the common ancestor and back again to the other parent, and applying the formula

$$F_x = \sum (1/2)^n (1 + F_A) \qquad (8.26)$$

The following table will be helpful in calculating F:

n	1	2	3	4	5	6	7	8	9
$(1/2)^n$	0.5000	0.2500	0.1250	0.0625	0.0312	0.0156	0.0078	0.0039	0.0019

Linebreeding is a special form of inbreeding utilized for the purpose of maintaining a high genetic relationship to a desirable ancestor. Figure 8-12 shows a pedigree in which close linebreeding to B has been practiced so that A possesses more than 50% of B's genes. Individual D possesses 50% of B's genes and transmits 25% to C. Individual B also contributes 50% of his genes to C. Hence, C contains 50% + 25% = 75% B genes and transmits half of them

(37.5%) to A. Individual B also contributes 50% of his genes to A. Therefore, A has 50% + 37.5% = 87.5% of B's genes.

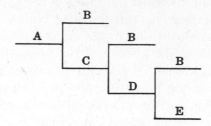

Fig. 8-12. Pedigree exemplifying close linebreeding.

(b) Based on Phenotypic Similarity. Positive phenotypic assortative mating is seldom practiced in its purest form among the selected individuals, i.e., mating only "look-alikes" or those with nearly the same selection indices. However, it can be used in conjunction with random mating; a few of the best among the selected group are "hand-coupled," artificially cross-pollinated, or otherwise forced to breed.

> **EXAMPLE 8.18**
> A beef cattle rancher may maintain a small "show string" in addition to a commercial herd. The few show animals would be closest to the ideal breed type (conformation of body parts, size for age, color markings, shape of horns, etc.) and would be mated like-to-like in hopes of generating more of the same for displaying at fairs and livestock expositions. The rest of the herd would be randomly mated to produce slaughter beef. Some of the cows from the commercial herd might eventually be selected for the show string; some of the young bulls or cows of the show string might not prove to be good enough to save for show and yet perform adequately as members of the commercial herd.

Both inbreeding and positive phenotypic assortative mating tend to reduce genetic heterozygosity, but the theoretical end results are quite different.

> **EXAMPLE 8.19**
> As a model, consider a quantitative trait governed by two loci, each with a pair of alleles both additive and equal in effect. Inbreeding among the five phenotypes would ultimately fix one of four homozygous lines (*AABB*, *AAbb*, *aaBB*, *aabb*). Positive phenotypic assortative mating would fix one of two lines (*AABB* and *aabb*).

The rate at which heterozygous loci can be fixed (brought to homozygosity) in a population can be greatly accelerated by combining a system of close inbreeding with the additional restriction of positive phenotypic assortative mating; in other words, they must also "look" alike.

3 NEGATIVE ASSORTATIVE (DISSORTATIVE) MATING

(a) Based on Genetic Relatedness. When a mating involves individuals that are more distantly related than the average of the selected group it is classified as a negative genetic assortative mating. Individuals sort into mating pairs based on dissimilarity. This may involve crossing individuals belonging to different families, or crossing different inbred varieties of plants, or crossing different breeds of livestock. It may occasionally involve crossing closely related species such as the horse and ass (donkey, burro) to produce the hybrid mule. The usual purpose of these "outcrosses" is an attempt to produce offspring of higher heterozygosity (which may lead to superior phenotypes) than that normally found in the parental populations.

Many recessives remain hidden in heterozygous conditions in noninbred populations, but as homozygosity increases in an inbred population there is a greater probability that recessive traits, many of which are deleterious, will begin to appear. One of the consequences of inbreeding is a loss in vigor (i.e., less productive vegetatively and reproductively) that commonly accompanies an increase in homozygosity (inbreeding depression). Crosses between inbred lines usually produce a vigorous hybrid F_1 generation. This increased reproductive rate of heterozygous individuals has been termed **heterosis** or **hybrid vigor**. The genetic basis of hybrid vigor is still a subject of controversy, largely centered around three ideas. First, hybrid vigor may result from the action and interaction of dominant growth or fitness factors present in the hybrid (**dominance theory**; see Problem 8.3). Second, heterozygosity may produce hybrids that express phenotypes even more extreme or valuable than their parents (**overdominance theory**). And finally, some heterozygotes may have an advantage over homozygotes (**heterozygote advantage theory**). For example, individuals that are homozygous for the sickle-cell anemia mutation have sickle-cell anemia (i.e., the homozygote is disadvantaged due to the disease state); however, individuals carrying this allele (e.g., heterozygotes) have greater resistance to malaria than sickle-cell mutation homozygotes (see Natural Selection and Evolution section in Chapter 9).

Phenotypic variability in the hybrid generation is generally much less than that exhibited by the inbred parental lines (Fig. 8-13). This indicates that the heterozygotes are much less subject to environmental influences than the homozygotes. Geneticists use the term "canalized" to indicate that the organism's development is highly regulated genetically. Another term often used in this connection is **homeostasis**, which signifies the maintenance of a "steady state" in the development and physiology of the organism within the normal range of environmental fluctuations.

A rough guide to the estimation of heterotic effects (H) is obtained by noting the average excess in vigor that F_1 hybrids exhibit over the midpoint between the means of the inbred parental lines (Fig. 8-13).

$$H_{F_1} = \overline{X}_{F_1} - (1/2)(\overline{X}_{P_1} + \overline{X}_{P_2}) \qquad (8.27)$$

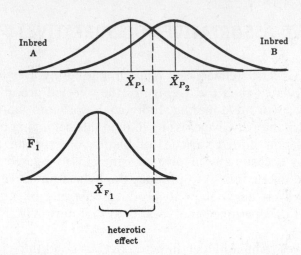

Fig. 8-13.   Hybrid vigor in the progeny from crossing inbred lines.

The heterosis exhibited by an F_2 population is commonly observed to be half of that manifested by the F_1 hybrids.

(b) Based on Phenotypic Dissimilarity. When intermediate phenotypes are preferred, they are more likely to be produced by mating opposite phenotypes. For example, general-purpose cattle can be produced by crossing a beef type with a dairy type. The offspring commonly produce an intermediate yield of milk and hang up a fair carcass when slaughtered (although generally not as good in either respect as the parental types). The same is true of the offspring from crossing an egg type (such as the Leghorn breed of chicken) with a meat type (such as the Cornish). Crossing phenotypic opposites may also be made to correct specific defects.

EXAMPLE 8.20
Brahman cattle have more heat tolerance and resistance to certain insects than European cattle breeds. Brahmans are often crossed to these other breeds in order to create hybrids with the desirable qualities of both parental populations.

EXAMPLE 8.21
Sometimes wild relatives of agriculturally important crops may carry genes for resistance to specific diseases. Hybrids from such crosses may acquire disease resistance, and successive rounds of selection combined with backcrossing to the crop variety can eventually fix the gene or genes for disease resistance on a background that is essentially totally that of the cultivated species.

SOLVED PROBLEM 8.15
Calculate the inbreeding coefficient for A in the following pedigree.

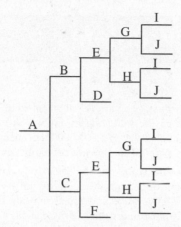

Solution: First we must convert the pedigree to an arrow diagram.

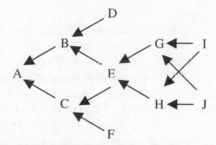

There is only one pathway from B to C and that goes through ancestor E. However, ancestor E is himself inbred. Note that the parents of E are full sibs; i.e., G and H are 50% related (see Example 8.17). By formula (8.23),

$$F_E = (1/2)R_{GH} = 1/2(0.5) = 0.25$$

The inbreeding coefficient of A is given by equation (8.26),

$$F_A = (1/2)^n(1 + F_{ancestor}) = (1/2)^3(1 + 0.25) = 0.156$$

where n is the number of arrows connecting the individual (A) through one parent (B) back to the common ancestor (E) and back again to the other parent (C).

SOLVED PROBLEM 8.16

The average plant heights of two inbred tobacco varieties and their hybrids have been measured with the following results: inbred parent (P_1) = 47.8 in, inbred parent (P_2) = 28.7 in, F_1 hybrid ($P_1 \times P_2$) — 43.2 in. (a) Calculate the amount of heterosis exhibited by the F_1. (b) Predict the average height of the F_2.

Solution:

(a) The amount of heterosis (formula (8.27)) is expressed by the excess of the F_1 average over the midpoint between the two parental means.

$$\text{Heterosis} = \overline{X}_{F_1} - \frac{1}{2}(\overline{X}_{F_1} + \overline{X}_{P_1}) = 43.2 - \frac{1}{2}(47.8 - 28.7) = 43.2 - 38.25 = 4.95 \text{ in}$$

(b) As a general rule, the F_2 shows only about half the heterosis of the F_1: 1/2(4.95) = 2.48. Hence, the expected height of F_2 plants = 38.25 + 2.48 = 40.73 in.

Supplementary Problems

POLYGENIC TRAITS

8.17. Beginning at some arbitrary date, two varieties of wheat were scored for the length of time (in days) to heading, from which the following means were obtained: variety X = 13.0 days, variety Y = 27.6 days. From a survey of 5,504,000 F_2 progeny, 86 were found to head out in 13 days or less. How many pairs of factors are probably contributing to early flowering?

8.18. Suppose that the average skin color on one racial population is 0.43 (measured by the reflectance of skin to red light of 685-nm wavelength); the average skin color of a population racially distinct from the first is 0.23; and racial hybrids between these two populations average 0.33. If about 1/150 offspring from hybrid (racially mixed) parents have skin colors as extreme as the average of either race, estimate the number of segregating loci in the hybrid parents that contribute to skin color variability in their offspring.

THE NORMAL DISTRIBUTION

8.19. From a sample of 10 pig body weights determine (a) mean body weight, (b) sample standard deviation (s), (c) the weight that will probably be exceeded by $2\frac{1}{2}\%$ of this population. Pig weights: 210, 215, 220, 225, 215, 205, 220, 210, 215, 225.

8.20. Suppose six pairs of genes were contributing to a quantitative trait in a cultivated crop. Two parental lines with averages of 13,000 lb/acre and 7000 lb/acre produced an intermediate hybrid F_1 with a variance of 250,000 lb^2. Estimate the standard deviation of the F_2 by formula (8.6).

8.21. Two strains of mice were tested for susceptibility to a carcinogenic drug. The susceptible strain had an average of 75.4 tumorous lung nodules, whereas the resistant strain failed to develop nodules. The F_1 from crossing these two strains had an average of 12.5 nodules with a standard deviation of ±5.3; the F_2 had 10.0 ± 14.1 nodules. Estimate the number of gene pairs contributing to tumor susceptibility by use of formula (8.6).

TYPES OF GENE ACTION

8.22. Calculate the metric values of the parents and their F_1 hybrids in the cross $AA\ B'B'\ CC\ D'D'$ × $A'A'\ BB\ C'C'\ DD$ assuming (a) additive gene action where unprimed alleles contribute 3 units each to the phenotype and primed alleles contribute 6 units each, (b) primed alleles are fully dominant to unprimed alleles; at a given locus, genotypes with one or two primed alleles produce 12 units and the recessive genotype produces 6 units.

HERITABILITY

8.23. Let V_i = phenotypic variance between identical twins, V_f = phenotypic variance between fraternal twins, and heritability $(h^2) = (V_f - V_i)/V_f$. Given the following differences in intelligence quotients (IQ) of 20 pairs of twins (all females, reared together, and identically tested at the same age), estimate the heritability of IQ.

Identical twins	6	2	7	2	4	4	3	5	5	2
Fraternal twins	10	7	13	15	12	11	14	9	12	17

8.24. Suppose that population A has a mean IQ of 85 and that of population B is 100. Estimates of heritability of IQ in both populations are relatively high (0.4 to 0.8). Explain why each of the following statements is false.

(a) Heritability estimates measure the degree to which a trait is determined by genes.

(b) Since the heritability of IQ is relatively high, the average differences between the two populations must be largely due to genetic differences.

(c) Since population B has a higher average IQ than population A, population B is genetically superior to A.

8.25. Flower lengths were measured in two pure lines, and their F_1 and F_2 and backcross progenies. To eliminate multiplicative effects, logarithms of the measurements were used. The phenotypic variances were $P_1 = 48$, $P_2 = 32$, $F_1 = 46$, $F_2 = 130.5$, $B_1(F_1 \times P_1) = 85.5$, and $B_2(F_1 \times P_2) = 98.5$. (a) Estimate the environmental variance (V_E), the additive genetic variance (V_A), and the dominance genetic variance (V_D). (b) Calculate the degree of dominance. (c) Estimate the narrow sense heritability of flower length in the F_2.

8.26. Let r_1, = phenotypic correlation of full sibs, r_2 = phenotypic correlation of half sibs, r_3 = correlation of offspring with one parent, r_4 = correlation of monozygotic twins, and r_5 = correlation of dizygotic twins. In the following formulas, determine the values of x and/or y:

$$(a)\ h^2 = x(r_1 - r_2) \quad (b)\ h^2 = xr_1 - yr_3 \quad (c)\ h^2 = x(r_4 - r_5)$$

8.27. In the following table, Y represents the average number of bristles on a specific thoracic segment of *Drosophila melanogaster* in four female offspring and X represents the number of bristles in the mother (dam) of each set of four daughters.

Family	1	2	3	4	5	6	7	8	9	10
X	9	6	9	6	7	8	7	7	8	9
Y	8	6	7	8	8	7	7	9	9	8

(a) Calculate the daughter-dam regression. (b) Estimate the heritability of bristle number in this population assuming $s_X = s_Y$.

8.28. A flock of chickens has an average mature body weight of 6.6 lb. Individuals saved for breeding purposes have a mean of 7.2 lb. The offspring generation has a mean of 6.81 lb. Estimate the heritability of mature body weight in this flock.

8.29. Yearly wool records (in pounds) are taken from a sample of 10 sheep: 11.8, 8.4, 9.5, 10.0, 10.9, 7.8, 10.8, 8.5, 11.8, 10.5. (a) Calculate the range within which approximately 95% of the sheep in this population are expected to be found. (b) If the additive genetic variance is 0.60, what is the heritability estimate of wool production in this breed?

8.30. Determine (a) the dominance variance and (b) the environmental variance from the following information: heritability [formula (8.9)] = 0.3, phenotypic variance = 200 lb^2, total genetic variance = 100 lb^2, and epistatic variance is absent.

8.31. Thickness of backfat in a certain breed of swine has been estimated to have a heritability of 80%. Suppose the average backfat thickness of this breed is 1.2 in and the average of individuals selected from this population to be the parents of the next generation is 0.8 in. What is the expected average of the next generation?

8.32. The average yearly milk production of a herd of cows is 18,000 lb. The average milk production of the individuals selected to be parents of the next generation is 20,000 lb. The average milk production of the offspring generation is 18,440 lb. (a) Estimate the heritability of milk production in this population. (b) If the phenotypic variance of this population is 4,000,000 lb^2, estimate the additive genetic variance. (c) Between what two values is the central 68% of the original (18,000 lb average) population expected to be found?

8.33. About 1903, Johannsen, a Danish botanist, measured the weight of seeds in the Princess variety of bean. Beans are self-fertilizing and therefore this variety is a pure line. The weights in centigrams (cg) of a small but representative sample of beans are listed below.

19	31	18	24	27	28	25	30	29
22	29	26	23	20	24	21	25	29

(*a*) Calculate the mean and standard deviation for bean weight in this sample. (*b*) Calculate the environmental variance. (*c*) Estimate the heritability of bean weight in this variety. (*d*) If the average bean weight of individuals selected to be parents from this population is 30 cg, predict the average bean weight of the next generation.

SELECTION METHODS

8.34. The length of an individual beetle is 10.3 mm or 0.5 when expressed in "standardized" form. The average measurement for this trait in the beetle population is 10.0 mm. What is the variance of this trait?

8.35. Given the swine selection index $I = 0.14W - 0.27S$, where W is the pig's own 180-day weight and S is its market score. (*a*) Rank the following three animals according to index merit:

Animal	Weight	Score
X	220	48
Y	240	38
Z	200	30

(*b*) If differences in index score are 20% heritable, and parents score 3.55 points higher than the average of the population, how much increase in the average score of the progeny is expected?

8.36. A beef cattle index (I) for selecting replacement heifers takes the form $I = 6 + 2WW' + WG'$, where WW' is weaning weight in standardized form and WG' is weaning grade in standardized form. The average weaning weight of the herd = 505 lb with a standard deviation of ± 34.5 lb. The average weaning grade (a numerical score) is 88.6 with a standard deviation of ± 2.1. Which of the following animals has the best overall merit?

Animal	Actual Weaning Weight	Actual Weaning Grade
A	519	88
B	486	91

8.37. Suppose 360 ewes (female sheep) are available for proving sires. All ewes lamb; 50% of ewes lambing have twins. The 10 rams with the highest progeny test scores will be kept as flock sires. How much selection can be practiced among the progeny-tested individuals; i.e., what proportion of those tested can be saved if a test requires (*a*) 18 progeny, (*b*) 12 progeny, (*c*) 6 progeny?

8.38. During the same year, three dairy bulls were each mated to a random group of cows. The number of pounds of butterfat produced by the dams and their daughters (corrected to a 305-day lactation at maturity with twice daily milking) was recorded as shown below.

Bull	Dam	Dam's Record	Daughter's Record
A	1	600	605
	2	595	640
	3	615	625
	4	610	600
B	5	585	610
	6	590	620
	7	620	605
	8	605	595
C	9	590	590
	10	590	595
	11	610	600
	12	600	605

(*a*) Calculate the sire index for each of the three sires. (*b*) Which sire would you save for extensive use in your herd?

MATING METHODS

8.39. A is linebred to B in the following pedigree. Calculate the inbreeding coefficient of A.

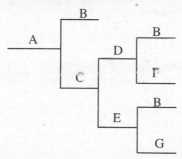

8.40. Calculate the inbreeding of A in the following. (*Hint*: There are nine pathways between B and C).

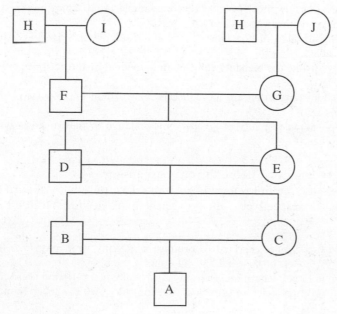

8.41. The yield of seed (in bushels per acre) and plant height (in centimeters) was measured on several generations of corn. Calculate by formula (8.27), (a) the amount of heterosis in the F_1 resulting from crossing the parental varieties with the inbreds, (b) the yield and height expectations of the F_2.

	Seed Yield	Plant Height
Parental varieties	73.3	265
Inbreds	25.0	193
F_1 hybrids	71.4	257

Review Questions

Vocabulary For each of the following definitions, give the appropriate term and spell it correctly. Terms are single words unless indicated otherwise.

1. The kind of phenotypic variation associated with quantitative traits. (One or two words.)
2. A bell-shaped distribution of continuous phenotypic variation. (One or two words.)
3. A squared standard deviation.
4. A type of allelic interaction in which the phenotype of a heterozygote is outside the phenotypic limits of the corresponding homozygotes.
5. The proportion of the phenotypic variance of a trait that is attributable to gene effects.
6. A statistic expressing how much (on average) one sample variable may be expected to change per unit change in some other variable. (Two words.)
7. A statistical measurement of how closely two sets of sample data are associated, having limits ± 1. (Two words.)
8. A method of estimating the breeding value of an individual by the performance or phenotype of its offspring. (Two words.)
9. The mating of individuals that are more closely related than the average of the population to which they belong.
10. The superior phenotypic quality of heterozygotes relative to that of homozygotes, commonly called "hybrid vigor."

Multiple-Choice Questions Choose the one best answer.
For problems 1–3, use the following information. Two pure lines of corn have mean cob lengths of 9 and 3 in, respectively. The polygenes involved in this trait all exhibit additive gene action.

1. Crossing these two lines is expected to produce a progeny with mean cob length (in inches) of (a) 12.0 (b) 7.5 (c) 6.0 (d) 2.75 (e) none of the above
2. If the variation in F_1 cob length ranges from 5.5 to 6.5 in, this variation is estimated to be due to segregation at (a) two loci (b) three loci (c) four loci (d) five loci (e) none of the above

3. If only two segregating loci contribute to cob length, and we represent the parental cross as
   AABB (9-in average cob length) × *aabb* (3-in average), the fraction of the F$_2$ expected to be 4.5
   in is (*a*) 1/8 (*b*) 1/16 (*c*) 3/32 (*d*) 3/16 (*e*) none of the above
4. If a mouse population has an average adult body weight of 25 g with a standard deviation of
   ±3 g, the percentage of the population expected to weigh less than 22 g is approximately (*a*)
   16 (*b*) 33 (*c*) 68 (*d*) 50 (*e*) 25
5. With reference to problem 4 above, if the genetic variance in mouse body weight is 2.7, the
   environmental variance is approximately (*a*) 22.3 (*b*) 6.3 (*c*) 0.3 (*d*) 3.3 (*e*) none of the
   above
6. In another population of mice, the total genetic variance of adult body weight is 4 g^2 and the
   environmental variance is 12 g^2. The broad estimate of heritability for this trait in this population is approximately (*a*) 0.15 (*b*) 0.20 (*c*) 0.25 (*d*) 0.33 (*e*) none of the above
7. If the correlation coefficient between body weight of full sibs is 0.15, then the heritability of this
   trait in this population is (*a*) less than 0.15 (*b*) 0.25 (*c*) 0.3 (*d*) 0.6 (*e*) none of the above

For problems 8-10, use the following information. A population of adult mice has a mean
body weight of 30 g. The average weight of mice selected for breeding purposes is 34 g. The
progeny produced by random mating among the selected parents average 30.5 g.

8. The selection differential (in grams) is (*a*) 0.5 (*b*) 4 (*c*) 3.5 (*d*) 2 (*e*) none of the above
9. The genetic gain (in grams) is (*a*) 0.5 (*b*) 4 (*c*) 3.5 (*d*) 2 (*e*) none of the above
10. The heritability estimate for adult body weight in this population is (*a*) 0.050 (*b*) 0.625 (*c*)
    0.125 (*d*) 0.250 (*e*) none of the above

Answers to Supplementary Problems

8.17. Eight pairs of factors
8.18. Three or four loci (pairs of alleles)
8.19. (*a*) 216 lb (*b*) ± 6.58 lb (*c*) 229.16 lb
8.20. ± 1000 lb/acre
8.21. 4.16 or approximately four gene pairs
8.22. (*a*) Both parents, F$_1$ = 36 units (*b*) Both parents = 36 units, F$_1$ = 48 units
8.23. 0.64
8.24. (*a*) Heritability estimates measure the proportion of the total phenotypic variation that is due
       to genetic variation for a trait among individuals of a population. There is no genetic varia-
       tion in a pure line (heritability = 0), but blood groups (for example) would still be 100%
       determined by genes. (*b*) Suppose that a group of identical twins were divided, one member of
       each pair to the two populations A and B. Each population would then have the same genetic
       constitution. If population A is not given equal social, educational, and vocational oppor-
       tunities with population B, then A might be expected to show lower average IQ. In other
       words, the average IQs of these populations would be reflective solely of nongenetic (envir-
       onmental) differences regardless of the heritability estimates made in each population. (*c*) The
       answer to part (*b*) demonstrates that the difference between phenotypic averages of two
       populations does not necessarily imply that one population is genetically superior to the
       other. Important environmental differences may be largely responsible for such deviations.

We could imagine that a pure line (heritability = 0) would be very well suited to a particular environment, whereas a highly genetically heterogeneous population with a high heritability for the same trait might be relatively poorly adapted to that same environment. In other words, the high heritabilities for IQ within populations A and B reveal nothing about the causes of the average phenotypic differences between them.

8.25. (a) $V_E = 42.0$, $V_A = 77.0$, $V_D = 11.5$ (b) 0.55 (c) 0.59

8.26. (a) $x = 4$ (b) $x = 4$, $y = 2$ (c) $x = 2$

8.27. (a) $b = 0.22$ (b) $h^2 = 2b = 0.49$

8.28. 0.35

8.29. (a) 7.17–12.83 lb (b) 0.3

8.30. (a) 40 lb^2 (b) 100 lb^2

8.31. 0.88 in

8.32. (a) 0.22 (b) 880,000 lb^2 (c) 16,000–20,000 lb

8.33. (a) $\overline{X} = 25$ cg, $s = \pm3.94$ cg (b) 15.53 cg^2; note that this is the square of the phenotypic standard deviation in part (a). In pure lines, all of the variance is environmentally induced. (c) $h^2 = 0$, since a pure line is homozygous; there is no genetic variability. (d) $\overline{X} = 25$ cg; no genetic gain can be made by selecting in the absence of genetic variability.

8.34. 0.36 mm^2

8.35. (a) Y = 23.34, Z = 19.90, X = 17.84 (b) 0.71 point

8.36. $I_A = 6.526$, $I_B = 6.041$; A excels in overall merit.

8.37. (a) 1/3 (b) 22.2% (c) 1/9

8.38. (a) A = 630.0, B = 615.0, C = 597.5 (b) Sire A

8.39. 0.25

8.40. 0.4297

8.41. (a) Heterosis for seed yield = 22.2 bushels/acre, for plant height = 28 cm (b) 60.3 bushels/acre, 243 cm

Answers to Review Questions

Vocabulary

1. continuous variation
2. normal (or Gaussian) distribution
3. variance
4. overdominance
5. heritability (broad definition)
6. regression coefficient
7. correlation coefficient
8. progeny test
9. inbreeding
10. heterosis

Multiple-Choice Questions

1. *c* 2. *e* (environmental) 3. *e* (1/4) 4. *a* 5. *b* 6. *c* 7. *c* 8. *b* 9. *a* 10. *c*

CHAPTER 9

Population Genetics and Evolution

The study of genetics at the population level is called **population genetics**. A **Mendelian population** may be considered to be a group of sexually reproducing organisms with a relatively close degree of genetic relationship (such as a species, subspecies, population, breed, variety, strain) residing within defined geographic boundaries wherein interbreeding occurs. If all the gametes produced by a **Mendelian population** are considered as a hypothetical mixture of genetic units from which the next generation will arise, we have the concept of a **gene pool**. Alleles occur in this pool at some frequency that is a function of how common they are in the population.

Generally, there is tremendous genetic variation within most natural populations. This variation occurs at the DNA and protein levels and is often expressed in different phenotypes in the population. Differences in DNA or protein sequences are referred to as **polymorphisms**.

The gene pool changes over time as new alleles enter or existing alleles exit the pool through events such as natural selection, migration, or mutation. These changes are the basis for **evolution**. Evolution can be observed on the level of alleles and their changing frequencies (**microevolution**) or on the larger scale of geologic time through the extinction as well as the emergence of new species (**macroevolution**).

Hardy-Weinberg Equilibrium

If we consider a pair of alleles A and a, we will find that the percentage of gametes in the gene pool bearing A or a will depend upon the genotypic frequencies of the parental generation whose gametes form the pool. For example, if most of the population were of the recessive genotype aa, then the frequency of the recessive allele in the gene pool would be relatively high, and the percentage of gametes bearing the dominant A allele would be correspondingly low.

When matings between members of a population are completely at random, i.e., when every male gamete in the gene pool has an equal opportunity of uniting with every female gamete, then the zygotic frequencies expected in the next generation may be predicted from a knowledge of the gene (allelic) frequencies in the gene pool of the parental population. That is, given the relative frequencies of A and a gametes in the gene pool, we can calculate (on the basis of the chance union of gametes) the expected frequencies of progeny genotypes and phenotypes. If p = percentage of A alleles in the gene pool and q = percentage of a alleles, then we can use the Punnett square to produce all the possible chance combinations of these gametes.

♀ ＼ ♂	p Ⓐ	q ⓐ
p Ⓐ	p^2 AA	pq Aa
q ⓐ	pq Aa	q^2 aa

Note that $p + q = 1$; i.e., the percentage of A and a gametes must add to 100% in order to account for all of the gametes in the gene pool. The expected genotypic (zygotic) frequencies in the next generation then may be summarized as follows:

$$(p + q)^2 = p^2 + 2pq + q^2 = 1.0$$
$$AA \quad Aa \quad aa$$

Thus, p^2 is the fraction of the next generation expected to be homozygous dominant (AA), $2pq$ is the fraction expected to be heterozygous (Aa), and q^2 is the fraction expected to be recessive (aa). All of these genotypic fractions must add to unity to account for all genotypes in the progeny population.

This formula, expressing the genotypic expectations of progeny in terms of the gametic (allelic) frequencies of the parental gene pool, is called the **Hardy-Weinberg** rule. This rule is named after G. H. Hardy and W. Weinberg who independently formulated it in 1908. It is important because, if a population conforms to the conditions on which this formula is based, there should be no change in the allele frequencies in the population from generation to generation. Should a population initially be in disequilibrium, one generation of random mating is sufficient to bring it into genetic equilibrium, and thereafter the population will remain in equilibrium (unchanging in allelic frequencies) as long as the Hardy-Weinberg conditions persist.

Several assumptions underlie the attainment of genetic equilibrium as expressed in the Hardy-Weinberg equation. When these conditions are met, the predictions from the rule are valid.

1. The population is infinitely large and mates at random (**panmictic**).
2. No selection is operative, i.e., each genotype under consideration can survive just as well as any other (no differential mortality), and each genotype is equally efficient in the production of progeny (no differential reproduction).
3. The population is closed, i.e., no immigration of individuals from another population into nor emigration from the population under consideration is allowed.
4. There is no mutation from one allelic state to another. Mutation may be allowed if the forward and back mutation rates are equivalent, i.e., *A* mutates to *a* with the same frequency that *a* mutates to *A*.
5. Meiosis is normal so that chance is the only factor operative in gametogenesis.

If we define **evolution** as any change in the allele frequencies of a population, then a violation of one or more of the Hardy-Weinberg restrictions could cause the population to move away from the equilibrium frequencies. Changes in gene frequencies can be produced by sampling errors most evident in a very small populations (**genetic drift**), by selection, migration, or mutation pressures, or by nonrandom assortment of chromosomes (**meiotic drive**). No population is infinitely large, spontaneous mutations cannot be prevented, selection and migration pressures usually exist in most natural populations, etc., so it may be surprising to learn that despite these violations of Hardy-Weinberg restrictions many populations do conform, within statistically acceptable limits, to equilibrium conditions between two successive generations. Changes too small to be statistically significant deviations from equilibrium expectations between any two generations can nonetheless accumulate over many generations to produce considerable alterations in the genetic structure of a population.

A **race** is a phenotypically, genetically, and usually geographically, isolated interbreeding population of a species. Races of a given species can freely interbreed with one another. Members of different **species**, however, are generally reproductively isolated to a recognizable degree. **Subspecies** are races that have been given distinctive taxonomic names. Varieties, breeds, strains, etc., of cultivated plants or domesticated animals may also be equated with the racial concept. Geographic isolation is usually required for populations of a species to become distinctive races. Race formation is generally a prerequisite to the splitting of one species into two or more species (**speciation**). Differentiation at many loci over many generations is generally required to reproductively isolate these groups by time of breeding, behavioral differences, ecological requirements, hybrid inviability, hybrid sterility, and other such mechanisms. Yet, clearly, single loci that determine traits significant for mate choice or recognition could yield a new species very rapidly.

Although alleles at a single autosomal locus reach equilibrium following one generation of random mating, gametic equilibrium involving two independently assorting genes is approached rapidly over a number of generations. At equilibrium, the product of coupling gametes equals the product of repulsion gametes.

EXAMPLE 9.1
Consider one locus with alleles A and a at frequencies represented by p and q, respectively. A second locus has alleles B and b at frequencies r and s, respectively. The expected frequencies of coupling gametes AB and ab are pr and qs, respectively. The expected frequencies of repulsion gametes Ab and aB are ps and qr, respectively. At equilibrium, $(pr)(qs) = (ps)(qr)$. Also at equilibrium, the disequilibrium coefficient (d) is $d = (pr)(qs) - (ps)(qr) = 0$.

For independently assorting loci under random mating, the disequilibrium value of d is halved in each generation during the approach to equilibrium because unlinked genes experience 50% recombination. The approach to equilibrium by linked genes, however, is slowed by comparison because they recombine less frequently than unlinked genes (i.e., less than 50% recombination). The closer the linkage, the longer it takes to reach equilibrium. The disequilibrium (d_t) that exists at any generation (t) is expressed as

$$d_t = (1 - r)d_{t-1}$$

where r = frequency of recombination and d_{t-1} = disequilibrium in the previous generation.

EXAMPLE 9.2
If $d = 0.25$ initially and the two loci experience 20% recombination (i.e., the loci are 20 map units apart), the disequilibrium that would be expected after one generation of random mating is $dt = (1 - 0.2)(0.25) = 0.2$. This represents $0.20/0.25 = 0.8$ or 80% of the maximum disequilibrium that could exist for a pair of linked loci.

SOLVED PROBLEM 9.1
In a population gene pool, the alleles A and a are at initial frequencies p and q, respectively. Prove that the gene frequencies and the zygotic frequencies do not change from generation to generation as long as the Hardy-Weinberg conditions are maintained.

Solution: Zygotic frequencies generated by random mating are

$$p^2(AA) + 2pq(Aa) + q^2(aa) = 1$$

All of the gametes of AA individuals and half of the gametes of heterozygotes will bear the dominant allele A. Then the frequency of A in the gene pool of the next generation is

$$p^2 + pq = p^2 + p(1-p) = p^2 + p - p^2 = p$$

Thus, each generation of random mating under Hardy-Weinberg conditions fails to change either the allelic or zygotic frequencies.

SOLVED PROBLEM 9.2

Prove the Hardy-Weinberg law by finding the frequencies of all possible kinds of matings and from these generating the frequencies of genotypes among the progeny using the symbols shown below.

Alleles		Genotypes		
A	a	AA	Aa	aa
Frequency: p	q	p^2	$2pq$	q^2

Solution: There are six kinds of matings (ignoring male-female differences) that are easily generated in a mating table.

		Male Parent		
		AA p^2	Aa $2pq$	aa q^2
Female Parent	AA p^2	p^4	$2p^3q$	p^2q^2
	Aa $2pq$	$2p^3q$	$4p^2q^2$	$2pq^3$
	aa q^2	p^2q^2	$2pq^3$	q^4

The matings $AA \times Aa$ occur with the frequency $4p^3q$. Half the offspring from this mating are expected to be AA $[(1/2)(4p^3q) = 2p^3q]$, and half are expected to be Aa (again with the frequency $2p^3q$). Similar reasoning generates the frequencies of genotypes among the progeny shown in the following table.

	Mating	Frequency	Genotypic Frequencies among Progeny		
			AA	Aa	aa
1	$AA \times AA$	p^4	p^4	—	—
2	$AA \times Aa$	$4p^3q$	$2p^3q$	$2p^3q$	—
3	$AA \times aa$	$2p^2q^2$	—	$2p^2q^2$	—
4	$Aa \times Aa$	$4p^2q^2$	p^2q^2	$2p^2q^2$	p^2q^2
5	$Aa \times aa$	$4pq^3$	—	$2pq^3$	$2pq^3$
6	$aa \times aa$	q^4	—	—	q^4

$$\text{Sums:} (AA) = p^4 + 2p^3q + p^2q^2 = p^2(p^2 + 2pq + q^2) \quad = p^2$$
$$(Aa) = 2p^3q + 4p^2q^2 + 2pq^3 = 2pq(p^2 + 2pq + q^2) \quad = 2pq$$
$$(aa) = p^2q + 2pq^3 + q^4 = q^2(p^2 + 2pq + q^2) \quad = q^2$$
$$\text{Total} = 1.00$$

SOLVED PROBLEM 9.3

At what allelic frequency does the homozygous recessive genotype (aa) become twice as frequent as the heterozygous genotype Aa in a Hardy-Weinberg population?

Solution: Let q = frequency of recessive allele, p = frequency of dominant allele. The frequency of homozygous recessives (q^2) is twice as frequent at heterozygotes ($2pq$) when

$$q^2 = 2(2pq)$$
$$= 4pq$$
$$= 4q(1-q)$$
$$= 4q - 4q^2$$
$$0 = 4q - 5q^2$$
$$0 = q(4 - 5q)$$

Therefore, either $q = 0$ (which is obviously an incorrect solution), or

$$4 - 5q = 0$$
$$5q = 4$$
$$q = 4/5 \text{ or } 0.8$$

Proof:

$$q^2 = 2(2pq)$$
$$(0.8)^2 = 4(0.2)(0.8)$$
$$0.64 = 0.64$$

Calculating Gene Frequencies

1 AUTOSOMAL LOCI WITH TWO ALLELES

(a) **Codominant Autosomal Alleles.** When codominant alleles are present in a two-allele system, each genotype has a distinctive phenotype. The numbers of each allele in both homozygous and heterozygous conditions may be counted in a sample of individuals from the population and expressed as a percentage of the total number of alleles in the sample. If the sample is representative of the entire population (containing proportionately the same numbers of genotypes as found in the entire population) then we can obtain an estimate of the allelic frequencies in the gene pool. Given a sample of N individuals of which D are homozygous for one allele ($A^1 A^1$), H are heterozygous ($A^1 A^2$), and R are homozygous for the other allele ($A^2 A^2$), then N = D + H + R. Since each of the N individuals is diploid at this locus, there are 2N alleles represented in the sample. Each $A^1 A^1$ genotype has two A^1 alleles. Heterozygotes have only one A^1 allele. Letting p represent the frequency of the A^1 allele and q the frequency of the A^2 allele, we have

$$p = \frac{2D + H}{2N} = \frac{D + \frac{1}{2}H}{N} \qquad q = \frac{H + 2R}{2N} = \frac{\frac{1}{2}H + R}{N}$$

This allows the translation of phenotype frequencies into gene (allele) frequencies.

SOLVED PROBLEM 9.4

In Shorthorn cattle, the genotype $C^R C^R$ is phenotypically red, $C^R C^W$ is roan (a mixture of red and white), and $C^W C^W$ is white. (*a*) If 108 red, 48 white, and 144 roan animals were found in a sample of Shorthorns from the central valley of California, calculate the estimated frequencies of the C^R allele and the C^W allele in the gene pool of the population. (*b*) If this population is completely panmictic, what zygotic frequencies would be expected in the next generation? (*c*) How does the sample data in part (*a*) compare with the expectations for the next generation in part (*b*)? Is the population represented in part (*a*) in equilibrium?

Solution:

(*a*)

Numbers	Phenotypes	Genotypes
108	Red	$C^R C^R$
144	Roan	$C^R C^W$
48	White	$C^W C^W$
300		

First, let us calculate the frequency of the C^R allele. There are 108 red individuals each carrying two C^R alleles; $2 \times 108 = 216$ C^R alleles. There are 144 roan individuals each carrying only one C^R allele; $1 \times 144 = 144$ C^R alleles. Thus, the total number of C^R alleles in our sample is $216 + 144 = 360$. Since each individual is a diploid (possessing two sets of chromosomes, each bearing one of the alleles at the locus under consideration), the total number of alleles represented in this sample is $300 \times 2 = 600$. The fraction of all alleles in our sample of type C^R becomes $360/600 = 0.6$ or 60%. The other 40% of the alleles in the gene pool must be of type C^W. We can arrive at this estimate for C^W by following the same procedure as above. There are $48 \times 2 = 96$ C^W alleles represented in the homozygotes and 144 in the heterozygotes; $96 + 144 = 240$; $240/600 = 0.4$ or 40% C^W alleles.

(*b*) Recall that panmixis is synonymous with random mating. We will let the frequency of the C^R allele be represented by $p = 0.6$, and the frequency of the C^W allele be represented by $q = 0.4$. Then according to the Hardy-Weinberg law, we would expect as genotypic frequencies in the next generation

$$p^2 = (0.6)^2 = 0.36 C^R C^R : 2pq = 2(0.6)(0.4) = 0.48 C^R C^W : q^2 = (0.4)2 = 0.16 C^W C^W$$

(*c*) In a sample of size 300 we would expect $0.36(300) = 108$ $C^R C^R$ (red), $0.48(300) = 144$ $C^R C^W$ (roan), and $0.16(300) = 48$ $C^W C^W$ (white). Note that these figures correspond exactly to those of our sample. Since the genotypic and gametic frequencies are not expected to change in the next generation, the original population must already be in equilibrium.

(b) Dominant and Recessive Autosomal Alleles. Determining the gene frequencies for alleles that exhibit dominance and recessive relationships requires a different approach from that used with codominant alleles. A dominant phenotype may have either of two genotypes, *AA* or *Aa*, but we have no way (other than by laboriously testcrossing each dominant phenotype) of distinguishing how many are homozygous or heterozygous in our sample. The only phenotype whose genotype is known for certain is the recessive (*aa*). If the population is in equilibrium, then we can obtain an estimate of *q* (the frequency of the recessive allele) from q^2 (the frequency of the recessive genotype or phenotype). Then the frequency of the dominant allele is

$$p = 1 - q \quad \text{or} \quad p = 1 - \sqrt{q^2}$$

SOLVED PROBLEM 9.5
White wool is dependent upon a dominant allele *B* and black wool upon its recessive allele *b*. Suppose that a sample of 900 sheep of the Rambouillet breed in Idaho gave the following data: 891 white and 9 black. Estimate the allelic frequencies.

Solution:

$$p^2(BB) + 2pq(Bb) + q^2(bb) = 1.0$$

If we assume the population is in equilibrium, we can take the square root of that percentage of the population that is of the recessive genotype (phenotype) as our estimator for the frequency of the recessive allele.

$$q = \sqrt{q^2} = \sqrt{9/900} = 0.1 = \text{ frequency of allele } b$$

Since $p + q = 1$, the frequency of allele B is 0.9.

(c) Sex-Influenced Traits. The expression of dominance and recessive relationships may be markedly changed in some genes when exposed to different environmental conditions, most notable of which are the sex hormones. In sex-influenced traits (Chapter 5), the heterozygous genotype usually will produce different phenotypes in the two sexes, making the dominance and recessive relationships of the alleles appear to reverse themselves. We shall consider only those sex-influenced traits whose controlling genes are on autosomes. Determination of allelic frequencies must be indirectly made in one sex by taking the square root of the frequency of the recessive phenotype ($q = \sqrt{q^2}$). A similar approach in the opposite sex should give an estimate of the alternate allele, p. Corroboration of sex influence is obtained if these estimates of p and q made in different sexes add close to unity. It is generally difficult to determine whether a trait is sex-influenced or sex-linked.

SOLVED PROBLEM 9.6
In the human population, an index finger shorter than the ring finger is thought to be governed by a sex-influenced gene that appears to be dominant in males and recessive in females. A sample of the males in this population was found to contain 120 short and 210 long index fingers. Calculate the expected frequencies of long and short index fingers in females of this population.

Solution: Since the dominance relationships are reversed in the two sexes, let us use all lowercase letters with superscripts to avoid confusion with either dominance or codominance symbolism.

	Phenotypes	
Genotype	Males	Females
$s^1 s^1$	Short	Short
$s^1 s^2$	Short	Long
$s^2 s^2$	Long	Long

Let $p = $ frequency of s^1 allele, $q = $ frequency of s^2 allele. $p^2(s^1 s^1) + 2pq(s^1 s^2) + q^2(s^2 s^2) = 1.0$. In males, the allele for long finger s^2 is recessive. Then

$$q = \sqrt{q^2} = \sqrt{210/(120 + 210)} = \sqrt{0.64} = 0.8 \qquad p = 1 - 0.8 = 0.2$$

In females, short index finger is recessive. Then $p^2 = (0.2)^2 = 0.04$ or 4% of the females of this population will probably be short fingered. The other 96% should possess long index fingers.

2 AUTOSOMAL LOCI WITH MULTIPLE ALLELES

If we consider three alleles, A, a', and a, with the dominance hierarchy $A > a' > a$, occurring in the gene pool with respective frequencies p, q, and r, then random mating will generate progeny with the following frequencies:

$$(p + q + r)^2 = \quad p^2 + 2pq + 2pr + q^2 + 2qr + r^2 = 1$$

Genotypes: $\underbrace{AA \quad Aa' \quad Aa}$ $\underbrace{a'a' \quad a'a}$ $\underbrace{aa}$

Phenotypes: A a' a

For ease in calculation of a given allelic frequency, it may be possible to group the phenotypes of the population into just two types.

EXAMPLE 9.3

In a multiple allelic system where $A > a' > a$, we could calculate the frequency of the top dominant allele A by considering the dominant phenotype (A) in contrast to all other phenotypes produced by alleles at this locus. The latter group may be considered to be produced by an allele a_x, which is recessive to A.

Let $p =$ frequency of allele A, $q =$ frequency of allele a_x
$q^2 =$ frequency of phenotypes other than A
$q = \sqrt{q^2}$
$p = 1 - q =$ frequency of gene A

Many multiple allelic series involve codominant relationships such as $(A^1 = A^2) > a$, with respective frequencies p, q, and r. More genotypes can be phenotypically recognized in codominant systems than in systems without codominance.

$$(p + q + r)^2 = p^2 + 2pr + 2pq + q^2 + 2qr + r^2 = 1$$

Genotypes: $\underbrace{A^1A^1 \quad A^1a}$ $\underbrace{A^1A^2}$ $\underbrace{A^2A^2 \quad A^2a}$ $\underbrace{aa}$

Phenotypes: A^1 A^1A^2 A^2 a

The use of this formula in calculating multiple allelic frequencies is presented in Solved Problem 9.7. Similar methods may be utilized to derive other formulas for calculating gene frequencies in multiple allelic systems with more than three alleles, but their computation becomes too involved for our purposes at the introductory level. Therefore, multiple allelic problems in this chapter will be mainly concerned with three alleles.

SOLVED PROBLEM 9.7

The ABO blood group system is governed by a multiple allelic system in which some codominant relationships exist. Three alleles, I^A, I^B, and i, form the dominance hierarchy $(I^A = I^B) > i$. (a) Determine the genotypic and phenotypic expectations for this blood group locus from a population in genetic equilibrium. (b) Derive a formula for use in finding the allelic frequencies at the ABO blood group locus. (c) Among New York Caucasians, the frequencies of the ABO blood groups were found to be approximately 49% type O, 36% type A, 12% type B, and 3% type AB. What are the allelic frequencies in this population? (*d*) Given the population in part (*c*) above, what percentage of type A individuals are probably homozygous?

Solution:

(*a*) Let p = frequency of I^A allele, q = frequency of I^B allele, r = frequency of i allele. The expansion of $(p + q + r)^2$ yields the zygotic ratio expected under random mating.

Genotypic Frequencies	Genotypes	Phenotypes (Blood Groups)
p^2	$I^A I^A$	A
$2pr$	$I^A i$	
q^2	$I^B I^B$	B
$2qr$	$I^B i$	
$2pq$	$I^A I^B$	AB
r^2	ii	0

(*b*) Let A, B, and O represent the phenotypic frequencies of blood groups A, B, and O, respectively. Solving for the frequency of the recessive allele i,

$$r = \sqrt{r^2} = \sqrt{O}$$

Solving for the frequency of the I^A allele,

$$p^2 + 2pr + r^2 = A + O; \quad (p+r)^2 = A + O; \quad p = \sqrt{A+O} - r = \sqrt{A+O} - \sqrt{O}$$

Solving for the frequency of the I^B allele $q = 1 - p - r$. Or, following the method for obtaining the frequency of the I^A allele,

$$q = \sqrt{B+O} - \sqrt{O}$$

Presenting the solutions in a slightly different form,

$$\underbrace{\sqrt{A+O} - \sqrt{O}}_{p} + \underbrace{\sqrt{B+O} - \sqrt{O}}_{q} + \underbrace{\sqrt{O}}_{r} = 1.0$$

$$p = 1 - \sqrt{B+O}; \qquad q = 1 - \sqrt{A+O}; \qquad r = \sqrt{O}$$

(*c*) Frequency of allele i

$$\sqrt{O} = \sqrt{0.49} = 0.70 = r$$

Frequency of I^B allele

$$1 - \sqrt{A+O} = 1 - \sqrt{0.36 + 0.49} = 0.08 = q$$

Frequency of allelle I^A

$$1 - \sqrt{B+O} = 1 - \sqrt{0.12 + 0.49} = 0.22 = p$$

Check: $p + q + r = 0.22 + 0.08 + 0.70 = 1.00$

(*d*)
$$p^2 = I^A I^A = (0.22)^2 \qquad = 0.048$$
$$2pr = I^A i = 2(0.22)(0.7) = \underline{0.308}$$
$$= 0.356 = \text{total group A individuals}$$

Thus, $48/356 = 0.135$ or 13.5% of all group A individuals in this population are expected to be homozygous.

3 SEX-LINKED LOCI

(a) Codominant Alleles. Data from both males and females can be used in the direct computation of sex-linked codominant allelic frequencies. Bear in mind that in organisms with an X-Y mechanism of sex determination, the heterozygous condition can only appear in females. Males are hemizygous for sex-linked genes.

EXAMPLE 9.4

In domestic cats, black melanin pigment is deposited in the hair by a sex-linked gene; its alternative allele produces yellow hair. Random inactivation of one of the X chromosomes occurs in each cell of female embryos. Heterozygous females are thus genetic mosaics, having patches of all-black and all-yellow hairs called tortoise-shell pattern. Since only one sex-linked allele is active in any cell, the inheritance is not really codominant, but the genetic symbolism used is the same as that for codominant alleles.

Sex	Phenotypes		
	Black	Tortoiseshell	Yellow
Females	$C^b C^b$	$C^b C^y$	$C^y C^y$
Males	$C^b Y$	—	$C^y Y$

Let p = frequency of C^b, q = frequency of C^y.

$$p = \frac{2(\text{no. of black females}) + (\text{no. of tortoiseshell females}) + (\text{no. of black males})}{2(\text{no. females}) + \text{no. males}}$$

$$q = \frac{2(\text{no. of yellow females}) + (\text{no. of tortoiseshell females}) + (\text{no. of yellow males})}{2(\text{no. female}) + \text{no. males}}$$

SOLVED PROBLEM 9.8

The genetics of coat colors in cats was presented in Example 9.4: $C^B C^B$ females or $C^B Y$ males are black, $C^Y C^Y$ females or $C^Y Y$ males are yellow, $C^B C^Y$ female are tortoiseshell (blotches of yellow and black). A population of cats in London was found to consist of the following phenotypes:

	Black	Yellow	Tortoiseshell	Totals
Males	311	42	0	353
Females	277	7	54	338

Determine the allelic frequencies using all of the available information.

Solution: The total number of C^B alleles in this sample is $311 + 2(277) + 54 = 919$. The total number of alleles (X chromosomes) in this sample is $353 + 2(338) = 1029$. Therefore, the frequency of the C^B allele is $919/1029 = 0.893$. The frequency of the C^Y allele would then be $1 - 0.893 = 0.107$.

(b) Dominant and Recessive Alleles. Since each male possesses only one sex-linked allele, the frequency of a sex-linked trait among males is a direct measure of the allelic frequency in the population, assuming, of course, that the allelic frequencies thus determined are representative of the allelic frequencies among females as well.

SOLVED PROBLEM 9.9

White eye color in *Drosophila* is governed by a sex-linked recessive gene w and wild-type (red) eye color is produced by its dominant allele w^+. A laboratory population of *Drosophila* was found to contain 170 red-eyed males and 30 white-eyed males. (*a*) Estimate the frequency of the w^+ allele and the w allele in the gene pool. (*b*) What percentage of the females in this population would be expected to be white-eyed?

Solution:

(*a*)

Observed No. of Males	Genotypes of Males	Phenotypes of Males
170	w^+Y	Wild type (red eye)
30	wY	White eye
200		

Thus, 30 of the 200 X chromosomes in this sample carry the recessive allele w.

$$q = 30/200 = 0.15 \text{ or } 15\% \; w \text{ alleles}$$
$$p = 1 - q = 1 - 0.15 = 0.85 \text{ or } 85\% \; w^+ \text{ alleles}$$

(*b*) Since females possess two X chromosomes (hence two alleles), their expectations may be calculated in the same manner as that used for autosomal genes.

$$p^2(w^+w^+) + 2pq(w^+w) + q^2(ww) = 1.0 \text{ or } 100\% \text{ of the females}$$

$q^2 = (0.15)^2 = 0.0225$ or 2.25% of all females in the population are expected to be white-eyed.

Testing a Locus for Equilibrium

In cases where dominance is involved, the heterozygous class is indistinguishable phenotypically from the homozygous dominant class. Hence, there is no way of checking the Hardy-Weinberg expectations against observed sample data unless the dominant phenotypes have been genetically analyzed by observation of their progeny from testcrosses. Only when codominant alleles are involved can we easily check our observations against the expected equilibrium values through the chi-square test (Chapter 2).

The number of variables in chi-square tests of Hardy-Weinberg equilibrium is not simply the number of phenotypes minus 1 (as in chi-square tests of classical Mendelian ratios). The number of observed variables (number of phenotypes $= k$) is further restricted by testing their conformity to an expected Hardy-Weinberg frequency ratio generated by a number of additional variables (number of alleles, or allelic frequencies $= r$). We have $(k - 1)$ degrees of freedom in the number of phenotypes, $(r - 1)$ degrees of freedom in establishing the frequencies for the r alleles. The combined number of degrees of freedom is $(k - 1) - (r - 1) = k - r$. Even in most chi-square tests for equilibrium involving multiple alleles, the number of degrees of freedom is the number of phenotypes minus the number of alleles.

SOLVED PROBLEM 9.10

A human serum protein called haptoglobin has two major electrophoretic variants produced by a pair of codominant alleles Hp^1 and Hp^2. A sample of 100 individuals has 10 Hp^1/Hp^1, 35 Hp^1/Hp^2, and 55 Hp^2/Hp^2. Are the genotypes in this sample conforming to the frequencies expected for a Hardy-Weinberg population within statistically acceptable limits?

Solution: First, we must calculate the allelic frequencies.

$$\text{Let } p = \text{frequency of } Hp^1 \text{ allele} = \frac{2(10) + 35}{2(100)} = 55/200 = 0.275$$

$$q = \text{frequency of } Hp^2 \text{ allele} = 1 - 0.275 = 0.725$$

From these gene (allelic) frequencies we can determine the genotypic frequencies expected according to the Hardy-Weinberg equation.

$$
\begin{aligned}
Hp^1/Hp^1 &= p^2 \\
&= (0.275)^2 \\
&= 0.075625 \\
Hp^1/Hp^2 &= 2pq \\
&= 2(0.275)(0.725) \\
&= 0.39875 \\
Hp^2/Hp^2 &= q^2 \\
&= (0.725)^2 \\
&= 0.525625
\end{aligned}
$$

Converting these genotypic frequencies to numbers based on a total sample size of 100, we can do a chi-square test.

Genotypes	Observed (o)	Expected (e)	Deviation (o−e)	$(o-e)^2$	$(o-e)^2/e$
Hp^1/Hp^1	10	7.56	2.44	5.95	0.79
Hp^1/Hp^2	35	39.88	−4.88	23.81	0.60
Hp^2/Hp^2	55	52.56	2.44	5.95	0.11
Totals	100	100.00	0		$\chi^2 = 1.50$

$\text{df} = k$ phenotypes $\quad r$ alleles $= 3 - 2 = 1; p = 0.2 - 0.3$ (Table 2-3).

This is not a significant χ^2 value, and we may accept the hypothesis that this sample (and hence presumably the population from which it was drawn) is conforming to the equilibrium distribution of genotypes.

SOLVED PROBLEM 9.11

One of the "breeds" of poultry has been largely built on a single-gene locus, that for "frizzled" feathers. The frizzled phenotype is produced by the heterozygous genotype $M^N M^F$. One homozygote $M^F M^F$ produces extremely frizzled birds called "woolies." The other homozygous genotype $M^N M^N$ has normal plumage. A sample of 1000 individuals of this "breed" in the United States contained 800 frizzled, 150 normal, and 50 wooly birds. Is this population in equilibrium?

Solution:

Let p = frequency of the M^F allele = $\dfrac{2(50) + 800}{2(1000)}$ = 0.45

q = frequency of the M^N allele = $1 - 0.45 = 0.55$

Genotypes	Equilibrium Frequencies		Calculations		Expected Numbers
$M^F M^F$	p^2	$(0.45)^2$	$= 0.2025(1000)$	$=$	202.5
$M^F M^N$	$2pq$	$2(0.45)(0.55)$	$= 0.4950(1000)$	$=$	495.0
$M^N M^N$	q^2	$(0.55)^2$	$= 0.3025(1000)$	$=$	302.5

Chi-square test for conformity to equilibrium expectations gives the following results:

Phenotypes	o	e	(o−e)	(o−e)^2	(o−e)^2/e
Wooly	50	202.5	−152.5	23,256	114.8
Frizzle	800	495.0	+ 305.0	93,025	187.9
Normal	150	302.5	−152.5	23,256	76.9
					$\chi 2 = 379.6$

df = 1; $p < 0.01$ (Table 2-3)

This highly significant chi-square value will not allow us to accept the hypothesis of conformity with equilibrium expectations. The explanation for the large deviation from the equilibrium expectations is twofold. Much artificial selection (by people) is being practiced. The frizzled heterozygotes represent the "breed" type and are kept for show purposes as well as for breeding by bird fanciers. Such breeders dispose of (cull) many normal and wooly types. Natural selection is also operative on the wooly types because they tend to lose their feathers (loss of insulation) and eat more feed just to maintain themselves, are slower to reach sexual maturity, and lay fewer eggs than do the normal birds.

Natural Selection and Evolution

Genetic diversity in a population of eukaryotic organisms is mainly generated by three mechanisms: mutation, independent assortment of alleles, and recombination. Relative to a given environment, some of these genetic variants will be better able to survive and reproduce than other variants. The genotype of an individual determines its various phenotypes, some of which will contribute to the overall **fitness** of the individual. Fitness is the ability of an individual to survive and reproduce (i.e., pass its genetic information to the next generation). Fitness is generally measured relative to others in a population. **Natural selection**, or the processes of nature such as the environment or other external factors (e.g., food type and availability), acts by choosing or "selecting" individuals with the highest fitness in a population. Thus, favorable gene combinations (**adaptive**) tend to be perpetuated while less adaptive ones tend to be eliminated from the population. **Evolution** is the process whereby natural selection acts on the genetic diversity of a population in this manner.

EXAMPLE 9.5

A population of bacterial pathogens infecting a person will contain a certain level of genetic diversity. That diversity may include a gene or an allele that allows one individual bacterial cell (or part of the population) to be resistant to an antibiotic. When the environment of the population includes the antibiotic, multiplication of the individual containing the resistance gene (higher fitness) will occur over those that do not have this beneficial allele (and, thus, are less fit). Thus, the population will evolve into one that is generally drug resistant from one that was not. The antibiotic, in this case, is the environmental condition that selects the resistant phenotypes of individuals in the population.

1 MOLECULAR EVOLUTION

Evolution at the molecular level is driven primarily by mutation and selection. Mutations and recombination provide the genetic variation from which natural selection "selects" or favors transmission of the most adaptive gene combinations. Mutations can fall into three distinct classes based on their effects on the phenotype: (1) detrimental, (2) neutral, or (3) beneficial. Most **detrimental mutations** will tend to be eliminated in a population due to selection against individuals harboring this allele. Some deleterious mutations may survive in a population if they have a beneficial function in a heterozygous state. For example, the sickle-cell hemoglobin allele produces a defective hemoglobin protein; it is codominant with the wild-type hemoglobin allele. When homozygous, this mutant allele causes sickle-cell disease, and, often, early death (decreased fitness). However, individuals that are heterozygous for this mutation are resistant to infection by the malarial parasite, *Plasmodium*. This advantage tends to perpetuate the detrimental allele in a gene pool where malaria is endemic. **Neutral mutations**, those that have neither advantageous nor nonadaptive phenotypic effects, are generally not affected by natural selection and either survive or do not survive in populations as a result of **genetic drift**. Neutral mutations are those that alter an amino acid codon in such a way as to not change the amino acid coded for (e.g., GCU changed in the third position to GCC; both codons specify alanine) or in such a way that a functionally equivalent amino acid is substituted (e.g., one basic amino acid is substituted for another). **Beneficial mutations** or alleles generally survive due to the increased fitness afforded to individuals harboring these alleles. Ultimately, these alleles tend to become **fixed** in the population. During fixation, one allele replaces another (less fit) allele. Most mutations are deleterious and therefore tend to be lost due to selection. Therefore, most genetic diversity in a population is likely to be neutral or of little fitness consequence.

By comparing DNA or protein sequences from different individuals, evolutionary changes at the molecular level can be observed. In general, the more differences that two sequences contain, the more distantly related they are. For example, one would expect the sequence of a gene in a bacterium to be very different from a homologous gene in a human. More specific terms describing homologous gene relationships are **paralog** and **ortholog**. A paralog (or **paralogous gene**) is a homologous gene within a species, while an ortholog (or **orthologous gene**) refers to a homologous sequence in different species. Paralogs arise by gene duplication and differential mutations, whereas orthologs arise from a common ancestor during speciation. In addition, orthologs may or may not carry out similar functions. From these types of sequence comparisons, **phylogenetic trees** (Fig. 9-1) can be built. These trees show the relatedness among different groups of organisms (commonly different species) based upon similarities in DNA or amino acid sequences of a gene.

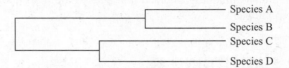

Species A
Species B
Species C
Species D

Fig. 9-1. A simplified example of a phylogenetic tree. The length of the branches represents distances of divergence between alleles. In this example, species C is more closely related to species D (at this locus) than to species A or B. Species A and B share a more recent common ancestor than do species C and D.

Ribosomal genes are typically highly conserved and are thus commonly used as genes for observing evolution over long periods of time. Other genes, such as cytochrome genes or fibrinoprotein genes, are less conserved and can be used to observe evolutionary change over shorter periods of time. Noncoding regions, such as introns, are often used for short-scale (ecological) time periods.

2 MACROEVOLUTION

Evolution of organisms living today (**extant**) can be observed in the fossil record. From this, a basic timeline of evolution can be constructed, a simplified version of which follows. The Earth is believed to have come into existence about 4.5 billion years ago. At about 4.2 billion years ago, its surface was covered with oceans. The evolution of complex, multicellular organisms is proposed to have begun with the formation of rudimentary "cells" about 3.7 billion years ago. These ancestral "cells" were membrane delimited and probably possessed the ability to reproduce and transmit genetic information. RNA is now thought to be the first information molecule that evolved in Earth's early environment. This is due to the discovery that some RNA molecules can act as catalysts (**ribozymes**). It is generally accepted that these ancestors gave rise to the three domains of extant organisms: the **Bacteria** (the Eubacteria, Chapter 10), the **Archaea** (the Archaebacteria, Chapter 10), and the **Eukarya** (all eukaryotes). Single-celled organisms were prevalent on Earth until about 0.7 billion years ago when there was a rapid expansion of multicellular organisms known as **metazoans**. Metazoans showed a tremendous diversity in body plan. From that time, diversity has been greatly reduced primarily due to several extinctions of some major groups, now known only from fossils. More recent extinctions, such as the great dinosaur extinction 65 million years ago, show that life continues to replace itself. Predominant forms of life have limited "life spans." One of the key patterns of macroevolution is a **stasis**, or long-term dominance of a single group, punctuated (interrupted) by rapid extinctions that are followed by the emergence of new groups. The emergence of *Homo sapiens* from primate relatives is thought to have occurred 6 million years ago. The DNA of our closest primate relative, the chimpanzee, is 99% similar to ours and has a chromosome karyotype that is also strikingly similar.

Supplementary Problems

HARDY-WEINBERG EQUILIBRIUM

9.12. At what allelic frequency is the heterozygous genotype (*Aa*) twice as frequent as the homozygous genotype (*aa*) in a Hardy-Weinberg population?

9.13. There is a singular exception to the rule that genetic equilibrium at two independently assorting autosomal loci is attained in a nonequilibrium population only after a number of generations of random mating. Specify the conditions of a population that should reach genotypic equilibrium after a single generation of random mating.

9.14. Let the frequencies of a pair of autosomal alleles *A* and *a* be represented by p_m and q_m in males and by p_f and g_f in females, respectively. Given $q_f = 0.6$ and $q_m = 0.2$, (*a*) determine the equilibrium gene frequencies in both sexes after one generation of random mating, and (*b*) give the genotypic frequencies expected in the second generation of random mating.

9.15. The autosomal gametic disequilibrium in a population is expressed as $d = 0.12$. The two loci under consideration recombine with a frequency of 16%. Calculate the disequilibrium (*d* value) that existed in the gamete pool of (*a*) the previous generation, and (*b*) the next generation.

9.16. The frequency of a sex-linked allele is 0.4 in males and 0.8 in females of a population (XY sex determination) not in genetic equilibrium. Find the equilibrium frequency of this allele in the entire population.

9.17. A laboratory population of flies contains all females homozygous for a sex-linked dominant allele and all males hemizygous for the recessive allele. Calculate the frequencies expected in each sex for the dominant allele in the first three generations of random mating.

9.18. For two independently assorting loci under Hardy-Weinberg conditions, (*a*) what is the maximum value of the disequilibrium coefficient (*d*)? (*b*) Specify the two conditions in which a population must be in order to maximize *d*.

9.19. Given gene *A* at frequency 0.2 and gene *B* at frequency 0.6, find the equilibrium frequencies of the gametes *AB*, *Ab*, *aB*, and *ab*.

CALCULATING GENE FREQUENCIES

Autosomal Loci with Two Alleles

Codominant Autosomal Alleles

9.20. A population of soybeans is segregating for the colors golden, light green, and dark green produced by the codominant genotypes $C^G C^G$, $C^G C^D$, $C^D C^D$, respectively. A sample from this population contained 2 golden, 36 light green, and 162 dark green. Determine the frequencies of the alleles C^G and C^D.

9.21. The MN blood group system in humans is governed by a pair of codominant alleles L^M and L^N. A sample of 208 Bedouins in the Syrian Desert was tested for the presence of the M and N antigens and found to contain 119 group M ($L^M L^M$), 76 group MN ($L^M L^N$), and 13 group N ($L^N L^N$). (a) Calculate the gene frequencies of L^M and L^N. (b) If the frequency of $L^M = 0.3$, how many individuals in a sample of size 500 would be expected to belong to group MN?

Dominant and Recessive Autosomal Alleles

9.22. The ability of certain people to taste the chemical phenylthiocarbamide (PTC) is governed by a dominant allele T, and the inability to taste PTC by its recessive allele t. If 24% of a population is homozygous taster and 40% is heterozygous taster, what is the frequency of t? (*Hint:* Use the same method as that employed for codominant alleles for greatest accuracy.)

9.23. Gene A governs purple stem and its recessive allele a produces green stem in tomatoes; C governs cut-leaf and c produces potato-leaf. If the observations of phenotypes in a sample from a tomato population were 204 purple, cut : 194 purple, potato : 102 green, cut : 100 green, potato, determine the frequency of (a) the cut allele, (b) the allele for green stem.

9.24. An isolated field of corn was found to be segregating for yellow and white endosperm. Yellow is governed by a dominant allele and white by its recessive allele. A random sample of 1000 kernels revealed that 910 were yellow. Find the allelic frequency estimates for this population.

9.25. The R locus controls the production of one system of antigens on the red blood cells of humans. The dominant allele results in Rh-positive individuals, whereas the homozygous recessive condition results in Rh-negative individuals. Consider a population in which 85% of the people are Rh-positive. Assuming the population to be at equilibrium, what is the gene frequency of alleles at this locus?

9.26. What is the highest frequency possible for a recessive lethal that kills 100% of its bearers when homozygous? What is the genetic constitution of the population when the lethal allele reaches its maximum?

9.27. Dwarf corn is homozygous recessive for gene d, which constitutes 20% of the gene pool of a population. If two tall corn plants are crossed in this population, what is the probability of a dwarf offspring being produced?

9.28. A metabolic disease of humans called phenylketonuria is the result of a recessive gene. If the frequency of phenylketonurics is 1/10,000, what is the probability that marriages between normal individuals will produce a diseased child?

Sex-Influenced Traits

9.29. Baldness is governed by a sex-influenced trait that is dominant in men and recessive in women. In a sample of 10,000 men, 7225 were found to be nonbald. In a sample of women of equivalent size, how many nonbald women are expected?

9.30. The presence of horns in some breeds of sheep is governed by a sex-influenced gene that is dominant in males and recessive in females. If a sample of 300 female sheep is found to contain 75 horned individuals, (a) what percentage of the females is expected to be heterozygous, (b) what percentage of the males is expected to be horned?

Autosomal Loci with Multiple Alleles

9.31. The genetics of the ABO human blood groups is presented in Solved Problem 9.7. (a) A sample of a human population was blood-grouped and found to contain 23 group AB, 441 group O, 371 group B, and 65 group A. Calculate the allelic frequencies of I^A, I^B, and i. (b) Given the gene frequencies $I^A = 0.36$, $I^B = 0.20$, and $i = 0.44$, calculate the percentage of the population expected to be of groups A, B, AB, and O.

9.32. The color of screech owls is under the control of a multiple allelic series: G^r (red) $> g^i$ (intermediate) $> g$ (gray). A sample from a population was analyzed and found to contain 38 red, 144 intermediate, and 18 gray owls. Calculate the allelic frequencies.

9.33. Several genes of the horse are known to control coat colors. The A locus apparently governs the distribution of pigment in the coat. If the dominant alleles of the other color genes are present, the multiple alleles of the A locus produce the following results: A^+ = wild-type (Przewalski) horse (bay with zebra markings), A = dark or mealy bay (black mane and tail), a^t = seal brown (almost black with lighter areas), a = recessive black (solid color). The order of dominance is $A^+ > A > a^t > a$. If the frequency of $A^+ = 0.4$, $A = 0.2$, $a^t = 0.1$, and $a = 0.3$, calculate the equilibrium phenotypic expectations.

Sex-Linked Loci

9.34. A genetic disease of humans called hemophilia (excessive bleeding) is governed by a sex-linked recessive gene that constitutes 1% of the gametes in the gene pool of a certain population. (a) What is the expected frequency of hemophilia among men of this population? (b) What is the expected frequency of hemophilia among women?

9.35. Color blindness in humans is due to a sex-linked recessive gene. A survey of 500 men from a local population revealed that 20 were color blind. (a) What is the gene frequency of the normal allele in the population? (b) What percentage of the females in this population would be expected to be normal?

9.36. The white eyes of *Drosophila* are due to a sex-linked recessive gene and wild type (red eyes) to its dominant allele. In a *Drosophila* population the following data were collected: 15 white-eyed females, 52 white-eyed males, 208 wild-type males, 365 wild-type females (112 of which carried the white allele). Using all the data, calculate the frequency of the white allele.

TESTING A LOCUS FOR EQUILIBRIUM

9.37. A pair of codominant alleles governs coat colors in Shorthorn cattle: $C^R C^R$ is red, $C^R C^W$ is roan, and $C^W C^W$ is white. A sample of a cattle population revealed the following phenotypes: 180 red, 240 roan, and 80 white. (a) What is the frequency of the C^R allele? (b) What is the frequency of the C^W allele? (c) Does the sample indicate that the population is in equilibrium? (d) What is the chi-square value? (e) How many degrees of freedom exist? (f) What is the probability that the deviation of the observed from the expected values is due to chance?

9.38. A blood group system in sheep, known as the XZ system, is governed by a pair of codominant alleles (X and X^Z). A large flock of Rambouillet sheep was blood-grouped and found to contain 113 X/X, 68 X/X^Z, and 14 X^Z/X^Z. (a) What are the allelic frequencies? (b) Is this population conforming to the equilibrium expectations? (c) What is the chi-square value? (d) How many degrees of freedom exist? (e) What is the probability of the observed deviation being due to chance?

9.39. The frequency of the T allele in a human population $= 0.8$, and a sample of 200 yields 90% tasters of a chemical called PTC (Problem 9.22) (T-) and 10% nontasters (tt). (a) Does the sample conform to the equilibrium expectations? (b) What is the chi-square value? (c) How many degrees of freedom exist? (d) What is the probability that the observed deviation is due to chance?

9.40. In poultry, the autosomal gene F^B produces black feather color and its codominant allele F^W produces splashed-white. The heterozygous condition produces Blue Andalusian. A splashed-white hen is mated to a black rooster and the F_2 was found to contain 95 black, 220 blue, and 85 splashed-white. (a) What F_2 ratio is expected? (b) What is the chi-square value? (c) How many degrees of freedom exist? (d) What is the probability that the observed deviation is due to chance? (e) May the observations be considered to conform to the equilibrium expectations?

⁇⁇ Review Questions

Vocabulary For each of the following definitions, give the appropriate term and spell it correctly. Terms are single words unless indicated otherwise.

1. The breeding structure of a population when each gamete has an equal opportunity of uniting with any other gamete from the opposite sex. (One or two words.)
2. The total genetic information possessed by the reproductive members of a population of sexually reproducing organisms. (Two words.)
3. An important model of population genetics. (Three words.)
4. The condition of a locus that does not experience a change in allelic frequencies from one generation to the next.
5. An interbreeding group of organisms sharing a common gene pool. (Two words.)
6. Changes in gene frequencies due to sampling errors in very small populations. (Two words.)
7. A deviation from Hardy-Weinberg expectations at any specific time in a population.
8. A phenotypically and/or geographically distinctive group, composed of individuals inhabiting a defined geographic and/or ecological region, and possessing characteristic phenotypic and gene frequencies that distinguish it from other such groups.
9. The general hallmark criterion that demarcates one biological species from another. (Two words.)
10. Any change in the genetic composition of a population, such as a change in allele frequency.

Multiple-Choice Questions Choose the one best answer.

For problems 1–4, use the following information. Snapdragon flowers may be red ($C^r C^r$), pink ($C^r C^W$), or white ($C^W C^W$). A sample from a population of these plants contained 80 white, 100 pink, and 20 red-flowered plants.

1. The frequency of the red allele (C^r) in this sample is   (a) 0.10   (b) 0.20   (c) 0.30   (d) 0.45   (e) none of the above
2. The percentage of pink-flowered plants expected on the basis of the Hardy-Weinberg equation is approximately   (a) 35   (b) 45   (c) 50   (d) 55   (e) none of the above
3. A chi-square test of the sample data against the Hardy-Weinberg expectations produces a chi-square value of   (a) 1.96   (b) 2.43   (c) 2.87   (d) 3.02   (e) 3.11
4. Refer to Table 2-3 to answer this question. Assuming the sample is representative of its population in problem 3 above, it may be said that (a) the chi-square test is significant and the sampled population is not in genetic equilibrium, (b) the chi-square test is nonsignificant and the sampled population is not in genetic equilibrium (c) the chi-square test is significant and the sampled population is in genetic equilibrium (d) the chi-square test is nonsignificant and the sampled population is in genetic equilibrium (e) the chi-square value is significant, thereby invalidating the test.

For problems 5–7, use the following information. Black pelage is an autosomal dominant trait in guinea pigs; white is the alternative recessive trait. A Hardy-Weinberg population was sampled and found to contain 336 black and 64 white individuals.

5. The frequency of the dominant black gene is estimated to be   (a) 0.60   (b) 0.81   (c) 0.50   (d) 0.89   (e) none of the above
6. The percentage of black individuals that is expected to be heterozygous is approximately (a) 46   (b) 57   (c) 49   (d) 53   (e) none of the above
7. The probability that a black male crossed to a white female would produce a white offspring is approximately   (a) 0.12   (b) 0.14   (c) 0.16   (d) 0.18   (e) none of the above

For problems 8–10, use the following information. Yellow body color in *Drosophila* is governed by a sex-linked recessive gene; wild-type color is produced by its dominant allele.

8. A sample from a Hardy-Weinberg population contained 1021 wild-type males, 997 wild-type females, and 3 yellow males. The percentage of the gene pool represented by the yellow allele is estimated to be   (*a*) 0.04   (*b*) 0.16   (*c*) 0.21   (*d*) 0.42   (*e*) none of the above
9. If the frequency of the yellow allele is 0.01, the percentage of wild-type females expected to carry the yellow allele is   (*a*) 1.98   (*b*) 1.67   (*c*) 2.04   (*d*) 2.76   (*e*) none of the above
10. If the frequency of the yellow allele is 1.0 in females and 0 in males, the frequency of that allele in males of the next generation is expected to be   (*a*) 1.0   (*b*) 0.5   (*c*) 0.33   (*d*) 0.67   (*e*) none of the above

Answers to Supplementary Problems

9.12. 0.5

9.13. All individuals are *AaBb*.

9.14. (*a*) $p_m = p_f = 0.6$, $q_m = q_f = 0.4$ (*b*) $AA = 0.36$, $Aa = 0.48$, $aa = 0.16$

9.15. (*a*) $1/7 = 0.143$ (*b*) 0.1008

9.16. $2/3 = 0.67$

9.17. Males: (1) = 1.0, (2) = 0.5, (3) = 0.75; females: (1) = 0.5, (2) = 0.75, (3) = 0.625

9.18. (*a*) 0.25   (*b*) 1/2 *AABB* : 1/2 *aabb* or 1/2 *aaBB* : 1/2 *AAbb*

9.19. $AB = 0.12$, $Ab = 0.08$, $aB = 0.48$, $ab = 0.32$

9.20. $C^D = 0.9$, $C^G = 0.1$

9.21. (*a*) $L^M = 75.5\%$, $L^N = 24.5\%$ (*b*) 210

9.22. $\iota = 0.56$

9.23. (*a*) $C = 0.30$ (*b*) $a = 0.58$

9.24. $Y = 0.7$, $y = 0.3$

9.25. $R = 0.613$, $r = 0.387$

9.26. 0.5; all individuals are heterozygous carriers of the lethal allele.

9.27. $\left(\dfrac{2pq}{1-q^2}\right)^2 \left(\dfrac{1}{4}\right) = \dfrac{p^2q^2}{(1-q^2)^2} = \dfrac{1}{36} = 0.0277778$

9.28. $\left[\dfrac{2(0.99)(0.01)}{(0.99)^2 + 2(0.99)(0.01)}\right]^2 (0.25) = 0.01\%$

9.29. 9775

9.30 (*a*) 50%   (*b*) 75%

9.31 (*a*) $I^A = 0.05$, $I^B = 0.25$, $i = 0.70$ (*b*) A = 44.6%, B = 21.6%, AB = 14.4%, O = 19.4%

9.32. $G^r = 0.1$, $g^i = 0.6$, $g = 0.3$

9.33. 64% wild type, 20% dark bay, 7% seal brown, 9% black

9.34. (*a*) 1/100   (*b*) 1/10,000

9.35. (*a*) 0.96   (*b*) 99.84%

9.36. $w = 0.19$

9.37. (*a*) 0.6   (*b*) 0.4   (*c*) Yes   (*d*) 0   (*e*) 1   (*f*) 1

9.38. (*a*) $X = 0.75$, $X^Z = 0.25$ (*b*) Yes   (*c*) 0.7   (*d*) 1   (*e*) 0.3–0.5

9.39. (*a*) No   (*b*) 18.75   (*c*) 1   (*d*) < 0.001

9.40. (*a*) 1/4 black : 1/2 blue : 1/4 splashed-white   (*b*) 4.50   (*c*) 1   (*d*) 0.01–0.05   (*e*) No

Answers to Review Questions

Vocabulary

1. random mating or panmixis
2. gene pool
3. Hardy-Weinberg rule
4. equilibrium
5. Mendelian population
6. genetic drift
7. disequilibrium
8. race or subspecies
9. reproductive isolation
10. evolution

Multiple-Choice Questions

1. *e* (0.35)   2. *b* (45.5%)   3. *a*   4. *d*   5. *a*   6. *b*   7. *e* (0.286)   8. *e* (0.29)   9. *a*   10. *a*

CHAPTER 10

Genetics of Bacteria

Characteristics of Bacteria

Each cellular organism is classified as either prokaryote or eukaryote (acceptable alternative spellings are procaryote and eucaryote). A **prokaryote** is generally a single-celled organism whose DNA is not confined within a true nucleus. A eukaryote can be single-celled or multicelled and has its genetic material isolated from the rest of the cell by a nuclear membrane. All bacteria are prokaryotes. There are two main prokaryotic groups—the **Eubacteria** (true bacteria) or **Bacteria** and the **Archaebacteria** (ancient bacteria) or **Archaea**. These two groups are distinguished based on their cellular structures and DNA sequences. The Eubacteria contains most of the organisms we typically identify as bacteria, such as the common gut microbe *Escherichia coli,* and the causative agent of strep throat, *Streptococcus pyogenes*. The Archaea contains organisms that have been hypothesized to be evolutionarily older, or at least distinct, from the Eubacteria, such as the methanogens (bacteria that produce methane). All other forms of life (fungi, plants, animals) are eukaryotes in the **Eukarya** group. Bacteria are part of the larger category of organisms called **microorganisms**. This category includes bacteria as well as some smaller eukaryotes, such as the fungi, protozoans, and algae.

Most bacterial cells have a wall surrounding their plasma membrane. The wall contains a unique chemical called **peptidoglycan** (also called **murein**). There are two main types of bacteria based on the structure of their walls, and, in part, on the amount of peptidoglycan in their walls. The **Gram-positive** bacteria have a thick peptidoglycan layer, while the **Gram-negative** bacteria have a thinner peptidoglycan layer with an extra outer membrane. The antibiotic **penicillin** is effective against bacteria because it disrupts the synthesis of peptidoglycan.

Bacteria do not reproduce sexually (i.e., by formation of haploid gametes produced by meiosis, and fusion of gametes to form diploid zygotes) and do not use mitosis as a mechanism for cellular division. The bacterial chromosome does not condense, it has no centromere, and no spindle develops. Instead, the circular bacterial chromosome is replicated and, as the cell elongates, new cell wall material is laid down and the chromosome copies move apart in a process called **binary**

fission. Bacteria can divide much more rapidly than eukaryotes (once every 20 min under ideal conditions, in contrast to 24–48 h or longer for many eukaryotic cells). Bacteria are typically about 1 μm wide and 1–5 μm long. Some bacteria can be as wide as 50 μm and several millimeters long, but this is rare. As mentioned earlier, bacteria are primarily single-celled organisms; however, there are exceptions. The Actinomycetes contain bacterial species that form filamentous, multicellular structures and, more recently, many different types of bacteria previously known only as single-celled organisms have been observed to form multicellular, slimy structures called **biofilms**. Biofilms are difficult to destroy due to difficulty in penetrating their multilayer, complex structure. They have been observed to accumulate in pipes used for food processing, medical catheters, or in patient tissues—resulting in difficult-to-treat diseases. In fact, symptoms of cystic fibrosis are thought to be due, in part, to biofilm accumulation in the patient's lungs.

Most of the genetic information of a bacterial cell resides in a single, circular, double-stranded DNA molecule, commonly referred to as the **bacterial chromosome**, in a region of the cell called the **nucleoid**. Some bacterial DNA is complexed with basically charged proteins to form a kind of bacterial chromatin analogous to the association of histones and other chromosomal proteins with DNA in eukaryotic chromatin. Some bacteria may also contain small, self-replicating DNA circles called **plasmids**. There seldom are any membrane-bound organelles in bacterial cells.

SOLVED PROBLEM 10.1

Under optimal conditions, some bacteria can divide every 20 min. Suppose each cell has a mass of 2×10^{-9} milligrams (mg). The mass of Earth is approximately 5.97×10^{27} grams (g). Determine the time (in hours) required for the progeny of a single cell dividing without restriction at the above rate to equal the weight of Earth.

Solution: At time zero we have one cell; 20 min later we have two cells; at 40 min there are four cells; at 60 min there are eight cells; etc. Thus, there are three cell divisions per hour. The number of cells at any hour, t, is 2^{3t}. The mass of each cell in grams is 2×10^{-9} mg $\times 10^{-3}$ mg/g $= 2 \times 10^{-12}$ g. The number of cells equivalent to the weight of Earth is

$$(5.97 \times 10^{27} \text{ g})/(2 \times 10^{-12}) = 2.98 \times 10^{39} = 2^{3t} \text{ hours (h)}$$

from which $3t \log 2 = \log 2.98 + \log 10^{39}$,

$$t = \frac{\log 2.98 + 39}{3 \log 2} = \frac{39.475}{3(0.301)} = 43.7 \text{ h}$$

Culture Techniques

Elemental requirements for most life on this planet include carbon, hydrogen, nitrogen, oxygen, phosphorus and sulfur, plus some mineral elements, such as magnesium and zinc. The elements are used by cells to form the basic molecular building blocks (e.g., proteins, carbohydrates, nucleic acids, lipids) of the cell. Cells also require an energy source, such as sugars or light from the sun, to help them generate the needed power for cellular reactions. Bacteria grown in the laboratory in an aqueous solution of nutrients and energy sources are referred to as a bacterial **culture**. Bacteria can also be grown in a **broth** or liquid culture medium or on a medium that is made solid by the addition of **agar**. Agar is a complex carbohydrate isolated from algae that forms a solid matrix similar to gelatin. In addition, and unlike gelatin, most bacteria cannot digest agar. Solid media of this type are often poured into flat, circular containers called **Petri dishes** or into test tubes to make a **slant**. The inoculation of bacteria onto an agar surface is called **plating**. When a dilute sample from a culture is plated, each bacterial cell reproduces itself into a cluster of thousands of cells called a **bacterial colony** or clone that is visible to the naked eye. Barring mutation, all members of a colony are genetically identical. The number of cells in a culture can be estimated by plating.

> **EXAMPLE 10.1**
> Suppose that 0.1 ml of a 10^6-fold dilution of a bacterial culture is plated on nutrient agar and 200 colonies develop. If 0.1 ml produces 200 colonies, I ml should produce 10 times as many colonies. Furthermore, since the original culture was diluted 10^6, it must contain 10^6 more bacteria than the diluted sample. Thus, the cell density of the original culture is estimated to be $200 \times 10 \times 10^6 = 2 \times 10^9$ cells per milliliter.

When an undiluted sample of a dense culture is plated, the colonies are so numerous that they form a **lawn** of solid bacterial growth over the entire surface of the agar. Rare mutants can be easily isolated from such a lawn by several techniques to be explained later in this chapter.

Bacterial Phenotypes and Genotypes

Bacteria exist in a number of morphological forms: **bacilli** (rod-shaped), **cocci** (spherical), **spirilla** (spiral), **spirochetes** (helical), and **branched**. Because they are so small, individual bacterial cells are rarely studied in genetics. However, bacterial colonies are large enough to examine macroscopically and often exhibit variations in size, shape or growth habit, texture, color, and response to nutrients, dyes, drugs, antibodies, and viral pathogens (bacterial viruses, called bacteriophage or phage). Some bacteria can grow on **minimal media** containing a carbon and energy source (e.g., glucose), a nitrogen source, a source of sulfur, a few inorganic salts, and water. Bacteria that can grow on such a simple, "unsupplemented" medium are said to be **prototrophic**. If any other organic substance must be added to minimal medium to obtain growth, the bacteria are said to be **auxotrophic**. A medium that contains all the organic nutrients (amino acids, nucleotides, etc.) that could be required by any auxotrophic cell is called **complete medium**.

Five major types of phenotypic changes are commonly produced by bacterial mutations:

1. A change from prototrophy to auxotrophy or vice versa; i.e., the loss or recovery of the ability to produce products of biosynthetic pathways. For

example, a mutation that produces a defect in the gene that specifies the enzyme that converts glutamic acid to glutamine would cause the cell to be dependent on the environment for glutamine.

2. The loss or recovery of the ability to use alternative nutrients. For example, a mutation in the gene for the enzyme that converts the sugar lactose into glucose and galactose renders the cell incapable of growing in a medium where lactose is the only carbon source. These kinds of mutations that are involved in catabolic (degradative) reactions are independent of prototrophy or auxotrophy.

3. A change from drug sensitivity to drug resistance or vice versa. For example, most bacteria are sensitive to the antibiotic streptomycin, but resistant strains can be produced by mutation.

4. A change from phage sensitivity to phage resistance or vice versa. For example, a mutation in the bacterial receptor for the phage would render the cell resistant to infection.

5. The loss or recovery of structural components of the cell surface. For example, one pneumococcus strain may possess a polysaccharide capsule, whereas another strain may not have a capsule.

The symbols used to represent bacterial phenotypes and genotypes conform to the following rules: phenotypic symbols consist of three roman letters (the first letter is capitalized) with a superscript "+" or "−" to denote presence or absence of the designated character, and "*s*" or "*r*" for sensitivity and resistance, respectively. Genotypic symbols are lowercase with all components of the symbol italicized.

EXAMPLE 10.2

If the cell can synthesize its own leucine, its phenotype is symbolized Leu$^+$ and its genotype, *leu$^+$*. The substance that characterizes the phenotype in this case (leucine) is symbolized Leu. The genotype that is auxotrophic for leucine is *leu* or *leu$^-$*, and the phenotype in this case is Leu$^-$ (unable to grow without leucine supplementation). If more than one gene is required to produce the substance, the three-letter symbol would be followed by an italicized letter, such as *leuA*, *leuB*, etc. The genotype for resistance to the antibiotic drug penicillin is *pen^r* or *pen-r*; Pen^r or Pen-r is the corresponding phenotype. In partial diploids, the two haploid sets are separated by a diagonal line; thus, *leu$^+$/leuA$^-$*.

Genetically different members of the same bacterial species are sometimes recognized as different **strains** if the differences are small, or as different **varieties** if the differences are substantial.

EXAMPLE 10.3

One of the most thoroughly studied bacterial species is *Escherichia coli*, or *E. coli*. Strains are designated by adding an unitalicized capital letter or number after the species name, thus *E. coli* B, *E. coli* S, etc. The three most commonly used strains of *E. coli* are *E. coli* B (host for phages of the T series), *E. coli* C (host for the single-stranded DNA phage ϕX174), and *E. coli* K12 (harboring the lambda prophage). Note that variants within a strain are indicated by adding a number after the strain letter.

Isolation of Bacterial Mutants

There are several different mutant bacterial phenotypes that can be used to identify mutations effecting various functional pathways. When the mutant phenotype has an advantage over the wild type under a specified set of conditions, a **selection** scheme is employed to isolate the mutant. For example, it is relatively easy to isolate phage- or drug-resistant mutant clones by plating the bacteria on a medium that contains a **selective agent**, such as a phage or drug. Only resistant cells will form colonies on such plates. Prototrophic mutants can be isolated from an auxotrophic culture by plating on minimal medium; only prototrophic colonies would grow on such a plate. Isolation of auxotropic mutants is more difficult because they do not have a growth advantage over wild type. Generally, either an **enrichment** or a **screening** scheme must be used in this case. There are at least four methods for isolating auxotrophic mutants from prototrophic cultures, each of which is enhanced by first treating the culture with a mutagenic agent to increase the mutation rate.

1. In the **delayed enrichment method**, a diluted culture is plated on minimal medium and then covered with an agar layer of the same medium. The plate is incubated and the locations of prototrophic colonies are marked on the plate. A layer of nutrient medium is then added, and the nutrient is allowed to diffuse through the minimal agar. After another incubation period, the appearance of any new colonies may represent auxotrophs that could only grow after supplementation.
2. The **limited enrichment method** is a simplification of the delayed enrichment method. Bacteria are plated on minimal medium containing a very small amount of nutrient supplementation. Under such conditions, auxotrophic bacteria will undergo limited growth until the supply of nutrients is exhausted, and hence will form small colonies. Prototrophic bacteria will continue to grow and produce large colonies.
3. Penicillin interferes with the development of the bacterial cell wall only in growing cells, causing them to rupture. In the **penicillin enrichment technique**, the bacterial culture is exposed to penicillin in minimal medium. Growing prototrophic cells die, whereas auxotrophic cells cannot grow and therefore are not killed. The culture is then plated on nutrient medium without penicillin. Only auxotrophic colonies should form on the plate.
4. In the **replica plating technique** (Fig. 10-1), the bacteria are first plated on nutrient agar and allowed to form colonies. A sterile velvet pad is then pressed onto the surface of this "master plate." The nap of the velvet picks up representatives from each colony on the plate. The pad is then pressed onto the sterile surface of one or more "replica plates" containing minimal medium. Only prototrophic colonies grow on the replica plate. Colonies on the master plate that are not represented on the replica plate may be auxotrophs. These colonies can be picked from the master plate to form a "pure" auxotrophic culture. The purity of cultures must be constantly monitored to remove newly formed mutants.

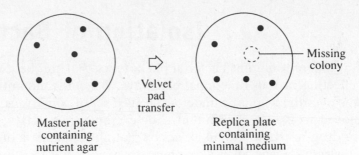

Fig. 10-1. The replica plating technique. The colony missing on the replica plate is likely to be an auxotrophic mutant. In reality, there would be about 100 colonies on each plate so that screening hundreds or thousands of such plates might be necessary to find a particular auxotrophic mutant.

SOLVED PROBLEM 10.2

The discipline of bacterial genetics began in 1943 when S. E. Luria and M. Delbrück published a paper entitled "Mutations of bacteria from virus sensitivity to virus resistance." Before this time, it was not known if the heredity of bacteria changed adaptively in specific ways as a consequence of exposure to specific environments, or whether specific mutants existed in the population prior to an environmental challenge. The former idea was Lamarckian, the latter was neo-Darwinian. Luria and Delbrück found that there was great variation from one trial to another in the number of *E. coli* that were resistant to lysis by phage T1. In order to determine which of the two hypotheses was correct, they devised the following "fluctuation test." Twenty 0.2-ml "individual cultures" and one 10-ml "bulk culture" of nutrient medium were incubated with about 10^3 *E. coli* cells per milliliter. The cultures were incubated until they contained about 10^8 cells per milliliter. The entire 0.2 ml of each individual culture was spread on a nutrient agar plate heavily seeded with T1 phages. Ten 0.2-ml samples from the bulk culture were also treated in similar fashion. After overnight incubation, the total number of T1-resistant (Ton^r) bacterial cells was counted; the results are presented in the following table. What inferences can be drawn from this "fluctuation test"?

Individual Cultures		Samples from Bulk Culture	
Culture Number	Number of Resistant Colonies	Sample Number	Number of Resistant Colonies
1	1	1	14
2	0	2	15
3	3	3	13
4	0	4	21
5	0	5	15
6	5	6	14
7	0	7	26
8	5	8	16
9	0	9	20
10	6	10	13
11	107		
12	0		
13	0		
14	0		
15	1		
16	0		
17	0		
18	64		
19	0		
20	35		

Solution: Variances for each experiment can be calculated from the square of formula (8.2) or (8.3); the individual cultures have a variance of 714.5, whereas the variance of samples from the bulk culture is 16.4.[1] In a Poisson distribution, the mean and the variance are essentially identical; hence, the variance/mean ratio should be near unity (1.0). The variance/mean ratio for the bulk culture samples is 16.4/16.7 = 0.98 or nearly 1.0, as expected from a random distribution of rare events. The samples from the bulk culture collectively serve as a control for the individual cultures. The same ratio for the individual cultures, however, is 714.5/11.3 = 63.23, indicating that there are extremely wide fluctuations of the numbers of Ton^r cells in each culture around the mean. If resistance to the T1 phages occurs with a given probability only after contact with the phages, then each culture from both the individual and batch experiments should contain approximately the same average number of resistant cells. On the other hand, if Ton^r mutants occurred prior to contact with the phages, great variation around the mean is expected from one individual culture to another because some will incur a mutation

[1] When the probability of a rare event (e.g., a specific mutation) is relatively small and the sample size is relatively large, the binomial distribution (Example 2.26) is essentially the same as a Poisson distribution, but is much easier to solve by the latter.

early and others late (or not at all) during the incubation period. This experiment argues in favor of the mutation hypothesis and against the induced resistance hypothesis.

Certain mutations, such as that to phage resistance, are preadaptive in that their selective advantage only becomes manifest when phages are in the environment as a selective agent; in this case, T1-sensitive bacteria (Ton^s) are killed by T1 phages, allowing only the few Ton^r cells to survive and multiply. Phage resistance depends upon altering the structure of the bacterial receptor sites to which T1 phages normally attach. Immunity to superinfection by a specific phage is based upon production of a repressor of phage replication by a lysogenic cell.

SOLVED PROBLEM 10.3

Two triple auxotrophic strains of *E. coli* are mixed in liquid medium and plated on complete medium, which then serves as a master for replica plating onto six kinds of media. From the position of the clones on the plates and the ingredients in the media, determine the genotype for each of the six clones. The gene order is as shown.

$$thr^- leu^- thi^- bio^+ phe^+ cys^+ \times thr^+ leu^+ thi^+ bio^- phe^- cys^-$$

Gene symbols are abbreviated as follows:

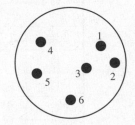

thr = threonine	*bio* = biotin
leu = leucine	*phe* = phenylalanine
thi = thiamin	*cys* = cystine

Master Plate (complete medium)

Replica plates: Each dish contains minimal medium plus the supplements listed below each dish.

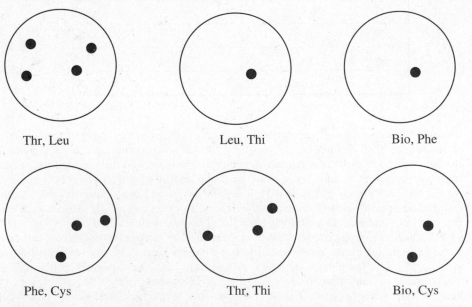

Thr, Leu	Leu, Thi	Bio, Phe
Phe, Cys	Thr, Thi	Bio, Cys

Solution: Clone 1 grows when supplemented with Thr and Leu or Thr and Thi, but not with Leu and Thi. Therefore, this colony is auxotrophic for Thr alone (*thr$^-$ leu$^+$ thi$^+$ bio$^+$ phe$^+$ cys$^+$*).

Clone 2 appears only on the plate supplemented with Phe and Cys. This is a double auxotrophic colony of genotype *thr$^+$ leu$^+$ thi$^+$ bio$^+$ phe$^-$ cys$^-$*.

Clone 3 appears on all replica plates and therefore must be prototrophic ($thr^+ leu^+ thi^+ bio^+ phe^+ cys^+$).

Clone 4 grows only when supplemented with Thr and Leu and therefore must be a double auxotroph of genotype $thr^- leu^- thi^+ bio^+ phe^+ cys^+$.

Clone 5 and clone 1 always appear together on the replica plates and therefore have the same genotype.

Clone 6 can grow in the presence of Phe and Cys or Bio and Cys. The common factor is Cys, for which this strain is singly auxotrophic ($thr^+ leu^+ thi^+ bio^+ phe^+ cys^-$).

Bacterial DNA Replication and Cell Division

The circular chromosome of bacteria presents special problems for replication. Circular chromosomes usually have a single site, called the **origin**, or **ori**, site, at which replication originates. By contrast, many ori sites exist on each chromosome of eukaryotes. Once the replication process starts, it usually proceeds bidirectionally from the ori site to form two **replication forks**.

As the two strands of a right-handed, double-helical, circular DNA unwind during replication, the molecule tends to become **positively supercoiled** or overwound, i.e., twisted in the same direction as the strands of the double helix. These supercoils would interfere with further replication if they were not removed. **Topoisomerases** are a group of enzymes that can change the topological or configurational shape of DNA. **DNA gyrase** is a bacterial topoisomerase that makes double-stranded cuts in the DNA, holds on to the broken ends so they cannot rotate, passes an intact segment of DNA through the break, and then reseals the break on the other side (Fig. 10-2). This action of DNA gyrase quickly removes positive supercoils and momentarily relaxes the DNA molecule into a more energetically stable state. However, with the expenditure of energy, DNA gyrase normally pumps **negative supercoiling** or underwinding (twisting in a direction opposite to the turns of the double helix) into relaxed DNA circles so that virtually all DNAs in both prokaryotes and eukaryotes naturally exist in the negative supercoiled state. Relaxed circles and positively supercoiled DNA exist only in

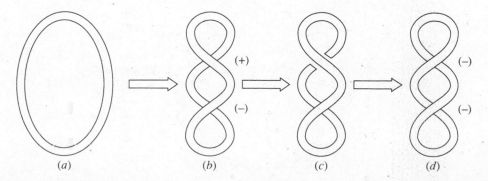

$$(a) \qquad (b) \qquad (c) \qquad (d)$$

Fig. 10-2. A proposed mechanism whereby DNA gyrase "pumps" negative supercoiling into DNA. A relaxed, covalently closed, circular DNA molecule (*a*) is bent into a configuration for strand passage (*b*). DNA gyrase makes double-stranded cuts (*c*), holds on to the ends, passes an intact segment through the break, and reseals the break on the other side (*d*).

the laboratory. Localized regions of DNA transiently and spontaneously unwind to single-stranded "bubbles" and then return to their former topology as hydrogen bonds between complementary base pairs are broken and reformed by thermal agitation. This process is sometimes referred to as "breathing." The strain of underwinding is thus momentarily relieved in a superhelix by an increase in the number, size, and duration of these bubbles. An equilibrium normally exists between these supercoiled and "bubbled" states. More bubbles form as the temperature increases.

At each replication fork, an enzyme called **helicase** unwinds the two DNA strands. **Single-stranded, DNA-binding (SSB) proteins** protect the single-stranded regions in the replication forks from forming intrastrand base pairings that could cause a tangle of partially double-stranded segments that would interfere with replication. The enzyme **primase** synthesizes short RNA primers using a region on each strand as a template. Primers are required for DNA polymerase to begin extending the new DNA strand. Three DNA polymerase enzymes (referred to as **pol I**, **pol II**, and **pol III**) have been found in *E. coli*. Pol III is the principal replicating enzyme. Gaps left by pol III are filled by pol I, and DNA ligase seals the nicks. The function of pol II is not well established, although it is known that it is not involved in RNA primer replacement. In addition to their 5′ to 3′ synthetic activity, both pol I and pol III have 3′ to 5′ exonuclease activity, which plays an "editing" or "proofreading" role by removing mismatched bases inserted by error during chain polymerization. Pol I also has 5′ to 3′ exonuclease activity by which it normally removes primers and replaces them with complementary DNA sequences after polymerization has begun. About halfway through the above replication process, the replicative intermediate molecule looks like the Greek letter theta (θ). This type of replication is therefore referred to as theta replication (Fig. 10-3).

Another type of bacterial replication is used to transfer a linear DNA molecule during bacterial conjugation or for the production of linear phage genomes. A nick occurs in one strand of a DNA double helix, creating free 3′-OH and 5′-P

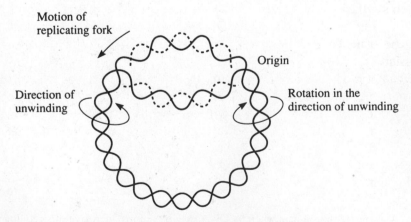

Fig. 10-3. Theta (θ) replication. Newly synthesized DNA is indicated by broken lines. Overwinding of the unreplicated segment (caused by unwinding of the daughter branches) is removed by the nicking action of DNA gyrase.

termini. Helicase and SSB proteins establish a replication fork. No primer is necessary because a strand with a free 3'-OH is available for elongation by DNA polymerase III as the leading strand. Simultaneously with replication of the leading strand, the template for the lagging strand is displaced. The displaced strand is discontinuously replicated to produce Okazaki fragments in the usual way (see Fig. 3-11). The result of this replication model is a circle with a linear tail, resembling the Greek letter sigma (σ). Hence, this model is called **sigma replication** or **rolling-circle replication** (Fig. 10-4). The circle may revolve several times, creating concatemers or covalently connected, linear repetitions of bacterial genomes. An endonuclease enzyme makes cuts at slightly different positions on each DNA strand of the concatemer to create genome-sized segments containing "sticky ends" (single-stranded complementary ends). The linear genomes circularize by base pairing of the sticky ends. DNA ligase seals each gap to create covalently closed (circular), double-stranded DNA molecules.

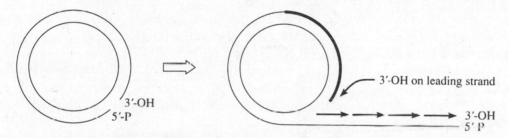

Fig. 10-4. Rolling-circle or sigma (σ) replication. Newly synthesized DNA is indicated by heavy lines.

A replicating bacterial chromosome is thought to be attached to invaginations of the cell membrane at each replication fork. After DNA replication, the cell elongates by growth of the sector between the two attachment points, causing the two chromosomal replicas to move apart. A septum of new cell membrane and wall is then synthesized between the two chromosomes, creating two progeny cells (Fig. 10-5). The "passing down" of DNA from parent to progeny cell is called **vertical gene transfer** and the overall process of bacterial cell division is called **binary fission**.

When bacteria are growing exponentially, most cells contain two to four identical chromosomes in various stages of replication. If a mutation occurs during replication, the new copy of DNA will be slightly different from the parental template. If this mutation occurs in a coding region that results in a defective protein, the effect of the mutation (i.e., the phenotype) will only be observable once the new cell has depleted its level of wild-type protein. This is a phenomenon known as **phenotypic lag**.

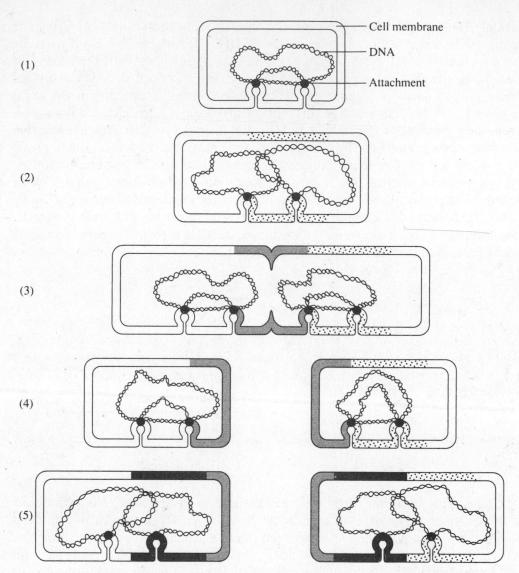

Fig. 10-5.   A model for segregation of bacterial DNA ("chromosome") replicas. (1) A circular DNA molecule is attached to invaginations of the cell membrane at two points. The DNA is a theta structure (about half replicated). (2) Replication is complete. (3) Cell division is beginning via growth of the membrane region shown in medium gray. Both daughter chromosomes are already partially replicated. (4) Cell division is complete. (5) Daughter cells are midway through the next generation. The membrane attachment points have moved to the center of the cell as a result of growth of new membrane (dark gray) after cell division. (After G. S. Stent and R. Calendar, *Molecular Genetics,* 2nd ed., W. H. Freeman and Company, New York, 1978.)

EXAMPLE 10.4

Resistance to a specific bacteriophage can be acquired by mutation of a gene responsible for the phage receptor on the cell's surface. Resistance cannot be fully realized until the receptor sites (synthesized under the direction of the former phage-sensitive genotype) have been completely diluted out through successive cell divisions. If even one receptor remains on the mutant cell, it is still susceptible to phage infection. Thus, many cell generations may be required before a phage-resistant mutation can be fully expressed in a progeny cell.

Bacterial Transcription

In bacteria, all RNA molecules (mRNAs, rRNAs, and tRNAs) are synthesized by the same enzyme, **RNA polymerase**. The complete, functional enzyme (**holoenzyme**) consists of five different polypeptide chains: β, β', α, ω, and σ. Each gene or genes to be transcribed has a region preceding it called the **promoter**, where the RNA polymerase complex binds to begin transcription. The **sigma (σ) subunit** is involved in initiation. It recognizes and binds to two different sequences of DNA in the promoter region: at 10 bases upstream (toward the $5'$ end) of the first base to be copied is the **Pribnow box** and at 35 bases upstream is the **-35 sequence**. The first base to be transcribed is generally labeled the $+1$ base and any sequence in front of or upstream of this base is given a negative number. This is why the sequence located 35 bases upstream is known as the -35 sequence, and the Pribnow box is often referred to as the -10 box. The binding of sigma to this region of DNA helps to align the remainder of the enzyme (called the **core enzyme**). The most common sequence found in Pribnow boxes of various genes, called the **consensus sequence**, is TATAAT and the consensus of the -35 region is TTGACA. RNA chains usually initiate with $5'$ adenosine triphosphate (pppA) or guanosine triphosphate (pppG). After transcription has begun, sigma factor dissociates from the core enzyme and may then associate with the same or another core enzyme to get it started on a promoter. "Strong" promoters have Pribnow box and -35 sequences that closely match the consensus sequence, allowing sigma to bind more effectively. This favors the production of many copies of RNA. Promoters that have Pribnow box and -35 sequences that are further from the consensus are "weak" and foster only low levels of transcription. The transcriptional activity of RNA polymerase may be blocked or its affinity for a promoter may be increased by the attachment of other specific DNA-binding proteins near the transcription initiation site. Some DNA sequences that precede (**leaders**) or follow (**trailers**) the gene may also control transcription. In addition, these sequences may play a role in regulating translation.

During elongation of the RNA molecule, ribonucleoside triphosphates of the bases A, U, G, and C pair with complementary bases in the sense strand of DNA and are then connected with $3'-5'$ phosphodiester bonds by RNA polymerase. Thus, the RNA molecule grows from its $5'$ end toward its $3'$ end ($5' \rightarrow 3'$). The DNA double helix unwinds ahead of the advancing RNA polymerase to expose more of the sense strand for extending the RNA chain. The DNA reforms its double-helical shape after the enzyme has passed a given region.

RNA polymerase in bacteria stops transcription of an RNA chain at DNA sequences known as **terminators** [Fig. 10-6(*a*)], and dissociates from the DNA. To function properly, some terminators require an accessory protein called **rho (*ρ*) factor**. The *ρ*-independent terminators have a diad symmetry in the double-stranded DNA, centered about 15–20 nucleotides before the end of the RNA, and have about six adenines in the sense strand that are transcribed into uracils at the end of the RNA. The RNA transcript of the diad symmetry folds back on itself to form a hairpin structure, ending with approximately 6 uracils [Fig. 10-6(*b*)]. Because an RNA-DNA hybrid consisting of polyribo-U and polydeoxy-ribo-A is very unstable, the RNA chain is quickly released from the DNA duplex. The *ρ*-dependent terminators lack this poly-A region. It is thought that even a weak hairpin structure causes RNA polymerase to pause, allowing *ρ* factor to attach to the terminator and cause dissociation of the RNA and RNA polymerase.

In bacteria, one single RNA transcript often contains the coding regions for multiple genes. This type of RNA is called **polycistronic** or **polygenic** (Fig. 10-7).

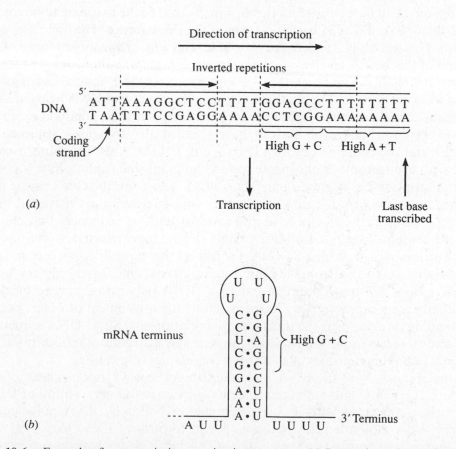

Fig. 10-6. Example of a transcription termination sequence. (*a*) Inverted repeats at the end of a gene are transcribed into (*b*), an mRNA that folds into a stem-and-loop structure that dislodges RNA polymerase from its DNA template. (After David Freifelder, *Molecular Biology*, Jones and Bartlett Publishing Company, Inc., Sudbury, MA, 1987.)

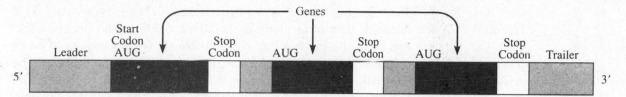

Fig. 10-7. Example of a polycistronic mRNA molecule. Untranslated regions are stippled or white; coding regions are dark.

The $70S^2$ bacterial ribosome consists of two major subunits: a larger 50S subunit and a smaller 30S subunit. The 50S subunit contains two rRNA molecules (23S and 5S); the 30S subunit contains a single 16S rRNA molecule. All three rRNAs are transcribed into a single 30S pre-rRNA transcript containing a **leader** sequence at the 5′ end, a **trailer** sequence at the 3′ end, and nonfunctional spacer regions between the three rRNA sequences. Likewise, all tRNAs are derived by nuclease digestion from longer pre-tRNA primary transcripts containing from one to as many as seven different tRNAs. Many bacterial genes with related functions are transcribed into a polycistronic mRNA under the control of one promoter; these genes are said to make up an **operon**.

Bacterial Translation

No membrane separates the DNA from the ribosomes in a bacterial cell. Hence, as soon as the 5′ end of an mRNA is transcribed, ribosomes can begin translation. In other words, transcription and translation are coupled processes in bacteria.

Translation occurs in three major steps: (1) initiation, (2) elongation, and (3) termination. A nucleotide sequence called the **Shine-Dalgarno sequence** (AGGAGG is the consensus sequence) in the leader of an mRNA molecule is complementary to a sequence at the 3′ end of 16S rRNA and thereby serves as a binding site for ribosomes. The initiation codon near the 5′ end of an mRNA molecule is 5′-AUG, coding for the amino acid methionine. In bacteria, a formyl group (CHO) becomes attached to the amino group of the methionine after it has become attached to its tRNA molecule. A **deformylase** enzyme removes the formyl groups from some polypeptide chains soon after their synthesis commences. In other cases, an **aminopeptidase** enzyme removes the terminal methionine (or part of the amino terminal end), so that not all functional bacterial proteins have formyl-methionione or methionine at their N termini.

Initiation of protein synthesis begins with the formation of a complex involving the 30S ribosomal subunit, guanosine triphosphate (GTP), and three protein **initiation factors** (IF1, IF2, IF3). In the next step, formylated methionyl-tRNA and the mRNA attach to the IF-30S-GTP complex, forming a 30S initiation

---

[2] S = Svedberg unit; a sedimentation coefficient for molecules in an ultracentrifuge. The S value tends to increase with the molecular weight of the molecule, but the geometry of the molecule also may be influential. Note that S units are not additive; i.e.; 50S subunit + 30S subunit = 70S for the complete bacterial ribosome, not 80S.

complex. Then IF3 is released, the 50S subunit is added, GTP is hydrolyzed, and IF1 and IF2 are released. The final complex is called a 70S initiation complex. Only the initiator region of the mRNA can simultaneously form two sets of base-pairing interactions (16S rRNA-mRNA and mRNA-fMet tRNA). In this way, the AUG start codon is distinguished from other AUG codons downstream in the mRNA. Base pairing between the leader of the mRNA and the 16S rRNA somehow dissociates after formation of the 70S initiation complex, so that elongation can begin.

The elongation phase requires GTP, three protein elongation factors (EF-Tu, EF-Ts, and EF-G), and a **peptidyl transferase** activity (now thought to be carried out by catalytic RNA molecules—ribozymes—within the ribosome). Two molecules of GTP are hydrolyzed for each amino acid added to the growing polypeptide chain. EF-Tu and EF-Ts cyclically interact to align each amino-acyl-tRNA complex (AA-tRNA) for effective codon-anticodon base pairing. EF-G (also called **translocase**) and GTP form a complex that mediates the movement of peptidyl-tRNA from the A site to the P site. Hydrolysis of GTP to guanosine diphosphate (GDP) is required for entry of the ribosome into the next elongation cycle. Several ribosomes may simultaneously translate the same mRNA. A collection of ribosomes of this kind is called a **polyribosome**.

Termination of translation in *E. coli* requires at least two protein **release factors**. RF1 recognizes the mRNA stop codons UAG and UAA; RF2 recognizes UGA and UAA. These RFs cause peptidyl transferase to transfer the completed polypeptide chain to water instead of to an AA-tRNA. After chain termination, the 30S and 50S ribosomal subunits dissociate from the mRNA and are then free to recycle into new initiation complexes on the same or different mRNA templates.

If a translational product is to be transported across the plasma membrane to the cell's exterior, the protein usually contains 15–30 additional amino acids (called a **signal sequence**) at its N terminus. The signal sequence is rich in uncharged, usually hydrophobic, amino acids that become anchored in the membrane, whereas the remainder of the polypeptide chain is extruded through the membrane as it is synthesized. A **signal peptidase** then cleaves the signal sequence to release the protein from the cell.

Genetic Recombination

There are three basic mechanisms by which DNA can be transferred from one mature, independent bacterial cell to another. These processes are known generally as **horizontal gene transfer**. If the donor DNA is incorporated, or recombined, into the genome of the recipient cell, a **recombinant** organism with one or more new phenotypes may result.

1 TRANSFORMATION

Bacterial **transformation** is the transfer of naked DNA that generally originates from one bacterial cell into a different cell. When a bacterial cell ruptures (in a process called cellular **lysis**), its circular DNA becomes released into the environment. The efficiency of transformation depends upon primarily on the competence of the cell. **Competence** is the ability of a cell to incorporate naked DNA. Not all bacterial species are capable of competence and those that are only become competent during a restricted part of their life cycle. During the competent state, the cell produces one or more proteins called "competence factors" that (1) modify the cell wall so it can bind exogenous (foreign) DNA fragments and (2) help take in and incorporate the foreign DNA. As the double-stranded DNA fragments penetrate the cell wall, one of the strands is degraded. Any fragment of DNA that has been transferred (by transformation or some other method) from a donor cell to a recipient cell is referred to as an **exogenote**; the **endogenote** is the native DNA of the recipient cell. A bacterial cell that has received an exogenote is initially diploid for part of its genome, and is said to be a **merozygote**. However, single-stranded exogenotes are unstable and will usually be degraded unless they are integrated into the endogenote. Any process of genetic exchange that transfers only part of the genetic material from one cell to another is called **meromixis**. It is thought that the single-stranded exogenote of transformation becomes coated with a protein (such as the RecA-protein of *E. coli*) that aids the exogenote to find a complementary region on the endogenote, to invade the double helix, to displace one of its strands, and to base-pair with the other strand. The displaced strand is enzymatically removed as the endogenote replaces it by homologous base pairing (a phenomenon known as **branch migration**). Trimming enzymes remove the free ends (either donor or recipient) and ligase seals the nicks. Once the exogenote is integrated into the endogenote and the displaced strand is degraded, the cell is no longer a merozygote. DNA size and concentration are also factors in transformation efficiency.

If the exogenote contains an allele of the endogenote, the resulting recombinant double helix would contain one or more mismatched base pairs, and is referred to as a **heteroduplex**. If progeny cells are to receive the new allele, mismatch repair must occur by excising a segment of the endogenote strand and using the exogenote strand as a template for its replacement. Since incorporation of the exogenote into the endogenote requires homologous recombination, the donor cell would normally belong to either the same species as the recipient cell or to a closely related one. Two or more closely linked genes may reside on the same transforming piece of DNA. If two or more genes are incorporated together into the endogenote, the recipient cell would be **cotransformed**. The frequency of cotransformation is a function of the linkage distance between the respective genes.

EXAMPLE 10.5

In 1928 Fred Griffith discovered the first example of bacterial transformation. *Streptococcus pneumoniae* is a bacterium that causes human pneumonia and can also kill mice. The virulent strain of this bacterium contains a polysaccharide capsule that tends to resist destruction by immune cells of the host species. A nonvirulent strain of the pneumococcus does not have a capsule. The virulent strain forms colonies with smooth borders on nutrient agar plates and is thus designated the S (smooth) strain; the nonvirulent strain forms colonies with rough borders and is designated the R (rough) strain. When mice were injected with both heat-killed S strain and live R strain, they died, and live S-strain bacteria were recovered from their bodies. Griffith did not know how to explain these results, but he called the process "transformation" and named the responsible substance "**transforming principle**." Later studies by Avery and others demonstrated that the transforming principle was naked DNA. In the Griffith experiment, the exogenote from the S strain contained the gene responsible for capsule formation. When this exogenote was incorporated into the endogenote of the R strain the transformant cells had the ability to make the capsule and thus became virulent S-type cells.

2   CONJUGATION

Bacterial **conjugation** involves the temporary union of two cells of opposite mating type, followed by unidirectional transfer of some genetic material through a cytoplasmic bridge from the donor cell to the recipient cell, and then disunion of the cells (**exconjugants**). An **episome** is an extrachromosomal genetic element that may exist either as an autonomously replicating circular DNA molecule or as an integrated DNA sequence within the host chromosome (e.g., phage lambda). A **plasmid** was originally defined as a small, circular DNA molecule that replicates autonomously of the bacterial chromosome and is incapable of integration into the bacterial chromosome. Plasmids often carry genes for antibiotic resistance that confer a selective advantage upon their host cells when antibiotics are in their environment. Currently, however, the word "plasmid" is often used for both episome and plasmid. Most plasmids are between 1/10 and 1/100 the size of the bacterial chromosome.

In some strains of *E. coli*, there is an episomal fertility factor (also known as a **sex plasmid** or **F plasmid**). Strains that carry an F plasmid are called males, designated F^+, and can manufacture a protein called **pilin** from which a conjugation tube, or **pilus**, can be constructed. Contraction of the pilus connecting two cells brings the conjugating cells into close contact. Cells that do not have an F plasmid are called females and are designated F^-. When an F^+ cell conjugates with an F^- cell, replication of the F plasmid is initiated. One strand of the F plasmid is broken, and replication by the rolling-circle mechanism (Fig. 10-4) causes the 5' end of the broken strand to enter the recipient cell through the pilus (Fig. 10-8), where it is copied into a double-stranded DNA molecule. The other strand of the F plasmid in the donor cell also replicates simultaneously so the donor cell does not lose its F plasmid (it remains F^+). The recipient cell thus becomes F^+.

An F plasmid has little homology with the bacterial chromosome, so homologous recombination between these two DNA circles rarely occurs. In approximately 1 in 10^5 cells, however, a nonhomologous recombination event causes F to become integrated into a site on the bacterial chromosome. Thus, F is an episome that can exist chromosomally or extrachromosomally. A cell with F integrated into its chromosome is called an **Hfr** (*h*igh *f*requency of *r*ecombination) cell. It is

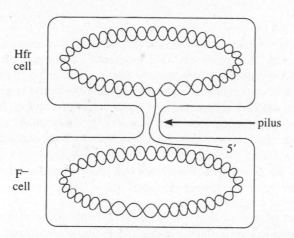

Fig. 10-8.   Transfer of a single strand of DNA from an Hfr (donor, male) cell to a recipient F⁻
(female) cell via the rolling-circle mechanism of replication. The same mechanism is used
to transfer the circular DNAs of plasmids.

so designated because many chromosomal genes can now be transferred from
donor to recipient with high frequency. In *E. coli,* integration of the F factor is
known to occur (in either of two orientations) at about 10 specific sites on the
chromosome. The integration site and the orientation of the integrated F determines the order with which chromosomal genes will be transferred during conjugation (Fig. 10-9). DNA replication begins in the Hfr cell at the F locus in such a
way that a small part of F is at the beginning of the donated segment; normal
bacterial genes then follow in sequence, and finally the remaining portion of the F
is replicated last. About 90 min is required for *E. coli* to transfer its entire genome
at 37°C under laboratory conditions. Thermal agitation of molecules (Brownian
movement) usually causes the pilus to rupture before DNA transfer is complete, so
a complete F particle is rarely recovered in the recipient cell. For this reason, the
F⁻ recipient cell usually remains F⁻ after conjugation with an Hfr cell. The presence of the exogenote in the recipient cell activates a recombination system that
causes genetic exchange to occur. In order to recover these recombinant bacteria,
it is convenient to use an Hfr cell that is sensitive to some antibiotic and a recipient
cell that is resistant to that same antibiotic. The locus of the antibiotic-resistance
gene in the recipient cell ideally should be so far from the origin of the exogenote
that the pilus will almost always rupture (stopping conjugation) before that locus
can be transferred.

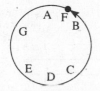   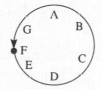

This Hfr strain transfers   This Hfr strain transfers   This Hfr strain transfers
its markers in the order   its markers in the order   its markers in the order
BCDEGAF.   GABCDEF.   DEGABCF.

Fig. 10-9.   Three examples of Hfr strains. Each strain transfers its genes in a unique sequence.
Arrows indicate the directions of transfer.

EXAMPLE 10.6

Suppose that an Hfr strain is able to synthesize leucine and is sensitive to the antibiotic streptomycin (*leu⁺*, *str-s*) and that a recipient F⁻ strain is unable to synthesize leucine, but is resistant to streptomycin (*leu⁻*, *str-r*). Mixing of these two strains allows conjugation to occur. Recombinants are selected by plating the mixture on minimal medium (without leucine) containing streptomycin. The Hfr strain cannot grow in the presence of streptomycin. The recipient strain cannot grow unless it has received the *leu⁺* gene by conjugation. Only the recombinants of genotype *leu⁺*, *str-r* will form colonies on the plate. In this case, *leu⁺* is the selected marker; *str-r* is the counterselective marker that prevents growth of any cell other than a recombinant on this type of medium.

The integration of F factor with the host chromosome is a reversible process. Normally, the excision event that releases the F factor from the host chromosome involves a crossover at the same position at which it was integrated. Occasionally, however, the excision event is aberrant, and the released F factor contains one or a few bacterial genes that were close to the integrated F (Fig. 10-10). Such a cell is symbolized **F′**. This cell thus has a deletion in part of its chromosome; the missing material is present in the F plasmid. If the chromosomal genes in the plasmid are essential genes, the F′ cell becomes dependent upon both the host chromosome and the F plasmid for its survival. Normal binary fission of an F′ cell produces progeny that are also F′. However, when an F′ cell conjugates with an F⁻ recipient cell, the recipient cell becomes partially diploid for the small piece of chromosomal material carried in the F′ particle. Because the F′ plasmid in this new F′ cell contains genetic material that is homologous with a segment of the host chromosome, there is a higher probability that this DNA may become integrated at the homologous region. Furthermore, the new F′ cell can conjugate with F⁻ recipients and transfer the F′ plasmid information to all its exconjugants with high frequency. This process is also referred to as **sex-duction** or **F-duction**.

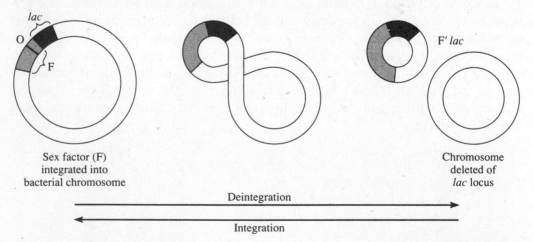

Fig. 10-10. Formation of an F genote (F′). O is the origin of chromosome transfer when the sex or fertility factor F is integrated into the bacterial chromosome. If the lactose locus (dark segment) is adjacent to the F locus (gray region) an F′ *lac* genote can be formed by deintegration (breakage and reunion). Not drawn to scale; the F plasmid is approximately 1/40, the length of the bacterial chromosome.

Plasmids carry genetic information for their own replication, but usually none that is essential for the life of the cell in normal environments. However, plasmids may carry one or more genes that confer selective advantage to the cell in certain environments, such as in the presence of antibiotics. Plasmids that carry genes for resistance to one or more substances normally toxic to the host are designated **R plasmids** or **R factors**. Some strains of *E. coli* carry a colicinogenic plasmid (Col) that can synthesize colicin, a protein that kills closely related bacterial strains that lack the Col plasmid. The best known of the bacterial plasmids are designated F, R, and Col.

3   TRANSDUCTION

The transfer of genetic material from one bacterium to another, using a bacterial virus (**bacteriophage, phage**) as a vector, is called **transduction**. This aspect of bacterial gene transfer is discussed in Chapter 11. However, the basic principle of a donor strain that provides DNA to a recipient strain is the same as in the previously discussed mechanisms. The primary difference is that the DNA is transferred through a bacteriophage intermediate.

Regulation of Bacterial Gene Activity

Within any cell, not all its genes are active at the same time. Some gene products need to be continuously synthesized, whereas others are necessary only during certain phases of the life cycle or perhaps only when particular environments are encountered. Even when genes are "turned on," the quantity of proteins they specify may need to be controlled. Some proteins need to be synthesized in large amounts and others only in small amounts. Therefore, the activity of virtually all genes needs to be regulated in one or more ways to make the most efficient use of the energy available to the cell. These regulatory mechanisms over gene expression may act at one or more levels. Regulation may occur at the level of the gene itself by controlling the timing and/or rate of transcription. Other control mechanisms may operate during translation. After translation, some proteins must be modified to become functional.

The transcriptional activity of genes may be unregulated if their products are needed regardless of environmental conditions. Such products are said to be synthesized **constitutively**. The quantity of products from these "unregulated" genes can vary, however, depending upon the relative affinities of their promoters for RNA polymerase. For those proteins that are required only under certain conditions, their genes would usually be governed by one or more **regulatory proteins**. Regulatory proteins usually do not have an enzymatic function, but instead interact with the DNA in or near the promoter of a gene to regulate transcription.

There are two basic types of regulatory proteins: repressors and activators. A **repressor** protein binds to a site called the **operator** within an operon. The attachment of a repressor protein to an operator prevents transcription of all structural genes in the same operon. A gene with this form of regulation is said to be under **negative control**. Proteins required for the expression of an operon are called **activators**. They may bind to **initiator or activator sites** that are located within an operon's promoter or, in the case of **enhancer sites,** they may bind at sequences far from the operon. When the binding of a regulatory protein to an initiator or enhancer site stimulates transcription of structural genes in the operon, a **positive control** mechanism is said to be at work.

The stimuli to which regulated genes respond may vary from relatively small molecules (e.g., sugars, amino acids) to relatively large substances (e.g., in eukaryotes, a complex of a steroid hormone and its protein receptor). A substance that turns on gene transcription is referred to as an **inducer**, whereas a substance that turns transcription off is said to be a **corepressor**. **Inducible genes** are usually involved in **catabolic** (degradative) reactions, as in the breakdown of a polysaccharide into simple sugars. **Repressible genes** are usually involved in **anabolic** (synthetic) reactions, as in the construction of amino acids from simpler precursors. Thus, there are two main types of transcriptional controls: (1) negative and (2) positive. Negative control can either be inducible or repressible and positive control is known only to be inducible. In addition, genes may be regulated by global controls or by post-translational mechanisms discussed in sections 4 and 5 in the following text.

1 NEGATIVE, INDUCIBLE CONTROL

The prototype of negative control by way of an inducible operon is the **lactose operon** of *E. coli*. β-galactosidase is an enzyme with dual functions. Its primary function is to catabolize lactose to glucose and galactose. Its secondary function is to convert the 1-4 linkage of glucose and galactose (in lactose) to a 1-5 linkage in **allolactose**. This enzyme is not normally present in high concentrations when lactose is absent from the cell's environment. Shortly after adding lactose to a medium in which glucose is absent, the enzyme begins to be produced. A transport protein called **galactoside permease** is required for the efficient transport of lactose across the cell membrane. This protein also appears in high concentration after lactose becomes available in the medium. The wild-type lactose operon (Fig. 10-11) consists of a regulatory gene (*lacI*) and an operon containing a promoter sequence (*lacP*), an operator locus (*lacO*), and three structural genes for β-galactosidase (*lacZ*), permease (*lacY*), and transacetylase (*lacA*; an enzyme whose function in lactose metabolism remains unresolved). Mutations at each of these loci have been found.

EXAMPLE 10.7
Some of the alleles of the lactose system are listed below. The *lacP* sequence contains a CAP-cAMP binding site (see section 4, page 348) and an RNA polymetase binding site (*p*).

Promoter Alleles
p^+ = wild-type promoter; normal affinity for RNA polymerase
p^- = mutant promoter cannot bind RNA polymerase; none of the structural genes in the lactose operon are transcribed

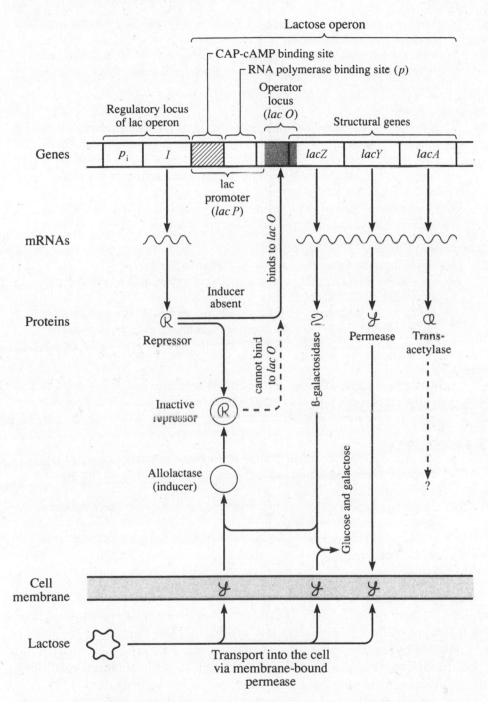

Fig. 10-11. Diagram of major elements controlling the lactose operon.

p^s = increased affinity for recognition by RNA polymerase; elevates the transcriptional level of the operon; s = "super promoter".

$p^{i\ cr}$ = affects the CRP-cAMP binding site to reduce the level of expression of lactose operon genes below 10% of wild type; $i\ cr$ = insensitive to catabolite repression

Operator Alleles

O^+ = in the absence of repressor, this operator "turns on" the structural genes in its own operon; i.e., the $lacZ^+$ and $lacY^+$ alleles in the same segment of DNA (cis position) can produce proteins; this operator is sensitive to the repressor; i.e., repressor will "turn off" the synthetic activity of the structural genes in the lactose operon

O^c = a constitutive operator that is insensitive to repressor and permanently "turns on" the structural genes in the lactose operon

Galactosidase Alleles

Z^+ = makes β-galactosidase if its operon is "turned on" or "open"

Z^- = a missense mutation that makes a modified, enzymatically inactive product called lacCZ protein

Z^{-ns} = results in the destruction of the polycistronic message downstream from the mutation so that there is no expression of any of the downstream lactose operon genes (a polar mutation); ns = nonsense

Permease Alleles

Y^+ = makes β-galactoside permease if its operon is "turned on"

Y^- = no detectable permease is formed regardless of the state of the operator; probably a nonsense mutation

Regulator Alleles

I^+ = makes a diffusible repressor protein that inhibits synthetic activity in any o^+ operon in the absence of lactose; in the presence of lactose, repressor is inactivated

I^- = a defective regulator that is unable to produce active repressor due to a nonsense or missense mutation

I^s = makes a "superrepressor" that is insensitive to lactose and inactivates any O^+ operon

There is some overlap in the promoter and operator sites of the *lac* system; in some other operons the operator locus may be totally embedded in the promoter. The regulator gene constitutively produces a repressor protein at low levels because it has an inefficient promoter. Its synthesis is unaffected by the level of lactose in the cell. The normal promoter of the lac operon, by contrast, binds RNA polymerase very efficiently. In the absence of lactose (**noninduced conditions**), an active repressor protein (produced by *lacI*) binds to the operator. RNA polymerase can neither bind to the promoter nor "read through" the operator sequence because repressor protein occupies that region. Hence, transcription of all three structural genes in the *lac* operon is prevented.

When lactose is present (**induced conditions**), it is transported inefficiently into the cell because only a few molecules of permease would normally be present. Inside the cell, some of the lactose would be converted to allolactose by β-galactosidase. Allolactose is the inducer of the *lac* operon. It binds to the repressor protein and causes a conformational change in the protein that alters the site by which it binds to the operator. This conformational change in a protein as a consequence of binding to another molecule is called an **allosteric transformation**. The allolactose-repressor complex can no longer bind to the operator, and it falls off the DNA. RNA polymerase can now read through the operator to transcribe the structural genes in the operon. The increased amount of permease now transports lactose across the membrane in large quantities, and the sugar is then digested by β-galactosidase. When lactose becomes depleted from the medium, newly synthesized repressor proteins will not be coupled with allolactose, so they can bind to the operator and shut off transcription of the structural genes in the operon. Furthermore, allolactose can reversibly bind to repressor protein, so that under low levels of lactose in the cell allolactose would tend to dissociate from repressor-allolactose complexes. Even when the *lac* operon is repressed, occasionally the repressor protein will diffuse from the operator momentarily. RNA polymerase may then be able to "sneak" past the open operator and synthesize a molecule of polycistronic mRNA, thus accounting for the very low levels of permease and β-galactosidase that are normally present in the cell. Bacterial mRNA molecules have a very short half-life (only a few minutes), so synthesis of proteins stops very soon after a cell is repressed. Proteins, on the other hand, are much more stable, but they would be diluted out with each subsequent cell division.

EXAMPLE 10.8

Bacteria of genotype $I^+O^+Z^+Y^+$ grown on media devoid of lactose will produce neither galactosidase nor permease because I^+ makes repressor substance that inactivates the O^+ operator and "turns off" the synthetic activity of structural genes Y^+ and Z^+ in its own operon.

EXAMPLE 10.9

Partial diploids can be produced in bacteria for this region of the chromosome. Cells of the genotype $I^-O^+Z^+Y^-/I^+O^cZ^-Y^+$ will produce the lacCZ protein constitutively (i.e., either with or without the presence of lactose inducer) because the allele O^c permanently "turns on" the genes in its operon (i.e., those in cis position with O^c). β-Galactosidase will be produced only inductively because in the presence of lactose (inducer), the diffusible repressor substance from *lacI*$^+$ will be inactivated and allow the structural gene *lacZ*$^+$ in cis position with the operator O^+ to produce enzyme.

The wild-type operon of the regulatory gene (*lacI*) in the lactose system consists of just a promoter (p_i) and the structural gene for the repressor protein (I^+). Its wild-type promoter is very inefficient, and only a few molecules of lac-repressor protein exist in the cell. In the operons of most regulatory genes in other systems, however, an operator locus is adjacent to its promoter, and **autoregulation** is possible. The repressor proteins made by these operons bind to their own operators to terminate transcription when the concentrations of their respective repressor molecules are elevated.

2   NEGATIVE, REPRESSIBLE CONTROL

An example of a repressible operon under negative control is found in the tryptophan system of *E. coli* (Fig. 10-12). The amino acid tryptophan is synthesized in five steps, each step mediated by a specific enzyme. The genes responsible for these five enzymes are arranged in a common operon (the ***trp* operon**) in the same order as their enzymatic protein products function in the biosynthetic pathway. The regulatory gene for this system constitutively synthesizes a nonfunctional protein called **aporepressor**. When tryptophan is in oversupply, the excess tryptophan acts as a **corepressor** by binding to the aporepressor. This forms a functional repressor complex. The functional repressor binds to the *trp* operator and coordinately represses transcription of all five structural genes in the operon. The promoter and operator regions overlap significantly, and binding of active repressor and RNA polymerase are thus competitive. When tryptophan is in low concentration, tryptophan dissociates from the aporepressor, and the aporepressor protein falls off the operator. RNA polymerase then synthesizes the polycistronic mRNA for all five enzymes of the tryptophan pathway.

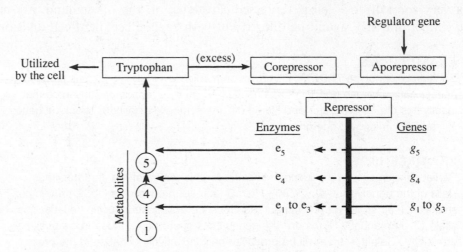

Fig. 10-12.   Negative repressible control in the tryptophan system of *E. coli.*

A secondary regulatory mechanism also exists in the tryptophan operon. At the $5'$ end of the polycistronic mRNA of this operon, 162 bases precede the coding segments for the five enzymes. This region is called a **leader sequence**. Part of this sequence is transcribed into a leader peptide of 14 amino acids, the function of which is unknown. There are two adjacent tryptophan codons in the leader peptide. When tryptophan is present in excess, transcription of the rest of the *trp* operon is prevented because RNA polymerase generates a transcription termination sequence; this phenomenon is known as **attenuation**. A model (Fig. 10-13) that explains attenuation assumes that (when tryptophan is abundant) movement of bacterial ribosomes follows closely behind the movement of RNA polymerase as it synthesizes mRNA, and all intramolecular base pairing is prevented in the mRNA segment in contact with the ribosome. In the experimental absence of ribosomes, only the leader mRNA is transcribed and no translation occurs [Fig. 10-13(*a*)]. Leader segments 1 and 2 become folded into stem and loop A by

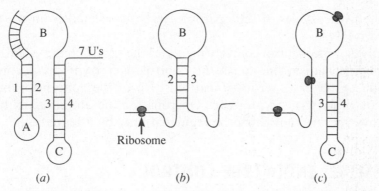

Fig. 10-13. Three kinds of stem and loop structures may form in the leader segment of a bacterial mRNA transcribed from the tryptophan operon of bacteria. (*a*) In the experimental absence of ribosomes, or when the cell is starved for the first few amino acids of the leader peptide (e.g., methionine, lysine, arginine, isoleucine), ribosomes will not be moving along the mRNA behind RNA polymerase. Hence, complementary leader sequences 1 and 2 base pair to form a stem (1,2) and loop A structure. Pairing of sequences 3 and 4 also forms a stem (3,4) and loop C structure that functions as a transcription termination signal, causing RNA polymerase to dissociate from the DNA before any of the "downstream" genes of the tryptophan operon can be transcribed. (*b*) If only the concentration of tryptophan is low, ribosomes will stall in sequence 1 at each tryphophan codon. This allows RNA polymerase to move ahead and transcribe sequences 2 and 3, which base pair to form a stem (2,3) and loop B structure. If sequence 3 first pairs with 2, it cannot pair with 4, which is synthesized later. The 2-B-3 structure thus acts as an "antiterminator," permitting RNA polymerase to transcribe the remainder of the leader and all of the genes of the tryptophan operon. Other ribosomes can attach to ribosome-binding sites preceding each of the coding regions on the polycistronic mRNA of the tryptophan operon and thereby initiate translation of all five proteins of the operon. When all amino acids are abundant (*c*), ribosomes follow behind RNA polymerase, covering sequence 1 with ribosomes before sequence 2 is synthesized. Thus, the antiterminator (2-B-3) cannot form. Sequences 3 and 4 are synthesized and form the stem and loop structure 3-C-4 before ribosomes can begin to translate 3.

complementary base pairing, whereas segments 3 and 4 fold into stem and loop C that acts as a transcription termination signal. As RNA polymerase synthesizes the seven uracils that follow segment 4, these uracils and the adjacent paired 3-4 region of mRNA (having just folded into stem and loop C) form a terminator signal that causes RNA polymerase to prematurely dissociate from the DNA before it can transcribe any of the DNA coding segments for the five enzymes of the *trp* operon.

When only the concentration of activated trp-tRNAs is low [Fig. 10-13(*b*)] ribosomes begin to translate region 1, thereby preventing pairing of regions 1 and 2. However, the ribosome tends to stall momentarily (especially at the pair of tryptophan codons), and this allows pairing of regions 2 and 3 to form a B stem-and-loop structure (called an **antiterminator**); regions 3 and 4 are thereby prevented from forming the C termination signal, and RNA polymerase is allowed to continue transcription on into the *trp* operon.

If activated trp-tRNAs are abundant [Fig. 10-13(*c*)], ribosomes follow so closely behind RNA polymerase that the antiterminator B structure cannot form, and therefore the terminator C structure does form. Thus, only the leader peptide (but none of the five enzymes of the operon) can be translated from the prematurely terminated mRNA.

The repressor mechanism coarsely regulates the tryptophan system, whereas the attenuation mechanism fine-tunes the control over tryptophan concentrations. Attenuation of the *trp* operon is also sensitive to the concentrations of several amino acids other than tryptophan. Operons for the amino acids histidine and leucine, however, are thought to be regulated only by attenuation.

3 POSITIVE, INDUCIBLE CONTROL

An example of a positive, inducible regulatory mechanism is found in the arabinose operon of *E. coli*. Arabinose is a sugar that requires three enzymes (coded by genes *araB*, *araA*, *araD*) for its metabolism. Two additional genes are needed to transport arabinose across the cell membrane, but they are located at a distance from the BAD cluster coding for the catabolic enzymes. The regulatory gene *araC* is close to the promoter for the BAD cluster. The protein product of the *araC* gene (AraC) is a repressor of the BAD cluster when the substrate arabinose is absent. However, when arabinose is present, it binds to the repressor (AraC), forming an **activator complex** that facilitates the binding of RNA polymerase to the promoter, thus inducing transcription of the operon. The preceding story is a gross over-simplification of the complexity that is already known about the regulation of the arabinose system. For example, **cyclic adenosine monophosphate (cAMP)** and **catabolite activator protein (CAP**; also known as **cyclic AMP receptor protein, CRP)** are also involved in the regulation of the arabinose system. The action of these last two molecules in the phenomenon of **catabolite repression** is discussed in the next section.

4 GLOBAL REGULATION OR MULTIPLE CONTROLS

A genetic locus may be regulated by more than one mechanism. When glucose is available, there is no need to catabolize other sugars, and the genes coding for these other sugar-catabolizing enzymes can be turned off. For example, if glucose is absent and lactose is present in the medium, the *lac* operon would be induced. But if glucose is present, induction of the *lac* operon does not occur. This phenomenon was originally termed the glucose effect; it is now known as **catabolite repression**. A complex of two molecules, namely, **cAMP** and **CAP**, acts as the activator in catabolite repression. Within the lac promoter (Fig. 10-11), there is a site for binding of a cAMP-CAP complex. RNA polymerase only binds effectively to the promoter if cAMP-CAP complex is also bound to this site. As the level of glucose increases within the cell, the amount of cAMP decreases and less cAMP-CAP complex is available to activate the *lac* operon. CAP is produced at low levels by its own genetic locus. The enzyme **adenylate cyclase** (adenylcyclase) converts adenosine triphosphate (ATP) to cyclic adenosine monophosphate (cAMP). Adenylate cyclase can become activated to first messenger status by the interaction of specific cell receptors with their target molecules; the cAMP thus produced (second messenger) can then regulate a battery of genes coordinately.

5 POSTTRANSLATION CONTROL

The expression of genes can be regulated after proteins have been synthesized (posttranslation control). **Feedback inhibition** (or **end-product inhibition**) is a regulatory mechanism involving inhibition of enzymatic activity. The end product of a synthetic pathway (usually a small molecule such as an amino acid) may combine loosely (if in high concentration) with the first enzyme in the pathway. This union does not occur at the catalytic site of the enzyme, but it does modify the tertiary or quaternary structures of the enzyme and hence inactivates the catalytic site. This allosteric transformation of the enzyme blocks its catalytic activity and prevents overproduction of the end product of the pathway and its intermediate metabolites.

EXAMPLE 10.10

The end product isoleucine in *E. coli*, when present in high concentration, unites with the first enzyme in its synthetic pathway and thus inhibits the entire pathway until isoleucine returns to normal levels through cellular consumption. Intermediates in the biosynthetic pathway are in numbered boxes; e = enzyme; g = gene.

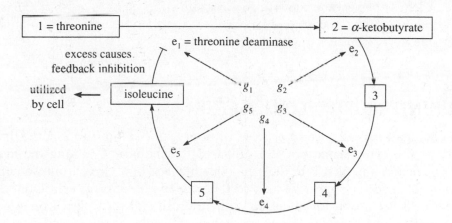

Mapping the Bacterial Chromosome

1 INTERRUPTED CONJUGATION

When Hfr and F^- cultures are mixed, conjugation can be stopped at any desired time by subjecting the mixture to the shearing forces of a Waring blender, which breaks the conjugation bridge. The sample is diluted immediately and plated on selective media, incubated, and then scored for recombinants. In addition to the selected marker, an Hfr strain must also carry a distal auxotrophic or sensitivity marker that prevents the growth of Hfr cells on the selective medium and thereby allows only recombinant cells to appear. This technique is called **counterselection**. Because of the polarity with which the Hfr chromosome is transferred, the time at which various genetic markers appear in the recipient indicates their linear organization in the donor chromosome. At a given temperature, the transfer of the first half of the Hfr chromosome proceeds at a relatively uniform rate. Therefore, the time of entry of different markers into a recipient (F⁻) cell is a function of the physical distance between them. Because of errors introduced by experimental manipulations, this method is best suited for markers that are more than 2 min apart.

EXAMPLE 10.11
An Hfr strain carrying the prototrophic markers a^+, b^+, c^+ is mixed with an F⁻ strain carrying the auxotrophic alleles a, b, c. Conjugation was interrupted at 5-min intervals and plated on media that revealed the presence of recombinants.

Time (min)	Recombinants detected
5	ab^+c
10	ab^+c^+
15	$a^+b^+c^+$

The order of the genes in the Hfr donor strain is ori — b^+ — c^+ — a^+; b is less than 5 time units from the origin (ori); c is less than 5 time units from b; a is less than 5 time units from c.

2   UNINTERRUPTED CONJUGATION

When conjugation is allowed to proceed without artificial interruption, the time of rupture of the cytoplasmic bridge is apparently randomized among the mating pairs. The nearer a marker is to the origin (leading end of donor chromosome), the greater its chances of appearing as a recombinant in a recipient cell. Donor and recipient cells are mixed for about an hour in broth and then placed on selective media that allows growth of F recombinants only for a specific marker. Counterselection against Hfr must also be part of the experimental design. The counterselective marker should be located as distally as possible from the selected marker so that unselected recombinants will not be lost by its inclusion. The frequencies with which unselected markers appear in selected recombinants are inversely related to their distances from the selected marker, provided they lie distal to it. Obviously, any unselected marker between the selected marker and the origin of the chromosome will always be transferred ahead of the selected marker. Proximal markers more than 3 time units apart exhibit approximately 50% recombination, indicating that the average number of exchanges between them is greater than 1. Just at the point where gross mapping by conjugation becomes ineffective, i.e., for markers less than 2 time units apart, recombination mapping becomes very effective, permitting estimation of distances between closely linked genes or between mutant sites within the same gene. Distances between genes can be expressed in three types of units: (1) time units, (2) recombination units, or (3) chemical units.

EXAMPLE 10.12

If 1 min of conjugation is equivalent to 20 recombination units in *E. coli*, and the entire chromosome is transferred in 100 min, then the total map length is 2000 recombination units. If 10^7 nucleotide pairs exist in the chromosome, then 1 recombination unit represents $10^7/2000 = 5000$ nucleotide pairs.

3   RECOMBINATION MAPPING

Virtually all of the opportunities for recombination in bacteria involve only a partial transfer of genetic material (meromyxis) and not the entire chromosome. One or more genes have an opportunity to become integrated into the host chromosome by conjugation, depending upon the length of the donor piece received. Exogenotes must become integrated if they are to be replicated and distributed to all of the cells in a clone. Only a small segment of DNA is usually integrated during transformation or transduction. Thus, if a cell becomes transformed for two genetic markers by the same transforming piece of DNA (double transformation), the two loci must be closely linked. Similarly, if a cell is simultaneously transduced for two genes by a single transducing phage DNA (cotransduction), the two markers must be closely linked. The degree of linkage between different functional genes (intergenic) or between mutations within the same functional gene (intragenic) may then be estimated from the results of specific crosses.

In merozygotic systems where the genetic contribution of the donor parent is incomplete, an even number of crossovers is required to integrate the exogenote into the host chromosome (endogenote).

EXAMPLE 10.13

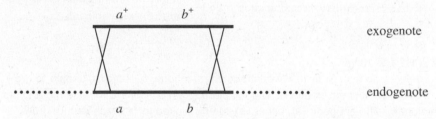

Prototrophic recombinants must integrate the exogenote from somewhere left of the *a* locus to right of the *b* locus. Two crossovers (an even number) are required for this integration.

EXAMPLE 10.14

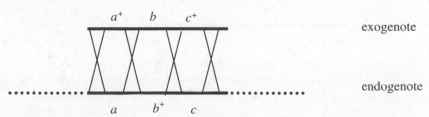

A prototrophic recombinant in this example requires a quadruple (even number) crossover for integration of all wild-type genes.

The total number of progeny is unknown in merozygotic systems so that recombination frequency cannot be expressed relative to this base. Therefore, recombination frequencies must be made relative to some standard that is common to all crosses. For example, the number of prototrophic recombinants produced by crossing two mutant strains can be compared with the number emerging from crossing wild type by mutant type. However, many sources of error are unavoidable when comparing the results of different crosses. This problem can be circumvented by comparing the number of prototrophic recombinants with some other class of recombinants arising from the same cross.

EXAMPLE 10.15

Ratio test for different functional genes. Suppose we have two mutant strains, a and b, where the donor strain (a^+b) can grow on minimal medium supplemented with substance B, but the recipient strain (ab^+) cannot do so.

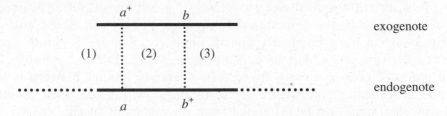

Crossing over in regions (1) and (2) produces prototrophic recombinants (a^+b^+) able to grow on unsupplemented medium. If the medium is supplemented with substance B, then a^+b recombinants arising by crossing over in regions (1) and (3) can grow in addition to the prototrophs.

$$\text{Standardized recombination ratio} = \frac{\text{number of prototrophs}}{\text{number of recombinants}}$$

EXAMPLE 10.16

Intragenic ratio test. Consider two intragenic mutations, b_1, and b_2, unable to grow in medium without substance B. The recipient strain contains a mutation in another functionally different gene (a), either linked or unlinked to b, which cannot grow unless supplemented by substances A and B.

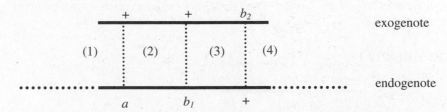

On unsupplemented medium, only prototrophs arising through crossovers in regions (1) and (3) appear. On medium supplemented only by substance B, recombinants involving region (1) and any of the other three regions can survive.

$$\text{Standardized recombination ration} = \frac{\text{number of colonies on unsupplemented medium}}{\text{number of colonies on B-supplemented medium}}$$

SOLVED PROBLEM 10.4
A strain of *E. coli* unable to ferment the carbohydrate arabinose (*ara⁻*) and unable
to synthesize the amino acids leucine (*leu⁻*) and threonine (*thr⁻*) is transduced by a
wild-type strain (*ara⁺ leu⁺ thr⁺*). Recombinants for leucine are detected by plating
on minimal medium supplemented with threonine. Colonies from the transduction
plates were replicated or streaked onto plates containing arabinose. Out of 270
colonies that grew on the threonine-supplemented plates, 148 could also ferment
arabinose. Calculate the amount of recombination between *leu* and *ara*.

Solution:

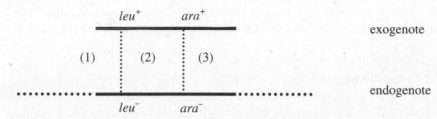

In order for a transductant to be *leu⁺ ara⁺*, crossing over in regions (1) and (3) must occur;
for *leu⁺ ara⁻* to arise, crossing over in regions (1) and (2) must occur.

$$\text{Standardized recombination ratio} = \frac{\text{no. of } leu^+ ara^-}{\text{no. of } leu^+} = \frac{270 - 148}{270} = 0.45 \text{ or } 45\%$$

4   ESTABLISHING GENE ORDER

Mapping small regions in microorganisms has revealed that multiple crossovers
often occur with much greater than random frequency, a phenomenon called
"localized negative interference." The only unambiguous method for determining
the order of very closely linked sites is by means of three-factor reciprocal crosses.
Suppose that the location of gene *a* is known to be to the left of gene *b* but that the
order of two mutants within the adjacent *b* gene is unknown. Reciprocal crosses
will yield different results, depending upon the order of the mutant sites.

EXAMPLE 10.17
Assume the order of sites is *a-b₁-b₂*.

Original Cross:

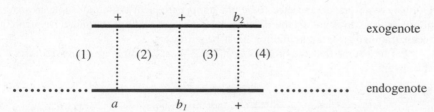

Reciprocal Cross:

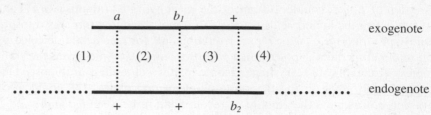

In the original cross, prototrophs (+ + +) can be produced by crossovers in regions (1) and (3). In the reciprocal cross, prototrophs arise by crossovers in regions (3) and (4). The numbers of prototrophs should be approximately equivalent in the two crosses.

SOLVED PROBLEM 10.5

Several $lacZ^-$ mutants, all lacking the ability to synthesize β-galactosidase, have been isolated. A cross is made between Hfr ($lacZ_1^-$ ade^+ str^s) × F⁻ ($lacZ_2^-$ ade^- str^r) where ade^- = adenine requirement, str^s and str^r = streptomycin sensitivity and resistance, respectively. Many of the ade^+ exconjugant clones were able to ferment lactose, indicating β-galactosidase activity. Only a few ade^+ clones from the reciprocal cross Hfr ($lacZ_2^-$ ade^+ str^s) × F⁻ ($lacZ_1^-$ ade^-str^r) were able to ferment lactose. What is the order of the $lacZ_1$ and $lacZ_2$ mutants relative to the ade locus?

Solution: First, assume that the order is $lacZ_2$-$lacZ_1$-ade. In the first mating, four crossovers are required to produce a streptomycin-resistant prototroph able to ferment lactose ($lacZ_2^+$ $lacZ_1^+$ ade^+ str^r).

(a)

The reciprocal mating requires only two crossovers to produce a prototroph able to ferment lactose.

(b)

Double crossovers are expected to be much more frequent than quadruple crossovers. The above scheme does not fit the data because the first mating was more frequent than the reciprocal mating. Our assumption must be wrong.

Let us assume that the order is $lacZ_1$-$lacZ_2$-ade. The first cross now requires a double crossover.

(c)

The reciprocal cross requires four crossover events.

(d)

Hfr exogenote

F⁻ endogenote

The reciprocal cross is expected to be much less frequent under this assumption and is in agreement with the observations.

5   COMPLEMENTATION MAPPING

An F particle that carries another bacterial gene other than the sex factor produces a relatively stable F^+ merozygote. These partial diploids can be used for complementation tests of mutants affecting the same trait (see Example 3.7).

EXAMPLE 10.18

An Hfr strain of *E. coli* is unable to ferment lactose ($lacZ_1^-$) and can transfer the $lacZ_1^-$ gene through conjugation to a mutant ($lacZ_2^-$) recipient, forming the heterogenote $lacZ_2^-/(F\text{-}lacZ_1^-)$. If $lacZ_1^-$ and $lacZ_2^-$ are mutations in the same gene (i.e., functional alleles), then complementation does not occur and only mutant phenotypes are produced. If $lacZ_1^-$ and $lacZ_2^-$ are mutations in different genes, complementation could produce wild types able to ferment lactose.

SOLVED PROBLEM 10.6

Six mutations are known to belong to three genes. From the results of the complementation tests, determine which mutants are in the same gene.

1	2	3	4	5	6	
0	+	+				1
	0		+	+		2
		0	+	0		3
			0		+	4
				0	+	5
					0	6

+ = complementation
0 = noncomplementation
blank = not tested

Solution: Mutations 3 and 5 are in the same gene, since they fail to complement each other. Mutations 1 and 3 are in different genes, since they do complement each other. We will arbitrarily assign these to genes A and B.

(3, 5)	1	
Gene A	Gene B	Gene C

Mutations 1 and 2 are in different genes, but we do not know whether 2 is in A or C. However, 5 and 2 complement and therefore 2 cannot be either in gene A or B and thus must be in C.

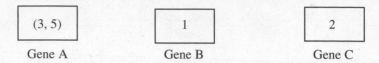

Mutations 3 and 4 complement; thus, 4 must be in either B or C. But 2 and 4 also complement; thus, 4 cannot be in C and must reside in B.

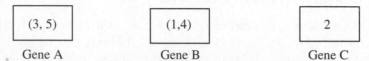

Mutation 6 cannot be in A since it complements with 5. Thus, 6 is either in B or C. Since 6 and 4 complement, they are in different genes. If 6 cannot be in A or B, it must be in C. The mutants are grouped into genes as shown below.

(3, 5)	(1,4)	(2, 6)
Gene A	Gene B	Gene C

Intragenic complementation may sometimes be possible when the enzyme product is composed of two or more identical polypeptide chains. Experimental evidence has shown that an *in vitro* mixture of inactive enzymes from some complementing mutants can "hybridize" to produce an enzyme with up to 25% normal activity. Mutants that fail to complement with some but not all other mutants are assumed to overlap in function. A complementation map can be constructed from the experimental results of testing all possible pairs of mutants for complementary action in bacterial merozygotes or in fungal heterokaryons. A complementation map cannot be equated in any way with a crossover map, since the gene is defined by different criteria. A complementation map tells us nothing of the structure or location of the mutations involved. Complementation maps are deduced from merozygotes or heterokaryons; crossover maps are deduced from recombination experiments.

EXAMPLE 10.19

Three mutants map by complementation as follows:

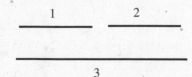

This indicates that mutants 1 and 2 are complementary and do not overlap in function. Hence, 1 and 2 are nonallelic mutations by this criterion. Mutant 3 fails to complement with either 1 or 2 and hence must overlap (to some degree) with both 1 and 2. Hence, 3 is functionally allelic with both 1 and 2.

6   MAPPING BY DELETION MUTANTS

A deletion in some segment of a functional gene cannot recombine with point mutations in that same region even though two point mutations at different sites within this region may recombine to produce wild type. Another distinctive property of deletion mutants is their stability; they are unable to mutate back to wild type. The use of overlapping deletions can considerably reduce the work in fine structure analysis of a gene.

EXAMPLE 10.20
Determining the limits of a deletion. Suppose that a series of single mutants (I, 2, 3, 4) has already been mapped as shown below:

A deletion that fails to recombine with point mutants 1 and 2 but does produce wild type with 3 and 4 extends over region X. A deletion that yields no recombinants with 3 and 4 has the boundaries diagrammed as Y. A deletion mutant that produces wild type only with point mutants 1 or 4 has the limits of Z.

EXAMPLE 10.21
Assigning point mutations 1–4 to deletion regions R, S, and T.

Given the deletions R, S, and T as shown above, the point mutation that recombines to give wild type with deletions S and T, but not with R, is 1. Number 3 is the only one of the four mutants that fails to recombine with one of the three deletions.

Supplementary Problems

ISOLATION OF BACTERIAL MUTANTS

10.7. Approximately 10^8 *E. coli* cells of a triple auxotrophic strain ($arg^-lys^-ser^-$) are plated on complete medium to form a bacterial lawn. Replica plates are prepared containing minimal medium supplemented by the amino acids arginine, lysine, and serine. What is the genotype of the colonies that appear on the replica plates?

Master Plate   Replica Plates

Medium contains: arg, lys, ser   arg & lys   arg & ser   lys & ser
                                            (1)   (2)   (3)

10.8. Two triple auxotrophic bacterial strains are allowed to conjugate in broth, diluted, and plated onto complete agar (master plate). Replica plates containing various supplements are then made from the master. From the position of each clone and the type of media on which it is found, determine its genotype.

$$met^-thr^-pan^-pro^+bio^+his^+ \times met^+thr^+pan^+pro^-bio^-his^-$$

Symbols   Master Plate

met = methionine
thr = threonine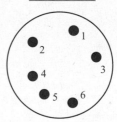
pan = pantothenic acid
pro = proline
bio = biotin
his = histidine

Replica plates: Each dish contains minimal medium plus the supplements shown at the bottom.

   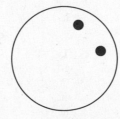   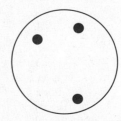

met, thr   met, pan   thr, pan

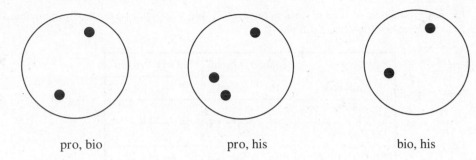

pro, bio   pro, his   bio, his

10.9. A bacterial strain unable to synthesize methionine (met^-) is transduced by a phage from a bacterial strain unable to synthesize isoleucine (ile^-). The broth culture is diluted and plated on minimal medium supplemented with isoleucine. An equivalent amount of diluted broth culture is plated on minimal medium. Eighteen clones appeared on the minimal plates and 360 on the isoleucine plates. Calculate the standardized recombination ratio.

10.10. DNA damage (mutation) is an essential initiation event for a cell to transform into a cancerous state, but it is not the only event causing cancer (see Chapter 11). Therefore, DNA-damaging agents (mutagens) are only potential carcinogens (agents causing cancer). Most chemical carcinogens are not biologically active in their original form; they must first be metabolized to carcinogenic metabolites. The **Ames test** (named after the inventor, Bruce Ames) is one of the standard tests for a quantitative estimate of the mutagenic potency of a chemical. This test employs an auxotrophic strain of *Salmonella typhimurium* that cannot make the amino acid histidine (his^-). To increase the sensitivity of the tester strain (1) it carries a mutation that makes the cell envelope more permeable to allow penetration of the test chemicals, (2) its capacity for excision repair is eliminated so that most of the primary lesions remain unhealed, and (3) a genetic element that makes DNA replication more error prone is introduced via a plasmid. Rat liver extract is added to a minimal medium culture plate coated with a thin layer of these bacteria. A disk of filter paper is impregnated with the chemical to be tested; the paper is placed in the center of the plate. After 2 days of incubation, the number of colonies are counted. (*a*) What events are being scored by the colony counts? (*b*) Why was mammalian liver extract added to the test? (*c*) Diagram the expected distribution of colonies on a plate containing a known carcinogen. Explain why this distribution develops. (*d*) Suppose that the test chemical (e.g., nitrosoguanidine) is mixed with the bacteria prior to plating at two dosages (low and high). A control is run simultaneously with these two doses. Diagram the expected distribution of colonies on these three plates.

BACTERIAL DNA REPLICATION AND CELL DIVISION
10.11. Several lines of evidence suggest that the circular chromosome of *E. coli* has two replication forks. The length of one whole unreplicated chromosome is 1300 μm (about 500 times longer than the *E. coli* cell). There are ten base pairs per one complete turn of the DNA double helix, equivalent to 34 Å or $3.4 \times 10^{-3}\mu$m. (*a*) How many nucleotide base pairs are in the *E. coli* DNA complement or genome? (*b*) If the *E. coli* genome is replicated in 40 min at 37°C by two replicating forks, how many revolutions per minute (rpm) must the parental double helix make to allow separation of its complementary nucleotide strands during replication?

REGULATION OF BACTERIAL GENE ACTIVITY
10.12. In addition to the $lacI^+$ allele, producing repressor for the lactose system in *E. coli* and the constitutive $lacI^-$ allele, a third allele I^s has been found, the product of which is unable to combine with the inducer (lactose). Hence, the repressor ("superrepressor") made by I^s remains unbound and free to influence the operator locus. (*a*) Order the three alleles of the *lacI* locus in descending order of dominance according to their ability to influence the lactose operator. (*b*) Order the four alleles of the promoter locus (*p*) in descending order of dominance according to their ability to bind RNA polymerase. (*c*) Using + for production and 0

for nonproduction of the enzymes permease (P) and β-galactosidase (β-gal), complete the following table. (*Hint*: See Example 10.7.)

Genotype	Inducer Absent		Inducer Present	
	P	β-gal	P	β-gal
(1) $I^+ O^+ Y^+ Z^+$				
(2) $I^- O^+ Y^+ Z^+$				
(3) $I^s O^+ Y^+ Z^+$				
(4) $I^+ O^c Y^+ Z^+$				
(5) $I^- O^c Y^+ Z^+$				
(6) $I^s O^c Y^+ Z^+$				

10.13. For each of the following partial diploids, determine whether enzyme formation is constitutive or inductive: (*a*) I^+/I^+, O^+/O^+ (*b*) I^+/I^+, O^+/O^c (*c*) I^+/I^+, O^c/O^c (*d*) I^+/I^-, O^+/O^+ (*e*) I^-/I^-, O^+/O^+.

10.14. In the lactose operon of *E. coli*, *lacY*$^+$ makes permease, an enzyme essential for the rapid transportation of galactosides from the medium to the interior of the cell. Its allele *lacY*$^-$ makes no permease. The galactoside lactose must enter the cell in order to induce the *lacZ*$^+$ gene to produce the enzyme β-galactosidase. The allele *lacZ*$^-$ makes a related but enzymatically inactive protein called lacCZ. Predict the production or nonproduction of each of these products with a normal operator O^+ by placing + or 0, respectively, in the table below.

Genotype	Inducer Absent			Inducer Present		
	P	β-gal	lacCZ	P	β-gal	lacCZ
(1) $I^+ Y^+ Z^-$						
(2) $I^+ Y^+ Z^+$						
(3) $I^- Y^+ Z^+$						
(4) $I^+ Y^- Z^+$						
(5) $I^- Y^- Z^-$						
(6) $I^- Y^- Z^+$						

P = permease enzyme; β-gal = beta-galactosidase enzyme; lacCZ = lacCZ protein.

10.15. In genotype (1) of the table below, the *lacI*$^-$ allele allows constitutive enzyme production by the *lacY*$^+$ and *lacZ*$^+$ genes in an operon with a normal operator gene, O^+. The action of *lacI*$^-$ might be explained by one of two hypothesis: (1) *lacI*$^-$ produces an internal inducer, thus eliminating the need for lactose in the medium to induce enzyme synthesis; *lacI*$^+$ produces no internal inducer, or (2) *lacI*$^+$ produces a *repressor* substance that, in the absence of lactose inducer, blocks enzyme formation, but in the presence of lactose inducer the repressor becomes inactivated to allow enzyme synthesis; *lacI*$^-$ produces no repressor. (*a*) Assuming dominance of the *lacI*$^-$ allele under the first hypothesis in an *E. coli* partial diploid of the constitution *lacI*$^+$/*lacI*$^-$, would internal inducer be produced? (*b*) Under the conditions of part (*a*), would enzymes be produced constitutively or inductively in a wild-type *lac* operon? (*c*) Assuming dominance of the *lacI*$^+$ allele under the second hypothesis in a partial diploid of the constitution *lacI*$^+$/*lacI*$^-$, would repressor be produced? (*d*) Under the conditions of part (*c*), would enzymes be produced constitutively or inductively in a wild-type *lac* operon? (*e*)

From the pattern of reactions exhibited by genotypes (2) and (3) in the table below, determine which of the two hypotheses is consistent with the data. (*f*) Is the repressor substance diffusible, or can it only act on loci in cis position with the *lacI* locus? How can this be determined from the information in the table?

Partial Diploid Genotype	Lactose Absent		Lactose Present	
	P	β-gal	P	β-gal
(1) $O^+Y^-Z^-I^-/O^+Y^+Z^+I^-$	+	+	+	+
(2) $O^+Y^-Z^-I^-/O^+Y^+Z^+I^+$	0	0	+	+
(3) $O^+Y^-Z^-I^+/O^+Y^+Z^+I^-$	0	0	+	+

10.16. List two kinds of single mutations that can change the function of an operator.

10.17. A bacterial mutation renders a cell incapable of fermenting many sugars (e.g., lactose, sorbitol, xylose) simultaneously. The operons of genes specifying the respective catabolic enzymes are wild type (unmutated). Offer an explanation for this phenomenon.

10.18. Shown below is a hypothetical biosynthetic pathway subject to feedback inhibition; letters represent metabolites; numbers represent enzymes. Identify the enzymes that are most likely to be subject to feedback inhibition and their inhibitor(s). *Note*: The inhibitor may consist of more than one metabolite.

MAPPING THE BACTERIAL CHROMOSOME

10.19. A cross is made between the streptomycin-resistant (*str*^{r}) F$^-$ strain of genotype *gal*$^-$ *thr*$^-$ *azi*^{r} *lac*$^-$ *Ton*^{r} *mal*$^-$ *xyl*$^-$ *leu*$^-$ and the prototrophic Hfr strain that has opposite characters. After 60 min of contact, samples are transferred to plates with minimal medium plus streptomycin. The original mixture is in the ratio 2×10^7 Hfr to 4×10^8 F$^-$. The percentages of each Hfr gene transferred are: 72% *Ton*^{s}, 0% *mal*$^+$, 27% *gal*$^+$, 91% *azi*^{s}, 0% *xyl*$^+$, 48% *lac*$^+$. (*a*) How many F$^-$ cells exist in the original mixture for every Hfr cell? (*b*) What is the counterselective agent that prevents Hfr individuals from obscuring the detection of recombinants? (*c*) In what order are these genes probably being transferred by the Hfr strain?

10.20. Four Hfr strains of *E. coli* are known to transfer their genetic material during conjugation in different sequences. Given the time of entry of the markers into the F$^-$ recipient, construct a genetic map that includes all of these markers and label the time distance between adjacent gene pairs.

| Strain 1. | Markers: | *arg - thy - met - thr* |
| | Time (min): | 15 21 32 48 |

| Strain 2. | Markers: | *mal - met - thi - thr - try* |
| | Time (min): | 10 17 22 33 57 |

| Strain 3. | Markers: | *phe - his - bio - azi - thr - thi* |
| | Time (min): | 6 11 33 48 49 60 |

| Strain 4. | Markers: | *his - phe - arg - mal* |
| | Time (min): | 18 23 35 45 |

10.21. Two mutants at the tryptophan locus, trp_A^- and trp_B^-, are known to be close to a cysteine locus (*cys*). A bacterial strain of genotype $cys^+ trp_A^-$ is transduced by phage from a bacterial strain that is $cys^- trp_B^-$. The reciprocal cross is also made wherein the strain $cys^- trp_B^-$ is transduced by phage from a strain that is $cys^+ trp_A^-$. In both cases, the numbers of prototrophic recombinants are equivalent. Determine the order of the tryptophan mutants relative to the cysteine marker.

10.22. Five point mutations (*a–e*) were tested for wild-type recombinants with each of the five deletions shown in the topological map below. The results are listed in the table below (+ = recombination, 0 = no recombination). Determine the order of the point mutations.

Deletions

	1	2	3	4	5
a	0	0	+	+	+
b	+	+	+	0	+
c	0	0	+	+	0
d	0	+	0	0	0
e	0	+	0	0	+

Review Questions

Vocabulary For each of the following definitions, give the appropriate term and spell it correctly.

1. Descriptive of all mutant strains of bacteria requiring supplementation to minimal medium for their growth.
2. A colony of genetically identical bacterial cells.
3. The Greek letter representing an intermediate structure of the bacterial genophore when it is halfway through bidirectional replication from a single origin.
4. A solid mass of bacterial cells covering the surface of nutrient agar medium in a Petri dish.
5. A form of bacterial recombination requiring cell-to-cell contact.
6. A form of genetic recombination between bacterial cells that is mediated by a bacteriophage.
7. A partially diploid exconjugant.
8. A small, circular DNA molecule capable of replication independent of the genophore in its bacterial host.
9. The bacterial topoisomerase that pumps negative supercoiling into relaxed DNA circles.
10. The protein that unwinds double-stranded DNA at each replication fork.

Multiple-Choice Questions Choose the one best answer.

1. Which of the following is classified as a prokaryote? (*a*) protozoa (*b*) yeast (*c*) bacteria (*d*) algae (*e*) more than one of the above

2. Which of the following is found in prokaryotes? (*a*) mitochondria (*b*) histones (*c*) actin and myosin (*d*) formylated methionine (*e*) more than one of the above

3. The physiologically receptive state in which a bacterial cell is able to be transformed is called (*a*) sensitized (*b*) activated (*c*) competence (*d*) lysogenic (*e*) inducible.

4. Which of the following is not a method for gene transfer in bacteria? (*a*) translocation (*b*) conjugation with Hfr transfer (*c*) transformation (*d*) transduction (*e*) conjugation with an F' cell.

5. Which of the following modes of cell division or DNA replication is not used by bacteria? (*a*) binary fission (*b*) rolling circle (*c*) theta replication (*d*) mitosis (*e*) bidirectional replication

6. The distances between bacterial genes, as determined from interrupted conjugation experiments, are measured in units of (*a*) recombination (*b*) nucleotide pairs (*c*) minutes (*d*) micrometers (*e*) percentage of genome

7. The subunit of RNA polymerase involved in initiation of bacterial transcription is designated (*a*) alpha (α) (*b*) beta (β) (*c*) gamma (γ) (*d*) sigma (σ) (*e*) rho (ρ)

8. A Shine-Delgarno sequence (*a*) serves as a binding site for bacterial ribosomes (*b*) is involved in bacterial transcription (*c*) forms part of mRNA trailers (*d*) serves as a recognition site for stopping RNA synthesis (*e*) serves as a recognition site for termination of translation

9. The type of gene regulation governing the lactose operon of *E. coli* that involves the *lac I* gene product is best described as (*a*) positive, inducible (*b*) negative, repressible (*c*) attenuation (*d*) feedback inhibition (*e*) negative, inducible

10. The ineffectiveness of many antibiotics today is most closely associated with (*a*) bacteriophages (*b*) F plasmids (*c*) R plasmids (*d*) catabolite repression (*e*) bacterial transformations

Answers to Supplementary Problems

10.7. Plate 1 contains a mutation to *ser*$^+$; plate 3 contains a mutation to *arg*$^+$.

10.8.

Clone	Genotype of Clone					
	met	thr	pan	pro	bio	his
1	+	+	+	+	+	+
2	+	−	−	+	+	+
3	−	+	−	+	+	+
4	+	+	+	+	+	−
5	+	+	+	−	+	+
6	+	−	+	+	+	+

10.9. 20% recombination

10.10. (*a*) Back mutations (reverse mutations) from *his*$^-$ to *his*$^+$. (*b*) It supplies the mammalian metabolic functions that are usually required to convert a chemical into its carcinogenic metabolites. (*c*) After 2 days, most of the *his*-bacteria have died for lack of histidine. Back-mutation rates are expected to be proportional to concentration of the chemical that forms a radially diminishing concentration gradient around the paper disk. Close to the disk there is a zone in which no cells grow because of toxic levels of the chemical. Beyond this zone there may be so many *his*⁺ revertants that the cells almost form a continuous lawn. At the periphery are a few larger clones (because they are isolated) representing spontaneous *his*⁺ mutants that have not been exposed to the chemical.

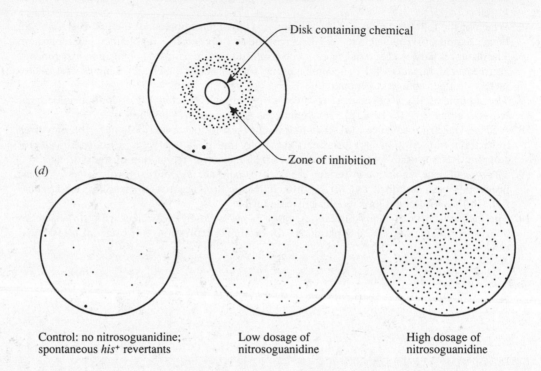

(*d*)

Control: no nitrosoguanidine; spontaneous *his*⁺ revertants

Low dosage of nitrosoguanidine

High dosage of nitrosoguanidine

10.11.

(*a*) $\dfrac{(1300\ \mu\text{m})(10\ \text{bp/turn})}{(3.4 \times 10^{-3}\,\mu\text{m/turn})} = 3.9 \times 10^6$ or 3900 kilobase pairs

(*b*) rate of chain growth $= \dfrac{3900\ \text{kb}}{2(2400\ \text{s})} = 0.8$ kilobase/second

$\dfrac{800\ \text{bp/s}}{10\ \text{bp/rev}} \times 60\ \text{s/min} = 80\ \text{rev} \times 60\ \text{min}^{-1} = 4800\ \text{rev/min}$

10.12. (*a*) $I^s,\ I^+,\ I^-$　(*b*) $p^s,\ p^+,\ p^{i\ cr},\ p^-$

(*c*)

	Inducer Absent		Inducer Present	
	P	**β-gal**	**P**	**β-gal**
(1)	0	0	+	+
(2, 4, 5, 6)	+	+	+	+
(3)	0	0	0	0

10.13. (*a*), (*d*) = inductive; (*b*), (*c*), (*e*) = constitutive

10.14

	Inducer Absent			Inducer Present		
	P	β-gal	lacCZ	P	β-gal	lacCZ
(1)	0	0	0	+	0	+
(2)	0	0	0	+	+	0
(3)	+	+	0	+	+	0
(4)	0	0	0	0	+	0
(5)	0	0	+	0	0	+
(6)	0	+	0	0	+	0

All + answers for parts (4), (5), and (6) are due to new level
of products by sneak synthesis.

10.15. (a) Yes (b) Constitutively (c) Yes (d) Inductively (e) Note that in genotypes (2) and (3) the $lacI^-$ allele fails to produce enzymes in the absence of the external inducer (lactose); it fails to exhibit dominance. Therefore, the first hypothesis is incorrect. Under the second hypothesis, $lacI^+$ is dominant and produces a repressor that (in the absence of lactose inducer) blocks enzyme synthesis as seen in genotypes (2) and (3). (f) Genotype (3) has $lacY^+$ and $lacZ^+$ on one DNA molecule and $lacI^+$ on a different DNA molecule. Yet in the absence of lactose inducer, the repressor made by $lacI^+$ still prevents the production of enzymes by $lacY^+$ and $lacZ^+$. Therefore, the repressor must be able to act at a distance (it behaves as a diffusible substance) on genes that are either on the same DNA molecule (cis position) or on a different DNA molecule (trans position).

10.16. (1) A change that prevents repressor binding. (2) Modifications that increase repressor binding so that operons cannot be derepressed even when inducer is bound.

10.17. The mutation could be in the gene for adenyl cyclase or in the gene for catabolite activator protein (CAP).

10.18. I inhibits 7; J inhibits 9; G alone or (I, J) together inhibits 5; E inhibits 3; enzyme 1 could be inhibited by (I, J, E), (G, E), (I, J, C), or (C, G).

10.19. (a) 20 (b) Streptomycin; Hfr is streptomycin-sensitive (str^s). (c) Origin-(thr^+ leu^+)-azi^s-Ton^s-lac^+-gal^+-str^s-(mal^+ xyl^+). The genes for synthesizing the amino acids threonine and leucine must have entered first, otherwise none of the other recombinants could survive on unsupplemented medium. *Note:* The order of markers within parentheses has not been determined.

10.20. *arg - thy - mal - met - thi - thr - azi - bio - try - his - phe - arg*
      6   4   7   5   11   1   15   8   14   5   12

10.21. $cys\text{-}trp_B\text{-}trp_A$

10.22. $a\text{-}c\text{-}d\text{-}e\text{-}b$

Answers to Review Questions

Vocabulary

1. auxotrophic
2. clone
3. theta (θ)
4. lawn
5. conjugation
6. transduction
7. merozygote
8. plasmid
9. DNA gyrase
10. helicase

Multiple-Choice Questions

1. *c* 2. *d* 3. *c* 4. *a* 5. *d* 6. *c* 7. *d* 8. *a* 9. *e* 10. *c*

CHAPTER 11

Viruses, Transposable Elements, and Cancer

Viruses

Viruses are noncellular entities that are obligate, intracellular parasites. They require a host cell in order to reproduce. In addition, they have the following characteristics:

1. Viruses have only one kind of nucleic acid (either DNA or RNA), whereas cells have both kinds.
2. Viruses have no protein-synthesizing system of their own (i.e., they have no ribosomes); they have no energy-conversion system of their own (i.e., they do not metabolize food to generate ATP).
3. Viruses are not contained by a lipid membrane of their own making (although some viruses become surrounded by an **envelope** of modified host membrane as they leave the cell). They have no internal membranes.
4. Viruses are not affected by antibiotics, although their host cells might be.
5. Viruses have no cytoskeleton or means of motility other than diffusion.
6. Viruses do not "grow" in the classical sense of increasing in mass; i.e., once the virus is formed, it does not increase in size.

A fully formed virus is called a **virion**, and it has its genetic material protected within a protein coat known as a **capsid**. The individual protein subunits that make up the capsid are called **capsomeres** (see example in Figure 11-6). The basic viral unit of nucleic acid and protein coat is called a **nucleocapsid**. Viruses are much smaller than either prokaryotic or eukaryotic cells; their sizes are typically in the nanometer (nm) range (20–300 nm). A nanometer is 1/1000 of a micrometer (μm). Viruses cannot be seen under a light microscope; thus, specialized microscopes such as electron microscopes must be used. Viral genome sizes are also smaller than cellular genomes; the smallest is around 500 kb and the largest is about 5000 kb. Thus, viral genomes do not code for a large number of genes. They typically encode genes required for infection (e.g., enzymes), genome replication (e.g., polymerases), virion production (e.g., capsid proteins), host cell destruction (e.g., the enzyme lysozyme), and others required for any special aspect of their lifestyle. Viruses can have a linear or circular genome of single-stranded or double-stranded nucleic acid. Viral genomes can be composed of DNA or RNA. Most viruses have only one molecule of nucleic acid in their genome, but some have more than one.

Viruses are highly symmetrical in their structure. They can have a **helical symmetry** (in which the viral protein coat forms a long helix), an **icosahedral symmetry**, an **enveloped** structure, or a **complex** structure. An icosahedron is a spherical shape with 20 equilateral, triangular faces. Protein coat subunits (capsomeres) spontaneously assemble to construct each face of the icosahedron. Enveloped viruses have their nucleocapsid core surrounded by a phospholipid bilayer that they obtain as they are extruded from the host cell they infected. These viruses are more common in the animal world. Complex viruses are usually composites of several structures (e.g., bacterial viruses have an icosahedral head and a long, helical tail with tail fibers) or they have unusual, asymmetrical structures.

Cells that are susceptible to viral infection have specific **receptors** on their surfaces to which each virus can attach. Cells without these receptors would be refractory to infection by viruses. Viruses are specialized to infect particular hosts and host cell types. Generally, a particular type of virus does not infect more than one host, species (or strain) but occasionally this does occur. For example, the human immunodeficiency virus (**HIV**) is thought to have originally infected non-human primates in rain forest areas, but it most likely obtained a mutation that allowed it to jump between species, resulting in the human disease, acquired immunodeficiency syndrome, **AIDS**.

Bacteriophages

Viruses that infect bacteria are called **bacteriophages** or simply **phages**. The plural form "phages" is used when referring to different species (e.g., lambda and T4 are both phages). When referring to one or more virions of the same species, the word **phage** is used; thus, a bacterial cell may be infected by one or more lambda phage. The most commonly studied phages have a roughly spherical icosahedral capsid to which a tail is attached. The tail may be long or short, contractile or noncontractile. Other kinds of phage have tailless heads or filamentous structures. The genetic material of most phages is double-stranded DNA (dsDNA), although some single-stranded DNA (ssDNA), single-stranded RNA (ssRNA), and double-stranded RNA (dsRNA) phages are known. Enveloped forms are rare. Other characteristics that are useful for classification include molecular weight, genomic base composition (G + C content), antigenic specificities of the capsid, and species (or strains) of susceptible host cells (**host range**). **Host restriction** is the ability of a bacteriophage to replicate in only certain strains of bacteria.

EXAMPLE 11.1
Several bacterial species synthesize a site-specific endonuclease enzyme that can digest any foreign DNA containing the specific nucleotide sequence that constitutes the recognition site of the enzyme. According to the restriction and modification model proposed by W. Arber, such a bacterium would also contain a methylase enzyme to *modify* (by methylation) these same sequences in its own DNA, and thus protect it from digestion by endogenous endonuclease. Foreign DNA from a different source, such as an infecting phage particle, would not have these recognition sites methylated and hence would be destroyed (and thus *restricted* from surviving in that strain) by the host's endonuclease.

The nucleic acid from a single phage particle typically infects a bacterial cell, replicates itself many times, produces viral proteins to make numerous viruses, and ruptures the cell to release several hundred progeny phage. Repetitions of this reproductive process can cause a turbid bacterial culture to rapidly become clear owing to lysis of the host cells. The various types of viral reproductive processes are sometimes referred to as "life cycles"; this is a misnomer because viruses are not considered to have all of the properties attributed to living cells. If a dilute solution of phage is plated on a confluent growth of bacterial cells ("lawn") on nutrient agar in a Petri dish, a cleared area, or "hole," will develop around each position where a phage particle was deposited. These holes, referred to as **plaques**, contain millions of progeny phage that have been released from lysed cells. By counting the number of plaques on a plate, and knowing the amount and dilution of the phage suspension added to the plate, one can estimate the total number of phage particles or the **phage titer** in the original phage solution.

SOLVED PROBLEM 11.1
A prepared solution of bacteriophage has an unknown titer (phage number). An aliquot, 0.1 ml, is used to inoculate 1 ml of a susceptible *E. coli* culture (This amount of *E. coli* is enough to grow into a lawn on the appropriate type of culture medium.) The inoculated culture is then mixed with molten **top agar** (agar that is poured on top of another type of culture medium) and poured on top of a solid culture medium in a petri plate. The top agar is allowed to harden and the plated bacteria are incubated for 12–18 hours. After incubation, a lawn of bacteria is observed to contain many plaques. The number of bacteriophage plaques is counted and found to equal 220. Determine the infective phage titer in the original phage solution.

Solution: Each plaque can be assumed to be the result of a single bacteriophage infecting a single *E. coli* cell. This assumption may not be valid if too many phage particles are used for the inoculation. However, let us assume that one plaque results from one phage infecting a single cell. Thus, each plaque represents one bacteriophage particle, known as a **plaque-forming unit (pfu)**. In order to find the number of infective bacteriophage particles in the original culture, simply divide the counted pfu number by the volume of the phage solution used to inoculate the bacterial culture, as follows:

$$\frac{220 \text{ pfu}}{0.1 \text{ ml}} = 2200 \text{ pfu/ml phage solution}$$

If a dilution of the phage solution, such as 1 : 100, was used as an inoculum, this dilution would have to be taken into account in the above calculation, as follows:

$$\frac{220 \text{ pfu}}{0.1 \text{ ml}} \times 100 = 2.2 \times 10^5 \text{ pfu/ml phage solution}$$

1 BACTERIOPHAGE LIFE CYCLES

Most phages (such as phage T4 that infects *E. coli*) have only a **lytic life cycle** in which they kill the host cell in the production of progeny phage. Such phages are said to be **virulent**. A few phages (such as phage lambda that also infects *E. coli)* have a **lysogenic life cycle** in which they may either temporarily camouflage themselves and act as a **temperate phage** (**nonvirulent**) or enter a lytic cycle.

(a) Lytic Cycles. The first step in the life cycle of **phage T4** (Fig. 11-1) involves the adsorption of a virion to a specific receptor site on the surface of the host cell. Any cell lacking this receptor would be immune to infection by T4. Each type of phage can usually infect only one species of bacteria, and in some cases only a particular strain or strains of that species; the number of such cell types in which a phage can carry out its lytic cycle constitutes its host range. Following adsorption, T4 injects its DNA through its tail into the host cell. The empty phage capsid remains outside the bacterium as a **ghost** (so named because of the empty appearance of the head in electron micrographs). The filamentous phage, **M13**, is able to penetrate the cell wall and then has its nucleic acid released by host-cell enzymes that digest the coat proteins.

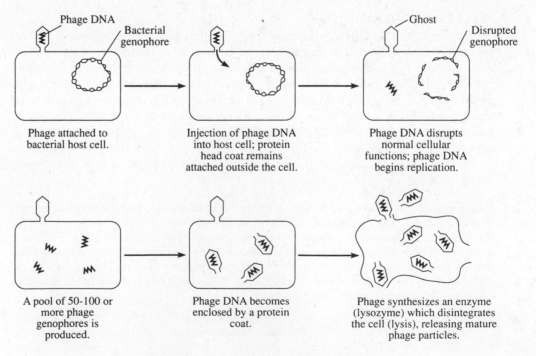

Fig. 11-1. Lytic life cycle of T-even phages.

Once the naked phage DNA is inside the cell, different phages may use different strategies to produce progeny particles. Generally, however, the phage DNA initially is transcribed by the host's RNA polymerase into "early mRNAs." Later mRNAs may be synthesized by a phage RNA polymerase that was made from an early mRNA; or perhaps the bacterial RNA polymerase becomes modified to transcribe phage genes preferentially or exclusively. These mRNAs become translated into catalytic (enzymatic), regulatory, and structural proteins. The regulatory proteins of the phage control the timing at which various phage genes become active. The structural proteins form heads, tails, and other protein parts of the complete phage particle as needed. The phage enzymes mediate replication of many copies of the phage genome, further transcription, and sometimes even the destruction of the host's DNA.

EXAMPLE 11.2

Phage T4 specifies the enzyme hydroxymethylase that modifies the cytosine bases in its own DNA to **5-hydroxymethylcytosine**. Such modified bases are resistant to degradation by host-cell nucleases, making the phage more successful during infection.

Several different mechanisms are known for packaging phage DNAs into their protein coats. In *E. coli* phage T4, rolling-circle replication of its double-stranded DNA produces long, tandemly linked series (concatemers) of phage genomes. It is thought that the end of the concatemer enters the head, followed by enough DNA to fill the head. The concatemer is then cleaved at a nonspecific site. Since the DNA capacity of the head is greater than the length of one phage genome (monomer), the gene order will be different in each linear fragment cut from the concatemer. Terminal regions will be present twice within each monomer (terminally redundant). Since each phage monomer cut from a concatemer begins at a different gene sequence, they collectively form a cyclically permuted set (Fig. 11-2).

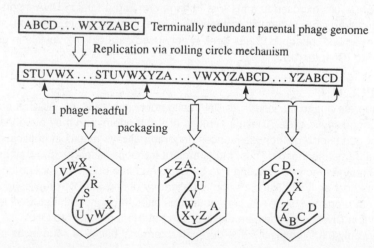

Fig. 11-2. Production of a terminally redundant and cyclically permuted set of progeny phage by cutting (vertical arrows) the linear tail produced by sigma replication according to the "headful rule." Terminal redundancy can be present without cyclic permutation in some phages that do not use this mechanism.

In phage **lambda (λ)**, the circular genome is replicated early in the lytic cycle to increase the number of templates for transcription and further replication. Later in the cycle, rolling-circle replication provides the genomes for packaging into the heads of progeny phage. Lambda genomes are also cut from a concatemer, but unlike phage T4, the cuts are made at base-specific sequences known as **cos sites** (for cohesive site). Linear phage genomes always end with single-stranded termini because they are cut from the concatemer at the cos site by a sequence-specific **terminase** or **Ter system**. Ter-cutting requires that two cos sites or one cos site and a free cohesive end (1/2 cos) be present on a single concatemeric DNA molecule. A modified lambda genome that is 79–106% the length of a normal λ phage genome will still be cut by the Ter system and become packaged into phage heads. This is an important property of λ that makes it useful as a vector for genetic cloning (see Chapter 12).

After assembly of the phage capsids is completed, a lytic protein called **lysozyme** is usually produced that ruptures the cell and releases the progeny phage in a typical **burst size** of 50–300 infective particles per cell. Most virulent phages follow the general lytic life cycle outlined above. Some exceptions, however, are known.

EXAMPLE 11.3

E. coli phage M13 is filamentous and contains a circular, single-stranded DNA molecule. Among all known phages reproducing vegetatively, M13 is the only one that neither kills nor lyses its host cell. Infective progeny phage leave the cell by budding from its surface without causing cell damage. Upon infection, the entire phage particle penetrates the cell wall by being absorbed at the end of a sex (F) pilus. The entry of coat proteins into the cell is another feature unique to this phage. One genetically engineered strain of M13 (M13mp7) contains the promoter (*lacP*), operator (*lacO*), and β-galactosidase gene (*lacZ*) of the *E. coli* lactose operon. Insertion of a foreign DNA segment into *lacZ* inactivates the gene and no enzyme is produced. Lac⁻ bacterial cells infected with wild-type M13mp7 would be able to ferment lactose. On EMB agar, lactose-fermenting colonies would appear dark purple. Cells exposed to a M13mp7 phage carrying a foreign DNA insert in *lacZ* would be unable to ferment lactose; therefore, they grow into colorless colonies. Like lambda, phage M13 has been widely used as a cloning vehicle in genetic engineering (Chapter 12).

EXAMPLE 11.4

The temperate phage *Mu* inserts its DNA obligatorily into its *E. coli* host chromosome during its lytic cycle. These insertions are at random, and they often inactivate host genes or regulatory sequences. Insertion always results in duplication of a terminal target sequence. Thus, Mu is a giant transposon (page 387) that has acquired phage functions enabling it to be packaged into phage coats and to escape its host by lysis. Transposition is obligatory during Mu DNA replication. Insertion of progeny Mu DNA occurs at various sites throughout the lytic cycle. Various host DNA sequences are always found at the termini of Mu DNA. However, only Mu DNA inserts; the duplicated terminal bacterial sequences are not inserted.

(b) Lysogenic Cycles. There are two types of lysogenic cycles. In the most common type, typified by *E. coli* phage lambda (λ), the phage DNA becomes integrated into the host chromosome. In the other type, represented by *E. coli* **phage P1**, the phage DNA does not integrate into the host chromosome, but somehow replicates in synchrony with it as a plasmid. Both the integrated and plasmid forms of phage DNA are called **prophage**.

The establishment of an integrated lambda prophage occurs in four major steps:

1. Linear phage DNA is injected into the host bacterial cell; the phage DNA is circularized by base pairing of its terminally redundant tails.
2. Some early phage genes are transcribed to produce a few molecules of a repressor protein and an **integrase** enzyme. The repressor then turns off transcription of phage genes.
3. The phage DNA is usually integrated or inserted at a specific site into the host chromosome as a prophage with the aid of integrase.
4. The bacterium survives and multiplies; the prophage is replicated along with the host chromosome.

The mechanism whereby an infected cell is switched to either the lytic or the lysogenic cycle is not well understood. Many details of the lysogenic cycle are known for phage lambda, but they are too complex to be presented here. However, two conditions seem to favor the establishment of the lysogenic cycle of a temperate phage: (1) depletion of nutrients in the growth medium and (2) high **multiplicity of infection** (**MOI**)—i.e., many adsorbed phages per bacterium. Phage can carry out the lytic cycle only in cells that are actively metabolizing. When nutrients are depleted, bacteria degrade their own mRNAs and proteins before they become dormant. When nutrients become available to an uninfected dormant bacterium, it can again resume growth. A phage-infected cell that becomes dormant interrupts the lytic cycle, and usually loses the ability to produce phage. The cell dies. On the other hand, if the cell can become **lysogenized** (containing a prophage), both the phage and the bacterium can survive a dormant period, and the potential for production of phage by induction persists (Fig. 11-3).

If a lysogenic bacterium sustains damage to its DNA, it would be advantageous for the prophage to **deintegrate** from the bacterial chromosome, enter the lytic cycle, produce progeny phage, and leave that cell. When bacterial DNA is damaged, a protease (**RecA protein**) of the SOS repair mechanism is activated. This protease cleaves the lambda repressor that has kept the prophage in its inactive state. The prophage DNA becomes **derepressed**, an **excisionase** enzyme is synthesized, and the prophage deintegrates from the host chromosome to enter the lytic cycle. This is the process known as **prophage induction**. If ultraviolet radiation has damaged the host DNA, the ensuing prophage induction is termed **UV induction**. When a nonlysogenic F^- bacterial cell receives, by conjugation, a prophage from a lysogenic Hfr donor, the recipient cell dies by induction of the lytic phage cycle. This form of prophage induction is termed **zygotic induction**.

In the lysogenic cycle of phage **P1**, the prophage is not integrated into the bacterial chromosome. Upon entry into the cell, P1 DNA circularizes and is repressed. It remains as a free, supercoiled, plasmidlike molecule, and replicates once with each cell division so that each daughter cell receives one copy of the prophage. The mechanism for this orderly assortment is not well known.

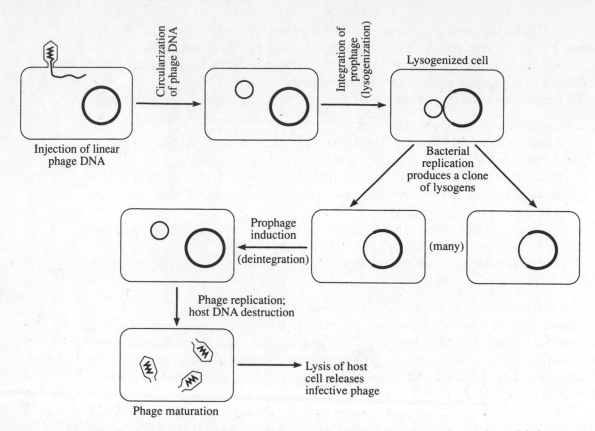

Fig. 11-3. Lambda-type lysogenic cycle. Thin lines represent phage DNA; thick lines represent bacterial chromosome.

2 TRANSDUCTION

Transduction is the virus-mediated transfer of DNA from a donor bacterial cell to a recipient cell. There are two types of transduction: specialized and generalized. In both cases, bacterial DNA is incorporated into the genome of a mature virus that then goes on to infect a different bacterial host. In this process, the bacterial DNA is transferred to a new recipient cell. Generally, specialized or generalized tranducing phage are defective due to the integration of additional DNA into their genomes. However, they must be normal enough to infect a new cell.

(a) Specialized Transduction. Specialized transduction occurs when a specific region of the bacterial chromosome becomes integrated into a mature viral particle. There are four distinguishing characteristics of specialized transduction: (1) the only bacterial genes that can be transduced are those very near the site at which the prophage is integrated, (2) only λ-type prophage are involved, (3) it results from defective excision of the prophage from the host chromosome, and (4) recombinant progeny bacteria may be partial diploids. The only site at which lambda phage integrates into the host chromosome (Fig. 11-4) is between the genes for galactose fermentation (*gal*) and biotin synthesis (*bio*). The head of the phage can only contain a limited amount of DNA, so if the prophage deinte-

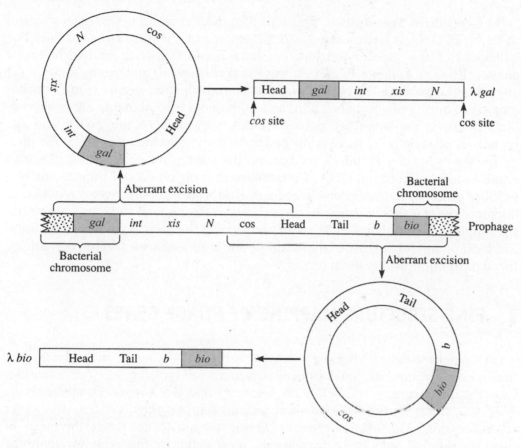

Fig. 11-4.   Formation of λ*bio* (or λ*gal*) by aberrant excision of the prophage from the bacterial chromosome (stippled).

grates abnormally from the host chromosome (taking some bacterial DNA in place of its own DNA), only the *gal* or *bio* genes could be transduced. Thus, all transducing lambda phages are defective in part of their own genome and cannot replicate on their own. A lambda phage that transduces the galactose genes is therefore called λ*gal* or λ*dg* (*d* = defective; *g* = galactose). If a *gal$^-$* cell is infected by λ*dg* (bearing the gene *gal$^+$*), integration of the defective prophage into the host chromosome produces (for the *gal* locus) a partial diploid recombinant chromosome. Aberrant excision of the prophage is usually a rare event, so restricted transduction is an event of low frequency in nature. However, high-frequency transduction can be attained under laboratory conditions. If a bacterial cell is doubly infected with a wild-type lambda phage and a λ*dg* phage, the wild-type phage can supply the functions missing in the defective phage, and the progeny will contain about equal numbers of both types. When the lysate is used for transduction, the process is referred to as **high-frequency transduction**. In many cases, because of its defective genome, λ*dg* fails to be integrated into the host chromosome (and therefore is not replicated). At each division, only one of the two progeny cells contains the defective phage genome; this process is called **abortive transduction.**

(b) Generalized Transduction. This type of transduction occurs when any region of bacterial DNA is incorporated into the genome of a mature virus particle. The hallmarks of this type of transduction have been extensively studied in the **P phages** (P1 of *E. coli* and P22 of *Salmonella typhimurium*) and are as follows: (1) any bacterial gene can be transduced, (2) the transduction results from a packaging error during phage maturation, and (3) haploid recombinants are produced. Since there is no homology between DNA sequences of the phage and the sequences of their host, there is no preferential site at which the prophage integrates. Any gene can be transduced because the head of the phage can package an entire headful of bacterial DNA. **Cotransduction** is the process of transducing two or more genes via the same defective phage. Reciprocal crossing over is required to integrate the transduced genes, so recombinant bacteria tend to be haploid rather than diploid. The endogenote segment replaced by the exogenote fails to replicate for lack of an ori (origin of replication) site and becomes lost in the culture through dilution or digestion.

3 FINE-STRUCTURE MAPPING OF PHAGE GENES

(a) Complementation Mapping. Because so little of the phage or bacterial genome consists of nonfunctional sequences, virtually all crossing over occurs within, rather than between, genes. Before the discovery that DNA is genetic material, the gene was thought to be the smallest genetic unit by three criteria: mutation, recombination, and function. Semour Benzer set out to determine the limits of these operational units by performing the most definitive fine-structure mapping ever performed on a phage gene. He chose to investigate the *r*II region of phage T4. When wild-type (r^+) T4 infects its host *E. coli*, it produces relatively small plaques. Many mutations have been found that have a shorter life cycle than the wild type, thus yielding larger plaques. These "rapid lysing" mutants may be classified into three phenotypic groups, depending on their ability to lyse three strains of *E. coli* (Table 11-1). The sites of *r*I, *r*II, and *r*III map at noncontiguous locations in the T4 genome. The *r*II region is about 8 recombination units long, representing approximately 1% of the phage DNA. Complementation tests were used to determine that the *r*II region consists of two genes.

Table 11-1. Distinguishing Characteristics of Rapid Lysing Mutants of Phage T4 in Comparison with Wild Type

Strain of Phage T4	Plaque Type on *E. coli* Strains		
	B	S	K
r^+	Wild	Wild	Wild
*r*I	Large	Large	Large
*r*II	Large	Wild	No plaques
*r*III	Large	Wild	Wild

EXAMPLE 11.5

If two *r*II mutants are added to strain K12 in sufficient numbers to ensure that each cell is infected with at least one of each mutant, one of two results is observed. If all the *E. coli* K12 cells lyse after one normal propagation cycle (20–30 min), we may infer that the mutants were in different functional units (genes). Each mutant was making a different polypeptide chain, and the two chains "cooperated" to allow a normal-sized burst to occur. On the other hand, if the two *r*II phages contain a mutation in the same gene, progeny phage can only be produced by genetic recombination with a frequency dependent upon how closely the two point mutations are linked. In any event, only a few of the cells would be expected to lyse by this mechanism in the same period of time. Thus, the results of complementation are easily distinguished from those of recombination. Benzer found that all of the point mutations in the *r*II region mapped into two genes (A and B).

(b) Deletion Mapping. Benzer also found that about 10% of his more than 2000 *r*II mutants did not backmutate to wild type because they were deletions of various lengths. By infecting cells with two different deletions, wild-type recombinants could be produced if the deletions did not overlap. Of course, wild-type recombinants cannot be produced if the two deletions overlap to any extent. Thus, by a series of crosses Benzer was able to draw a topological map in which deletions were shown to overlap or not overlap. The lengths of the deletions or the degree of overlap or non-overlap is arbitrary at this point, although they can be determined by crosses with point mutations. Having obtained a topological map, it is then possible to assign one or more point mutations to a relatively small segment of a gene by crossing them with deletion mutants. A point mutation cannot recombine with a deletion over the same site. This principle allowed Benzer to group mutants within relatively small regions of each gene. He did not attempt to order the point mutations within each of these small segments, but he did subject them to recombination tests to ascertain identity or nonidentity. This was accomplished by doubly infecting strain B with a pair of *r*II point mutants (e.g., *r*IIa and *r*IIb) in broth, and then allowing them to lyse the culture. The total number of progeny phage can be estimated by plating dilutions of this lysate on *E. coli* B and counting the resulting plaques. Wild-type recombinants are scored by plating the lysate on strain K. For every wild-type (r^+IIa, r^+IIb) plaque counted on strain K, we assume that an undetected double mutant (*r*IIa, *r*IIb) reciprocal recombinant was formed.

$$\text{Recombination percentage} = \frac{2(\text{number of plaques on K})100}{\text{number of plaques on B}}$$

The smallest reproducible recombination frequency that Benzer observed between two sites in the *r*II region was about 0.02%, corresponding to approximately 1/400 of a genetic region whose total length is only 8 recombination units. Thus, it was concluded that the genes of the *r*II region each contained hundreds of possible mutation sites, and that recombination could occur between the closest of these mutant sites. Benzer reasoned that the smallest distance within which recombination occurs (**recon**) might be as small as adjacent nucleotide pairs. The smallest bit of DNA that when mutated could cause a phenotypic effect (**muton**) was found to be as small as five nucleotide pairs or smaller. Since Benzer's observations, the muton has been shown to be a single nucleotide base pair.

A surprising finding of this work was that the point mutants were not randomly located in the *r*II region; a few locations (called **hot spots**) in both genes had many more mutations than elsewhere (over a hundred in a couple of positions vs. about 1 to 10 elsewhere).

SOLVED PROBLEM 11.2

In an attempt to determine the amount of recombination between two mutations in the *r*II region of phage T4, strain B of *E. coli* is doubly infected with both kinds of mutants. A dilution of $1 : 10^9$ is made of the lysate and plated on strain B. A dilution of $1 : 10^7$ is also plated on strain K. Two plaques are found on K, 20 plaques on B. Calculate the amount of recombination.

Solution: In order to compare the numbers of plaques on B and K, the data must be corrected for the dilution factor. If 20 plaques are produced by a (10^9) dilution, the lesser dilution (10^7) would be expected to produce 100 times as many plaques.

$$\text{Recombination percentage} = 2(\text{no. of plaques on K})(100)/(\text{no. of plaques on B})$$
$$= 2(2)(100)/20(100)$$
$$= 0.2\%$$

SOLVED PROBLEM 11.3

Seven deletion mutants within the A gene of the *r*II region of phage T4 were tested in all pairwise combinations for wild-type recombinants. In the table below of results, $+$ = recombination, 0 = no recombination. Construct a topological map for these deletions.

	1	2	3	4	5	6	7
1	0	+	0	0	+	0	0
2		0	0	0	+	+	0
3			0	0	+	+	0
4				0	+	0	0
5					0	0	0
6						0	0
7							0

Solution: If two deletions overlap to any extent, no wild-type recombinants can be formed.

Step 1. Deletion I overlaps with 3, 4, 6, and 7 but not with 2 or 5.

```
        1                 2, 5
  |_____|       |_____|

        3, 4, 6, 7
  |_____|
```

Step 2.   Deletion 2 overlaps with 3, 4, and 7 but not with 1, 5, or 6.

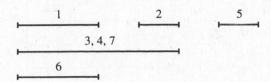

Step 3.   Deletion 3 overlaps with 1, 2, 4, and 7 but not with 5 or 6.

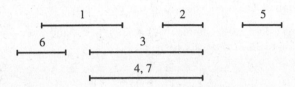

Step 4.   Deletion 4 overlaps with 1, 2, 3, 6, and 7 but not with 5.

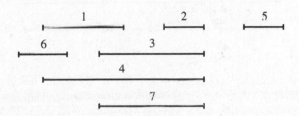

Step 5.   Deletion 5 overlaps with 6 and 7 but not with 1, 2, 3, or 4. To satisfy these conditions we will shift 5 to the left of 1 and extend 7 into part of region 5 so that the overlaps in step 4 have not changed. Now segment 5 can overlap 6 and 7 without overlapping 1, 2, 3, or 4.

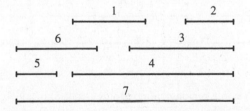

Step 6.   Deletion 6 overlaps with 1, 4, 5, and 7 but not with 2 or 3. No change is required.
Step 7.   Deletion 7 overlaps with all other regions, just as it was temporarily diagrammed in step 5. This completes the topological map. Although the overlaps satisfy the conditions in the table, we have no information on the actual lengths of the individual deletions.

SOLVED PROBLEM 11.4
Five point mutations (*a–e*) were tested for wild-type recombinants with each of the seven deletion mutants in Problem 11.3. Determine the order of the point mutations and modify the topological map accordingly.

	1	2	3	4	5	6	7
a	0	+	0	0	+	+	0
b	+	+	+	0	+	0	0
c	+	+	+	+	0	0	0
d	0	+	+	0	+	0	0
e	+	0	0	0	+	+	0

Solution: A deletion mutation cannot recombine to give wild type with a point mutant that lies within its boundaries. The topological map was developed in Problem 11.3.

Step 1. Mutant *a* does not recombine with 1, 3, 4, and 7 and therefore must be in a region common to all of these deletions.

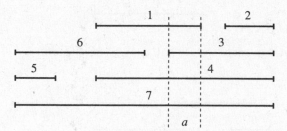

Step 2. Mutant *b* does not recombine with 4, 6, and 7 and thus lies in a region common to these three deletions. As the topological map stands in step 1, a point mutant could not be in regions 4, 6, and 7 without also being in region 1. Therefore, this information allows us to modify the topological map by shortening deletion 1, but still overlapping deletion 6. This now gives us a region in which *b* can exist.

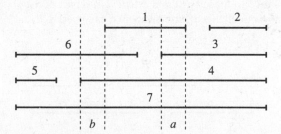

Step 3. Mutant *c* lies in a region common to deletions 5, 6, and 7. Mutant *d* lies in a region common to deletions 1, 4, 6, and 7. Mutant *e* lies in a region common to deletions 2, 3, 4, and 7. Thus, the order of these point mutations is *c-b-d-a-e* and the topological map is modified as shown in step 2.

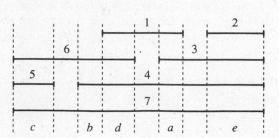

SOLVED PROBLEM 11.5
Propose a procedure for establishing the location of a lambda prophage with respect to other bacterial genes.

Solution: Cross a donor Hfr cell (lysogenic for lambda) with a nonlysogenic F⁻ recipient. Once the lambda genome has been transferred via conjugation to the F⁻ recipient cell, the cell often will die by lysis (a phenomenon known as zygotic induction). Using the interrupted mating technique (blender treatment), it should be possible to determine the point at which the frequencies of origin-proximal recombinants decrease with time. At this point, all of the lambda genome has been donated and can therefore be related on the temporal map to the location of other bacterial genes. Since recipient cells contain no repressor (and no unbound repressor is likely to be transferred during conjugation), the exogenous prophage has a good chance of entering the lytic cycle because relatively high levels of repressor are required to establish the lysogenic state within the F⁻ recipient cell. A lysogenic cell usually contains sufficient unbound ("cytoplasmic") repressor to inhibit superinfection by one or a few lambda phages, but a nonlysogenic cell has no immunity to infection by exogenous lambda phages or lambda prophages. Hence, in the latter case, zygotic induction of exogenous lambda prophages has a good chance of occurring.

An alternative procedure for mapping the location of prophage lambda is by mating a nonlysogenic Hfr strain with an F⁻ strain lysogenic for lambda and studying the loss of immunity to superinfection by lambda through recombination. However, this would be much more laborious than the previous procedure.

Eukaryotic Viruses

Eukaryotic viruses differ in many respects from virulent bacteriophages; some of the more obvious differences are outlined in Table 11-2.

Table 11-2. Some Differences between Virulent Bacteriophages and Eukaryotic Viruses

Bacteriophages	Eukaryotic Viruses
20–60-min life cycle	Typically 6–48-h life cycle
DNA virus = 50–1000 burst size	500–10^5 virions per cell per generation
Fraction of burst size capable of infection is very high (at or near 1.0)	Fraction of progeny virions capable of infection is very low (10^{-4} to 10^{-1})
Life cycle terminates in cell death (M13 is an exception); all phage simultaneously released	Some infected cells die; others continue to grow and release progeny viruses continuously
Cessation of synthesis of host DNA, mRNA, and protein is an early event after infection	Impairment of host functions usually occurs in late states of infection
Viral genome typically injected, with capsid remaining outside cell	No tailed forms; viral DNA is never injected into host cells; virion is typically taken up inside cell

(continued)

Table 11-2. *(continued)*

Bacteriophages	Eukaryotic Viruses
No segmented genomes	Some single-stranded RNA viruses have segmented genomes; double-stranded RNA viruses always have segmented genomes and are isocapsidic
No 5′ capping	5′ end of each ssRNA virion (+) strand or viral mRNA is capped
No poly-A tails	3′ end of ssRNA animal viruses have poly-A tail
Phage proteins rarely penetrate the host cell (phage M13 is an exception)	Viral proteins commonly enter the cell; removal of the capsid (uncoating) is an early event in infection
No replicative intermediate	Replicative intermediate required for ssRNA virions
Many rounds of recombination common with multiple infections	Genetic recombination between viral genomes rare
No known highly mutable forms	Some highly mutable forms; e.g., HIV-1, influenza virus
Restriction sites specifically methylated	No RE sites specifically methylated

1   ANIMAL VIRUSES

Some animal viruses cause relatively mild diseases such as the common cold; others cause more severe problems or even life-threatening conditions such as rabies, AIDS, and cancer. Much of what is known about the molecular biology of eukaryotic cells has come from studying their viruses.

Viral capsids are usually constructed from only one or a few types of proteins and thus do not require much coding information. The viruses that infect animal cells have been classified into four morphological types: (1) naked icosahedral (20 faces), (2) naked helical, (3) enveloped icosahedral, and (4) enveloped helical. **Naked viruses** have a protein capsid but no lipid envelope. An **envelope** is a portion of the host-cell membrane acquired as the virus leaves the cell by a **budding** process. The envelope is derived from the cell membrane in two steps. First, glycoproteins specified by the viral genome are inserted into the membrane. Then the virion capsid attaches to the cytoplasmic ends of the glycoproteins, causing the membrane to adhere to the capsid. The enveloped virus pinches off from the cell surface without creating a hole in the cell membrane. To infect another cell, an infective virus particle attaches to a specific receptor on the host-animal cell membrane, either by capsid proteins of a naked virion or by the viral glycoproteins extending from the surface of an enveloped virion. The attached virion is then engulfed by the host cell and the viral genome becomes **uncoated** (removal of the capsid) inside the cell.

The genomes of animal viruses may be DNA or RNA, single-stranded or double-stranded. Double-stranded DNA viral genomes may be either linear or circular. Circular double-stranded DNA may be covalently closed on one or both strands. All known double-stranded RNA viral genomes are segmented (i.e., consisting of multiple RNA molecules, each carrying a different set of genes). Some single-stranded RNA genomes are also known to be segmented. There are no known segmented DNA viruses or circular RNA genomes in eukaryotic viruses. Most DNA virus genomes are copied in the nucleus, using the host's RNA polymerase and other enzymes for capping, splicing, and adding poly-A tails in processing their transcripts. Most RNA viruses (except influenza virus) replicate in the cytoplasm.

The great diversity of animal viruses has been classified into 15–20 viral families based on characteristics such as type and structure of nucleic acid, virion morphology, and common antigenic determinants. Because of the dependent relationship of the viral genome on its mRNAs for replication, animal viruses have been recognized as falling into seven groups as summarized in Fig. 11-5.

> **EXAMPLE 11.6**
> All adenoviruses have a double-stranded DNA genome of about 36 kb and a naked icosahedral capsid of 252 subunits with prominent spikes at the vertices. Their mRNAs are transcribed directly from their genomes and their genomes replicate directly from their double-stranded DNA templates.

There are four major types of viral infection. The most common type is an **acute** or **lytic infection** that causes fairly rapid death of the host cell when it ruptures to release progeny virions. The second type involves the virus entering a dormant state similar to lysogeny of bacterial viruses (phages) as discussed earlier. This type of infection is called a **latent infection**. The third type involves virions that are slowly released from the cell surface without lysing or killing the host cell. This type of infection is termed **persistent infection**. Typically, as the virion buds from the cell, it acquires a lipid envelope of host membrane. Prior to becoming encapsulated, certain viral proteins have been synthesized and become incorporated into the host-cell membrane. Thus, the viral envelope contains viral proteins that enable the virion to attach to another host cell and spread the infection. The fourth type of viral infection involves transformation of a normal cell into a tumor cell, leading to **cancer**. This type of infection will be discussed later in the Cancer section of this chapter.

(a) DNA Viruses. A typical double-stranded DNA (dsDNA) virus attaches to a cell receptor and then is taken into the cell, where its capsid is removed (uncoated). The viral DNA is replicated using host-cell enzymes. The viral DNA is also transcribed by host enzymes into mRNAs, which in turn are translated (by the host's ribosomes and enzymes) into viral capsid proteins or (in some cases) into enzymes that favor viral DNA replication over that of host DNA. The capsomeres become organized into a capsid around the viral DNA to form progeny virions. The virions are released from the host cell by lysis or by budding. Deviations from this generalized life cycle exist in the dsDNA hepatitis B virus and in the ssDNA parvoviruses.

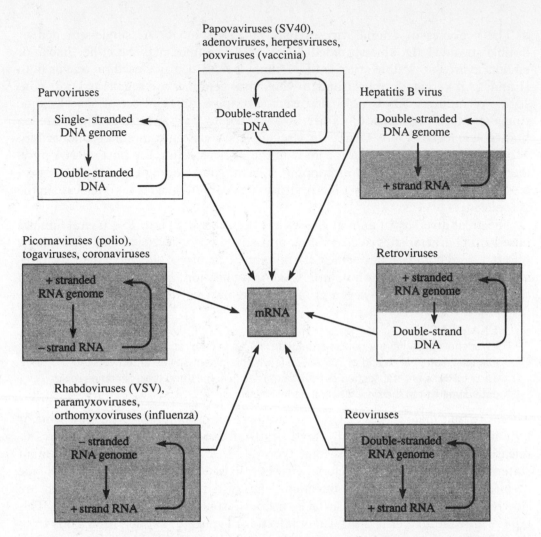

Fig. 11-5 A classification of animal viruses based on replication mechanism and the origin of mRNA. Arrows within each box indicate flow of information during replication. Arrows to mRNA originate at the template for mRNA synthesis. (After J. D. Watson et al., *Molecular Biology of the Gene*, 4th ed., Benjamin/Cummings Publishing Company, Inc., 1987.)

(b) RNA Viruses. Few host cells contain the enzymes necessary to replicate or repair RNA (rare exceptions are mentioned with viroids later in this chapter). Thus, the genes of RNA viruses have much higher mutation rates (e.g., 10^{-3} to 10^{-4}) than DNA viruses, and they must either code for these enzymes or carry these enzymes with them when they infect a host cell. RNA viruses with single-stranded genomes that function as mRNAs are said to have positive or plus (+) strand genomes specifying (minimally) the coat proteins and the enzyme(s) needed for replication. RNA viruses with negative or minus (−) strand genomes have DNA that is complementary to the genomic or mRNA strand, and so cannot be translated. Such viruses must, therefore, encode an RNA-dependent, RNA polymerase that can synthesize a (+)RNA strand from a (−)RNA template, and this enzyme must be packaged in the virion together with the viral RNA genome. For all RNA viruses except the retroviruses, double-stranded RNA is always an intermediate in viral RNA replication, even if the infective virion contains only single-stranded RNA (ssRNA). Double-stranded RNA is replicated in an analogous manner as DNA; i.e., each RNA strand serves as a template for making a complementary RNA strand. The viral enzyme that replicates viral RNA in this way is an RNA-dependent RNA polymerase called **RNA replicase**. **Retroviruses** contain a **reverse transcriptase** enzyme that copies their RNA genome into a DNA copy (**cDNA**). This cDNA can then be used for transcription or, in the case of retroviruses, can become integrated into the genome of the host cell.

Four model life cycles for the RNA viruses are easily recognized.

Model 1. If the viral RNA is double-stranded (dsRNA), the (+)strand is transcribed to produce RNA replicase. This enzyme not only replicates viral dsRNA [using both (+)strands and (−)strands as templates] to form dsRNA progeny genomes, but also makes many (+)copies using the (−)strands as templates. These extra (+)strands are required as mRNA templates for translating viral proteins in a relatively short period of time.

Model 2. If the viral RNA is a single (+)strand, the virion enters the host cell, becomes uncoated, and the (+)strand RNA is translated to produce an RNA replicase. The replicase then synthesizes a complementary (−)RNA strand using the (+)strand as a template, thereby forming a double-stranded RNA replicative intermediate. The (−)strands are needed as templates for the synthesis of (+)genomic strands of progeny virions. Some of the (+)strands are translated by the host cell's machinery into capsid proteins, membrane proteins, etc.

Model 3. If the viral RNA is a single (−)strand, it cannot serve as a translational template (mRNA) for making RNA replicase. Hence, this enzyme must be brought into the host cell along with the viral RNA. The RNA replicase uses the (−)strand as a template to produce a complementary (+)strand. More (−)strands are produced using the (+)strand(s) as templates and more (+)strands are produced using the (−)strand(s) as templates. The (+)strands serve as mRNAs for making viral proteins. The (−)strands then associate with the capsid proteins and RNA replicase to be packaged into progeny virions.

Model 4. Retroviruses contain single (+)strand RNA genomes that are not used as mRNA. They are first reverse transcribed into a cDNA copy. This cDNA integrates into the genome. The latent state begins. Upon activation, viral mRNAs are produced from the integrated viral DNA and viral-specific proteins are produced. Retroviruses are further discussed later in this chapter under the subject of cancer.

2 PLANT VIRUSES

Plant viruses exist in rod and polyhedral shapes. Most plant viruses have genomes consisting of a single RNA strand of the (+)type. The best-known plant virus is the rod-shaped tobacco mosaic virus (TMV; Fig. 11-6), which has a single-stranded (+)RNA genome of 6395 nucleotides. Some viruses with (+)genomes, however, cannot replicate unless the host cell is infected with two different virions. Such viruses are said to have **segmented genomes**. If the genomic fragments reside in different capsids, the virus is said to be **heterocapsidic**; if the fragments reside in the same capsid, the virus is said to be **isocapsidic**.

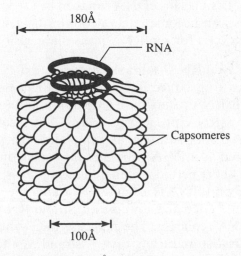

Fig. 11-6. Section of tobacco mosaic virus (3000 Å long) showing its single-stranded RNA genome and its associated capsomeres.

EXAMPLE 11.7
The heterocapsidic genome of the cowpea mosaic virus consists of two RNA chains, each encoding different proteins essential for replication. Each of these RNA chains is encapsulated into separate virions. The virus can only replicate in a host cell that has been infected by both kinds of virions.

Relatively few plant viruses have DNA genomes. There are only two classes of DNA plant viruses. The cauliflower mosaic virus belongs to the first class, which contains a double-stranded DNA genome in a polyhedral capsule. The second class of DNA plant viruses contains the geminiviruses (gemini = twins), characterized by a connected pair of capsids, each containing a circular, single-stranded DNA molecule of about 2500 nucleotides. The paired genomes may be identical in some viruses and markedly different in others.

Plant **viroids** have a very small RNA genome of 240–350 nucleotides in a single-stranded circle that can form extensive internal base pairing. This gives it essentially a stiff, double-helical structure and renders it resistant to digestion by ribonuclease enzymes that usually cut only at unpaired ribonucleotides. The genome is too small to code for any proteins. In the potato spindle tuber viroid there are no

AUG initiation codons, and frequent stop codons occur in all reading frames. At least some plant cells (unlike animal cells) are known to contain enzymes capable of replicating RNA. Viroids are not encapsulated in a protein coat. They do not pass through a DNA stage in their life cycle and are not integrated into the host chromosomes. Little else is known of their life cycles.

The variations within each class of plant and animal virus are too numerous to even mention here. Details of the life cycles of specific viruses can be found in more comprehensive texts.

Transposable Elements

Most genes reside at a specific locus or position on the chromosome. Some genes or closely linked sets of genes can mediate their own movement from one location to another and may exist in multiple copies (sometimes hundreds or thousands) dispersed throughout the genome. These elements have been variously called "jumping genes," "mobile elements," "insertion sequences," "cassettes," and "transposons." The formal name for this family of mobile DNA sequences is **transposable elements** and their movement is called **transposition**. Transposable elements were first discovered in corn and later in phages, bacteria, fungi, insects, viruses, and more complex animals.

Transposable elements can be of two different types. The first type is a relatively short DNA element that has the ability to "jump" or move to a new spot in the genome. These elements generally encode one or a few genes, one of which is an enzyme called **transposase** that is required for the movement. These elements are found in both prokaryotes and eukaryotes. The second type of transposable element is one that must go through an RNA intermediate (an mRNA) that is copied to DNA before this DNA is inserted into a new spot in the genome. These elements are termed **retrotransposons** or **retroposons** and are related to retroviruses. They are found primarily in eukaryotic organisms. Transposable elements can also be functional, meaning that they are able to transpose themselves, or nonfunctional, meaning that they cannot transpose on their own.

Transposition may result in mutation and is potentially a major source of genome diversity and change. Transposable elements can move adjacent DNA sequences with them, resulting in mutation or maybe spots for recombination events within a genome. If a transposon becomes inserted into the coding region of a gene, it interrupts the coding sequence and inactivates the expression of the gene. In addition, transposable elements may contain transcription and/or translation termination signals that block the expression of other genes downstream of the insertion site, thus affecting other genes downstream, as in an operon. This "one-way" mutational effect (or polarity) is referred to as a **polar mutation**. Transposition may also result in induction of oncogenic or cancer-promoting activities.

1 BACTERIAL ELEMENTS

The transposable elements of bacteria fall into two major classes. **Simple transposons** (also called **insertion sequences**, or **IS**) carry only the genetic information necessary for their transposition (e.g., the gene for the enzyme transposase). Simple insertion sequences have no known effects beyond transposition and inactivation of the gene (or operon) into which they may insert. **Complex** or **composite transposons** (**Tn**) contain additional genetic material unrelated to transposition, such as drug-resistance genes.

The hallmark of a simple transposon, whether it is known to transpose or not, is the presence of identical, **inverted terminal repeat** (**ITR**) sequences of 8–38 bp. Each type of transposon has its own unique inverted repeat. On either side of a transposon is a short (less than 10 bp) direct repeat (Fig. 11-7).

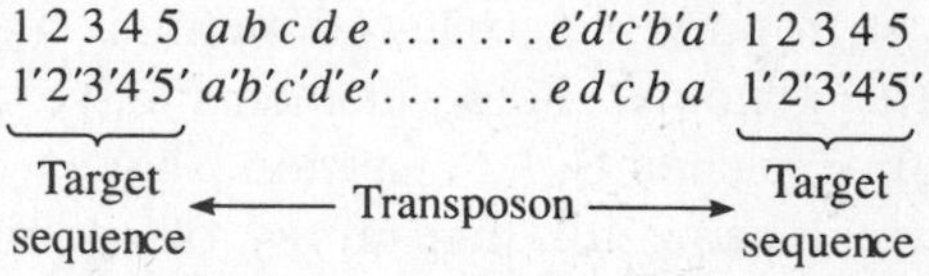

$$
\begin{array}{lll}
1\,2\,3\,4\,5 & a\,b\,c\,d\,e\ldots\ldots e'd'c'b'a' & 1\,2\,3\,4\,5 \\
1'2'3'4'5' & a'b'c'd'e'\ldots\ldots e\,d\,c\,b\,a & 1'2'3'4'5'
\end{array}
$$

Target
sequence ⟵—— Transposon ——⟶ Target
sequence

Fig. 11-7. Diagram of a simple transposon. Letters and numbers are used instead of nucleotides to make the repeats easier to read. Primed and unprimed symbols of the same kind (e.g., 3 and 3′, d and d′) represent complementary base pairs. The ends of a transposon consist of inverted repeats (represented by letters). Direct repeats of the target sequence (represented by numbers) flank the transposon. The dotted central region of simple transposons contains only genes necessary for translocation, but in complex transposons it may contain one or more additional genes.

If a transposon exists in multiple copies, these direct repeats are of different base composition at each site where the transposon exists in the chromosome; the inverted terminal repeats, however, remain the same for a given transposon. The sequence into which a transposable element inserts is called the **target sequence**. During insertion of a transposon, the singular target sequence becomes duplicated and thus appears as direct repeats flanking the inserted transposable element. The direct repeats are not considered part of the transposon. No homology exists between the transposon and the target site for its insertion. Many transposons can insert at virtually any position in the host chromosome or into a plasmid. Some transposons seem to be more likely to insert at certain positions (hot spots), but rarely at base-specific target sites. The enzyme(s) required for transposition is (are) encoded in the central region of the transposon. Transposons usually generate a high incidence of deletions in their vicinity because of imprecise excision that removes some adjacent sequences along with the transposon.

Two copies of a transposable element can transpose a DNA sequence between them. For example, in bacteria, Hfr cells are formed by the integration of the sex factor F into the host chromosome. An integrated F sequence is always flanked by two copies (in direct repeat) of one of the insertion sequences located in an F plasmid. This kind of complex transposon obviously may have effects in addition to those of the simple insertion sequences at their ends.

Some **complex transposons** carry one or more bacterial genes for antibiotic resistance in their central regions. Because transposons can shuttle in and out of plasmids as well as chromosomes, it is thought that multiple drug resistance, characteristic of R plasmids ("R" for resistance), developed in this way. Such plasmids are easily transferred by conjugation to antibiotic-sensitive bacteria and, with the aid of natural selection, very quickly spread resistance throughout a bacterial species within a patient. Transposons do not carry genes that are essential for survival under normal conditions, but in hostile environments (e.g., in the presence of antibiotics or an immune system) the genes carried by a transposon may make the difference between life and death of the bacterial cell.

Two models of transposition in prokaryotes have been proposed, on the basis of the fate of the donor site. The transposon might be excised from the donor site, leaving no copy of itself at the donor site (**conservative mode**). Alternatively, the transposon might be replicated, allowing one copy to transpose to another site and leaving an identical copy at the donor site (**replicative mode**). Only the replicative mode could produce multiple copies at various sites in the genome. In bacteria, the number of copies of a transposon appears to be regulated, seldom exceeding 20 copies per genome. In eukaryotes, however, the copy number can be very high.

2   EUKARYOTIC ELEMENTS

As mentioned earlier, eukaryotic cells can contain DNA elements that transpose on their own. A good example of this type of element is the *Ac-Ds* system in maize.

EXAMPLE 11.8

Mobile genetic elements ("jumping genes") were first discovered in maize by Barbara McClintock in the 1950s. Insertion of the controlling element *Ds* into or adjacent to a locus governing kernel color inhibits the production of color and results in a colorless phenotype. Excision of *Ds* reverses the effect and produces colored spots on a colorless background. The *Ds* elements occur in different sizes as deleted forms of a larger complete element called *Ac*. The *Ds* elements are non-autonomous because they remain stationary unless an *Ac* element is also present, whereas *Ac* elements are autonomous because they can move independently. Both *Ac* and *Ds* elements have perfect inverted repeats of 11 bp at their termini, flanked by 6- to 8-bp direct repeats of the target site. Thus, *Ac* and *Ds* are transposons. *Ac* need not be adjacent to *Ds* or even on the same chromosome in order to activate *Ds*. When *Ds* is so activated, it can alter the level of expression of neighboring genes, the structure of the gene product, or the time of development when the gene expresses itself, as a consequence of nucleotide changes inside or outside a given gene. An activated *Ds* element can also cause chromosomal breakage, which can yield deletions or generate a bridge-breakage-fusion-bridge cycle. Several other systems like the *Ac/Ds* system are now known in maize. Each has a target gene that is inactivated by insertion of a receptor element into it, and a distant regulator element that is responsible for the mutational instability of the locus. The receptor and regulator elements are both considered to be controlling elements of the target gene.

The other type of element found in eukaryotic systems is the retrotransposon. The retrotransposition process begins with transcription of DNA into mRNA. The RNA is then copied by the action of a reverse transcriptase enzyme into a DNA copy (cDNA). This cDNA then integrates into a new region of the genome. The reverse transcriptase is generally one of the genes encoded by the retrotransposon. Another commonly encoded gene is **integrase**, an enzyme required for integration of the cDNA into the genome. Retrotransposons exist in high copy number in mammalian genomes (up to 500,000 elements making up as much as 40% of the genome). They can be short (200–200 bp) or long (3,000–8,000 bp) DNA segments. Short elements are called **SINE**s, for *s*hort *in*terspersed *e*lements and the longer ones are termed **LINE**s for *l*ong *in*terspersed *e*lements.

EXAMPLE 11.9
In a human genome, there exist about 300,000 members of a SINE sequence (approximately 300 bp) that is cut by a base-specific DNase called *Alu*I. Members of this *Alu* family are related, but not identical in base sequence. Each member is flanked by direct repeats. Although transposition has not been observed for any member, the *Alu* family is thought to have evolved from a DNA copy of an RNA molecule that plays a role in protein synthesis. Since no function, essential or otherwise, has been attributed to this family, it may represent an example of "selfish DNA" whose function is to make copies of itself.

The most common LINE in mammals is the **L1** element. The *Ty* element in yeast and the *copia* element in *Drosophila* are other examples of retrotransposons. Many of these elements contain genes or processed **pseudogenes** related to cellular and retroviral genes. Pseudogenes are nonfunctional genes with sequence similarity to a functional gene found elsewhere in the genome of an organism.

Cancer

Cancer is a genetic disease resulting from multiple mutational events. These mutations alter the normal functioning of a cell so that it takes on the following characteristics: (1) it becomes **immortal**, i.e., it is capable of unlimited cell division, (2) it becomes **independent** from normal cellular controls that limit growth and division, and (3) it becomes **invasive** by spreading to other tissues, in a process called **metastasis**. One cancer cell is capable of dividing into a clonal mass of cells called a **tumor**. Tumors are not necessarily life-threatening (e.g., warts or galls) and may occur in both plants and animals. Cancers, however, are animal diseases characterized by uncontrolled cellular proliferation and spread of the abnormal cells into other tissues. Plants do not have cancers because their cell walls prevent metastasis of tumor cells. **Oncogenesis** is the process by which a normal cell becomes cancerous; **oncology** is the study of cancer. A **neoplasm** is a population of potentially cancerous cells growing out of control. If the neoplasm is confined to its place of origin and has no tendency to recur after removal, it is a **benign** neoplasm. If it metastasizes from its site of origin, it becomes a life-threatening **malignant** neoplasm.

Some oncogenic mutations may be inherited, and/or be induced by environmental exposure to mutagens that damage DNA. Oncogenic mutations may also occur spontaneously. A **carcinogen** is any agent (e.g., mutagenic chemicals, ionizing radiations, and certain viruses) that can promote a cancerous state. Aside from the irritant fibers of asbestos, all carcinogens are thought to be mutagenic (causing damage to DNA), but not all mutagens are carcinogenic.

EXAMPLE 11.10

Xeroderma pigmentosum is a genetic syndrome characterized by extreme sensitivity to ultraviolet light and the tendency to develop multiple skin cancers. It is inherited as an autosomal recessive trait that produces a defective enzyme. Individuals with this genotype are unable to repair ultraviolet-induced DNA damage. This disease provides strong evidence that cancer originates in cells that have sustained permanent damage to DNA.

EXAMPLE 11.11

A type of blood cell cancer known as chronic myelogenous leukemia is associated with a reciprocal translocation involving the tip of the long arm of chromosome 9 and a portion of the long arm of chromosome 22. The chromosome 22 that bears a piece of chromosome 9 is called a Philadelphia chromosome (Example 7.23). A cellular protooncogene called *c-abl* (normally located on chromosome 9) becomes activated to oncogenic status when translocated to chromosome 22. A homologus gene (*v-abl*) exists in the highly oncogenic Abelson murine (mouse) leukemia virus.

Cancer is generally conceded to involve at least two major steps. The first step, termed **initiation**, results from a single exposure to a carcinogen. The second step, called **promotion**, involves one or more exposures to the same initiator or even to unrelated substances called **promoters** that complete the conversion of a cell to the neoplastic state. In general, the time interval between the exposure to an initiator and a promoter is not critical. However, the order of application is critical; the individual must be exposed to the initiator first, followed by the promoter. The promotion stage is a gradual process, often requiring many weeks in rodents and years in humans. Phorbol esters are among the most well-known promoters. Further heritable changes of an unknown nature are thought to be experienced during the promotion phase. Some substances [e.g., benzo(*a*)pyrene and polycyclic hydrocarbons, at relatively high doses] can both initiate and promote tumor formation.

Many carcinogens must undergo chemical alteration within the exposed individual before they are capable of becoming active. This process is called **metabolic activation**. Enzymes in various tissues, especially those in the liver, are responsible for converting the inactive **precarcinogens** into active carcinogens (see Supplementary Problem 10.10). Not all species have the enzymes necessary to convert a given precarcinogen into a carcinogen; hence, such a species would not be susceptible to induction of cancer by that substance.

There are two different types of mutations that can lead to cancerous growth. The first type is a "gain-of-function" (typically dominant) mutation in a gene that normally promotes cell growth and division. Cell division is a very complicated process that is likely to be controlled by many different genes (some stimulatory, others inhibitory). Interference with the timing at which these genes act, the amount of gene product produced, or the activity of the gene product could lead to cancer. Most cells in the body are terminally differentiated and are not actively dividing; thus, cell division genes are generally turned off. If they become inappropriately activated, uncontrolled cell division may result. These types of cancer-causing genes are called **oncogenes**. Normal cellular counterparts of oncogenes are called **protooncogenes**. Some of these genes are carried by **oncogenic viruses** and are sometimes designated as **v-oncogenes** (for viral). V-oncogenes can be linked to potent promoters that lead to their inappropriate and high-level expression, which can lead to deregulated cell division. The second type of mutation is a "loss-of-function" (typically recessive) mutation in a **tumor suppressor** gene. Tumor suppressor genes have normal roles as suppressors of cell growth and division. These are the genes that are turned on in inactive cells. If inappropriately turned off, the cell may enter the cell cycle and begin division.

Oncogenes can be grouped into five classes based on the nature of their protein products: (1) altered peptide hormones, (2) altered cell receptors, (3) altered G-proteins involved in signal transduction, (4) altered protein kinases, and (5) altered DNA regulatory proteins (transcription factors). A genetic locus on human chromosome 17 is usually associated with colorectal cancer; most of these cancer cells lose one copy of the gene and the other copy has a single base pair mutation. Cancers like this that are associated with the absence of at least one normal gene copy are hypothesized to develop because the normal gene encodes a **tumor suppression factor**.

EXAMPLE 11.12

The **Jun** and **Fos** transcription factors work together to activate transcription of genes that promote cell cycle progression. **V-Jun** and **v-Fos** (v stands for viral oncogene) contain deletions and point mutations that make them insensitive to control signals that would deregulate their activities. Thus, these dominant mutations result in constant activation of cell cycle genes. Even if one wild-type copy of the gene is present inside the cell (i.e. **c-Jun**; c stands for cellular), the unresponsive v-Jun protein will still act in a deregulated manner.

EXAMPLE 11.13

One of the first tumor suppressor genes found is the **RB**, or **retinoblastoma**, gene. Retinoblastoma is a cancer of the retina that usually occurs in early childhood. Normal individuals have two wild-type copies of the RB gene on chromosome 13. Individuals that are homozygous for mutations in both RB genes develop cancer. Heterozygous individuals are carriers and are predisposed to cancer if their wild-type copy becomes mutated or lost through a somatic event. The protein encoded by the RB gene is involved in preventing cells from entering the G_1/S transition of the cell cycle, keeping them from dividing. When normal RB function is lost, cell division occurs inappropriately, leading to cancer.

1   STUDYING CANCER *IN VITRO*

Cells can be grown *in vitro* using culture techniques referred to as **cell culture** or **tissue culture** in order to facilitate study. Fibroblast cells (responsible for the formation of extracellular fibers such as collagen in connective tissue) are the easiest cell type to grow in tissue culture; other cell types are more challenging to grow. After placing some cells into a glass or treated plastic flat-bottomed vial containing a nutrient-rich medium, the cells settle and attach to the bottom of the container and begin to grow. This is called a **primary culture**. The cells may divide a few times, but they eventually reach a crisis period in which most of them die. After maintaining the few survivors for many months in fresh medium, some cells may start to grow again, producing an **established cell line**. Cells of an established line have become immortalized and can (given fresh nutrients and removal of waste products) continue to divide indefinitely. However, if they are not continually subcultured, most of the cells stop growing when they have formed a confluent monolayer on the bottom of the container. Their growth is arrested by **contact inhibition** or **density-dependent growth**. If forced to grow for many generations at high density, or if treated with carcinogens, some of the cells undergo **neoplastic transformation** and lose contact inhibition. Such transformed cells are thought to resemble or be identical to a tumorigenic condition *in vivo*. Growth of transformed cells causes them to break out of a cultured monolayer and pile up on each other, forming a **focus** (**foci**, plural). Transformed cells can often be grown indefinitely in culture and have several other important properties that distinguish them from normal cells:

1. They can grow in cell suspension; they no longer require surface contact for growth and may lose their affinity for attachment to substrates. *In vivo*, this property fosters metastasis.
2. They require less supplementation in the nutrient medium.
3. They have disorganized microfilaments (part of the cytoskeleton), and therefore have a tendency to take on a spherical shape.
4. They may concentrate certain molecules to high levels.
5. Tumor antigens may appear on the cell surface.
6. They often form tumors when injected into an animal of the same species from which they were derived.
7. Their chromosome number usually exceeds the normal diploid number (aneuploidy).
8. If neoplastically transformed by either a DNA or RNA virus, the cell always contains integrated viral DNA.

One of the most important uses of established cell lines has been in the production of monoclonal antibodies by use of the **somatic cell hybridization** technique. A myeloma is a plasma cell tumor that grows well *in vitro*. Plasma cells are mature lymphocytes that secrete a single kind of antibody. It is possible to fuse a mouse myeloma cell (defective in its ability to make antibodies) with a mouse plasma cell to produce a **hybridoma** able to multiply indefinitely in cell culture and also able to secrete a single kind of antibody (monospecific). Commonly, the plasma cells are derived from spleens of mice that have been immunized with a specific antigen. If the myeloma cells are mutant with regard to their ability to make the enzyme HGPRT, they cannot grow in HAT medium, but they can grow if fused with normal plasma cells that can make this enzyme. The plasma cells, however, usually grow so poorly in HAT medium that they either die or are rapidly outgrown by the hybrid cells. The hybridoma clones that survive in HAT medium can then be assayed for antibodies reactive with the immunizing antigen. Once the desired clones are found, they can be frozen for later use or propagated indefinitely in cell culture or by injection into **syngeneic** mice (genetically identical to the plasma cell source) to produce monoclonal antibody-secreting tumors.

Many different plasma cell clones are usually stimulated to respond to the same antigen, each clone producing an antibody that is reactive to a different component or determinant of the antigen (e.g., to different parts of the same antigenic protein molecule). Even antibodies of different clones that recognize the same antigenic determinant may differ in their antigen-binding strengths and in the degree to which they cross-react with related determinants. The monoclonal antibodies produced by a given hybridoma are identical in all respects and can be economically made in virtually unlimited quantities. They are in great demand for a variety of diagnostic, medical, industrial, and research purposes.

2   STUDYING CANCER *IN VIVO*

To become malignant, a transformed cell must undergo several further changes in order to metastasize *in vivo*. Some cells of a tumor must burrow their way into a blood vessel or into a vessel of the lymphatic system and then, at some other location, must reverse the process and burrow out into a tissue again. Basement membranes underlie the epithelial cells from which the common cancers are derived, and these membranes consist of a complex of proteins, including collagen IV, laminin, and fibronectin. They also surround the smooth muscles in blood vessel walls. Metastatic cells must produce new protease enzymes to digest the basement membrane (e.g., type IV collagenase, transin). Solid tumors must recruit a rich network of blood vessels to supply them with nutrients for their growth. The process that stimulates formation of these blood vessels is called **angiogenesis**. Tumor cells are known to produce angiogenic factors that enhance the growth of blood vessels toward the tumor. To form a new tumor, the replicated metastasized cells must regain the ability to clump together. In some tumors this has been attributed to the presence of large amounts of a sugar-binding protein on the surface of the tumor cells. During all of this movement and tumor reestablishment, cancer cells have had to undergo additional mutations that allow them to avoid being destroyed by the killer cells and/or antibodies of the immune system (e.g., a mutation might alter the proteins in the cell membrane that normally mark the cell for destruction by the immune system).

3   ONCOGENIC VIRUSES

Some viruses are known to carry oncogenes that trigger neoplastic transformation; they are called **oncogenic viruses**. Among the vertebrates, about 50 oncogenic viruses have been found to contain DNA and about 150 contain RNA. Several families of DNA viruses contain oncogenic viruses, but among the RNA viruses only some of the retroviruses produce tumors. **Retroviruses** are so named because they contain an enzyme, **reverse transcriptase**, that synthesizes DNA from an RNA template. This activity is unusual in that most cells only synthesize DNA from DNA, not RNA. Table 11-3 displays some of the major differences between DNA and RNA oncogenic viruses.

Table 11-3.   A Comparison of the Two Major Classes of Oncogenic Viruses

DNA Tumor Viruses	Oncogenic Retroviruses
Some of these viruses cause a **productive infection** (producing progeny virions) in cells of one species (**permissive cells**) and a tumor in another species (**nonpermissive cells**)	Usually cause tumors in most species in which they can cause a productive infection
Infection of most nonpermissive cells by these viruses is abortive; very few of these cells become cancerous	Most infected permissive cells are tumor cells
Prophage integration involves loss of viral genes; progeny virions cannot then be produced	Integration of viral DNA is obligatory for the production of virions
Some of these viruses contain oncogenes that encode essential early proteins for viral replication	All oncogenes of these viruses are nonessential for production of progeny virions
No virus-induced protein kinases are known in DNA tumor viruses	Some of these viruses produce virus-induced protein kinases

EXAMPLE 11.14
The Rous sarcoma virus (RSV) is one of the best understood retrovirus. Upon entry into a host cell, reverse transcriptase, contained within the RSV virion, produces a double-stranded DNA (dsDNA) from the single-stranded virion RNA. This molecule then circularizes and becomes integrated into the host chromosome as a provirus. Progeny virion RNA is synthesized from the provirus by host-cell RNA polymerase II. The provirus is rarely excised. Its presence does not seem to inhibit cell division, so daughter cells inherit the provirus and continue to produce active virions. In contrast to a lambda phage lysogen, the RSV provirus does not make a repressor, and progeny virions are produced continuously without the necessity of deintegration of the provirus.

Unlike bacterial lysogeny, where all phage genes except the one responsible for repression of lytic functions are silenced, genes of the proretrovirus (viral DNA integrated into the host chromosome) are transcribed to produce proteins, some of which are involved in the induction of cancer, and others of which are involved in replication of viral RNA genomes. The integration of viral dsDNA into a host chromosome is an essential step in the life cycle of all oncogenic viruses. The retroviruses are enveloped (membrane-bound) virions containing a single plus (+) strand RNA genome and an RNA-dependent DNA polymerase called reverse transcriptase. This enzyme synthesizes a minus (−) DNA strand using the viral

(+)RNA genomic strand as a template. The same enzyme then degrades the viral RNA and synthesizes a complementary (+)DNA strand using the (−)DNA strand as a template, thereby forming a dsDNA replicative intermediate. The viral dsDNA is then integrated into a host chromosome in the same manner as DNA oncogenic viruses.

Oncogenic viruses cause cancer by two general mechanisms: (1) insertional inactivation and (2) oncogenes. In insertional mutagenesis, the viral DNA causes a mutation simply by becoming integrated into the host's DNA. Some of these mutations might inactivate cancer-suppressor genes. Alternatively, by inserting near a host gene involved in initiation of the normal cell cycle, the activity of that gene might be stimulated to overproduction of its product (e.g., a growth factor).

Many retroviruses contain oncogenes that are identical or very similar (perhaps differing by only one or a few nucleotides) to normal cellular genes involved in control of the cell cycle (called protooncogenes). It is generally believed that retroviruses have, in the course of their evolution, acquired their oncogenes from these normal (probably essential) cellular counterparts, the protooncogenes. These former cellular protooncogenes may become viral oncogenes by integrating into the viral genome in such a way as to be regulated by a powerful viral promoter, causing overproduction of a normal or near-normal growth factor, and resulting in excessive cell proliferation. Alternatively, some of these retroviral oncogenes code for kinase enzymes that phosphorylate specific amino acids in proteins. Normal host-cell kinases phosphorylate proteins at their serine or threonine residues. Retroviral kinases, however, phosphorylate tyrosine residues. Some host-cell growth factors normally stimulate cell division by causing the phosphorylation of tyrosine in the same proteins activated by retroviral kinases. Other oncogenes code for DNA-binding proteins and growth factor receptors, the overproduction or untimely production of which may lead to uncontrolled cell division.

EXAMPLE 11.15
Rous sarcoma virus (RSV) is a retrovirus containing an oncogene v-src (for virus, sarcoma-producing) that can transform cells. All vertebrates possess DNA sequences similar to v-src, and these are called c-src (cellular origin). The product of v-src is a phosphoprotein (pp) enzyme, namely, phosphokinase, called pp60-v-src [60 = 60,000 daltons (Da) molecular weight]. Most cellular protein kinases phosphorylate the amino acid serine or threonine, but pp60-v-src is tyrosine-specific. Phosphorylation can activate some proteins and inactivate others. Thus, one kinase may affect several proteins in different ways. The number of such proteins affected by pp-60-v-src and their normal functions in control of cell division are not yet known.

EXAMPLE 11.16
The oncogene v-sis, carried by simian sarcoma virus, encodes a protein similar to the platelet-derived growth factor (PDGF) made by the cellular protooncogene c-sis. It is believed that the excess PDGF produced by the virus overwhelms the normal controls on cell division.

Supplementary Problems

BACTERIOPHAGES

11.6. Six deletion mutants within the A gene of the *r*II region of phage T4 were tested in all pairwise combinations for wild-type recombinants. In the following table, + = recombination, 0 = no recombination. Construct a topological map for these deletions. (*Hint*: see Solved Problem 11.3.)

	1	2	3	4	5	6
1	0	0	0	0	0	0
2		0	0	0	0	+
3			0	+	0	0
4				0	+	+
5					0	+
6						0

11.7. Phage MS2 is a single-stranded RNA virus of *E. coli*. After infecting a cell, the phage RNA (the "plus" strand) is made into a double-stranded replicative intermediate form ("plus-minus") from which "plus" RNA is synthesized. The "minus" strands when isolated are not infective. Phage φX174 is a single-stranded DNA virus of *E. coli*. When injected into a bacterium, the same events as described for MS2 occur, but the "minus" strands when isolated are infective. Devise a reasonable hypothesis to account for these observations.

11.8. The DNA of bacteriophage T4 contains approximately 200,000 nucleotide pairs. The *r*II region of the T4 genome occupies about 1% of its total genetic length. Benzer has found that about 300 sites are separable by recombination within the rII region. Determine the average number of nucleotides in each recon.

11.9. The molecular weight of DNA in phage T4 is estimated to be 160×10^6. The average molecular weight of each nucleotide is approximately 400. The total genetic map of T4 is calculated to be approximately 2500 recombination units long. With what frequency are r^+ recombinants expected to be formed when two different *r* mutants (with mutations at adjacent nucleotides) are crossed?

11.10 A number of mutations were found in the *r*II region of phage T4. From the recombination data shown in the table below, determine whether each mutant is a point defect or a deletion (+ = recombination, 0 = no recombination). Two of the four mutants have been known to undergo backmutation; the other two have never been observed to backmutate.

	1	2	3	4
1	0	0	0	+
2		0	+	0
3			0	+
4				0

11.11. *Escherichia coli* strain B is doubly infected with two *r*II mutants of phage T4. A 6×10^7 dilution of the lysate is plated on *E. coli* B. A 2×10^5 dilution is plated on strain K. Twelve plaques appeared on strain K, 16 on strain B. Calculate the amount of recombination between these two mutants.

11.12. A nonlytic response usually is observed in lysogenic (λ) *E. coli* cells when conjugated with nonlysogenic Hfr donors or in crosses of Hfr (λ) $\times$ F$^-$(λ). The donated prophage is almost never inherited by the recombinants. Lysis is very anomalous in crosses of Hfr (λ) $\times$ F$^-$. Explain these observations.

11.13. Temperate phages such as lambda sometimes produce turbid plaques on lambda-sensitive indicator cells; virulent phages that cannot lysogenize always produce clear plaques on cells of their host range. (*a*) Offer an explanation for the turbid plaques. (*b*) Some lambda mutants produce only clear plaques. What genetic locus is most likely mutant in these cases?

11.14. When bacterial DNA is damaged by a mutagenic agent, excision repair normally operates to repair the lesion. This process is less than 100% efficient, however, so that some residual lesions remain unrepaired. If these lesions delay replication of DNA, an error-prone "SOS repair" system becomes operative, involving activation and increased production of a multifunctional protein called RecA protein (for "recombination"). RecA protein interferes with cell partition, resulting in elongation of cells into filaments. RecA protein also cleaves lambda repressor; this repressor must remain intact for the virus to remain dormant as a prophage. *E. coli* strain B is lysogenic for lambda; strain A is not lysogenic for lambda. This knowledge led Moreau, Bailone, and Devoret to devise a "prophage induction test" or "inductest" for potential carcinogens. Lysogenic strain B of *E. coli* is made defective in its excision repair system and genetically modified to make the cell envelopes permeable to a wide variety of test chemicals. This special strain is mixed with indicator strain A and rat liver extract; the mixture is then plated; the medium is covered with a thin layer of indicator bacteria interspersed with a few lysogenic bacteria. The test chemical is applied to a filter paper disk and placed in the center of the plate for a "spot test." (*a*) After incubation, how is DNA damage assayed? (*b*) Why is strain A required as an indicator? (*c*) What advantage does an inductest have over an Ames test? (*d*) Explain the selective advantage of lysogenic induction. (*e*) Genetic engineers have spliced the gene for galacto-kinase into a bacterial chromosome, thereby creating an organism for assaying mutagens by an enzymatic activity test. Where was this gene inserted into the chromosome and how does the system work?

11.15. The single-stranded phage ϕX174 of *E. coli* contains 5386 nucleotides coding for 11 proteins with a combined molecular weight of 262,000 bp. (*a*) If an average amino acid has a molecular weight of 110, by how many amino acids is the coding capacity of the phage exceeded? (*b*) How can ϕX174 code for more proteins than it has coding triplets? (*c*) Several animal viruses make more proteins than for which they seem to have coding triplets. Suggest some ways by which they might accomplish this feat if a single reading frame is used.

TRANSPOSABLE ELEMENTS

11.16. A given transposable element becomes duplicated at a fairly constant (although usually low) rate. Therefore, over evolutionary time, the descendants of a bacterial cell might be expected to contain thousands of copies of such a transposon. However, the number of copies of bacterial transposons is very low (usually only one or two per cell). (*a*) Offer an explanation for this low copy number. (*b*) Why have most bacterial transposons been isolated from plasmids rather than from the bacterial chromosome?

11.17. Transposition of a particular transposable element is found to be dependent on reverse transcriptase activity. Propose a mechanism for its transposition.

11.18. How might a transposition event result in oncogenesis?

EUKARYOTIC VIRUSES

11.19. Give at least two mechanisms whereby RNA viruses produce mRNA.

11.20. With regard to retroviruses: (*a*) specify their defining characteristic, (*b*) name the enzyme contained in their virions and list three biochemical activities of that enzyme, (*c*) identify the template for synthesis of retroviral mRNA, (*d*) identify the cellular location of their replication, (*e*) specify those attributes suggesting that their DNA-insertion mechanism is related to transposition.

11.21. The life cycles of eukaryotic viruses and bacteriophages have many similarities, including the establishment of new replication and transcription systems, regulation of gene action (e.g., early vs. late transcription), and synthesis of large quantities of structural proteins. There are certain aspects of viral life cycles, however, that are not (or only rarely) found in the life cycles of phage. Specify some of these unique aspects.

CANCER

11.22. Many tests of the oncogenic potential of viruses have been made with an established mouse cell line called NIH 3T3 because it had been used for many years to study viral transformation and chemical carcinogens. If cancer is a multistep process, how can the introduction of a single active viral oncogene transform these cells into cancerous cells?

11.23. Some slow-transforming viruses (such as avian leukemia virus) do not contain oncogenes. Offer an explanation as to how they might transform cells.

11.24. Cellular protooncogenes usually contain introns; viral oncogenes do not. (*a*) Propose a scenario for the origin of a viral oncogene. (*b*) Why is it more probable that oncogenes originate in cells rather than in viruses?

11.25. A virus is suspected to be involved in the development of breast cancer in certain strains of mice. The virus is transmitted through the milk to the offspring. In crosses where the female carries the "milk factor" and the male is from a strain that is free of the factor, about 90% of the female progeny develop breast cancer prior to 18 months of age. The virus usually does not initiate cancer development in the infected mouse until she enters the nursing stage, and then only in conjunction with a hormone (estrone) from the ovaries. Males from a virus-infected strain are crossed with females from a virus-free strain. (*a*) Predict the proportion of the offspring from this cross that, if individually isolated from weaning to 18 months of age, will probably exhibit breast cancer. (*b*) Predict the proportion of offspring from this cross that will probably exhibit breast cancer if housed in a group from weaning to 18 months of age. (*c*) Answer part (*a*) when the reciprocal is made. (*d*) Answer part (*b*) when the reciprocal cross is made.

11.26. Another case in which a disease is acquired through the milk (see Problem 11.25) is hemolytic anemia in newborn horses. A mare may produce two or three normal offspring by the same stallion and the next foal may develop severe jaundice within about 96 h after birth and die. Subsequent matings to the same stallion often produces the same effect. Subsequent matings to another stallion could produce normal offspring. It has been found that if nursed for the first few days on a foster mother, the foal will not become ill and develops normally. Evidently, something in the early milk (colostrum) is responsible for this syndrome. If the foal should become ill, and subsequently recovers, the incompatibility is not transmitted to later generations. (*a*) How might this disease be generated? (*b*) How is the acquisition of this disease different from that of breast cancer in mice?

Review Questions

Vocabulary For each of the following definitions, give the appropriate term and spell it correctly. Terms are single words unless indicated otherwise.

1. The protein coat of a virus.
2. The name for all viruses that infect bacterial cells.
3. The inability of some bacteriophage to replicate in certain hosts. (Two words.)
4. The type of viral life cycle characterized by plaque production.
5. Phage DNA integrated into a bacterial genome.
6. Portion of a host-cell membrane that is acquired by a eukaryotic virus as it leaves the cell.
7. An adjective descriptive of the type of viral infection wherein virions are slowly released from the cell without lysing or killing the host cell.
8. The class of RNA viruses that produce cDNA as part of their life cycle.
9. Mobile DNA elements.
10. Cancer-causing genes.

Multiple Choice Questions Choose the one best answer.

1. Which of the following is not characteristic of eukaryotic viruses? (*a*) only one kind of nucleic acid per virion (*b*) inhibited by antibiotics (*c*) replicate independently of cells (*d*) pass through bacterial filters (*e*) more than one of the above
2. Lambda phage can transduce bacterial genes only at or near the gene concerned with (*a*) repressor synthesis (*b*) immunity repressor (*c*) CAP (*d*) lactose fermentation (*e*) galactose fermentation
3. Viruses generally only infect one or a few particular species. This is called (*a*) host restriction (*b*) host range (*c*) cell specificity (*d*) viral range (*e*) host modification
4. Restriction-modification systems of bacteria exist to (*a*) protect bacteria from invading foreign DNA (*b*) promote conjugation (*c*) help the bacterial chromosome replicate (*d*) encourage recombination of new genetic material (*e*) promote complementation
5. A prophage is involved in (*a*) lytic cycle (*b*) oncogenesis (*c*) transposition (*d*) lysogeny (*e*) plaque formation
6. Specialized transduction is best characterized by (*a*) the transfer of a specific naked DNA sequence into a recipient cell (*b*) the transfer of an F plasmid to a recipient cell (*c*) the transfer of a particular region of the bacterial chromosome to a recipient cell via a phage vector (*d*) the transfer of a specific gene sequence through a sex pilus (*e*) the induction of oncogenesis
7. Which of the following enzymes is required for most transposition events: (*a*) DNA polymerase (*b*) telomerase (*c*) transposase (*d*) reverse transcriptase (*e*) RNA polymerase
8. Among all known phages reproducing vegetatively, the only one that neither kills nor lyses its host cell is (*a*) M13 (*b*) T4 (*c*) Mu (*d*) P1 (*e*) ϕX174
9. Cells that have been transformed into tumor cells exhibit the following characteristic(s). (*a*) If cells are transformed by an oncogenic virus, the virus must be integrated into host DNA. (*b*) They do not require surface contact to grow in cell culture. (*c*) They may form tumors when injected into an animal of the same species from which they were derived. (*d*) Their chromosome number often exceeds the normal diploid number. (*e*) All of the above are characteristics of transformed cells
10. Which of the following statements regarding oncogenic retroviruses is incorrect? (*a*) They can generate tumors in at least some species in which they can cause a productive infection. (*b*) Infective retroviruses acquire an envelope of host membrane as they exit the cell. (*c*) Integration of viral DNA into host DNA is obligatory for the production of progeny viruses. (*d*) All oncogenes of these viruses are nonessential for the production of progeny

virions. (*e*) All genes of the proretrovirus are silenced except for the one responsible for repression of lytic functions.

Answers to Supplementary Problems

11.6.

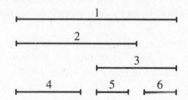

11.7. The DNA "minus" strand can serve as a template and can utilize the bacterial enzyme DNA polymerase for replication. The "minus" strand of RNA does not code for the enzyme that replicates RNA (RNA synthetase), and this enzyme is absent in uninfected bacteria. The "plus" strand of RNA carries the coded instructions for this enzyme and acts first as mRNA for enzyme synthesis. In the presence of the enzyme, single-stranded RNA can form a complementary strand and becomes a double-helical replicative form.

11.8. Approximately 7 nucleotides per recon.

11.9. 0.00625% of all progeny are expected to be r^+ recombinants.

11.10. m_1 and m_2 are overlapping deletions; m_3 is within m_1; m_4 is within m_2.

11.11. 0.5% recombination

11.12. Repressor is already present in the cytoplasm of the $F^-(\lambda)$ recipient cell. It binds to the operators that prevent prophage induction (vegetative reproduction of the phage) in either the $F^-(\lambda)$ or the Hfr (λ) donated chromosome segment. Early enzymes are therefore not produced, and recombination (leading to the inheritance of the donated lambda genes) cannot occur. Nonlysogenic F^- cells do not contain repressor. There are so few repressor molecules in a lysogenized cell that it is unlikely that free repressor would be bound to the newly synthesized donor fragment that moves almost immediately through the pilus into the F^- recipient. When the prophage from Hfr (λ) enters the F^- cell, a race occurs between the production of lambda repressor and an early protein of vegetative phage development. The outcome of this race is not predictable; hence, lysis is unpredictable in such crosses.

11.13. (*a*) Turbid plaques are due to secondary growth of lysogenized bacteria derived from the lambda-sensitive indicator strain. (*b*) Mutations of the gene coding for the lambda repressor of lytic activity are likely to produce an inactive (nonfunctional) repressor; hence, these mutants cannot lysogenize the indicator strain.

11.14. (*a*) DNA damage activates RecA protein, which then cleaves lambda repressor and opens up the viral genome for replication (induction). The cell bursts and releases viruses that infect and lyse the indicator strain A, causing plaques (holes) to appear in the bacterial "lawn" surrounding the paper disk. (*b*) If a cell of strain B is induced to lysis, the viruses cannot multiply in other cells of the same strain because active lambda repressor is present in these cells as a product of their prophages. Therefore, a nonlysogenic strain (A) is required to indicate how many viruses have been induced by the chemical treatment. (*c*) The inductest can assay a potential carcinogen at doses that would kill the tester bacteria in an Ames test (giving a false-negative reaction). The Ames test only detects the rare backmutations of *his*⁻ to *his*⁺, whereas DNA damage at any site can initiate lysogenic induction (a mass effect, independent of cell survival by toxic chemicals). (*d*) If DNA of the host cell cannot replicate, the cell is likely to die. Under these conditions it would be advantageous for a prophage to enter the lytic cycle and thereby possibly infect a "healthier" cell (like a "rat leaving a sinking ship"). (*e*) The gene for galactokinase was inserted adjacent to (and under the control of) the lambda repressor. When the mutagen damages DNA, RecA protein is activated and cleaves the repressor; this opens the operon to RNA polymerase and allows synthesis of the enzyme galactokinase, the activity of which can be quantitated spectrophotometrically when supplied with its substrate.

11.15. (*a*) (262,000 molecular weight/110 molecular weight per amino acid) − (5386 bases/3 bases per codon) = 586.5 amino acids

(*b*) Some phage genes are overlapping; i.e., the same sequence can be transcribed in different reading frames. The only structural feature responsible for gene overlap is the location of each AUG start codon

(*c*) By alternative intron cleavage sites from the same primary transcript, two proteins could be produced that have the same N terminus but different C termini. A polyprotein could also be enzymatically cleaved in more than one way to produce different products.

11.16. (*a*) Most of the DNA in bacteria, unlike the DNA in eukaryotic cells, is coding information. There is relatively little DNA that is not serving some function. Thus, the movement of most transposons to a new location would inactivate one or more vital genes, causing cell death or weakening the cell so that it cannot compete with normal cells. (*b*) Plasmids rarely are essential to their host cells, and therefore could tolerate the integration of transposable elements without interfering with vital gene functions.

11.17. The transposable element must first be copied into RNA. This RNA is then copied by reverse transcriptase into cDNA. This double-stranded cDNA is then capable of integrating into the genome at a different site.

11.18. Transposition may result in several types of insertional and recombinational mutation events: insertion in a coding region of a protooncogene, resulting in loss of function; insertion into the promoter of a protooncogene, resulting in loss of function; insertion near the controlling regions of a gene and acting as a fortuitous promoter, resulting in altered gene regulation; various types of recombination events that would result in translocations or deletions, upon "jumping out" a transposon may take nearby gene sequences with it, creating a deletion.

11.19. (1) Minus-strand RNA viruses transport into the cell a replicase enzyme that synthesizes mRNA from the (−) strand template.

(2) Plus-strand RNA viruses (other than retroviruses) use their infective strand as a template for synthesizing mRNA using host RNA polymerase.

(3) Retroviruses use their (+) strand as a template for DNA synthesis, which is then transcribed into mRNA.

(4) Double-stranded RNA viruses bring a replicase into the host cell that copies double-stranded RNA and synthesizes a (+) strand that functions as mRNA.

11.20. (*a*) They replicate from a DNA intermediate.

(*b*) RNA-dependent DNA polymerase (reverse transcriptase). Enzyme functions: (1) converting the single-stranded viral RNA to a DNA-RNA hybrid, (2) digesting RNA from a DNA-RNA hybrid, and (3) copying a primed single-stranded DNA to form a double-stranded DNA.

(*c*) The double-stranded DNA that is formed by reverse transcription from retroviral RNA.

(*d*) The DNA-RNA hybrid is made in the cytoplasm. The hybrid is converted to double-stranded DNA and becomes inserted into a host chromosome. Messenger RNA is made from the proviral DNA in the nucleus by host RNA polymerase.

(e) Their integrated proviral DNA is terminated at each end by a long terminal repeat and a short inverted repeat (like a composite transposon), which in turn is flanked by a short, direct repeat (like a target sequence).

11.21. (1) Viral proteins usually enter the infected cell along with the viral genome.
(2) The RNA of some viruses is converted to DNA.
(3) The mRNA of viruses is processed just like the cellular mRNA of their eukaryotic hosts.
(4) Polyproteins are produced by some viruses.

11.22. An established cell line has already experienced one or more early steps (e.g., immortalization) in the induction of neoplastic transformation before exposure to the effects of the oncogene.

11.23. These viruses might become integrated into the host DNA near a cellular protooncogene and activate it, via a viral enhancer sequence, to become an oncogene.

11.24. (a) A retrovirus becomes integrated as a provirus adjacent to a cellular protooncogene. The provirus and the adjacent protooncogene are transcribed into a single transcript. The RNA transcript is processed to remove introns and becomes packaged into a viral capsid. The infective virus is released from the host cell and infects another cell. (b) Each of the steps in part (a) involves known genetic mechanisms. There is no known mechanism by which introns from viruses can be inserted into cellular protooncogenes.

11.25. (a) (b) None of the progeny is expected to develop breast cancer because noninfected females have nursed them. (c) None of the progeny is expected to develop breast cancer because, in isolation, the infected females could never produce a litter and subsequently enter a lactation period, a prerequisite for expression of the milk factor. (d) 50% females × 90% of females develop breast cancer = approximately 45%.

11.26. (a) This disease is similar to the Rh blood group system incompatibility between a human mother and her baby. In this case, antibodies are transferred to the offspring through the milk rather than across the placenta. (b) The particular stallion that is used has an immediate effect on the character. This is not true in the acquisition of breast cancer in mice. The incompatibility disease in horses cannot be transmitted to later generations, so there is no evidence of a specifically self-duplicating particle like the infective agent that causes cancer in mice.

Answers to Review Questions

Vocabulary

1. capsid
2. bacteriophage or phage
3. host restriction
4. lytic
5. prophage
6. envelope
7. persistent
8. retroviruses
9. transposons
10. oncogenes

Multiple Choice

1. e (b, c) 2. e 3. b 4. a 5. d 6. c 7. c 8. a 9. e 10. e

CHAPTER 12

Molecular Genetics and Biotechnology

History

Prior to discovery that the chemical nature of the gene was DNA, the "gene" was an abstract unit of heredity. This period in history is referred to as **classical** or **formal genetics**. Classical genetics has been extremely successful in elucidating many basic biological principles without understanding the physical nature of the gene. The era of **molecular genetics** followed the discovery of DNA structure in 1953 and the determination that the fundamental unit of heredity was DNA and not protein. In the late 1970s, yet another milestone in genetics was reached when researchers found that they could manipulate DNA molecules in a test tube (*in vitro*), essentially "cloning" the first gene. This discovery led to further molecular genetic advances including **recombinant DNA technology (genetic engineering)** and spawned the **biotechnology** industry known today.

The histories of most scientific disciplines are generally characterized by relatively long periods of stagnation punctuated by bursts of rapid progress. Most of these flurries of research are initiated by new technical developments. This is certainly true of biochemistry and molecular biology. At least three major areas of technology have been influential in this respect: (1) instrumentation and techniques, (2) radioactive and fluorescent labels, and (3) nucleic acid enzymology. Beyond these technological influences, the use of computer technology to process biological information (i.e., DNA nucleotide sequences or protein amino acid sequences) is driving new discoveries in molecular genetics and biotechnology today. In fact, several new fields within molecular genetics—**Bioinformatics**, **Genomics**, and **Proteomics**—have been created. Each of these fields integrates the use of computer hardware and software to store, process and analyze biological information, such as DNA or protein sequences.

1 INSTRUMENTATION AND TECHNIQUES

(a) Instrumentation. The **analytical ultracentrifuge** was developed in the 1920s by Theodor Svedberg. The sedimentation rate of a substance during ultracentrifugation is mainly a function of its density and secondarily of its shape. The unit of sedimentation (S, in honor of Svedberg) is an expression of these parameters. The centrifuge has been modified for isolating organelles such as nuclei, ribosomes, mitochondria, and chloroplasts. It can be used for determining the minimum number of kinds of macromolecules in a biological specimen and for estimating the molecular weights of macromolecules.

The **electron microscope** was invented in the 1930s, and eventually enabled the direct visualization not only of cellular substructures but also of viruses and macromolecules. Circular genetic maps of microorganisms have been shown by electron microscopy to have a corresponding circular physical structure. Multiple ribosomes attached to an mRNA molecule (polysomes) have also been visualized by this instrument.

Electrophoresis is a technique that separates molecules according to their shapes, net charges, and molecular weights in an electric field, usually on solid or semisolid support media such as paper or agarose. This technique was first used in 1949 to differentiate sickle-cell hemoglobin from normal hemoglobin. Protein sequence analysis later revealed that the difference in electrophoretic mobilities of these proteins was due to a single amino acid difference in the β-chains. Electrophoresis is now used in a variety of DNA and protein analysis applications. DNA sequencing results are usually performed on (poly)acrylamide gels; **agarose gels** are usually employed to isolate and estimate the size of DNA fragments; proteins are often analyzed for their size and charge using **acrylamide gels**. Electrophoresis has been extensively used to differentiate **isozymes**, i.e., proteins possessing the same enzymatic properties but differing in primary structure.

X-ray-diffraction data from crystalline materials has been critical for the elucidation of three-dimensional shapes of nucleic acids (e.g., DNA, tRNAs) and proteins (e.g., myoglobins, viral capsomeres, enzymes). This technique has been expanded to include the determination of multi-subunit molecules.

During the mid-1940s and early 1950s, various forms of **chromatography** were perfected, enabling molecules to be separated by differences in solubilities in organic solvents, electrical charge, molecular weight, and specific binding properties for the support medium, or combinations of these factors. Erwin Chargaff used paper chromatography to determine the base compositions of DNAs from various sources. He found that the molecular ratio of adenine was equivalent to that of thymine and the ratio of guanine equals that of cytosine. This was a vital clue for James Watson and Francis Crick in their search for the structure of DNA.

Automated equipment (i.e., robotics) is now available for doing many repetitive biochemical tasks. DNA synthesizers ("gene machines") can be programmed to make oligonucleotide sequences of any desired composition. Automated instrumentation is now available for isolating DNA, and sequencing DNA or protein fragments, as well as for performing enzyme assays and other tasks. Computer software has been developed to interpret data from electropherograms, and to search databases for similar or identical sequences (**bioinformatics**).

(b) Techniques. Several techniques have been developed to separate, rejoin, synthesize, or break nucleic acid molecules. Separation of the complementary chains of a DNA molecule is known as **denaturation**. DNA is denatured if placed in alkali (0.2N NaOH) or when boiled. The latter process is referred to as **melting**. Separation of DNA strands can be detected by spectrophotometric instruments; **optical density (OD)** or **absorbance** at 260 nm increases during the melting process. The temperature at which the increase in OD_{260} is 50% of that attained when strand separation is complete is known as the **melting temperature (T_m)**. Because G and C base-pair by three hydrogen bonds, whereas A and T base-pair by two hydrogen bonds, the higher the G-C content in DNA the higher the melting temperature. Melting is enhanced where there are clusters of A's and T's, and also when all the purines (A, G) are on one strand and all the pyrimidines (T, C) are on the other strand.

If DNA is boiled and then quickly cooled, the strands will remain single; if cooled slowly, complementary strands will base-pair and reform double-helical DNA molecules. This process is called **renaturation** or annealing. Hybrid DNA-RNA molecules can be produced by analogous processes from single strands. RNA can be totally hydrolyzed to nucleotides by exposure to high pH (alkali). This property can be used to purify DNA from a mixture of DNA and RNA. Single-stranded DNA will bind to membranous filters made of nitrocellulose; RNA will pass through such filters. However, if single-stranded RNA is complementary to nitrocellulose-bound single strands of DNA, it will form DNA-RNA hybrid molecules and be retained by such a filter.

There are two main methods for breaking long DNA molecules into fragments of suitable size for sequencing or for recombinant DNA techniques: (1) **mechanical shearing** and (2) **restriction endonuclease** treatment. If a solution of DNA is subjected to the stirring forces of a Waring blender or forced through a narrow tube or orifice, the ends of long DNA strands will usually move at different speeds; this stretches the DNA and tends to break it near the middle. This phenomenon is called shear degradation or shearing. The higher the stirring speed or velocity of flow through an orifice, the greater the shearing force. Shearing can also be achieved by exposure of a solution of DNA to ultrasound. The effectiveness of any shearing force increases with molecular size of the DNA, but decreases with concentration (because entanglement of DNA molecules reduces the effective stretching). Restriction endonuclease treatment employs the use of enzymes that cut, or break the bonds between, specific base-pair sequences of DNA. These enzymes will be discussed later in this chapter in the sections Nucleic Acid Enzymology, and Recombinant DNA Technology.

2   RADIOACTIVE AND FLUORESCENT LABELS

(a) Radioactive Labels. Radioactive elements have historically been used as highly sensitive labels for detecting minute amounts of specific macromolecules.

However, recent advances in fluorescent molecule technology (i.e., the development of fluorescent dyes to tag molecules of interest) have led to the replacement of many radioactive tracers by these fluorescent molecules. Fluorescent molecules are often safer to use, better for the environment, and easier to dispose of.

EXAMPLE 12.1
A. D. Hershey and M. Chase differentially labeled the nucleic acid and the protein components of T2 phages. They used radioactive ^{32}P in place of normal ^{32}P to label DNA; radioactive ^{35}S was used in place of normal ^{35}S to label protein (cysteine and methionine are two amino acids that contain sulfur). Since there is no phosphorus in phage proteins and no sulfur in nucleic acids, the fate of both viral components could be followed during the viral life cycle. After allowing the phages to become attached to sensitive *Escherichia coli* host cells, the mixture was subjected to the shearing forces of a Waring blender. The mixture was centrifuged to sediment the cells and then activity characteristic of each radionuclide was assayed in the pellet and in the supernatant fluid. All of the ^{32}P activity was found in the bacterial pellet and virtually all of the ^{35}S was found in the supernatant. ^{32}P was found in some progeny phages, but no ^{35}S was found. The inference is that phages inject their DNA into host cells. Blender treatment shears the phage tail fibers from receptor sites on host cells; the empty phage protein capsids (ghosts) are therefore left free in the supernate. Semiconservative replication from the infecting ^{32}P-labeled DNA caused some progeny phages to be released with one of the original radioactively labeled infecting strands. This experiment was the first to demonstrate that DNA, and not protein, is the genetic material in phages.

DNA labeled with radioactive nuclides can reveal its own presence in a photographic technique called **autoradiography** or **radioautography**. A preparation of DNA on a slide or on filter paper can be covered with a photographic film or emulsion. As the radionuclides undergo radioactive disintegration, or decay, they release charged particles and/or photons that cause a chemical reaction on the film. After development of the X-ray film, the location of the labeled DNA is revealed by dark spots.

Radioactively labeled thymidine can be used to differentiate DNA from RNA molecules because uracil usually replaces thymine in RNA. Tritium (^{3}H) is a radioactive isotope of hydrogen commonly used to label thymidine; the labeled nucleoside is called tritiated thymidine. By allowing thymine-deficient (*thy⁻*) *E. coli* to grow in the presence of **tritiated thymidine**, its DNA becomes radioactively labeled. The **half-life** of tritium is 12.46 years, meaning that its radioactivity decreases by one-half each 12.46 years. Tritium-labeled thymidine and uridine are often used to tag or label newly synthesized DNA and RNA molecules, respectively. Tritium is often the radioisotope of choice in radioautography because it emits an extremely weak beta particle when it undergoes radioactive disintegration or decay. In a medium of unit density, the average tritium beta particle will penetrate only 1 μm. Therefore, in autoradiographs of tritium-labeled cells, darkened grains of the photographic emulsion will be localized within 1 μm of the decaying atoms. It was by autoradiography that John Cairns discovered the theta intermediate of circular DNA replication in bacteria (see Fig. 10-3).

A radioactive isotope of phosphorus (^{32}P) is also widely used to label nucleic acids; it emits a strong beta particle and has a half-life of 14.3 days. It is thus more radioactive ("hotter") than tritium and can reveal itself in much lower amounts than can tritium during the same decay period. An instrument called a **scintillation counter** is used to detect radioactive disintegrations. High-energy γ-rays can be detected by a crystal scintillation counter. A liquid scintillation counter must be used to detect weaker beta particles, although it can also detect γ-rays.

Any organic substance can be labeled with radioactive carbon (^{14}C). This isotope emits a weak beta particle and has a relatively long half-life of 5730 years. All living organisms incorporate a predictable amount of ^{14}C while alive. After death, ^{14}C decays to ^{14}N at the predictable rate of its half-life. This knowledge allows the dating of organic remains from the time of death up to about 40,000 years before the present.

Radioactive iodine (^{125}I) has a half-life of about 60 days, emits γ-rays, and is used to label proteins. This isotope is easily coupled to the amino acid tyrosine. Radioactive sulfur (^{35}S) is used to label the amino acids cysteine and methionine; ^{35}S can also be used to label nucleic acids. ^{35}S is more desirable than ^{32}P for most **autoradiography** (detection of radioactive signals on photographic film) because it has a half-life of 87.1 days and emits a much weaker beta particle that gives sharper bands. It is much less hazardous to handle than ^{32}P and poses less waste-disposal problems. Quantitation of small amounts (nanograms or picograms per milliliter) of these proteins can be accomplished by sophisticated techniques such as competitive protein binding assays and radioimmunoassays.

Even nonradioactive isotopes (e.g., ^{15}N) have been useful in solving fundamental problems in molecular biology.

SOLVED PROBLEM 12.1

In 1953, Watson and Crick proposed that DNA replicates semiconservatively; i.e., both strands of the double helix become templates against which new complementary strands are made so that a replicated molecule would contain one original strand and one newly synthesized strand. A different hypothesis proposes that DNA replicates conservatively; i.e., the original double helix remains intact so that a replicated molecule would contain two newly synthesized strands. Bacterial DNA can be "labeled" with a heavy isotope of nitrogen (^{15}N) by growing cells for several generations in a medium that has ^{15}NH$_4$Cl as its only nitrogen source. The common "light" form of nitrogen is ^{14}N. Light and heavy DNA molecules can be separated by high-speed centrifugation (50,000 rpm $= 10^5 \times$ gravity) in a 6M (molar) CsCl (cesium chloride) solution, the density of which is 1.7 g/cm^3 (very close to that of DNA). After several hours of spinning, the CsCl forms a density gradient, being heavier at the bottom and lighter at the top. In 1957, Matthew Meselson and Franklin W. Stahl performed a density-gradient experiment to clarify which of the two replication hypotheses was correct. How could this be done, and what results are expected after the first, second, and third generations of bacterial replication according to each of these hypotheses?

Solution: Bacteria from ^{15}N-labeled culture are transferred into medium containing ^{14}N as the only source of nitrogen. A sample is immediately taken and its DNA is extracted and subjected to density-gradient equilibrium centrifugation. The DNA forms a single band relatively low in the tube where its density matches that of the CsCl in that region of the gradient. After each generation of growth and replication of all DNA molecules, DNA is again extracted and measured for its density. According to the semiconservative theory, the first generation of DNA progeny molecules should all be "hybrid" (one strand containing only ^{14}N and the other strand only ^{15}N). Hybrid molecules would form a band at a density intermediate between fully heavy and fully light molecules. The second generations of DNA molecules should be 50% hybrids and 50% totally light, the latter forming a band relatively high in the tube where the density is lighter. After three generations, the ratio of light : hybrid molecules should be 3 : 1, respectively. The amount of hybrid molecules should be decreased by 50% in each subsequent generation.

According to the conservative replication scheme, the first generation of DNA molecules should be 50% heavy : 50% light. The second generation should be 25% heavy : 75% light. The third generation should be 12.5% heavy : 87.5% light. No hybrid molecules should be detected. The results of the Meselson and Stahl experiment supported the semiconservative theory of DNA replication.

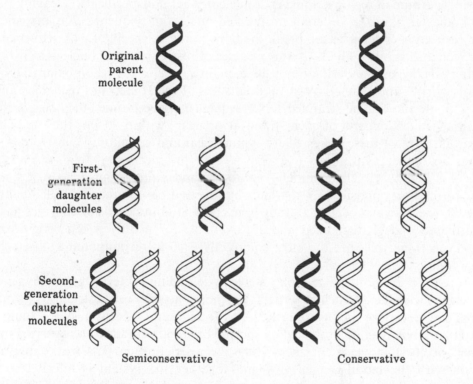

Original parent molecule

First-generation daughter molecules

Second-generation daughter molecules

Semiconservative   Conservative

(b) Fluorescent Labels. Fluorescent molecules are finding more uses in the biological sciences. These molecules contain a **fluorophore** that absorbs and emits light at particular wavelengths characteristic to the specific fluorophore. This emitted light can be detected and recorded by a number of different instruments (microscopes, scanners, spectrophotometers, etc). Fluorescent probes can be attached to a variety of biological molecules in order to detect and study them. DNA, proteins, and lipids, as well as specific molecules, such as actin and tubulin, can be labeled with a fluorescent molecule. It is possible to use two or more fluorescent probes in one experiment if their emitted light occurs at different wavelengths. For example, different molecules within a cell can each be labeled with a different fluorescent molecule and the activities and amounts of each molecule can be detected and compared at the same time. During DNA sequencing reactions, each nucleotide is labeled with a different fluorescent molecule, allowing strands terminating with a different nucleotide to be differentiated yet detected simultaneously (see section on DNA Sequencing later in this chapter).

3   NUCLEIC ACID ENZYMOLOGY

Nucleases are enzymes that hydrolyze, or break, the phosphodiester bonds that hold the nucleotides together. Those that remove terminal nucleotides, one at a time, are called exonucleases; those that break the sugar-phosphate backbone at nonterminal sites are called endonucleases. A **deoxyribonuclease (DNase)** attacks DNA molecules; a **ribonuclease (RNase)** degrades RNA molecules, especially at their single-stranded regions. Some endonucleases act nonspecifically, cleaving the phosphodiester bonds at different unspecified nucleotide sequences. Others, such as the **restriction endonucleases**, break the bonds only at specific DNA sequences, called recognition sites. There are three classes of restriction endonucleases: Types I, II and III. Types I and III do not have qualities useful for recombinant DNA technology (i.e., they cut DNA at random sites). **Type II enzymes** recognize and bind to a specific double-stranded DNA sequence. Once bound, they cleave the phosphodiester backbone of each strand at or near (within 20 bp) the site. The recognition sites of these enzymes are symmetrical and commonly consist of 4–8 bp that are either continuous (e.g., GAATTC) or interrupted (e.g., CANTG, where N is any base). The symmetry occurs around a midpoint, or axis of symmetry, formed on opposite DNA strands by inverted base sequences called **palindromes**. These enzymes occur naturally in bacteria, and hundreds of enzymes have been isolated and characterized and are available for use by genetic engineers. Thus, Type II enzymes are the most commonly used in recombinant DNA technology today.

Naturally, restriction enzymes play a role in defending bacteria against invasion by foreign nucleic acids, such as viruses. The host bacterial cell restriction enzyme systems can recognize the invading DNA as foreign and destroy it (see Example 11.1). Host DNA may be protected by specific base-pair modifications carried out by host-specific methylase enzymes. Thus, the restriction enzyme works in conjunction with the modifying enzyme and together the system is referred to a **restriction-modification system**. Nonhost, or foreign, DNA is recognized as such because its restriction sites are unmodified. Restriction endonucleases are named after the bacterial species or strain from which they were derived. For example, an enzyme from *Providentia stuartii*, will have the name *Pst* derived from the first letter of the genus (*P*) and the first two letters of the specific epithet (*st*). This name is italicized to honor the scientific name of the bacterium and can be followed by roman numerals to indicate that it is one of several enzymes isolated from that particular bacterial strain, e.g. *Pst*I, *Pst*II (the numerals are not italicized). Occasionally, a letter derived from the specific bacterial strain follows the name. For example, the enzyme *Eco*RI was derived from *E. coli* strain RY13 and *Hind*III was derived from *Haemophilus influenzae* strain Rd. If this is the case neither the strain letter nor the numerals are italicized.

EXAMPLE 12.2

The *Eco*RI restriction enzyme cuts bonds on the upper and lower strands within the palindromic DNA sequence at the arrows shown below.

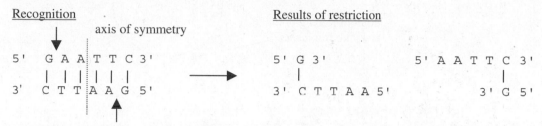

Notice that the 5′ to 3′ nucleotide sequence within the palindrome is the same starting from the 5′ (pronounced five prime) ends on both strands of the DNA (the palindrome). *Eco*RI cuts the DNA molecule in a staggered fashion, leaving "**overhang**," "**cohesive**" or "**sticky**" **ends**. In the above example, the overhangs are called 5′ overhangs because the unbonded nucleotides have a 5′ end. Other restriction enzymes can leave a 3′ overhanging end. Another restriction endonuclease (*Hae*III), derived from the bacterium *Haemophilus aegypticus*, snips DNA as shown below. Note that this enzyme cuts both strands at opposite bonds, leaving "**blunt**" **ends**.

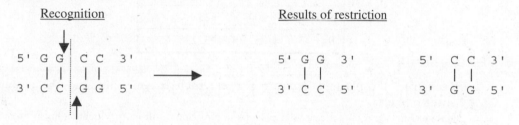

Restriction enzyme maps for any given DNA segment (linear or circular) can be constructed. Such maps show the location of various restriction endonuclease recognition sites on the DNA fragment. The sizes of the various restriction fragments can be expressed by molecular weight, but more commonly they are given in terms of number of base pairs (bp) or thousands of base pairs (kilobases, kb).

EXAMPLE 12.3

A linear 10,000-bp fragment of DNA has an unknown number of *Eco*RI restriction enzyme sites. You perform a restriction digestion of this fragment with *Eco*RI and obtain the following result: three fragments of sizes 5, 3, and 2 kb. These results suggest that there are two *Eco*RI sites present in this molecule. One possible arrangement of sites (arrows) is shown below. There are two other possible arrangements (not shown).

If this molecule were circular, as are bacterial **plasmids** commonly used in recombinant DNA technology, only two fragments (3 and 7 kb) would be obtained. The 7 kb fragment is a result of the joining of the 5 and 2 kb segments (see diagram below).

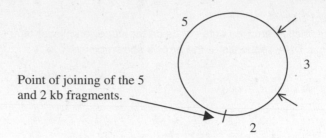

Point of joining of the 5
and 2 kb fragments.

SOLVED PROBLEM 12.2
Suppose that a circular plasmid contains 1000 bp. It is cut by three different restriction endonucleases, both singly and in pairs, with the results as shown below.

Enzyme(s)	Fragment Length(s)
A	1000
B	100, 300, 600
C	200, 800
A + B	50, 100, 300, 550
A + C	200, 375, 425
B + C	75, 100, 125, 225, 475

Reconstruct the plasmid, indicating where each enzyme cuts and the distances between all cuts.
Solution: Enzyme A cuts the circular plasmid at only one position, producing a linear molecule that is 1000 bp in length.

Enzyme B cuts the plasmid at three positions, producing fragments that are 100, 300, and 600 bp in length. Let us label these fragments B1, B2, and B3, respectively.

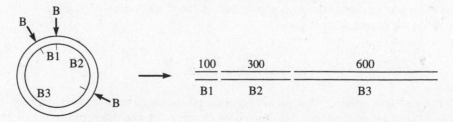

Enzyme C cuts the plasmid at two places, giving fragments that are 200 and 800 bp in length. Let us label them C1 and C2, respectively.

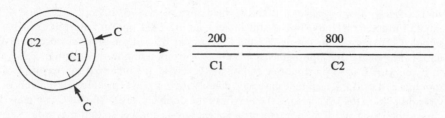

Digestion by both enzymes A and B produced four fragments of lengths 50, 100, 300, and 550 bp. Since the 100 and 300-bp fragments generated by enzyme B are still intact, enzyme A must have cut fragment B3 (600 bp) into two fragments of 50 and 550 bp. The location of this cut could be either closer to or farther from B2.

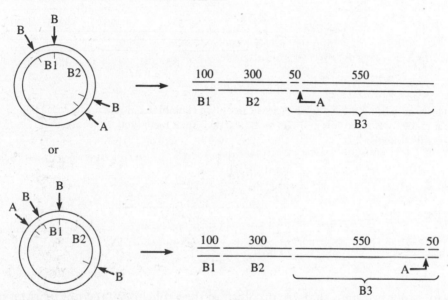

Digestion with both enzymes A and C produced three fragments of lengths 200, 375, and 425 bp. Obviously, the single cut by enzyme A must have been in fragment C2 (800 bp). The 425-bp fragment could be either to the left or right of A in the linear map.

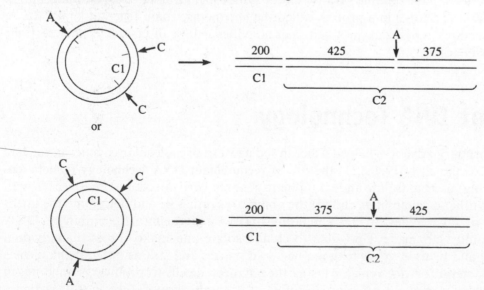

Double digestion with enzymes B and C yields five pieces: 75, 100, 125, 225, and 475 bp in length. Fragment B1 (100 bp) is still intact, but B2 (300 bp) and B3 (600 bp) have been degraded. Therefore, one cut by enzyme C occurs in B2 and the other cut occurs in B3. The sum of the 75- and 225-bp fragments is 300 bp, corresponding to the length of B2. Thus, one cut by enzyme C is 75 bp from one end of B2. Likewise, the other cut by enzyme C is 125 bp from one end of B3. Recall that the single digest with enzyme C produced fragments of 200 bp (C1) and 800 bp (C2). Therefore, the only way the double digest of B and C can make sense is to have the 75-bp and 125-bp fragments adjacent to one another so that they total 200 bp as in C1.

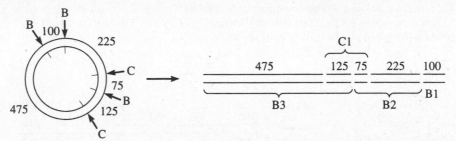

The results of all three double digests can only be combined in one meaningful way. Since the cut made by enzyme A is in fragment B3, 50 bp from a B cut, and far from either C cut, A must map as follows:

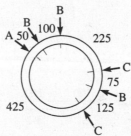

Many other enzymes that are involved in the replication, recombination, repair, modification, transcription, and translation of nucleic acids are utilized in recombinant DNA technology. In particular, DNA synthesis enzymes, known as **DNA polymerases**, have become very useful for synthesizing copies of DNA molecules *in vitro* (in a test tube) in a process called the **polymerase chain reaction**, or **PCR** (see later section). These enzymes and their activities will be discussed in later sections of this chapter.

Recombinant DNA Technology

Two primary advances helped usher in the new era of molecular genetics in the late 1970s to the early 1980s: (1) the use of **recombinant DNA technology (genetic engineering)** used to isolate and manipulate genes *in vitro* in order to endow cells with new synthetic capabilities and (2) the ability to synthesize and determine the linear order of nucleotides of DNA molecules (**DNA sequencing**). Recombinant DNA technology has made it possible to clone (isolate and make copies of individual genes) and transfer genes between bacterial species and strains or from eukaryotes into bacteria (or vice versa), causing the engineered cells to produce, sometimes in relatively large quantities, proteins of great economic importance such as enzymes (e.g., amylases, proteases), hormones (e.g., insulin, growth hormone), and **interferons** (lymphocyte proteins that prevent replication of many viruses). These proteins are made in such small quantities in human cells that the cost of their extraction and purification from blood or cadaver tissues has been very expensive, thus restricting their medical use in prophylaxis (prevention) and therapeutics (treatment) of disease. Many products are now produced successfully using genetic engineering techniques; such products include blood-clotting factors (e.g., tissue plasminogen activator, or TPA, used to activate the breakdown of blood clots and to prevent recurrence of blood clots in heart attack patients), and complement components (part of the immune system). In 1980, the United States Supreme Court decreed that new life forms created by genetic engineering could be patented, although this is not always true in other countries. This decision has contributed to the investment of large sums of money by private corporations into the development of many useful recombinant strategies.

There are many hopes for the future benefit of genetic engineering technology in helping solve some of society's problems. In the agricultural arena, soybeans, corn, and cotton have been genetically engineered or modified to contain added herbicide tolerance and pest-resistance genes. These crops are known as either **GM** (genetically modified) crops or **GMOs** (genetically modified organisms). For example, one type of GM corn (*Bt* corn) contains a gene from the bacterium *Bacilllus thuringiensis* that produces a protein that kills insect larvae. Larvae of the European corn borer can be costly pests for corn farmers. Having the pesticide gene in the corn eliminates the need to spray the crop with a pesticide. Other crops have been or are being engineered to contain genes that encode vaccines, added nutrients, and disease resistance, as well as resistance against cold, drought, and salinity. These few examples suffice to demonstrate the possibilities of this new technology and explain why there is such great excitement in the scientific, medical, agricultural, and pharmaceutical communities concerning its further development. While this technology is exciting and has proven useful, some have voiced concerns over potential environmental hazards, unknown risks to human health, and economic impact. These issues are still being investigated and must be resolved in order to make this technology useful *and* safe.

Most genetic engineering projects begin with the **cloning**, or isolation and amplification, of a gene. There are two major purposes for cloning a gene: (1) to obtain the sequence of its DNA (e.g., in order to discover how an abnormal gene differs structurally from its normal allele) and (2) to produce large amounts of a gene product (i.e., protein). Gene cloning generally involves three major steps: (1) A DNA fragment (or gene) of interest is isolated. (2) The fragment is spliced into a cloning vehicle called a **cloning vector**, such as a phage or a plasmid. (3) The vector is introduced into a host cell where it is replicated many times. A good cloning vector should be able to autonomously form many replicas within the host cell to greatly amplify its foreign DNA insert. The vector should have a single recognition site for several of the known restriction endonucleases so that it can be opened at only one position to receive a foreign piece of DNA. Ideally, a vector containing a foreign DNA insert should impart to its host cell some property (e.g., an antibiotic-drug-resistance gene) that distinguishes it from cells that do not contain an insert, so that the insert can be easily isolated.

Once the gene is cloned it can be analyzed (restriction mapped and its DNA sequence determined) and then transferred into a host cell or organism that will allow it to be expressed as a protein product. The host system might be a genetically engineered (**transgenic**) plant or a **mammalian cell culture**, yeast culture, or bacterial cell system that is used as a biological factory to produce large quantities of the desired product.

1   CUTTING AND LIGATING DNA MOLECULES

(a) **Enzymes that Create Cohesive Ends.** DNA cut by restriction enzymes that leave complementary cohesive ends can be rejoined when mixed together in a test tube. They are brought together by their complementary sequences and then an enzyme called **T4 DNA ligase** seals the nicks. A common "cutting" and "pasting" scenario is shown in Fig. 12-1, where one of the cut fragments is a linear piece of DNA (e.g., a piece of genomic DNA from a mammal) and the other is a circular bacterial plasmid. When these two fragments come together in the presence of DNA ligase, a stable, covalently closed, circular **recombinant DNA molecule** is created. This recombinant molecule is a hybrid or chimera of DNA from bacteria and a mammal.

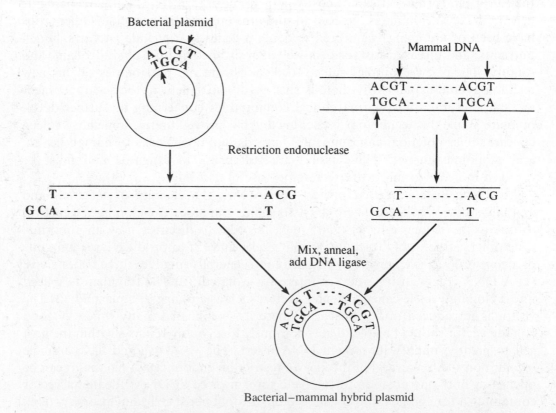

Fig. 12-1.   Formation of a recombinant DNA molecule by cutting a bacterial plasmid and mammalian DNA with the same restriction endonuclease.

(b) Joining Blunt-Ended DNA Fragments. DNA fragments that contain complementary cohesive ends can be joined by the action of the enzyme DNA ligase, as discussed earlier. This enzyme, originally isolated from the bacterial phage T4, catalyzes the formation of a phosphodiester bond between adjacent nucleotides. This enzyme can also be used to join blunt-ended fragments together, but there are several methods that can be employed.

(i) **Blunt-End Ligation.** DNA ligase can be used to directly join the blunt ends of double-stranded DNA (dsDNA) fragments. An advantage of this method is that it can join two defined sequences without introducing any additional material between them. The inability to control which pairs of blunt ends become joined is an obvious disadvantage (Fig. 12-2).

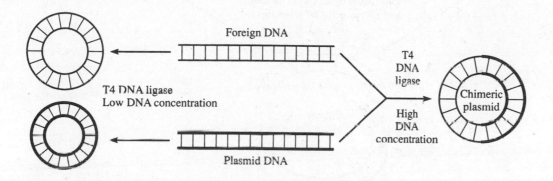

Fig. 12-2. Two possible results of blunt-end ligation via T4 DNA ligase.

Blunt-ended DNA fragments can also be modified at their ends to become overhangs. These ends can pair with other complementary ends and be sealed by DNA ligase. There are two main methods for modification of blunt-ends.

(ii) **Homopolymer Tails.** An enzyme called **terminal deoxynucleotidyl transferase** can add any available deoxynucleotides to the 3′ end of a single-stranded region of DNA without need of a template. If a string of A nucleotides are added onto the vector DNA and thymines (T) are added onto the foreign DNA, **homopolymer tails** are produced that should basepair. The gap (absence of one or more nucleotides in a DNA molecule) can be filled by DNA polymerase I using the other complete strand as a template. The nick (absence of a phosphodiester bond between adjacent deoxyribose sugars) can be sealed by **DNA ligase** (Fig. 12-3).

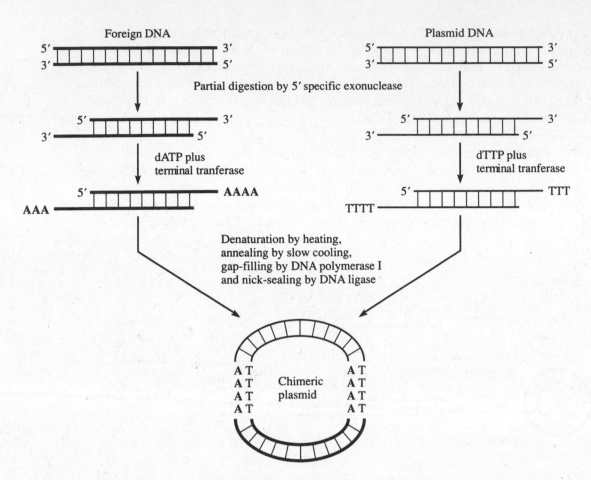

Fig. 12-3. Production of a recombinant (chimeric) plasmid by the formation of homopolymer tails.

(iii) **Linkers**. A third method employs short DNA segments, called **linkers** or **adapters**, that contain a restriction enzyme recognition site; these can be synthesized chemically. Linkers can be added covalently to the ends of a plasmid or to an insert by blunt-end ligation (Fig. 12-4). This method imposes no restriction on the choice of sites to generate the ends, yet allows retrieval of the insert by cleavage with the appropriate restriction enzyme.

It is also possible to convert cohesive ends into blunt ends by treatment with either an enzyme that will extend the shorter side of a 5′ overhang or one that will degrade or "chew back" the cohesive end of a 3′ overhang. Both of these methods employ the use of a DNA polymerase that either synthesizes new DNA in the case of extension or "filling in," or acts as a 3′ to 5′ nuclease in the case of degrading the 3′ overhang. In either case, the original restriction enzyme recognition site is generally destroyed. In the following diagram, N-N represents any conventional complementary base pair.

Extension:

```
5' N N N G      3'              5' N N N G A A T T  3'
   | | | |                         | | | | | | | |
3' N N N C T T A A 5'           3' N N N C T T A A  5'
```

Degradation:

```
5' N N N C T G C A 3'              5' N N N C  3'
   | | | |                            | | | |
3' N N N G      5'                 3' N N N G  5'
```

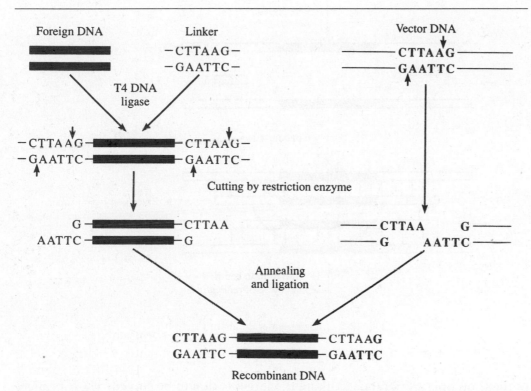

Fig. 12-4. Construction of a recombinant DNA via linkers carrying a recognition site for restriction enzyme *Eco*RI. Small arrows indicate sites of cleavage by *Eco*RI.

2 POLYMERASE CHAIN REACTION (PCR)

Cloning DNA segments for use in a variety of purposes (i.e., DNA sequencing) was formerly a relatively laborious *in vivo* process. However, in 1985, an *in vitro* technique, called the **polymerase chain reaction (PCR)**, was developed for making large amounts of any DNA sequence without the need for cloning using gene banks or libraries (see page 421). The PCR technique (Fig. 12-5) requires a pair of "**primers**" that are usually short (12–20 nucleotides long) pieces of chemically synthesized DNA (oligonucleotides), having nucleotide sequences specifically complementary

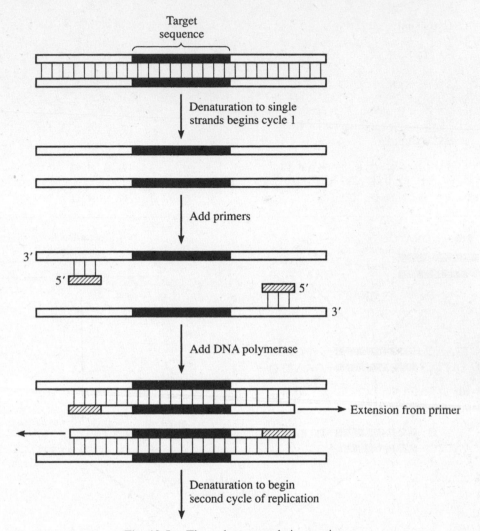

Fig. 12-5. The polymerase chain reaction.

to those on opposite strands flanking the target region to be copied. These primers thus define the ends of the DNA segment that will be duplicated. The original template source of DNA does not have to be highly purified, and even a very small amount of template can serve as the initiator for the PCR. For example, cheek cells can be scraped from the inside of a cheek and placed directly into the PCR reaction. The DNA sample is heated, allowing the complementary DNA chains to separate (**denaturation step**). The primers are then added together with a heat-tolerant DNA-polymerizing enzyme. The primers bind to the single-stranded chains during the cooling phase (**annealing step**), and the polymerizing enzyme extends the primer through the rest of the fragment (**extension step**), creating double-stranded DNA molecules. The process is then repeated; i.e., the mixture is reheated, and during the ensuing cooling phase the excess primers (or newly added primers) bind to template strands and become extended by the polymerizing enzyme to produce more double-stranded molecules. One cycle of denaturation, annealing, and extension can take several minutes, and about 20–30 cycles of heating plus DNA synthesis are normally run during gene amplification by the PCR (with heating and cooling times, this process can take several hours). After 20 cycles, a single DNA molecule can theoretically be amplified to about 1 million copies (called **amplicons**), and after 30 cycles to about 1 billion copies. With this quantity of DNA, nucleotide sequencing can be done easily.

Probes can also be used to locate genes or DNA segments of interest. Probes are commonly composed of a DNA sequence that is similar to the DNA or gene of interest. Because of the large quantity of DNA generated by the PCR, highly radioactive probes are not required, and the target segments can be detected by using nonradioactive probes or stains.

EXAMPLE 12.4
Human immunodeficiency virus type 1 (HIV-1) is the cause of acquired immunodeficiency syndrome (AIDS). Very few susceptible cells actually harbor the virus in an infected person. It is estimated that 1–10 copies of viral DNA per million cells can be detected through the use of the PCR and suitable probes.

3   GENE CLONING

A **library** is a collection of DNA fragments from a single source, such as a particular organism or a tissue. Libraries are created for the purpose of isolating a specific DNA fragment (or gene) of interest from a particular source. There are three general methods for achieving this by using recombinant DNA technology to (1) clone all of the genes of an organism into a **genomic DNA library**, (2) clone individual genes or DNA sequences of interest into a **cDNA library**, or (3) clone gene fragments into an **expression library**. Identification of recombinant clones in a genomic or cDNA library can be carried out using a DNA probe that contains complementary sequences to the gene fragment of interest while identification of expression library recombinants requires the use of a protein-specific antibody.

(a)   **Genomic DNA Library.** In **shotgun** experiments, genomic DNA from a donor organism is cut into many pieces by the same restriction endonuclease used to cleave the plasmid vector at its single restriction endonuclease site. The two kinds of fragments—various pieces of genomic DNA and the plasmid—are mixed *in vitro* and allowed to randomly rejoin by their complementary (cohesive, or "sticky") ends to form circles (see Fig. 12-1). If it is important that a DNA insert be connected in the proper orientation into a vector for expression of the foreign DNA, the vector and insert can be cleaved with two restriction enzymes that generate different cohesive tails at the ends of each fragment. The insertion can then occur in only one orientation (Fig. 12-6).

Each new recombinant molecule is designed to contain a different piece of the genomic DNA. Recipient bacterial cells are made permeable for transformation by the naked DNA of the plasmids by treatment with a cold calcium chloride solution. When plated on nutrient agar containing an antibiotic, each transformed cell will multiply many times to form a colony or clone, all cells of which contain

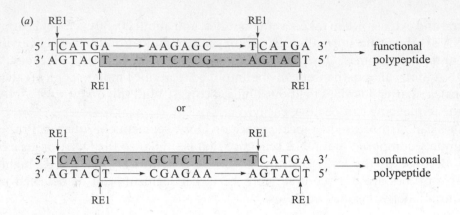

or

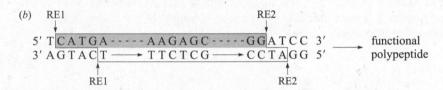

Fig. 12-6. A method for ensuring that a foreign DNA insert (box) will be joined to a cloning vector in the proper orientation for production of the desired protein. (*a*) If insert and vector are cleaved (vertical arrows) by the same restriction enzyme (RE1), the anticoding strand (serving as a transcriptional template for mRNA synthesis) may or may not be in the desired orientation with respect to the promoter, which is upstream (to the left) of the insert. Horizontal arrows indicate direction of transcription of the mRNA. Shaded strand indicates the desired antisense strand. (*b*) If the insert and the vector are cleaved by two different restriction enzymes (RE1 and RE2), only the desired orientation is possible.

multiple copies (sometimes hundreds) of the same chimeric plasmid. The transformed cells can grow on the medium containing the antibiotic since they contain a gene that confers resistance to the drug. The large set of clones that collectively contains all the donor's DNA (i.e., its genome) is known as a genomic DNA library. It should be noted that all the DNA is cloned, not just individual genes. Identification of the clone(s) containing the gene of interest will be discussed in section 5.

(b) Gene or cDNA Library. This method for isolating genes is much more specific than a genomic DNA library. Instead of cloning all of the donor's DNA, only specific genes (i.e., open reading frames) or gene fragments are cloned. If the desired protein is very small (15–20 amino acids) or a short segment of protein sequence is known, it is possible (using the genetic code) to chemically synthesize a corresponding DNA molecule. One of the earliest protein-coding sequences synthesized in this manner was that for the hormone somatostatin (14 amino acids; 42 base pairs in the DNA strand).

EXAMPLE 12.5

A portion of the protein sequence for somatostatin is shown below. The genetic code was used to perform **reverse genetics** (from protein to DNA) in order to figure out a hypothetical DNA sequence for this protein. Because of the degeneracy of the genetic code, the third base position in many mRNA codons may possibly contain more than one nucleotide. The potential nucleotides that can fill this position are indicated as letters in the 3rd position of the codon and below this position (see genetic code in Table 3.1). This synthetic piece of DNA is known as a **degenerate oligonucleotide**.

Amino acid sequence*	a	n	s	n	p	a	m	a	p	e	k
Codons (in DNA)**	GCA C G T	AAT C	TCA C G T	AAT C	CCA C G T	GCA C G T	ATG	CGA C G T	CCA C G T	GAA G	AAA G

*Single-letter amino acid code: a = ala, r = arginine, n = asparagine, d = aspartic acid, c = cysteine, g = glycine, e = glutamic acid, q = glutamine, h = histidine, i = isoleucine, l = leucine, k = lysine, m = methionine, f = phenylalanine, p = proline, s = serine, t = threonine, w = tryptophan, y = tyrosine, v = valine.

** The degenerate nucleotides are shown in the column underneath the third position in the codon.

Larger synthetic genes have been built using (1) PCR or (2) a combination of oligonucleotide synthesis, PCR, and restriction enzyme cutting and ligation. Most proteins, however, are too long to allow chemical synthesis of the corresponding DNA. In this case, it is possible to isolate the total population of mRNA molecules from those cells that are specialized to make the protein. For example, human insulin is produced only by the pancreas, even though the insulin gene is present in all human nucleated body cells. A purified preparation of insulin mRNA can be isolated from pancreatic cells. A synthetic oligonucleotide of thymidines (**oligo-dT**) is hybridized to the poly-A tail of the mRNA strand. The viral enzyme reverse transcriptase (RNA-dependent DNA polymerase) is added to make a single-stranded DNA copy (**cDNA or complementary DNA**) that ends in a hairpin loop (Fig. 12-7). The mRNA template is then destroyed with alkali or an RNase enzyme. The hairpin end of the remaining cDNA serves as a primer for extension synthesis of a complementary strand by **DNA polymerase I**. The loop is then removed by an enzyme called **S1 nuclease** to produce a double-stranded cDNA molecule. These blunt-ended molecules can now be spliced into a suitable vector and cloned as previously described. Alternatively, linkers can be added to the ends of the cDNA to facilitate cohesive end cloning. A library of cDNA fragments is called a **cDNA library**. The advantage to this kind of library is that when cloning eukaryotic genes, DNA sequences that represent only the coding (exon) information will be obtained since mature mRNA does not contain introns.

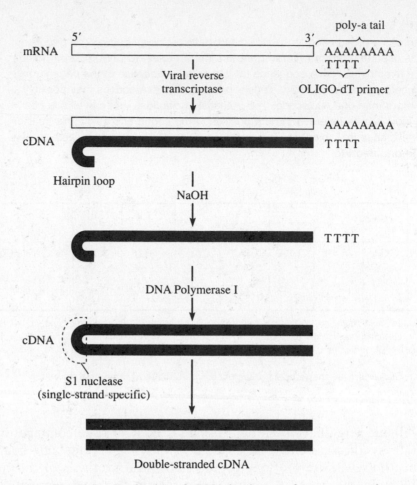

Fig. 12-7. Production of cDNA by the use of reverse transcriptase.

A specialized type of cDNA library called an **EST library** (for <u>E</u>xpressed <u>S</u>equence <u>T</u>ag) is used to make a fairly quick assessment of the variety of expressed genes in an organism or tissue. This is done by sequencing hundreds of cDNA library clones and then comparing that sequence information with known databases of gene sequences (see Bioinformatics section, p. 438).

(c) Expression Library. Expression libraries are built using vectors that contain very efficient regulatory sequences that allow genes to be highly expressed in host cells. These regulatory sequences are the promoters from host cell genes that are known to be efficiently expressed. Libraries of gene fragments can be created in expression vectors using the types of cloning procedures discussed for genomic libraries in section. These libraries are screened for clones able to produce the desired protein product. Alternatively, cloned genes can be fused to the strong promoters for expression of a useful gene product, such as insulin (see section on Production of Recombinant Gene Products in Industry, p. 437).

4   CLONING VECTORS

There are several choices for **cloning vectors**: (*a*) bacterial plasmids, (*b*) bacteriophage vectors, and (*c*) hybrid vectors.

   (a) Bacterial Plasmids. Bacterial plasmids are naturally occurring small, circular, extrachromosomal DNA molecules. They contain origins of replication so they are autonomously replicating (i.e., they can replicate separate from the chromosome). They naturally contain genes for antibiotic drug resistance, disease virulence factors, and gene transfer proteins. Molecular biologists have learned how to isolate and manipulate these plasmids to help in genetic engineering techniques.

   The larger the plasmid, the more inefficient it is as a cloning vector; larger plasmids are less easy to manipulate *in vitro* and less efficient in **transformation**. Transformation is the process by which foreign DNA (e.g., a plasmid) is taken up and incorporated into a bacterial host cell. Therefore, smaller (2–4 kb), nontransmissible plasmids that contain two different antibiotic-resistance genes are normally used as vectors. One of the original vehicles of this kind is the *E. coli* plasmid pBR322. It consists of 4363 bp and contains resistance genes for the antibiotics tetracycline and ampicillin. There is a single restriction endonuclease site for the restriction enzyme *Bam*HI in the entire pBR322 plasmid, and that site is within the tetracycline-resistance gene (*tet-r*). If both the donor DNA and the plasmid DNA are cut with *Bam*HI, the donor fragments can be spliced into the plasmid as described previously. The insertion of a foreign piece of DNA within the *tet-r gene* destroys the ability of this plasmid to confer resistance to tetracycline on the recipient bacterial cell, causing it to become sensitive to tetracyline (Tet-s). Recipient cells that are sensitive to both antibiotics are transformed with the gene library plasmids, some of which contain the donor DNA insert of interest. Three types of cells are produced from this transformation (see Fig. 12-8). Those cells that were not transformed remain Amp-s and Tet-s and will not grow on media containing either antibiotic; those that were transformed are Amp-r, but there are expected to be both Tet-s and Tet-r cells within this group. Tet-s cells contain plasmids with a DNA insert and Tet-r cells do not; i.e., their *tet-r* gene was not inactivated by the insertion of a foreign DNA molecule. One of the problems with plasmid vectors is that they can only accommodate smaller (< 5–8 kb) DNA inserts. Other vectors, such as bacteriophage λ derivatives and yeast artificial chromosomes, that can take larger DNA inserts (from 10 to over 50 kb) have been developed to overcome this problem.

   (b) Bacteriophage Vectors. Many bacteriophage vectors have been developed using lambda (λ) bacteriophage (for more on viruses, see Chapter 11). The central region of phage lambda (λ) contains genes involved in establishing and maintaining the lysogenic state, and hence is not essential for its lytic cycle. This region can be replaced with a foreign DNA insert, if it is of an appropriate size (up to 20 kb), and still allow the phage DNA to be packaged into phage heads. Large foreign inserts tend to be unstable in plasmids, so the two vectors complement one another. Furthermore, **transduction** (the transfer of genetic information to a bacterial cell via a viral intermediate) is a much more efficient process than transfor-

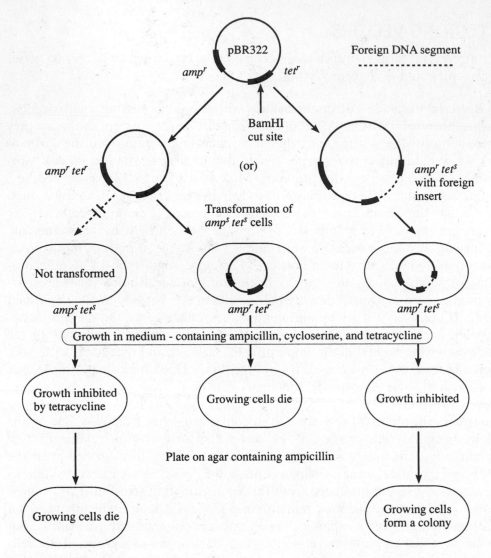

Fig. 12-8. Isolation of a recombinant plasmid using the pBR322 plasmid vector. Cycloserine kills any growing cells. Tetracycline inhibits growth without killing cells. Ampicillin kills growing cells that are Amp-sensitive.

mation, and it avoids the problem of the vector closing up without an insert. Genetically manipulated phage DNA without such an insert will not be packaged properly to become functional (infective) virions. The restriction enzyme *Eco*RI cuts lambda DNA at both ends of the nonessential region. The two essential end regions can be isolated by electrophoresis and ligated *in vitro* with foreign DNA cut by that same enzyme (Fig. 12-9).

Lambda-sensitive bacteria are grown on agar plates in high density to form a lawn of confluent growth. The artificially synthesized transducing phage are added in a concentration resulting in about 100 phage particles per plate, hence producing about 100 plaques of lysed bacterial cells per plate. Each plaque contains

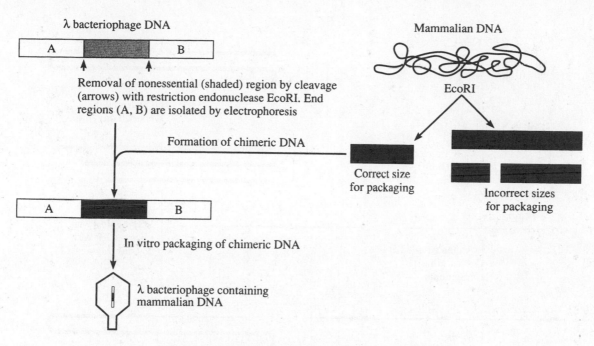

Fig. 12-9.   Cloning of foreign DNA in λ bacteriophage.

phage clones containing millions of identical phage genomes. Recombinant phage are produced when foreign DNA (e.g., mammalian DNA) is ligated to the manipulated phage (Fig. 12-9).

(c) Hybrid Vectors. Cosmids are plasmids into which have been inserted the *cos* sites (*co*hesive end *s*ites) required for packaging lambda DNA into its capsid. Cosmids can be perpetuated in bacterial cells or purified by packaging *in vitro* into phages. The main advantages of using cosmids are that inserts much longer than 15 kb can thereby be cloned and the ease of selecting a recombinant plasmid is greatly improved.

EXAMPLE 12.6
Plasmid ColE1 carries a gene for resistance to rifampicin (*rif-r*) and the *cos* sites of phage lambda, which can be recognized by the *cos*-site-cutting (Ter) system of *E. coli*. Cosmids such as this can function properly, provided that two *cos* sites are present and the *cos* sites are separated by no less than 38 kb and no more than 54 kb. Cleavage of ColE1 and foreign DNA by the restriction enzyme *Hind* III can be used to produce linear, recombinant molecules (Fig. 12-10). Transducing phage particles can be formed if the insert between the two *cos* sites is 38–54 kb in length. No particles are produced if no insert is made or if the insert is larger or smaller than that range. *In vitro* packaging (adding heads and tails) forms transducing particles containing cosmids with cohesive termini. Upon infection of a rifampicin-sensitive (Rif-s) cell with a transducing phage particle, the linear chimera becomes circularized and replicates using the ColE1 replication system. Plating cells on medium containing rifampicin selects for those cells containing the *rif-r* gene, the ColE1 region, and a foreign insert.

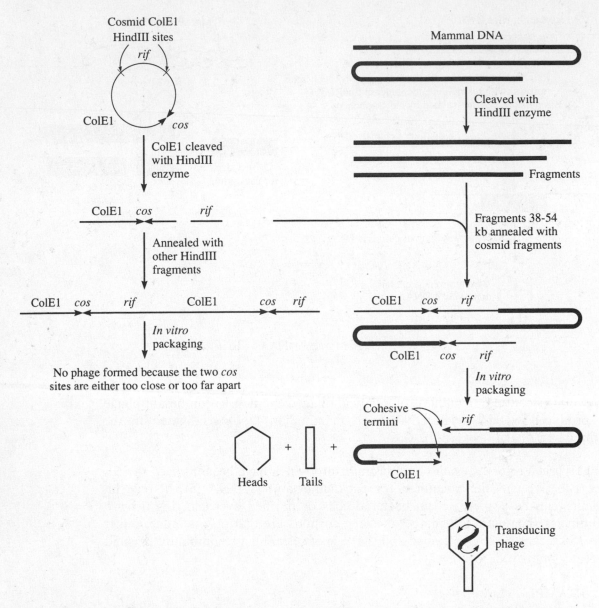

Fig. 12-10.   Formation of transducing phage via a cosmid. Thick lines represent mammal DNA; thin lines, cosmid DNA.

5   IDENTIFYING THE CLONE OF INTEREST

Finding a cell that contains the insert of interest among all the cells of a genomic library presents a major task.   The process involves screening all of the recombinant cells or phage obtained from the transformation or transduction to find the desired clone. Thus, it is called **screening**.   The likelihood of finding the desired gene fragment in a particular gene library can be estimated by the following formula, where N equals the number of recombinants required to screen, n equals the ratio of the organism's genome size relative to the average fragment size in the gene library, P equals the probability (i.e., $p = 0.95$ means there will be a 95% probability of finding the clone), and ln equals the natural log:

$$N = \frac{\ln(1 - P)}{\ln\left(1 - \dfrac{1}{n}\right)}$$

SOLVED PROBLEM 12.3

A researcher desires to clone the gene encoding a particular biosynthesis enzyme (E1) from the yeast *Saccharomyces cerevisiae*. Its genome size is 1.4×10^4 kb and the average size of the library fragments is 5 kb. The genomic library was created in vectors that were transformed into bacterial cells. With 95% probability, how many recombinant bacterial colonies will have to be screened in order to find this particular gene fragment?

Solution: First, solve for n (ratio of the organism's genome size relative to the average fragment size in the gene library).

$$n = \frac{1.4 \times 10^4 \text{ kb}}{5 \text{ kb}} = 2.8 \times 10^3$$

Then, N (the number of recombinant cells of a genomic library that need to be screened in order to find at least one containing the desired gene) can be calculated.

Step 1:

$$N = \frac{\ln(1 - 0.95)}{\ln\left(1 - \dfrac{1}{2.8 \times 10^3}\right)}$$

Step 2:

$$N = \frac{\ln(0.05)}{\ln(1 - 0.000357142)}$$

Step 3:

$$N = \frac{\ln(0.05)}{\ln(0.99964285)}$$

Step 4:

$$N = \frac{-2.995832}{-0.000357213} \approx 8387$$

Thus, nearly 8400 bacterial colonies obtained from transformation will have to be screened in order to find the DNA fragment of interest with 95% confidence. If the average gene library insert size were larger, one can see that the number of clones to be screened would decrease. In contrast, if the genome size were larger, many more colonies would have to be screened in order to achieve success. This formula can act as a guide for estimating the number of clones to screen.

(a) Screening a DNA library. The type of screening method used is dependent upon the cloning vector chosen. When using a plasmid vector, locating the desired DNA fragment can be accomplished by an *in situ* hybridization technique known as **colony hybridization** (Fig. 12-11). Recombinant colonies growing on nutrient agar plates are transferred to a piece of nitrocellulose or nylon membrane by

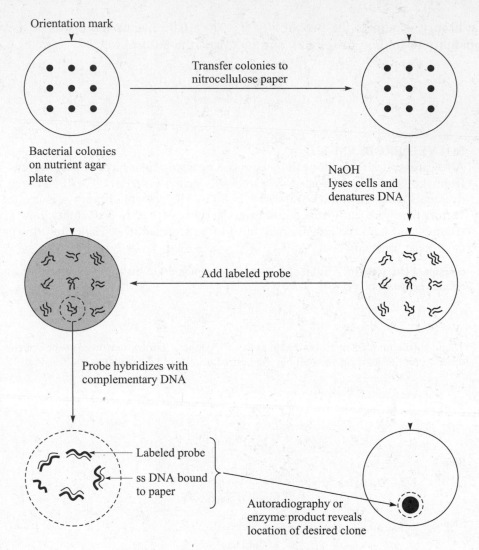

Fig. 12-11.   The colony hybridization technique.

pressing it into the plate, thereby transferring some cells from each colony (replica plating). The pattern of the colonies remains intact on the paper or membrane. The paper is then treated with a dilute sodium hydroxide solution to lyse the cells and denature the DNA to single strands. The cell contents are released, and its DNA binds tightly to the paper. Next, the sodium hydroxide is neutralized with acid. The paper is then covered with a solution containing a labeled DNA probe (a single-stranded piece of DNA that is composed of a sequence that is complementary to a portion of the gene of interest). The label can be a radioactive or fluorescent tag as discussed earlier. The probe hybridizes to complementary DNA sequences that exist on the paper, thereby becoming indirectly bound to the paper. The paper is washed to remove any probe that has not hybridized, and is then detected by an appropriate means (i.e., exposed to X-ray film for auto-radiography of radioactive probes). Detected spots correspond to colonies that contain the gene of interest. Cells from the corresponding clones can be grown in broth culture to any desired amount, thus producing a lot of plasmid. Plasmid DNA is easily isolated from the bacterial cells and the gene of interest can be liberated from the plasmid by digestion with the same restriction endonuclease that was used for its insertion. It can then be isolated from the larger plasmid DNA by electrophoresis.

For screening phage libraries, a similar technique can be used. The only difference is that recombinant plaques are "lifted" or transferred onto the nylon membrane. The library can then be screened using a complementary DNA probe to locate the rare clones containing the insert of interest.

(b) Screening an Expression Library. The most common procedures for detecting protein-secreting clones usually involve **antibodies** in a detection scheme called an immunoassay. Antibodies are protein molecules produced by immune cells that are capable of binding to particular sites on other proteins called **antigens**. Antibodies are produced in response to a foreign protein antigen introduced into vertebrate animals (e.g., mammals). Antibodies against one protein antigen are usually highly specific and they can be produced and then purified for use in immunoassays. A label or tag can be attached to antibodies to enable their detection once they have bound to a corresponding antigen. The most sensitive labels are radioactive isotopes used for **radioimmunoassay (RIA)** or enzyme labels used in **enzyme-linked immunosorbent assay (ELISA)**. The latter are often preferred because of the handling and disposal problems associated with radioactive materials.

> **EXAMPLE 12.7**
> Agar containing lysozyme and antibodies to a specific protein of interest is poured over bacterial colonies on a Petri plate and allowed to harden. Colonies lysed by lysozyme release their proteins. If the protein of interest is present, the antibodies will react with it and form a ring of precipitate around the colony.

6 SITE-DIRECTED MUTAGENESIS AND PROTEIN ENGINEERING

It is possible, by several techniques, to introduce one or more nucleotide alterations of known composition and location into specific genes or regulatory sequences. For example (Fig. 12-12), a plasmid carrying a gene of interest can be nicked at one position with an endonuclease. The plasmid DNA is then denatured and intact single-stranded circles are isolated. Short (13–30 bases) oligonucleotides of known complementary structure (either synthesized *de novo* or from cleavage by a restriction enzyme) can be made to have a mutant base at a desired site in the gene. This oligonucleotide is then renatured with the intact single-stranded circles to serve as a primer for *in vitro* replication of a strand that is not completely complementary to that of the plasmid strand. The replicated circles are sealed with DNA ligase. The covalently closed circles are isolated and used to transform bacteria. During *in vivo* replication, each strand of the plasmid serves as

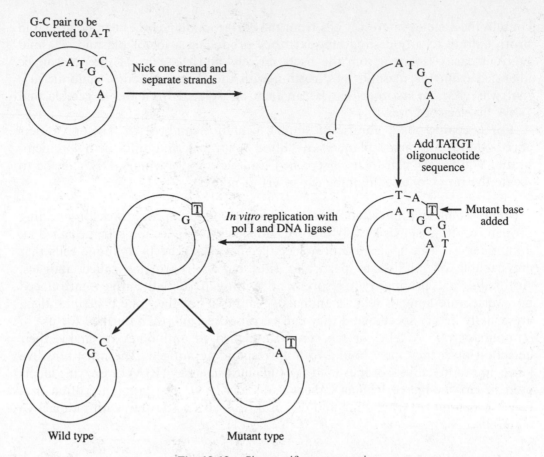

Fig. 12-12.   Site-specific mutagenesis.

a template for producing a progeny strand. Thus, some plasmids are produced with wild-type gene sequences and some with a single base-pair mutation at a known site. After isolation, these mutants can be identified and then evaluated for their effects on the functioning of the gene or regulatory sequence. This example is just one of the many different ways to introduce known and random mutations into DNA sequences. Biochemical and structural characteristics of proteins can be altered by making directed changes in the amino acid sequence. For example, protein stability, activity, temperature resistance, and pH optimum are just some of the qualities that can be changed using this technique. This is referred to as **protein engineering**.

7   POLYMORPHISMS

A **polymorphism** is the existence of two or more alleles at a locus in a population. Conventionally, a polymorphic element or locus is one at which the frequency of the most common allele is less than 0.99. Polymorphisms may exist minimally at three levels: (1) chromosome, (2) gene, or (3) restriction fragment length. DNA sequence polymorphisms can be as simple as a single nucleotide difference (known as **SNP** for **single nucleotide polymorphism**) or an insertion or deletion of a number of nucleotides (**indel**). Either of these types of polymorphisms can lead to differences in the abilities of restriction enzymes to recognize and cut a specific site. This type of polymorphic measure is called a **restriction fragment length polymorphism (RFLP)**. Chromosomal polymorphisms that are large enough to be detected under the light microscope may involve euploidy, aneuploidy, translocations, inversions, duplications, or deficiencies.

EXAMPLE 12.8

If the *Eco*RI restriction enzyme site, GAATTC, exists in three particular nucleotide positions in a gene sequence (allele 1) and mutation results in an alteration of the middle site to CAATTC (allele 2). The enzyme will no longer be able to recognize and cut the DNA strands at this position, resulting in a single 500-bp *Eco*RI fragment for allele 2 when compared with the two fragments (150 and 350 bp) predicted by cutting of allele 1. New restriction sites can also be created by mutation where there previously were none or different ones (see Example 12.9).

Allele 1:	GAATTC	GAATTC		GAATTC	
	1	150		500	bp
Allele 2:	GAATTC	CAATTC		GAATTC	
	1	150		500	bp

There are many uses for for RFLPs; however, they are particularly useful in medical and forensic genetics.

(a) Medical Genetics. The technique used to analyze RFLPs is **Southern blotting**, named after E. M. Southern, who first developed it. A restriction enzyme digest of an individual's DNA is electrophoresed (i.e., separated from other DNA sequences) on an agarose gel and then denatured to single strands. The single-stranded fragments are then transferred from the gel to nitrocellulose paper in the following manner. The gel is placed on normal filter paper that has been soaked in concentrated salt solution. The nitrocellulose paper is placed on top of the gel, with dry blotting paper and a weight on top of that. The salt solution moves through the gel, carrying the DNA fragments with it onto the nitrocellulose paper where they become trapped. The fragment pattern on the gel is thereby faithfully transferred onto the nitrocellulose. The fragment(s) of interest can then be located on the nitrocellulose by *in situ* hybridization with a radioactive DNA probe, followed by autoradiography. A similar technique, referred to as **northern blotting**, is used to identify RNA molecules that are similar to a probe sequence. Transfer of a protein electrophoresis pattern from a gel to a paper is called **western blotting**. In this case, the probe is usually a labeled antibody against the protein of interest.

EXAMPLE 12.9

The normal gene for the β-globin chain of human hemoglobin has a GAG codon for glutamic acid as the sixth amino acid from the N terminus. Individuals with sickle-cell anemia have a mutant, GTG for valine, at that same position. Since fetal hemoglobin does not contain β-globin chains, it is impossible to obtain fetal hemo-globin for prenatal analysis of this genetic disease. However, fibroblasts (which nor-mally do not make hemoglobin) contain the gene for β-chain of hemoglobin, and these cells can be retrieved by amniocentesis (see Human Cytogenetics section in Chapter 7, p. 242). The total DNA from fibroblast cells is digested with the restric-tion endonuclease *Mst*II and the fragments are separated by electrophoresis on an agarose gel. The DNA is then transferred onto a nitrocellulose membrane by Southern blotting, denatured to single strands, incubated with a radioactive β-globin gene probe, and autoradiographed. Only one DNA band of 1300 bp appears on the autoradiograph for normal hemoglobin (HbA), whereas two bands of lengths 200 and 1100 bp appear for sickle-cell hemoglobin (HbS). Hence, the GAG codon in the *β*-chain gene of HbA is not part of a recognition site for *Mst*II, but the muta-tion to GTG in HbS creates a new *Mst*II site.

(b) Forensic Genetics. Forensic genetics is concerned with legal matters and can be used to determine the identity or nonidentity of DNA from cells (e.g., blood, hair, semen) left at the scene of a crime with the DNA of cells of any suspect. It can also be used in cases of disputed parentage or for identifying the parentage of missing children. This branch of genetics utilizes a technique known as **DNA fingerprinting** to distinguish the DNA of a human from that of any other person. It depends on the fact that there are different numbers of tandem repetitive DNA sequences scattered throughout different human genomes. Any DNA sequence that exists in multiple copies strung together in various tandem lengths is referred to as a **minisatellite** or a **variable number of tandem repeats locus** (**VNTR locus**). The number, the pattern, and the length of these repeats are unique for each individual. Regardless of its length, each repeat contains a common (usually < 20 bp) core sequence that can be recognized by an appropriate radioactive probe. The DNA of an individual is extracted from a convenient sample of that person's cells (e.g., from white blood cells) and subjected to cleavage by one or more restriction endonucleases. The fragments are separated on an agarose gel, denatured to single strands, transferred to a nitrocellulose filter by Southern blotting, exposed to a radioactively labeled probe, and then autoradiographed. The number of bands that develop on the autoradiograph are unique for each individual.

DNA Sequencing

Two methods for determining the order of DNA nucleotides in a relatively short segment of a DNA molecule were originally developed about the same time in the late 1970s: (1) chemical cleavage (Maxam and Gilbert) and (2) enzyme (Sanger) sequencing. The enzyme method or primed synthesis method is the most commonly employed procedure and thus will be the only method discussed here. Several variations of this method have been developed, all of them employ-ing a DNA polymerase enzyme; only two of these variations will be discussed here.

1   FOUR FLUORESCENT LABELS

The fragment of DNA to be sequenced (target or template DNA) is most commonly contained within a plasmid or other type of cloning vector, or can be a purified PCR amplicon. A **DNA primer** that will specifically anneal to one of the 3' ends of the template strand is chemically synthesized. Such primers are relatively short (15–20 bp long), single-stranded oligonucleotide sequences. The two strands of the template DNA are denatured to single strands by heat and the primer molecules will bind to their complementary sequence on the desired template strand as the mixture cools (Figure 12-13). DNA polymerase is added together with a mixture of the four deoxyribonucleoside triphosphates [dATP, dCTP, dGTP, dTTP; together referred to as dNTPs (deoxyribonucleotides), Fig. 12-14(a)] and a mix of dideoxynucleoside triphosphates [ddATP, ddCTP, ddGTP and ddTTP; together referred to as ddNTPs (dideoxyribonucleotides), Fig. 12-14b)], each of which is labeled with a different fluorescent dye. These labeled dideoxynucleotides are called chain-extension terminators (**chain terminators**) because they have no 3' hydroxyl group with which to form internucleotide 3' to 5' phosphodiester linkages. Thus, once a ddNTP is added (bold type in Fig. 12-13) to an extending chain, further extension ceases. A suitable ratio of ddNTPs to normal dNTPs is chosen so that only a relatively small fraction of the growing extension products will be terminated by the addition of ddNTPs. The primed complexes are extended by DNA polymerase toward their 3' ends by random polymerization of each nucleotide from either dNTP or ddNTP pools, so that ideally every possible extension chain length should be produced. These reactions are now typically carried out using a PCR cycling format in a process called **cycle sequencing**. Polyacrylamide gel electrophoresis [Fig. 12-13(a)] is then applied under conditions that allow the separation of extension products different in length by a single nucleotide, in a single column or lane on the gel. Since DNA molecules are generally negatively charged, they migrate toward the anode (positive electrode) during electrophoresis. Thus, the smaller the extension product, the faster it will migrate through the gel. The last nucleotide in the extension product (indicated by the specific fluorescent dye) in the fastest-migrating band thus represents the 5' nucleotide of one strand in the target DNA and the last nucleotide in the extension product in the slowest-migrating band represents the 3' nucleotide of that strand. When each population of extension products of identical length passes a fluorescence scanner as a band, the fluorescent light emitted is recorded and analyzed by a computer to yield the entire sequence of that stretch of target DNA (**electropherogram**).

A typical sequencing gel can now yield 500–600 bp of sequence information per lane. In order to sequence a long stretch of target DNA (> 500–600 bp), multiple different sequencing reactions must be carried out, each beginning with a primer designed to anneal to and direct DNA synthesis from a different region of the DNA. Automated equipment can now accommodate almost 100 lanes in one electrophoresis run. That theoretically translates to approximately 50,000 bp of sequence data that can be generated in about 4–6 h. This kind of productivity requires lots of technical and robotic assistance, but it has made possible the relatively rapid sequencing of large genomes (such as the Human Genome).

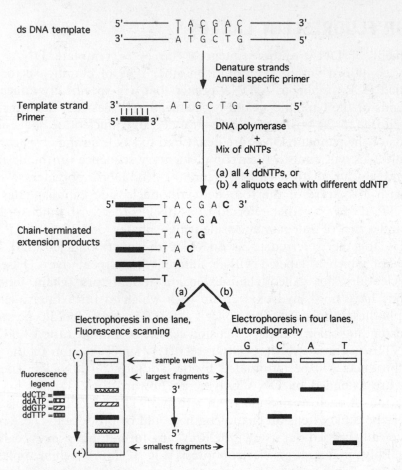

Fig. 12-13. Enzyme method of DNA sequencing. Chain-terminating ddNTPs are indicated by bold type. (*a*) Four fluorescent labels—one lane. (*b*) One radioactive label—four lanes. The DNA sequence read from either gel is 5′-TACGAC-3′.

2   ONE RADIOACTIVE LABEL

Alternatively, the template DNA molecules are divided into four reaction vessels and denatured to single strands. DNA polymerase and a mix of all four dNTPs are added, one of which (often dCTP) is radioactively labeled. Next, a different un-labeled ddNTP is added to each of the four reactions. Following chain extension, the products of the four reactions are electrophoresed separately in four lanes on the same gel [Fig. 12-13(*b*)]. Autoradiography reveals the position of the various sized, radioactively labeled extension products as dark bands on X-ray-sensitive film (**audioradiogram**), from which the nucleotide sequence of the target DNA can be directly read by eye. Again, the last nucleotide in the extension product (known by the lane in which the reaction was loaded) in the fastest-migrating band thus represents the 5′ nucleotide of one strand in the target DNA and the last nucleo-tide in the extension product in the slowest-migrating band represents the 3′

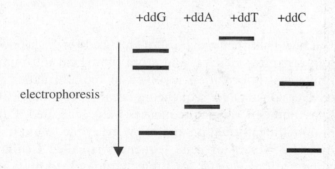

Fig. 12-14. Structure of a normal dNTP. (*a*) 2′-Deoxyribonucleoside-5′-triphosphate and a chain-terminating ddNTP. (*b*) 2′,3′-Dideoxyribonucleoside-5′-triphosphate.

nucleotide of that strand. This method requires more gel lanes per target DNA molecule so is much less efficient than using fluorescently labeled ddNTPs.

SOLVED PROBLEM 12.4

From the following electrophoretic banding pattern that represents a fragment sequenced by the one label, four-lane dideoxy (chain-termination) technique, determine the base sequence of the target DNA strand.

Solution: The sequence is read from the bottom of the gel to the top because the shortest fragments will travel faster and these are the ones closest to the 5′ end of the DNA being sequenced. Thus, the sequence is 5′-CGACGGT-3′. The target sequence is its complement: 3′-GCTGCCA-5′.

Production of Recombinant Gene Products in Industry

Proteins with industrial, agricultural, or pharmaceutical applications can be produced in a variety of microbial or cell culture systems. Several host systems have been developed: bacterial (*E. coli, Bacillus*), fungal (yeasts, *Aspergillus, Fusarium*), plant, mammalian, and insect cell culture systems. In order to produce a desired gene product in one of these systems, the gene must first be cloned. If it is a eukaryotic gene, a complete cDNA clone (from start ATG to final stop codon) must be obtained, particularly if expression will take place in a bacterial system. Bacteria do not possess the machinery to splice introns, and other systems may or may not correctly carry out intron splicing of genes from significantly different organisms. Furthermore, bacteria cannot carry out posttranslational modifications, such as glycosylation, that may be required for protein function. The cDNA clone is spliced behind the control of a host-specific promoter in a plasmid vector to guide transcription in the host system. For example, a cloned human gene would be spliced behind the promoter from the yeast alcohol oxidase (*AOX*) gene for expression in a yeast host-vector system. The *AOX* gene promoter is a strong promoter (i.e., produces many mRNA molecules) and can be upregulated by the presence of methanol. Once this recombinant expression plasmid is created, it is introduced into the cells of the host system. Stable recombinant cells can be grown in the lab and then used to produce the desired gene product. Expression of a gene from an organism in a cell of the same species is called **homologous gene expression**. When a gene is expressed in a completely different host-cell system, the term **heterologous gene expression** is used. This is the method employed to produce a variety of useful recombinant proteins, such as human insulin, blood-clotting factors, and human growth hormone.

Bioinformatics

Bioinformatics is a new field that was born of the need for high-powered computing ability to help organize, analyze, and store biological information. The primary types of biological information involved in bioinformatics are DNA and protein sequence data. Once DNA sequencing became technologically simple and automated, massive numbers of gene sequences were generated. Public databases were created to house the information and allow everyone to use it. The definitive database in the United States for gene sequences is called **GenBank**® which is administered by the National Center for Biotechnology Information (NCBI) and, as of June 2001, contains 12,973,707,065 nucleotide bases that are in 12,243,766 sequence records from thousands of different microbe, plant, and animal species. The database can be found at the NCBI website, http://www.ncbi.nlm.nih.gov/. There are additional databases of DNA sequences in Japan at the DataBank of Japan (DDBJ) and in Europe at the European Molecular Biology Laboratory (EMBL). All of these databases are cooperative systems.

Besides just storing biological information in the database, the database can be used to help analyze genes, their functions, and evolution. For example, if a gene is cloned and sequenced, this sequence can be used in a search called **BLAST**® against all known sequences (all 12 million ... and growing) to determine if (1) it has already been cloned or (2) it is related to an already known gene. If it is a new gene sequence, its relatedness to other sequences might help determine its possible biological function. Protein databases can also be searched.

EXAMPLE 12.10
Suppose you clone and sequence a DNA fragment obtained from a genomic library of *Xenopus* (frog) DNA. The sequence is entered into the BLAST® computer program at the NCBI web page and the top 10 matches obtained are shown below.

Sequences producing significant alignments:	Score
Rana dybowskii cdc2 kinase (*CDC2*) mR...	137
Rattus norvegicus cell division cycle cont...	96
R. norvegicus mRNA for *cdc2* promoter r...	96
Bos taurus cyclin-dependent kinase 1...	94
Chicken mRNA for *CDC2* protein kinase	74
Homo sapiens cyclin-dependent kinase 3 (CD...	68
H. sapiens mRNA *cdk3* for serine/threon...	68
C. japonica CDC2 gene exon 3	64

This data shown in the table provides several important pieces of information. First, it tells us the top 10 genes that our unknown DNA sequence is similar to in sequence. Next is information regarding the species to which each of these matching genes belongs (e.g., *Rattus norvegicus*). The score (based on an algorithm) is a relative measure of how similar the unknown, or query, sequence is to the identified sequence. The higher the number, the more similar the sequences.

Furthermore, the BLAST® program lines up the query sequence with each sequence in the database in an *alignment* (Fig. 12-15). The alignment shows the exact nucleotides that are similar by connecting them with a line. Those that are different are not connected. These types of alignments can give an estimate of gene relatedness, which can be inferred to represent some degree of evolutionary relatedness. All of the sequences in the database are annotated, so further information can be found. For example, most sequence annotations contain references to the research article wherein the information on the gene was published. This information can be accessed and a hypothesis regarding the function of the unknown gene, based on its similar relative (also known generally as a **homolog(ue)**; see section on Molecular Evolution in Chapter 9) can be made and then tested. It turns out that the gene family analyzed in this example is involved in regulation of the cell division cycle.

Query: 42 aaaatagagaagatcggagagggcacatatggggtcgtgtacaagggtcgtcacaaagca 101
           ||||||||||| || ||||| || || ||||| || ||||| ||||| | |||||| |
Sbjct: 133 aaaatagagaaattggagaaggtacctatggagttgtgtataagggtagacacaaaact 192

Fig. 12-15. A DNA sequence alignment of the query sequence and one retrieved as similar by the BLAST® program. These two sequences match each other at different locations, as indicated by the nucleotide sequence numbers shown at the left- and right-hand sides of each sequence. The vertical lines show identical nucleotide matches between the two sequences.

This example shows just one of the ways that bioinformatics can be used to help understand gene structure and function. Because DNA sequencing technology has advanced so rapidly, researchers are not just sequencing single genes but the genomes of entire organisms, ranging from bacteria and viruses to plants, insects, and humans. Most of this information is also being submitted to the public databases for use and analysis by scientists all over the world. Some of the information is being used by biotechnology and pharmaceutical companies to help develop better cures and treatments for diseases.

The Human Genome Project

A large collaborative effort to decode all of the 3 billion nucleotide base pairs of the human genome was launched in the mid-1980s. The international **Human Genome Project** effort is being funded by both government and industry sources. It was expected to be completed in the year 2003, the 50th year of the discovery of the DNA structure, and cost billions of dollars, but technological advances allowed the project to be completed a few years ahead of schedule. In an historical announcement on June 26, 2000 at the U.S. White House, leaders from industry (J. Craig Venter of Celera Genomics) and government (Francis Collins of the National Human Genome Research Institute) announced that the first draft of the human genome had been completed. It took 10 years to complete this first draft. The government-funded participants chose individual chromosomes to sequence while the private venture labs sequenced the entire genome in a large "shotgun" approach using computers to assemble the data into a genome-wide map. All-in-all, over 20 billion bases of sequence information has been gathered. These billions of bases overlap to provide a sequence map of the human genome. There is so much computer data that new computer hardware systems have been developed and storage space is measured in terms of **terabytes** (10^{15}), which is 1000 times larger than a gigabyte (10^{12}).

It is estimated that there are between 25,000 and 45,000 genes within the 3 billion base pairs that make up the human genome. Human genes can range in size from thousands to hundreds of thousands of base pairs (including exons and introns). For example, analysis of sequence data from chromosome 22 suggests that it contains over 800 genes, the largest of which exceeds 500,000 base pairs. Of the genes identified, only half (~ 400) have a hypothesized function, as discovered through sequence database comparisons. Some of the genes identified are responsible for at least 27 human disorders, including brain cancers and schizophrenia. **Gene families**, groups of genes that are similar, have been identified that appear to have originated from tandem duplication of genes and subsequent divergence by mutations. And that is just one of the 23 human chromosomes to be analyzed.

The completion of this human genome draft is predicted to be just the beginning of a new era in genetic and medical research. It is estimated that it will take decades to interpret and understand all of the information represented by the A's, T's, G's, and C's in our DNA. Not only will scientists begin to better understand gene and chromosome structure and function, but patterns and interactions between genes will emerge. This information also brings with it ethical concerns over the use of genetic information by government agencies, medical institutions, and insurance companies. There is much to discuss and discover as we enter this new age.

DNA Microarrays and Functional Genomics

A new focus of genetic research that has been driven by the availability of genome sequencing data is the identification of genes that are expressed (i.e., transcribed into mRNA molecules, then translated into proteins) in different cells or tissues under different conditions (presence or absence of a drug) or at different stages of normal or disease development. For example, one area of research is aimed at the elucidation of all genes expressed in cancer cells vs. normal cells. The term **tran-scriptome** is used to describe the set of all mRNA molecules in a particular cell or tissue type. Other methods in the emerging field of **proteomics** (the comparative study of protein expression in different cells or tissue types) are being designed to examine the total complement of proteins (referred to as the **proteome**) in a tissue or a population of cells at a given time or under particular circumstances. These methods will help researchers develop better therapies and possible cures for diseases such as cancer. The study of genome-wide gene expression (at the transcriptional and translational levels) is termed **functional genomics**.

DNA microarrays or **gene chips** were developed in the 1990s to determine the transcriptional status of thousands of genes in a single experiment (or a few) on a glass slide or small membrane. There are several ways to make a DNA microarray. One method begins with the PCR amplification of known gene sequences from an organism. A tiny amount (nanolitre, nl, 10^{-9} liter; 20–100 μm in diameter) of each PCR amplicon is then "printed" or spotted onto the surface of a glass slide in a grid fashion by a robotic arm called an "arrayer." A computer monitors the position of each DNA spot. Thousands of DNA spots can be printed on a single slide. For organisms with smaller genomes (e.g., most bacteria and yeasts), it is possible to print the DNA corresponding to every protein-coding sequence (i.e., gene) on one glass slide containing thousands of spots. Another method utilizes a special light-directed chemical synthesis process to build high-density DNA oligonucleotide arrays, one nucleotide at a time. Each oligonucleotide in the array corresponds to a different region of a gene. Gene chips or DNA microarrays may be used in a variety of different experiments designed to identify all of the transcripts present at a given time in a particular cell or tissue type.

EXAMPLE 12.11
Yeast cells carry out meiosis, the process of sexual reproduction (called sporulation in yeasts). In order to determine which genes are involved in this process, mRNA is isolated separately from nonsporulating and sporulating yeast cultures. The mRNA molecules are copied into cDNA molecules using the enzyme reverse transcriptase. During this process, the cDNAs from the nonsporulating culture are labeled with a green fluorescent tag (incorporated into a dNTP used in the reverse transcription reaction) and the cDNAs from the sporulating culture are labeled with a red fluorescent tag. The cDNAs are then mixed together and the solution is flooded onto a gene chip that contains spots of all the known yeast genes. The cDNAs will hybridize specifically with any homologous (i.e., similar) DNA on the chips. Any unbound cDNAs are then washed away and the bound fluorescently labeled cDNAs are quantitatively detected using a specially designed fluorescence microscope connected to a computer. Green spots indicate genes expressed in

nonsporulating cells, red spots indicate genes expressed in sporulating cells, and yellow spots (equal amounts of red and green label) indicate genes that are expressed under both conditions (see illustration below where green spots are indicated by solid circles, red spots by open circles, and yellow spots by gray circles). In this way, all of the genes expressed uniquely during the process of sporulation (those corresponding to the red spots) can be discovered. Relative levels of gene expression can also be determined.

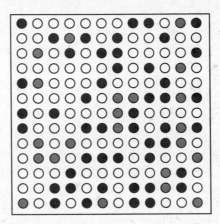

 # Supplementary Problems

12.5. The buoyant density (ρ) of DNA molecules in 6M CsCl solution increases with the molar content of G + C nucleotides according to the following formula:

$$\rho = 1.660 + 0.00098\,(G + C)$$

Find the molar percentage of (G + C) in DNA from the following sources: (*a*) *Escherichia coli*: $\rho = 1.710$ (*b*) *Streptococcus pneumoniae*: $\rho = 1.700$ (*c*) *Mycobacterium phlei*: $\rho = 1.732$.

12.6. Given two dsDNA molecules, the overall composition of which is represented by the segments shown below, determine which molecule would have the highest melting temperature. Explain.

(*a*) TTCAGAGAACTT
 AAGTCTCTTGAA

(*b*) CCTGAGAGGTCC
 GGACTCTCCAGG

12.7. The primary mRNA transcript for chicken ovalbumin contains seven introns (light, A–G) and eight exons (dark) as shown below.

| A | | B | C | D | E | | F | | G | |

If the DNA for ovalbumin is isolated, denatured to single strands, and hybridized with cytoplasmic mRNA for ovalbumin, how would the hybrid structure generally be expected to appear in an electron micrograph? *Note*: Double-stranded regions appear thicker than single-stranded regions.

12.8. About half the weight of RNA synthesized at any given time within a bacterial cell is rRNA. The 30S subunit of bacterial ribosomes contains one 16S rRNA molecule (1.5 kb); the 50S subunit contains one 23S rRNA (3 kb) and one small 5S rRNA (0.1 kb). Hybridization tests of 16S and 23S rRNAs with complementary single strands of DNA reveal that about 0.14% of DNA is coding for 16S rRNA and about 0.18% for 23S rRNA. Estimate the relative activity of rRNA genes as transcription templates compared with the average gene of the bacterial genome that gives rise to mRNA. *Note*: Assume that the amount of DNA allocated to 5S rRNA synthesis is negligible; likewise for all kinds of tRNAs.

12.9. Some bacterial proteins are normally secreted from the cell. If transgenic rat insulin protein could be attached by genetic engineering to such a secreted bacterial protein, it too might be secreted from the cell. Suppose that you are given an agar plate containing several recombinant bacterial clones known to contain the gene for insulin. Propose an autoradiographic method for identifying those clones that are secreting this protein. (*Hint*: antibodies can be attached to certain kinds of plastic in a way that leaves their antigen-combining sites free to react.)

12.10. Restriction endonuclease *Eco*RI makes staggered cuts in a 6-nucleotide DNA palindrome; restriction endonuclease *Hae*III cleaves at one point in the middle of a 4-nucleotide palindrome. If different aliquots of a purified DNA preparation are treated with these enzymes, which one would be expected to contain more restriction fragments? Explain (give the rationale for) your choice.

12.11. Only about 200 molecules of phage lambda repressor are made by bacteria when lambda is integrated at its specific attachment site between *E. coli* genes *gal* and *bio*. Some bacterial genes such as *lac* can be induced to produce more than 20,000 molecules of an enzyme per cell. If you could cut and splice genes and regulatory regions at your discretion, how would you design a bacterial cell for maximum synthesis of lambda repressor protein?

12.12. The goal is to clone a specific human gene. Human DNA is isolated and cut into ~ 15-kb fragments that are spliced into phage vectors (shotgun method). The phage are introduced into recipient bacterial cells. (*a*) How many kinds of cells exist? List their characteristics. (*b*) How many recombinant phage would have to be screened in order to find the DNA fragment of interest (the human genome is ~ 3×10^6 kb)? (*Hint*: See Solved Problem 12.3.)

12.13. The polymerase chain reaction (PCR) was originally performed with a DNA polymerase from the bacterium *E. coli*, a common inhabitant of the human gut (37°C). Each cycle of heating denatured the enzyme added during the previous cycle. In order to reduce costs and automate the PCR, another source of the enzyme had to be found. Where is the most likely place to find this alternative source?

12.14. Protein P is synthesized in relatively high amounts in the mouse pancreas. This protein has been isolated and purified and the sequence of 6 amino acids from the N-terminal end has been determined. If the gene for protein P is desired to be cloned for recombinant expression in a bacterial host system:

(*a*) How can a probe be prepared to identify the gene for protein P?
(*b*) Which is the best type of library (genomic, cDNA or expression) to construct for cloning this gene?

12.15. A cDNA library is being constructed from mouse pancreas tissue in order to clone the protein P gene.

(*a*) Outline the steps in cDNA production.
(*b*) Which type of vector is most appropriate to use in this case: plasmid pBR322 or phage lambda. Why?
(*c*) One method for cloning these cDNAs into a vector is to ligate short linker molecules onto the ends of each cDNA clone. The linkers each contain the recognition sequence for a

particular restriction enzyme. For example, the linker sequence, 5′ G<u>CTGCAG</u>C-3′ contains the *Pst*I restriction site (underlined). Now all the cDNA clones constructed contain *Pst*I sites on each end. Outline the steps involved in ligating these clones into a vector, such as plasmid pBR322 (assume that pBR322 has a unique *Pst*I site in the *amp*^{R} gene).

(d) After transforming *E. coli* with the plasmids, how can the cells that contain a recombinant plasmid (i.e., one containing any DNA insert) be identified?

(e) If one million ampicillin-sensitive, tetracycline-resistant clones are grown on nutrient agar plates, how are we going to detect the rare clone or clones that carry the gene for protein P?

(f) After selecting several clones identified as carrying the gene for protein P, recombinant cells from each clone are propagated to high density in nutrient broth. The recombinant plasmids (presumably containing a fragment of the gene for protein P) are then extracted and purified from the rest of the cellular DNA. How can the gene be isolated from the plasmid?

(g) How can we demonstrate that the gene we have isolated is indeed the one for protein P?

12.16. The gene for protein P is isolated and its DNA sequence is determined. It is found to contain 1000 bp of coding sequence and two *Eco*RI restriction enzyme sites, -GAATTC-, one at 150 bp internal to the start and one at 150 bp internal to the stop codon of the normal gene (see illustration below). A defective protein P has been discovered and its gene has also been cloned and sequenced. In the abnormal gene, one of these sequences has been changed to -GCATTC-.

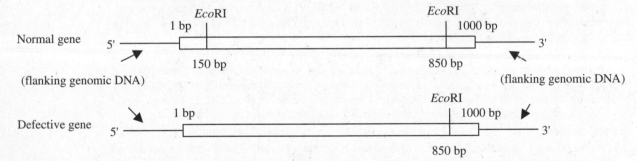

We want to find out if fetal cells contain the normal or the abnormal gene. So DNA from fetal cells is cleaved with *Eco*RI and the fragments are separated on an agarose gel and then transferred to a nylon membrane. A probe for the normal gene hybridizes to the internal 700-bp *Eco*RI fragment and will also hybridize with the abnormal gene sequence in this same region. The size of the fetal DNA fragments can be estimated by running DNA fragments of known sizes on the same gel. What band pattern is expected if the fetal cells contain the abnormal gene?

12.17. Diagram the electrophoretic pattern expected from a triple digest of the plasmid in Solved Problem 12.2 by restriction enzymes A + B + C.

Review Questions

Vocabulary For each of the following definitions, give the appropriate term and spell it correctly. Terms are single words unless indicated otherwise.

1. A technique that separates molecules according to their net charge in an electric field, usually on solid or semisolid support media such as paper or agarose.
2. Separation of complementary chains of a DNA molecule, usually by heating.
3. Reassociation of complementary single-stranded regions of DNA with DNA, or DNA with RNA.
4. Exposure of a photographic film to DNA labeled with a radioactive isotope.
5. Symmetrical sequences of nucleotide base pairs in double-stranded DNA that read the same on each strand from 5′ to 3′ .
6. Bacterial enzymes that break phosphodiester bonds in DNA at specific base sequences. (Two words.)
7. The random collection of a sufficiently large sample of cloned fragments of the DNA of an organism to ensure that all of that organism's DNA is represented in the collection. (Two words.)
8. An enzyme used to add deoxyribonucleotides to the 3′ ends of DNA chains without a template. (Two or three words.)
9. An *in vitro* technique for copying the complementary strands of a target DNA sequence simultaneously for a series of cycles until the desired amount is obtained. (Three words.)
10. The name of the product produced by reverse transcriptase enzyme from an mRNA template. (One or two words.)

Multiple-Choice Questions Choose the one best answer.

1. A radioisotope used to label proteins differentially from nucleic acids is (a) ^{32}P (b) ^{14}C (c) tritium (d) ^{35}S (e) ^{15}N
2. Which of the following single strands would be part of a palindrome in double-stranded DNA? (a) GAATTC (b) ATGATG (c) CTAATC (d) CCCTTT (e) none of the above
3. Which of the following is an enzyme used to form a phosphodiester bond in a nick between a 3′ end of one DNA chain and a 5′ end of another? (a) DNA polymerase (b) restriction endonuclease (c) DNA ligase (d) S1 nuclease (e) phosphodiesterase
4. Bacterial cells are rendered more permeable to uptake of plasmids by treatment with (a) heat (b) calcium chloride (c) alkali (d) a blender (e) ultrasound
5. The melting temperature of a DNA molecule is determined by using (a) electrophoresis (b) change in electrical conductivity (c) column chromatography (d) density-gradient ultracentrifugation (e) change in optical density
6. Which of the following is a desirable characteristic for a cloning plasmid? (a) a site at which replication can be initiated (b) one or more unique restriction endonuclease sites (c) one or more antibiotic-resistance or drug resistance genes (d) a highly active promoter (e) all of the above
7. The classical 1957 experiment of Meselson and Stahl was concerned with (a) mode of DNA replication (b) polymerase chain reaction (c) *in vitro* production of recombinant DNA molecules (d) synthesis of hybrid proteins (e) transduction via lambda phage
8. Many of the genes in lambda phage are clustered according to similarity of function. Which of these gene clusters could most likely be deleted and replaced with foreign DNA, making the recombinant phage a useful cloning vector? (a) nucleases to destroy host DNA (b) head capsomeres (c) phage-specific RNA polymerase (d) establishment and maintenance of lysogeny (e) tail proteins

9. Eukaryotic genes may not function properly when cloned into bacteria because of   (a) inability to excise introns   (b) destruction by native endonucleases   (c) failure of promoter to be recognized by bacterial RNA polymerase   (d) different ribosome binding sites   (e) all of the above

10. The DNA fingerprinting process involves   (a) chain terminators   (b) degenerate oligonucleotides   (c) VNTR loci   (d) RFLPs   (e) cDNA

Answers to Supplementary Problems

12.5.   (a) 51.02   (b) 40.82   (c) 73.47

12.6.   (b), because it has a higher (G + C)/(A + T) ratio

12.7.

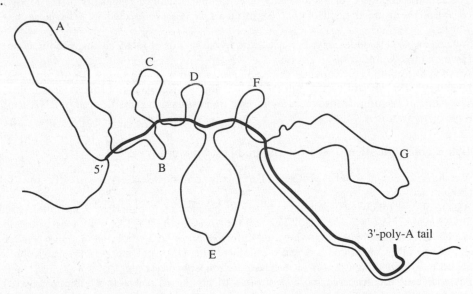

12.8.   Of all the RNAs that hybridize with DNA, the 16S and 23S rRNAs account for only 0.0014 + 0.0018 = 0.0032 or 0.32%. Since these rRNAs and mRNAs are about equally represented in a cell, the ratio 1/0.0032 = 312.5 expresses how much more active in transcription are the genes for the rRNAs.

12.9.   Attach anti-rat insulin antibodies to a plastic disk about the size of an agar plate. Impress the disk onto the plate and allow any secreted insulin to be specifically bound by the antibodies. Remove the plastic disk and expose it to radioactive anti-insulin antibodies, forming an "immunological sandwich" with the antigen (insulin) between two antibody molecules. Wash away any unattached radioactive antibodies and then make an autoradiograph of the disk. Images on the film can be used to identify the locations of insulin-secreting clones on the agar plate.

12.10.   More fragments are expected from *Hae*III because the probability of a specific four-base sequence is greater than the probability of a specific six-base sequence if the nucleotides are distributed along a chain in essentially a random order.

12.11. Insert the gene for lambda repressor protein immediately adjacent to the *lac* promoter in a plasmid. With no operator locus between these two genes, thousands of repressor molecules should be made per cell constitutively.

12.12. (*a*) Five kinds of cells: (1) bacteria that do not contain any plasmid DNA, (2) bacteria that took up the plasmid DNA, but do not contain any human DNA, (3) bacteria that contain the plasmid DNA with human DNA spliced in, but not the desired human gene sequence, (4) bacteria that contain plasmid DNA and a portion of the desired human gene, but not all of it, (5) bacteria that contain plasmid DNA and all of the desired human gene. (*b*) Approximately 600,000 recombinant phage, determined as follows.

$$n = \frac{3 \times 10^6 \, \text{kb}}{15 \, \text{kb}} = 2 \times 10^5$$
$$N = \frac{\ln(1 - 0.95)}{\ln\left(1 - \dfrac{1}{2 \times 10^5}\right)} = 599{,}140$$

12.13. Organisms that live in hot springs must have heat-stable enzymes. The first thermostable DNA polymerase used in automating PCR was isolated from a bacterium called *Thermus aquaticus* that normally lives in hot springs at a temperature of 70-80°C. Consequently, its polymerase is stable at temperatures near this point.

12.14. (*a*) The amino acid sequence can be used to construct a degenerate DNA probe (see Example 12.5) that can be labeled with radioactivity. (*b*) A cDNA library is most appropriate since this is a mammalian gene and it may contain introns. A genomic library will most likely allow the isolation of gene fragments, many of which may contain the introns. An expression library is more difficult to make, and useful if an antibody "probe" is being used.

12.15. (*a*) Steps in cDNA production: (1) isolate RNA from cells, (2) add reverse transcriptase enzyme, dNTPs and oligo-dT primers to make a first strand of cDNA, (3) degrade RNA, (4) add DNA polymerase I to make a second strand of cDNA, (5) clip the hairpin using S1 nuclease.

(*b*) Since the gene for protein P is 1000 bp or less in length, it is too small to clone into phage lambda, which requires fragments of approximately 15,000 bp. Thus, we should choose plasmid pBR322 and clone DNA fragments ranging from 2000 to 3000 bp. This will ensure that the entire gene will be found in one fragment, as well as its associated regulatory sequences.

(*c*) Cut the cDNAs and the pBR322 vector with *Pst*I enzyme, then ligate the cDNAs with the cut plasmid.

(*d*) By employing the replica plating technique, we can select cells that are ampicillin-sensitive and tetracycline-resistant. Only those cells that have incorporated the plasmid are tetracycline-resistant, and of these only the cells with an insert in the *amp*^{R} gene are ampicillin-sensitive.

(*e*) Blot each Petri plate with a nitrocellulose filter paper, thereby transferring some cells of each clone onto the filter for *in situ* hybridization. Lyse the cells and denature the DNA with a dilute sodium hydroxide solution. Single strands of the denatured DNA thus will stick to the filter. Flood the filter with the radioactive probe. After washing to remove any of the unhybridized probe, subject the filter to autoradiography. Select cells from the plate that correspond in position to the radioactive loci on the filter.

(*f*) Cut the plasmids with the *Pst*I enzyme. This should liberate the gene fragment, assuming that there are no *Pst*I sites within the gene fragment.

(*g*) The gene for protein P should hybridize specifically with the probe on a Southern blot. Also, by sequencing the gene fragment, it will be possible to ascertain the genetic code and then compare this to the original amino acid sequence obtained and any other previously cloned relatives of protein P using bioinformatics. If protein P has a measurable activity (e.g., enzymatic conversion of clear substrate into a color), the protein might be synthesized in an *in vitro* translation system and its activity checked. Finally, the gene might be cloned into an expression vector and the protein produced in *E. coli* (or another appropriate system). This protein could also be tested for activity.

12.16. After digestion with *Eco*RI, the normal gene for protein P should produce one band, 700 bp long. The mutant gene for an abnormal protein P should produce one band of at least 850 bp due to the loss of one of the *Eco*RI sites. The size of the band is dependent on where the next *Eco*RI site may reside in the neighboring flanking genomic DNA.

12.17. *Note*: The migration distance is not linear with increasing fragment length, but more like a log function.

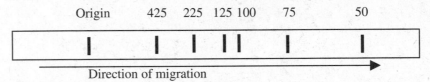

| Origin | 425 | 225 | 125 | 100 | 75 | 50 |

Direction of migration

Answers to Review Questions

Vocabulary
1. electrophoresis
2. denaturation or melting
3. renaturation or annealing
4. autoradiography
5. palindrome
6. restriction endonucleases
7. gene (DNA) library or genomic library
8. terminal (deoxyribonucleotidal) transferase
9. polymerase chain reaction
10. complementary DNA (cDNA)

Multiple-Choice Questions
1. *d* 2. *a* 3. *c* 4. *b* 5. *e* 6. *e* 7. *a* 8. *d* 9. *e* 10. *c*

CHAPTER 13

The Molecular Biology of Eukaryotes

The cells of fungi, protozoans, algae, plants, and animals contain a double-membraned organelle called the nucleus. The term eukaryote (eucaroyte is a variant spelling) refers to cells that contain such an organelle. The protoplasm between the nucleus and the plasma membrane constitutes the cell's cytoplasm. This division of the DNA from the rest of the cellular material is significant and dictates that the copying of DNA into RNA and the translation of RNA into protein are carried out in separate compartments, the nucleus and the cytoplasm, respectively. Other double-membrane-bound organelles exist in the cytoplasm, including mitochondria in both plants and animals and chloroplasts in plants. Bacteria (i.e., prokaryotes) do not have double-membraned organelles, but some may contain specialized membrane-bound structures made from cytoplasmic membrane invaginations or non-unit membranes, such as the **chlorosomes** involved in photosynthesis. Eukaryotic organisms can be single-cellular (i.e., yeasts) or multicellular (i.e., plants and animals) and their cells divide asexually by mitosis. Most undergo some form of sexual reproduction by the formation of gametes using meiosis followed by formation of a zygote that develops into a mature organism. Another important difference between prokaryotic and eukaryotic organisms is that eukaryotic genes are often interrupted by encoding sequences called **introns**. These introns are removed to connect the remaining, coding **exons** into a single mature mRNA (in a process called **splicing**) that is then translated into protein.

Because of the ease of growth, maintenance and manipulation, many single-celled prokaryotes, such as *E. coli*, and eukaryotes, such as yeasts, have been used for genetic studies of cellular function. These model systems have provided useful data regarding the information needed to carry out life. However, marked differences exist between prokaryotic and eukaryotic cells. This chapter will focus on the unique aspects of eukaryotic molecular cellular structure and function.

Genome Size and Complexity

As the new age of molecular genetics progresses, yielding more genome nucleotide sequences, the field of **comparative genomics** will provide new and enlightening information regarding genome function and evolution. Currently, over 73 microbial genomes have been sequenced, along with several insect (*Drosophila melanogaster*, mosquito), parasite (*Plasmodium*), worm (*Caenorhabditis elegans*), plant (*Arabidopsis*, corn, rice, potato), and mammal (human, mouse) genomes either completed or in progress. Information from some completed genomes is presented in Table 13-1.

Table 13-1.   Genomic Information of Various Organisms

Organism	Classification	Year Genome Completed	Mb in Genome	Approx. No. of Protein-Coding Genes
H. influenzae	Bacteria	1995	1.8	1700
E. coli	Bacteria	1997	4.6	4300
S. cerevisiae	Eukarya—yeast	1996	12.1	5900
C. elegans	Eukarya—nematode	1998	97	19,000
D. melanogaster	Eukarya—fly	2000	180	13,600
A. thaliana	Eukarya—plant	2000	100	25,000
H. sapiens	Eukarya—mammal	2001	3000	25,000–45,000

The genomes of eukaryotic organisms are typically much larger and more complex that those of prokaryotes. The human genome is estimated to contain between 25,000 and 45,000 protein coding genes and averages 3 billion base pairs (3000 Mb) in size, while many typical bacterial genomes, such as *E. coli*, encode only about 4000 genes in 4 million base pairs (megabases, Mb). Some plants and reptiles have even larger genomes; corn has a genome of 15,000 Mb. The complexity of the genome does not necessarily mirror the apparent complexity of the organism. For example, the total number of protein-coding genes in the *E. coli* genome is around 4000 while the multicelled fly genome contains just over three times that number (13,600). However, a fly seems outwardly to be more than three times as complex as a single-celled bacterium. Thus, organismal complexity is not simply a function of gene number. As organismal complexity increases, so does the size and complexity of gene structure. The average size of a bacterial gene is 1000 bp, while a mammalian gene can be as large as 600,000 bp (including exons and introns).

More complex organisms may have mechanisms that allow for the production of more than one protein from a single gene. An example of this type of mechanism is **alternative splicing**. Alternative splicing results in different proteins based on how the introns are spliced, or removed from the transcribed RNA to produce the mature mRNA. (See Regulation of mRNA processing section, p. 458.) Introns may not be spliced or spliced differently to produce unique mature mRNAs that, in turn, cause the translation of correspondingly unique proteins. It is estimated that the approximately 30,000 human genes are responsible for producing upwards of 100,000 (or more!) proteins.

Additionally, it is becoming apparent that as organisms become more complex, there is a higher degree of gene duplication in their genes and genomes. For example, *E. coli* is estimated to contain 1345 duplicated genes, while *Drosophila* contains 5536. Complexity in gene structure increases in terms of the number of domains present. A **domain** is a particular protein sequence element that generally is associated with a function (e.g., a DNA binding domain). The number of proteins in *Drosophila* that contain more than five domains (same or different) is around 100, while only 20 yeast proteins contain as many. Intron numbers also increase as organismal complexity increases; the yeast *S. cerevisiae* has a total of 220 introns, while *Drosophila* has 41,000.

Genes can be organized into families based on sequence similarity that generally translates to conservation in function. Several hundred gene families have been characterized, depending on the criteria. Families based on function such as transcription factors, DNA repair, protein kinases, transmembrane receptors, protein metabolism, etc., have been characterized in the fly and other organisms. Hundreds or thousands of genes may be placed in each gene family. For example, in the fly, nearly 4000 proteins have been categorized as cell growth and maintenance genes while only 57 are involved in DNA replication. Of the total number of genes identified from genome sequencing, anywhere from 30 to 50% do not have a hypothesized function. Function is generally assigned based on amino acid sequence similarity to proteins from other organisms whose function is known through previous genetic analysis (i.e., isolation of mutations and mutants).

Much of the eukaryotic genome contains **repetitive DNA** elements. Examples of repetitive DNA elements are telomeric sequences (at the ends of chromosomes), centromeric sequences, transposons, satellite DNA, rRNA and tRNA genes, and introns. For example, genes for rRNAs, tRNAs, and histones exist in tens, hundreds, or thousands of copies; and a few short DNA segments may be repeated over 10^5 times per genome (Table 13-2). Some genes may even contain highly repetitive DNA sequences. A recently characterized spider silk gene (*FLAG*) is composed of 11 repeating exons interspersed with nearly identical repeating introns. Each exon is ~ 1320 bp and is made up of virtually the same sequence. By contrast, prokaryotes have fewer repeated DNA sequences. The repetitive DNA of eukaryotes is just part of the genome that is generally known as noncoding. One estimate suggests that as much as 95% of the human genome does not code for proteins. This DNA has been referred to as "**junk DNA**" or "**selfish DNA**." The purpose of all this noncoding DNA has not been defined satisfactorily; however, it contains regulatory sequences, and ribosomal RNA and transfer RNA genes, as well as repetitive sequences. Additionally, noncoding DNA may play a role in generating genomic diversity necessary to fuel **genome evolution** and the generation of individual phenotypic diversity.

Table 13.2. Frequency Classes of Eukaryotic DNA Sequences

DNA Frequency Class	Number of Copies per Genome	Percentage of the Genome	Examples
Unique	1	10–80	Structural genes for ovalbumin, silk fibroin, hemoglobin
Middle repetitive	10^1–10^5	10–40	Genes for tRNA, rRNA, histones
Highly repetitive	$> 10^5$	0–50	Satellite sequences (5–300 nucleotides)

Organization of the Nuclear Genome

Functionally related bacterial genes are often clustered together in operons that produce **polycistronic mRNAs**. Eukaryotes have only **monocistronic cytoplasmic mRNAs** and their genes are not organized into operons. Many eukaryotic **repeated genes** that exist in multiple identical copies (e.g., genes for rRNAs, and tRNAs,) are clustered together on specific chromosomes, as components of multi-gene families. Other multiple-gene families may consist of a set of genes descended by duplication and mutation from one ancestral gene; they may be clustered together on the same chromosome or dispersed on different chromosomes. Such genes are usually coordinately controlled.

EXAMPLE 13.1

In humans, there are two families of hemoglobin genes. The alpha (α) family consists of a cluster of genes (including zeta [ζ], α_2, and α_1 on chromosome 16). The beta (β) family cluster on chromosome 11 includes epsilon (ε), gammas (γ^G, γ^A), delta (δ), and β. In addition, each family has one or more nonfunctional DNA sequences that are very similar to those of normal globin genes. These nonfunctional DNA gene-like sequences are referred to as **pseudogenes**. During the embryonic stage (less than 8 weeks) of development, the ζ- and ε-chains are synthesized. During the fetal period (8–41 weeks) the γ- and α-chains replace the embryonic chains. Beginning around birth and continuing for life, β-chains replace the gammas. A small fraction of adult hemoglobin has δ-chains in place of β-chains. The signals that control this switching on or off of the various hemoglobin genes is not known. The similarity in nucleotide structure of all these genes, however, suggests that early in evolution (perhaps 800 million years ago) a single ancestral globin gene began a series of duplications, followed by mutations and transpositions, to produce the two families and their multiple constituent genes and pseudogenes that exist today.

Gene Expression

1 TRANSCRIPTION

In contrast to the single RNA polymerase of prokaryotes, there are three such enzymes in eukaryotes, one for each major class of RNA. The enzyme that synthesizes rRNA is RNA polymerase I (pol I). The synthesis and processing of ribosomal RNA (rRNA) occurs in one or more specialized regions of the genome called **nucleoli**. Multiple copies of rRNA genes in tandem array are found in each nucleolus. The promoter regions for pol I lie upstream from the start site of transcription. A **Hogness box (TATA box)** lies within the promoter as the eukaryote analogue of the prokaryotic Pribnow box. The initiation of rRNA synthesis is highly species-specific; within a species, one or more proteins (essential for the transcription process) recognize promoters only in the rDNA of the same species.

RNA polymerase II (pol II) has its own specific initiation factors for synthesis of all eukaryotic mRNAs. Its promoters lie upstream from the start site of each gene, but the activity of the promoters may be increased by physically linked (i.e., in cis-position) DNA sequences called **enhancers**. Enhancers may function in either orientation, and may reside either within or upstream or downstream from their target genes (sometimes at great distances). The enhancing effect is mediated through sequence-specific DNA binding proteins. It is hypothesized that once the DNA binding protein attaches to the enhancer sequence, it causes the intervening nucleotides between the enhancer and the promoter to loop out and bring the enhancer into physical contact with the promoter of the gene it enhances. This loop structure then facilitates the attachment of RNA polymerase II molecules to the promoter of the transcribing gene.

The transcription termination signals for eukaryotic mRNA molecules are not known. RNA polymerase II continues elongating mRNA chains beyond the sequences found in mature mRNAs before termination occurs by an unknown mechanism. The transcript is then somehow specifically cleaved to form the correct 3' end.

Complex mechanisms (too involved to be presented here) ensure that the introns are removed from the pre-mRNA (primary transcript) and that the exons are spliced together in the proper order. Thereafter, the pre-mRNAs of eukaryotes undergo a number of covalent modifications before they are released from the nucleus as mature messenger molecules. The enzyme **poly-A polymerase** adds (without a template) a long stretch of adenine nucleotides to the 3' end of each pre-mRNA, forming a poly-A tail. Since only mRNA molecules (not rRNAs or tRNAs) have these tails, it might be suggested that they have something to do with translation. In contrast to most mRNAs, however, those for histone proteins in most species do not acquire poly-A tails. So the function of these tails remains a mystery. The 5' ends become "capped" with an unusual guanine nucleotide (3'-G-5'ppp5'-N-3'p). A methyl group is subsequently added to this backward guanine cap. Thus, both the 5' and 3' ends of most eukaryotic mRNAs possess free 2'- and 3'-OH groups on their terminal ribose sugars. Bacterial mRNAs contain specific ribosome binding sites in their leader sequences; eukaryotic mRNAs do not have these sites. Instead, a eukaryotic ribosome usually binds to the mRNA cap and then moves downstream along the mRNA until it encounters the first AUG initiation codon, and begins translation there.

Eukaryotic ribosomes, like their bacterial counterparts, consist of two major subunits, but they are more complex, existing as 40S and 60S subunits that together form an 80S complex. The rRNA components in more complex eukaryotes (e.g., angiosperms, vertebrates), having sedimentation coefficients of 18S, 5.8S, and 28S, are transcribed from 50 to 5000 identical genes tandemly arranged in that order into massive clusters located on one or more chromosomes as **nucleolar organizing regions (NORs)**. When active, these rRNA repeat units extend out from the main chromosome fiber as elongated threads (see Problem 13.23). When complexed with specific proteins involved in rRNA synthesis and processing, these clusters become visible under the light microscope as nucleoli, where the assembly of ribosomes begins. The number of NORs per haploid genome varies with the species from one to several. In *E. coli*, there are only seven copies of the rRNA genes. Very few bacterial genes exist in multiple copies, and even when they do the copy number is very small. As much as half of the eukaryotic primary rRNA transcript may be lost during processing of the mature rRNA molecule. Some of this loss is due to the removal of introns. In the protozoan ciliate *Tetrahymena thermophila*, *in vitro* rRNA transcripts appear to be self-splicing.

The eukaryotic genes encoding tRNAs generally also exist in multiple copies, from 10 to several hundred for each tRNA species per haploid genome. The identical genes within each tRNA family tend to be widely dispersed in species with relatively low numbers of tRNA gene copies. In organisms with more highly reiterated tRNA genes, they may form heteroclusters containing several kinds of tRNA genes. RNA polymerase III (pol III) is responsible for synthesizing not only all of the tRNAs but also 5S ribosomal RNA and other small RNAs. These transcripts are usually short (less than 300 nucleotides), with complementary end sequences that may allow formation of a stable base-paired stem. Sequences within the tRNA genes are required for transcription by pol III. Internal control regions (located inside the genes themselves) also direct termination of transcription by pol III. Thus, the same region can function both biosynthetically (in the gene) and structurally (in the RNA product).

2 TRANSLATION

The process of translating an mRNA into a polypeptide chain in eukaryotes is essentially the same as that in bacteria, but differs in several important ways. Whereas only three well-defined initiation factors are required for translation of *E. coli* mRNAs, many more are needed in eukaryotes. Eukaryotic initiation factors are designated eIFs to distinguish them from their bacterial counterparts. Other examples follow.

EXAMPLE 13.2
A tRNA^Met (symbolized Met-tRNA^Met when activated) brings an unformylated methionine into the first position on the ribosome. Hydrolysis of ATP to ADP is required for mRNA binding. The 40S ribosomal subunit is then thought to attach to the mRNA at its capped 5′ terminus, and then it slides along (consuming ATP) until it reaches the first AUG codon. Normally, only AUG is an efficient initiator codon in eukaryotes, whereas UUG, GUG, and AUU may also be used in *E. coli*.

EXAMPLE 13.3
Three different elongation factors (EFs) in eukaryotes replace those found in bacteria. However, a single termination factor (RF) replaces RF_1 and RF_2 of bacteria. RF recognizes all three stop codons (UAC, UAA, and UGA).

3 POSTTRANSLATION MODIFICATIONS

A nascent polypeptide chain may not become biologically active until after it has been modified in one or more specific ways, such as being enzymatically **phosphorylated**, **glycosylated**, or partly digested by **peptidase** enzymes. Phosphorylation involves the addition of one or more phosphate groups (Example 13.4) and glycosylation involves the addition of one or more carbohydrate groups to the protein sequence. Peptidase enzymes cleave the protein into smaller units (Example 13.5).

EXAMPLE 13.4
Protein kinases are enzymes that transfer terminal phosphate groups from ATP to specific amino acids on target proteins. **Phosphorylation** of these proteins may either raise or lower their biological activities. For example, the skeletal muscle enzyme glycogen synthetase is inactivated after phosphorylation, whereas phosphorylation of the enzyme glycogen phosphorylase increases its activity.

EXAMPLE 13.5
The hormone insulin is synthesized as a single-chain precursor (proinsulin) with little or no hormonal activity. Two internal cuts remove 31 amino acids from proinsulin, producing the two polypeptide chains of the functional dimer that are held together by disulfide bonds. Likewise, human growth hormone that circulates in blood is a "clipped" version of the pituitary form of that hormone.

Regulation of Gene Expression

Much less is known about gene regulation in eukaryotes than in prokaryotes. In contrast to bacteria, most eukaryotic cells (some algae, yeast, and protozoa are a few notable exceptions) are not free-living single cells. Multicellular eukaryotes usually show **cellular differentiation**. Differentiation allows cells to become specialized for certain tasks; e.g., liver cells are highly metabolic, muscle cells contract, nerve cells conduct impulses, red blood cells carry oxygen. The signals that cause eukaryotic cells to differentiate are largely endogenous (within the multicellular body). Eukaryotic cells cooperate with one another to maintain a fairly uniform internal environment despite variation in environmental conditions exterior to the organism; this regulatory phenomenon is known as **homeostasis**. Bacteria can turn their genes on or off repeatedly in response to various nutrients such as glucose or lactose in their environment. Switching genes on or off during development of eukaryotic cells, however, is usually a permanent change. Once a cell has started to differentiate, it can seldom be diverted to another developmental pathway.

Gene expression in eukaryotes involves six major steps that can each serve as a potential point for regulating protein production:

1. Uncoiling of nucleosomes
2. Transcription of DNA into RNA
3. Processing of the nuclear RNA (nRNA) or pre-mRNA
4. Transport of mRNA from nucleus to cytoplasm
5. Translation of mRNA into a polypeptide chain
6. Processing of the polypeptide chain into functional proteins

These six areas can be generally divided into three main points of control: transcriptional regulation (1, 2), regulation of RNA processing and (3, 4), and translational control (5, 6). Additional control strategies, such as multicopy genes, will also be discussed.

1 TRANSCRIPTIONAL REGULATION

Promoters and the genes that they control are generally adjacent; the promoter is the DNA sequence to which the RNA polymerase binds to begin transcription. Addition sites, called **enhancers**, may be several hundreds or thousands of base pairs either upstream or downstream from the promoters they stimulate. DNA binding proteins that bind to enhancer sequences are called **activators** or **repressors**. They interact with a series of other proteins that ultimately connects to the scaffold of proteins interacting at the promoter. Promoters and enhancers are also referred to as **cis-acting elements** because they are located on the same DNA strand as the gene they control. Enhancers can activate or repress the transcription of a gene. In yeast, enhancers are often referred to as **upstream activating elements**, or **UAS**s, because they are usually found upstream of a gene.

Unlike bacterial genes with similar functions, eukaryotic genes are not arranged under the control of one promoter. However, some genes are coordinately regulated despite their existence on completely different chromosomes. This coordinate regulation most likely occurs through a set of proteins called **transcription factors**. These proteins, encoded by separate and distinct genes, bind to specific DNA sequences in promoters to promote or repress the initiation of transcription. Thus, transcription factors belong to the class of proteins called **DNA binding proteins**. In this way, genes in different parts of the genome can still be coordinately controlled. These proteins are often referred to as **trans-acting factors** because they are encoded by genes at other locations in the genome (even on different DNA molecules). Most genes respond to more than one signal or combinations of signals, known as **combinatorial gene regulation**. Gene regulation can occur in response to either endogenous or exogenous signals.

(a) Exogenous Signals. Gene regulation in prokaryotes occurs mainly in response to exogenous signals such as the presence or absence of nutrients (e.g., glucose or lactose). Most gene regulation in eukaryotes occurs in response to endogenous signals produced by other cell types, but not exclusively so.

> **EXAMPLE 13.6**
> When plants are grown in darkness for several days they start to lose their green color (etiolation) because of loss of the enzymes that catalyze chlorophyll synthesis. Within a few hours after exposure of an etiolated plant to sunlight, more than 60 photosynthetic enzymes, chloroplast rRNA, and chlorophyll synthesis occur. A protein called **phytochrome** is covalently bound to a light-absorbing pigment. In the dark, phytochrome is inactive; in sunlight, it becomes activated and is thought to become a transcription factor for production of an unknown number of photosynthetic enzymes.

(b) Endogenous Signals. The best-known endogenous regulators of gene activity in eukaryotes are the **hormones**. These are substances produced by one cell type that have effects on other cell types. Hormones are usually transported throughout the organism (e.g., via the bloodstream in animals) but interact only with those cells that have the corresponding **receptors** on their cell surface. Some small hydrophobic molecules, such as steroids, may pass freely through the cell membrane and the hormone receptor could be in the cytoplasm or in the nucleus. The interaction of hormone and receptor eventually would cause a signal to be transmitted to the DNA at one or more specific sites to activate or repress the appropriate gene or set of genes.

> **EXAMPLE 13.7**
> Only the oviduct cells of the chicken respond to an injection of the steroid hormone estrogen by synthesizing ovalbumin mRNA. Other cell types fail to respond to estrogen because they lack the corresponding receptor. It is proposed that estrogen enters the cell by diffusion and binds to a cytoplasmic protein receptor. The hormone-receptor complex then migrates into the nucleus and initiates transcription of the ovalbumin gene.

A family of membrane proteins called **G proteins** are interposed between some signal molecules (e.g., hormones or neurotransmitters) and an "amplifier enzyme." If the hormone binds to a cell surface receptor, it induces a conformational change in the receptor. This change is transmitted through the cell membrane to a G protein, making it able to bind guanosine triphosphate (GTP); hence, the G in the name for these proteins. Binding of GTP causes a conformational change in the G protein that enables it to activate an amplifier enzyme. If the amplifier enzyme is adenyl cyclase, its activation results in the production of cyclic AMP (the second messenger). The cAMP can then regulate the activity of one or more genes coordinately.

Hormones might promote transcription by any of the following mechanisms:

1. The hormone could cause DNA to become uncoupled from histones (dissolution from nucleosomes) and thereby allow RNA polymerase to begin transcription.
2. The hormone might act as an inducer by inactivating a repressor molecule.
3. The hormone may bind directly to specific DNA sequences to facilitate binding of RNA polymerase or of a protein transcription factor.
4. The hormone may activate an effector protein (comparable to the CRP protein of the bacterial *lac* operon) so that the complex can bind to a site on the DNA and thereby stimulate binding of RNA polymerase.
5. The hormone could become attached to a protein already bound to DNA and thereby form an active complex that stimulates binding of RNA polymerase.

Modification of DNA nucleotides may play a role in regulating gene transcription. Some genes whose products are usually synthesized only in particular cell types (e.g., hemoglobin in erythrocytes; immunoglobulins in plasma cells) appear to be heavily **methylated** in cells that do not express the corresponding gene products and unmethylated in cells where those genes are expressed. The genes involved in general metabolism common to all cells (called **housekeeping genes**) are rarely methylated in or near their initiation regions. This mechanism of transcriptional regulation is carried out by proteins, such as the mammalian MBD1 protein, that bind preferentially to methylated cytosine residues at CpG DNA sequence islands. The binding or nonbinding affects transcription in a similar manner to other transcription factors.

2 REGULATION OF mRNA PROCESSING

Eukaryotic genes contain introns (noncoding regions) interspersed among the coding regions (exons). Part of the process that converts primary transcripts to mature mRNA molecules involves removal of the introns and splicing of the exons together. Variations in the excision and splicing jobs can lead to different mRNAs and, following translation, to different protein products. This type of gene regulation is called **alternative splicing**. Alternative splicing can play a significant role in developmental processes, such as sex determination in the fruit fly and production of immunoglobulin genes in mammals.

EXAMPLE 13.8
Immunoglobulins of class IgM have μ-type heavy chains that are produced in two varieties as a consequence of differences in intron/exon excision/splicing events. When processed in one way, the μ-chains are longer, ending with a group of hydrophobic amino acids at their carboxyl ends. This "water-fearing" tail tends to lodge in the lipid membrane and extends into the cytoplasm. The amino terminus extends outside the cell where it participates with an L chain to form an antigen-combining site (see Example 13.21). Thus, it becomes a cell receptor for a specific antigen. An alternative splicing step removes from the primary transcript the

sequence responsible for the hydrophobic tail. This shorter version of the μ-chain readily passes out of the cell and becomes part of the secretory antibody population found in the blood and other body fluids.

3 REGULATION OF TRANSLATION, PROTEIN STABILITY, AND ACTIVITY

There are three major methods by which eukaryotic cells are known to regulate protein production at the translational level: (1) by altering the half-life, or stability, of the mRNA, (2) by controlling the initiation and rate of translation, and (3) by modification of the protein after translation.

A typical eukaryotic mature mRNA consists of four major regions: (1) a 5' noncoding region (leader), (2) a coding region, (3) a 3' noncoding region (trailer), and (4) a poly-A tail. Each of the four segments may affect the half-life of mRNA molecules.

EXAMPLE 13.9
The mRNA transcribed from a normal human gene *c-myc* is relatively unstable, with a half-life of about 10 min. A mutant form of *c-myc* that is missing some of the 5' noncoding region of the normal *c-myc* produces an mRNA that is 3–5 times more stable than full-length mRNA.

EXAMPLE 13.10
Within the coding region of a histone gene, repositioning of the stop codon closer to the 5' end of its transcript not only produces abnormally short histone proteins but also at least doubles the half-life of the mutant mRNAs.

EXAMPLE 13.11
The mRNAs for human β-globin and δ-globin (Example 13.1) differ mainly in their 3' noncoding segments, yet δ-globin mRNA is degraded four times faster than β-globin mRNA.

EXAMPLE 13.12
Mature mRNA molecules do not normally exist as naked mRNAs but as ribonucleoprotein. One of the proteins normally bound to mRNAs is a **poly(A)-binding protein** (PABP). Experimental removal of PABP from normal mRNAs decreases their half-lives. Removal of poly-A tails from otherwise normal mRNAs greatly reduces their half-lives. Just how these changes in mRNA molecules influence their susceptibility to digestion by ribonuclease enzymes is not presently known.

EXAMPLE 13.13
Unfertilized sea urchin eggs store large quantities of mRNA complexed with proteins as ribonucleoprotein particles. In this inactive form it is called **masked mRNA**. Within minutes after fertilization, the mRNA somehow becomes "unmasked" and translation begins.

The 5′ untranslated regions (UTR) of mRNA molecules (leader sequence) can serve as regulators of translational initiation. This process is generally mediated by the presence or absence of a particular nutrient or metabolite. For example, there is a sequence present in the 5′ UTR of the human ferritin gene, called an iron-responsive element (IRE), that responds to the presence and absence of iron. Ferritin is a molecule involved in iron storage. When iron is absent, a protein, called the IRE-BP, can bind to the IRE sequence in the 5′ UTR of ferritin. This prevents efficient translation of the ferritin mRNA. However, when iron is present, the IRE-BP can no longer bind to the IRE and translation can proceed efficiently.

Posttranslational modifications, such as **ubiquitination**, can target proteins for proteolysis. **Ubiquitin** is a small protein that when covalently attached to target proteins signals their destruction by a complex of proteins known as the **proteosome**. Many genes involved in cell cycle regulation are quickly destroyed by this mechanism, allowing newly produced proteins to carry out the next step. Modifications such as phosphorylation are mechanisms that regulate protein activity that may lead to regulation of protein production. For example, some proteins are only active (i.e., can carry out their enzymatic or DNA binding capabilities) if they are phosphorylated on particular amino acid residues. Phosphorylation is carried out by enzymes called **kinases**. Phosphate residues can be removed by enzymes called **dephosphorylases**. In complex systems, there is often a cascade of kinases and dephosphorylases that activates a series of protein targets, leading ultimately to a transcription factor. The transcription factor then becomes activated (due to phosphorylation or dephosphorylation), resulting in the regulation of transcription of a particular gene or set of genes.

Another posttranslational control mechanism involves protein processing. Eukaryotes synthesize only monocistronic mRNAs, but the resulting single polypeptide chains may be cleaved into two or more functional protein components. A multicomponent protein such as this is termed a **polyprotein**.

EXAMPLE 13.14

A polyprotein called pro-opiomelanocortin is synthesized by the anterior lobe of the pituitary gland. A cut near the C (carboxyl) terminus first produces β-lipotropin. Then a cut near the N (amino) terminus produces adrenocorticotropic hormone (ACTH). In the intermediate lobe of the pituitary, β-lipotropin is further digested, releasing the C-terminal peptide β-endorphin; the ACTH is also cleaved to release α-melanotropin. Polypeptides that are destined to be released from the cell (after being processed in the Golgi apparatus) possess a **signal peptide**. This peptide usually consists of about 20 amino acids at or near the N terminus of a polypeptide chain. It serves to anchor the nascent polypeptide (as it is being synthesized) and its ribosome to the endoplasmic reticulum.

4 MULTICOPY GENES

The abundance of a species of RNA molecules may be regulated at the gene level by several mechanisms that serve to amplify the copy number of the gene. In order to understand some processes of selective gene amplification, a distinction must be made between germ-line genes (those that are passed on to offspring) and somatic genes (not hereditary). Evolution progresses by sequential modifications of pre-existing developmental patterns. Whatever mechanism that works initially to solve a biological problem tends to become so integrated into the overall developmental program with the passage of many generations that it cannot be changed thereafter. Thus, it is not surprising that different organisms may use quite different mechanisms to solve common biological problems.

EXAMPLE 13.15
In the ciliate protozoans, there are two kinds of nuclei: a polyploid somatic macronucleus (controlling all transcription during vegetative growth and asexual reproduction) and a haploid micronucleus containing the germ line. Fusion of haploid nuclei from conjugation of opposite mating types produces a diploid zygotic nucleus. The old macronucleus then degenerates and the zygotic nucleus divides to produce a new haploid micronucleus and an immature macronucleus. The macronuclear genome then becomes polyploid like the polytene chromosomes of *Drosophila*. The macronuclear chromosomes, however, become highly fragmented, and most of the fragments (up to 95% in some species) are degraded. The surviving fragments contain the genes required for vegetative growth and asexual reproduction. The mechanisms controlling this selective degradation and the distribution of surviving fragments into progeny cells during cell division are essentially unknown.

EXAMPLE 13.16
Amphibian oocytes contain a hundred to a thousand times more rRNA genes than are found in somatic cells, almost all of the increase being due to large numbers of extrachromosomal nucleoli. Each nucleolus contains one or more circular DNA molecules having 1–20 tandemly arranged rRNA genes coding for the 45S rRNA precursor. Most of these nucleolar circles are produced by the rolling-circle mechanism (as discussed in Chapter 10). These circles contain somatic genes and cannot replicate themselves. Extrachromosomal rRNA genes must be derived from the tandemly repeated rRNA germ-line genes.

Unlike the amphibian rRNA genes in Example 13.16, the chorion (eggshell) genes of *Drosophila* can be amplified without extrachromosomal replication. A large number of follicle cells surround the egg and produce the chorion. The genes encoding the chorionic proteins exist in two clusters (one on the X chromosome and one on an autosome). Only a single copy of each somatic gene is present. A developmentally controlled origin of replication, located within each gene cluster, is programmed to fire 3–6 times during interphase within the 5 h of choriogenesis. The process shown in Fig. 13-1 usually produces a 32- to 64-fold amplification of the chorion genes.

Development

The term **ontogeny** represents the development of an individual from zygote to maturity; **embryology** is the study of early ontogenic events. **Epigenesis** is the modern concept that development of differentiated cells, tissues, and organs occurs by cells acquiring new structures and functions while increasing in size

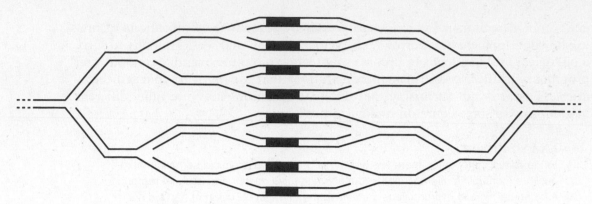

Fig. 13-1. Gene amplification of chorion genes (dark segments) in *Drosophila*. Three rounds of replication from a single replication origin near the chorion genes have occurred.

and complexity, as opposed to the now discredited notion that an organism develops simply by growth of tiny entities that are essentially fully formed in the fertilized egg (**preformation theory**).

1 DETERMINATION AND DIFFERENTIATION

After fertilization, the new diploid nucleus exists in a primarily maternal cytoplasm. This environment is critical for proper development of the embryo. There are mRNAs and proteins present, supplied by the maternal cytoplasm, that help guide the developing zygote. These mRNAs and proteins are encoded by **maternal effect genes** because mutations in these genes only have an effect if they originate in the mother. The nucleus begins to divide and **zygotic genes** are expressed. This further establishes local environments that are important for development. Ultimately, the cell number of the embryo begins to increase, and groups of cells are determined to become a particular organ or tissue (e.g., muscle cells that contract, neurons that transmit impulses, fibroblast cells that manufacture extracellular collagen or elastic fibers). This process is called cell fate **determination**. As subsequent gene expression programs ensue, fated cells become differentiated into their final functional state. Thus, development involves the **differentiation** of cells into specific types and tissues.

One of the primary mechanisms governing the developmental process is transcriptional control. It is thought that relatively few master control switches exist, and that they are arranged in a hierarchy, with early-acting genes controlling the expression of other genes that act at later times in development. The products of these master control genes that determine specific developmental pathways are called **morphogens**. Morphogens exert their effects through a gradient of their concentration. **Induction** is the determination of the developmental fate of one cell mass by another. This morphogenetic effect is brought about by a living part of an embryo (called an **inducer** or **organizer**) acting upon another part (**competent tissue**) via one or more morphogens. An undifferentiated cell may, under the influence of a mutant master control gene, follow a developmental pathway different from that which it normally would pursue (**transdetermination**), usually with bizarre (if not lethal) consequences.

Much of our understanding of development has come from studies of the fruit fly *Drosophila*. Many aspects of development are similar for eukaryotes in general (i.e., epigenesis, determination, and differentiation); however, some aspects are not shared. For example, maternal effect mutations have not been isolated in some genetic model systems (e.g., *Arabidopsis*) and development in *C. elegans* is highly dependent on cell-cell contact and communication, as opposed to gradients of morphogens.

EXAMPLE 13.17

In the fruit fly *Drosophila*, the genes that control development of its body plan can be grouped into three classes. **Maternal effect genes** are those genes of the mother that establish the organization of the egg through gradients of concentration. The embryo contains **segmentation genes** that establish the segmentation pattern of the fly. The embryo's **homeotic genes** switch on after the segmentation genes and establish the kind of structure that will develop in each body segment. Mutations in homeotic genes can cause a normal body part to develop in an abnormal location, e.g., a leglike antenna in *Drosophila*. A common conserved DNA sequence of about 180 bp (called a **homeobox**) is shared in most of the known homeotic genes and with at least some of the segmentation genes. The homeobox motif is usually repeated several times in one gene and gives rise to a 60 amino acid domain called a **homeodomain**. Homeodomains are very basic and have a helix turn helix motif that characterizes several well-known DNA binding proteins (e.g., CAP and lambda phage repressor). Homeotic genes act as transcriptional regulators that are thought to form a network of master control genes that switch on batteries of other genes whose activities specify the kind of body structure that will develop. Similar homeoboxes have been found in other invertebrates as well as in some vertebrates, including mammals, in which they are called *Hox* genes. But at present it is not known if they function in the same way as in *Drosophila*.

Maternal effect genes may provide certain substances or organize the egg cytoplasm in such a way that development of certain progeny phenotypes is essentially totally controlled by the maternal genotype rather than by the genotype of the embryo. Such effects may be ephemeral or may persist throughout the life of an individual. The substances that produce maternal effects are not self-perpetuating, and therefore must be synthesized anew for each generation of progeny by the appropriate maternal genotype.

EXAMPLE 13.18

A dominant gene *K* in the meal moth *Ephestia* produces a hormone-like substance called **kynurenine** that is involved in pigment synthesis. The recessive genotype *kk* is devoid of kynurenine and cannot synthesize pigment. Females of genotype *Kk* can produce *k*-bearing eggs containing a small amount of kynurenine. For a short time during early development, a larva may use this supply of kynurenine to develop pigment even though its own genotype might be *kk*. The color fades as the larva grows older because the maternally supplied kynurenine becomes depleted.

EXAMPLE 13.19

The direction in which the shell coils in the snail *Limnaea* can be dextral like a right-hand screw or sinistral like a left-hand screw. The maternal genotype organizes the cytoplasm of the egg in such a way that cleavage of the zygote will follow either of these two patterns regardless of the genotype of the zygote. If the mother has the dominant gene s^+, all her progeny will coil dextrally; if she is of the genotype *ss*, all her progeny will coil sinistrally. This coiling pattern persists for the life of the individual. See Solved Problem 13.2.

EXAMPLE 13.20

There are three important maternal effect genes in the fruit fly: *bicoid*, *hunchback*, and *nanos*. The bicoid and hunchback proteins are localized to the anterior portion of the egg and establish an anterior polarity in the egg. Bicoid is a transcription factor that activates the expression of *hunchback*. Hunchback, in turn, is a transcription factor that activates the transcription of other genes that are involved in the formation of head and thorax structures. Hunchback also represses the production of genes involved in posterior structures. The nanos protein is localized in the posterior of the egg. It acts as a translational repressor of *hunchback* mRNA. This suppresses the development of anterior features in the posterior of the egg and allows for expression of posterior feature regulators.

Depending on the signals it receives, the cell type and the species, a differentiated cell may or may not be able to **dedifferentiate** (revert to an unspecialized state). Differentiation is usually reversible at the nuclear level, as evidenced from nuclear transplantation experiments. However, fully differentiated cells often are incapable of replication. For example, spinal nerve cells, mature red blood cells (erythrocytes) that carry oxygen, and plasma cells that make antibodies can no longer divide. The undifferentiated **stem cells** from which mature blood cells are derived retain the capacity to replicate and differentiate into various blood cell types. Differentiation is seldom due to gain or loss of chromosomes or of genetic material (lymphocytes are a notable exception).

EXAMPLE 13.21

Antibodies are made by white blood cells (lymphocytes) known as plasma cells. Lymphoid stem cells in the bone marrow differentiate into B cells that can complete their maturation into antibody-secreting plasma cells after making specific contact with an antigen via their membrane receptor (an antibody molecule). An immunoglobulin (antibody) molecule is a tetramer composed of two identical heavy (H) polypeptide chains and two identical light (L) chains. There are five classes of immunoglobulin molecules (IgG, IgM, IgA, IgE, and IgD) based upon the structure of their heavy chains (γ, μ, α, ε, and δ, respectively). There are only two types of L chains (κ and λ). The carboxy ends of heavy and light chains possess amino acid sequences (called constant regions, designated "C") that are invariate within each H chain class and L chain type. The free amino ends of each chain differ in amino acid sequence and are referred to as variable ("V") regions. The V regions of an L chain and an H chain together form an antigen-binding site. The C_H region consists of three or four similar segments, presumably evolved by duplication of an ancestral gene, followed by subsequent mutational modifications; these similar segments are called "domains" and are labeled C_{H1}, C_{H2}, C_{H3}.

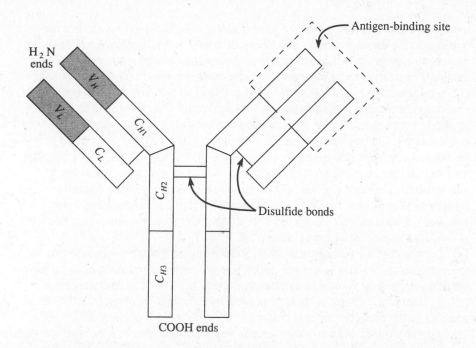

A mature plasma cell produces antibodies bearing a single class of H chain and a single class of L chain, hence also a single antigen-binding specificity. Although an individual may inherit different genes for the H and L chains, an unknown mechanism allows expression of only one gene for each of these chains, a phenomenon known as **allelic exclusion**. The first antibodies produced by a plasma cell are usually of class IgM. Later in that same cell, the same antigen-binding specificity may be associated with H chains of a different class (e.g., IgG or IgA). At any given time, however, a plasma cell is thought to synthesize primary mRNA H-chain transcripts of a single kind.

There are three immunoglobulin gene families: two for the light chain types (κ and λ) and one for the heavy chains, each on a different human autosome. Within each gene family, there usually are multiple DNA sequences (sometimes hundreds) coding for the V region of an immunoglobulin chain. There are also one or more sequences coding for the C region of that same chain. In embryonic lymphoid cells, the V and C segments of a gene family that ultimately code for a given immunoglobulin chain are not adjacent to one another, but are only loosely linked. As the cell matures into an antibody-secreting plasma cell, the V and C sequences become more tightly linked. The V and C sequences are "exons" that code for portions of the immunoglobulin polypeptide chain. An apparently random choice is made to connect one of the V exons to one of the C exons, and all unnecessary intervening material (exons or introns) is deleted (sometimes at the DNA level). Between the V and C regions there are a few J (for "joining") exons in both L- and H-chain families; the H-chain family also contains a few additional D (for "diversity") exons. These J and D segments contribute to hypervariable regions (also known as complementarity-determining regions, CDRs) that form part of an antigen-binding cavity in an immunoglobulin molecule. A light-chain V exon is joined to a J exon by a single recombination event. The V-J complex is then connected to a C exon at the level of mRNA by the standard RNA-splicing mechanism. Two recombination events are required to assemble a heavy-chain gene. The first event joins the J and D exons; the second event joins the V exon with the D-J complex to form a V-D-J complex. Since the joining of V-J or V-D-J is imprecise, this produces the phenomenon termed **junctional diversity** in the possible kinds of immunoglobulin chains. As a heavy-chain gene is being assembled, extra nucleotides (called **N regions**) can be inserted in a template-free fashion between the V-D or D-J segments. The random association of any V exon with any J or J-D complex is called **combinatorial translocation**. A V-D-J complex can be coupled to a C exon by either of two mechanisms. RNA splicing can connect the V-D-J group with one of the nearest C exons (either μ, or δ). Alternatively, the V-D-J group can be connected to more remote C exons (γ, α, or ε) by a third DNA recombination; this latter mechanism is known as **class switching**. A high level of point mutations in fully assembled antibody genes is another source of diversity called **somatic hypermutation**. In the formation of a tetrameric immunoglobulin molecule, two identical L chains can be associated with any two identical H chains, an option known as **combinatorial association**.

Estimates of the number of immunoglobulin components in the mouse are as follows. L chains (κ-type only) have 250 V and 4 J regions, and three sites for junctional diversity; total number of κL chains = $250 \times 4 \times 3 = 3000$. H chains have 250 V, 10 D and 4 J regions, plus three sites for junctional diversity at both V-D and D-J joints; total number of H chains = $250 \times 10 \times 4 \times 3 \times 3 = 90,000$. Combinatorial association of 3000 L chains with 90,000 H chains = 2.7×10^8 possible antibody molecules. This is an underestimate because it does not consider lambda L chains, N regions, somatic hypermutation, or the five classes (C exons) of the heavy chains. Moreover, in humans there are four different C_H exons for the four subclasses (unrelated to the number of C_H domains) of IgG, two each for IgM and IgA, and one each for IgD and IgE; for L chains there are four C_L exons for the four subtypes of the lambda family and one in the kappa family. Different C_H classes endow the immunoglobulin molecules with special effector functions such as complement binding (IgG and IgM), placental passage (IgG), secretion into body fluids (IgA), and binding to mast cells (IgE). Thus, the union of one kind of variable region with one kind of constant region in H chains contributes to an antibody-combining site that specifically binds antigen and also allows the immunoglobulin molecule to become biologically active.

SOLVED PROBLEM 13.1

A single cell, the fertilized egg, is **totipotent**; i.e., it has the capacity to produce a complete, normal adult individual. Repetitive mitotic divisions convert the zygote into the multicelled organism. During this cellular proliferation, many cells differentiate into types with different morphologies and physiological functions. These differences are associated with the different kinds of proteins made by these cells. For example, the protein hormone insulin is made only by the beta cells in the islets of Langerhans in the pancreas, whereas hemoglobin is made only by erythropoietic cells.

(a) Explain how different proteins are made by different cell types, given your knowledge of development process regulation.

(b) Are differentiated cells totipotent? Devise an experiment that might provide a positive answer to this question.

(c) In an experiment of the kind described in part (b), if the egg nucleus is exposed to ultraviolet light, a positive result might be due to failure of the radiation to destroy the native egg nucleus. Propose an experiment that might prove that this was not the cause of the positive result.

Solution:

(*a*) Different groups of genes are silenced or activated in each cell type, leading to a specific program of gene expression specific to that cell type. Translational regulation of mRNA molecules may also occur.

(*b*) Remove (by micropipette) or destroy (e.g., by radiation) the nucleus of a fertilized egg. Then transplant a diploid nucleus from a differentiated cell of the same species into the enucleated egg. If a complete, normal adult organism can develop from such an egg, then development in this species must be totipotent. We cannot generalize these results to all species because different species may not give similar results in such transplant experiments.

(*c*) Transplant a conspecific (same species) nucleus containing a genetic marker that differs from that of the recipient individual. If all cells of the resulting adult organism contain only the marker of the transplant, the native egg nucleus must have been destroyed by the ultraviolet light treatment.

2 APOPTOSIS

The phenomenon of programmed cell death is termed **apoptosis**. Apoptosis is necessary in several developmental processes to eliminate certain cells in an area where more specialized or different cell types will develop. For example, in insect morphogenesis, the cells of the larval stage must die and hydrolyze in order for entirely new cells with new functions to form. Apoptosis involves cell compaction, chromatin condensation, DNA fragmentation, and membrane blebbing. Dying cells ultimately fall apart and are engulfed by surrounding cells.

EXAMPLE 13.22

In the worm, *C. elegans*, 131 of the 1090 cells die predictably during the development of the mature organism. The programmed cell death of these cells is governed primarily by three genes. The *ced-3* and *ced-4* genes cause cells to die and the *ced-9* gene regulates their expression. Cells expressing *ced-9* survive and cells lacking *ced-9* die. Thus, *ced-9* blocks the expression of *ced-3* and *ced-4*.

The overexpression of genes homologous (similar) to *ced-9* in humans can lead to cancerous growth. Cancer is basically the dedifferentiation and division of terminally differentiated and normally nondividing cells. Genes whose inappropriate expression leads to cancer are called **oncogenes**. Inappropriate expression of genes similar to *ced-3* and *ced-4* can lead to degenerative diseases in which tissue cells are instructed to die early.

Somatic Nuclear Transfer and Cloning

Somatic nuclear transfer is a process by which the nucleus of a somatic (nongerm cell) is transferred into the cytoplasm of an enucleated oocyte to create a **clone** that is genetically identical to the organism from which the somatic cell nucleus was derived. The transfer can be done using a technique called **electrofusion**, where the enucleated oocyte and the specially treated somatic cell are forced to fuse with a jolt of electricity. The chimeric oocyte is then placed in a surrogate womb for embryonic development. Nuclear transfer was pioneered over 40 years ago in the frog, but was not successfully realized in mammals until 1996 with the cloning of a sheep named **Dolly**. Since the cloning of Dolly, somatic cell clones of cattle, goats, mice, and pigs have also been accomplished. The potential benefits of this technology is that cloned animals may be used for the production of therapeutic proteins or human-compatible tissues and organs. The production of therapeutic proteins would be accomplished by first creating a recombinant somatic cell nucleus (i.e., one that contains the gene encoding the desired therapeutic), then transferring this nucleus into the enucleated oocyte. This results in a **transgenic animal** clone.

> **EXAMPLE 13.23**
> The gene for human coagulation factor IX, missing in hemophiliacs, has been cloned into somatic sheep cells with genetic controls for expression in sheep milk. These somatic cells were used to create somatic cell sheep clones that expressed the protein in their milk. Current treatment for hemophiliacs is primarily by purifying factor IX from human blood. This is problematic in that it is expensive and has inherent disease risks (e.g., AIDS). Cloning proponents suggest that production of recombinant proteins in livestock milk may be cheaper and more easily regulated.

Human-compatible organs and tissues can potentially be created from a complicated technology involving either (1) genetic engineering of animals, such as pigs, to destroy problematic transplantation antigens on animal organs, or (2) by using embryonic stem cells to create human organs *in vitro* or in other species. The process of transferring an organ from one species into another is called **xenotransplantation**. Stem cells are **pluripotent** (i.e., have the potential to develop as multiple cell types) cells that are derived from various tissues; very early embryonic tissue yields totipotent embryonic stem cells. Currently, the ethics of stem cell research is a hot topic of debate and it is outlawed in many countries. The major controversy is centered on the creation of human embryos for research purposes. If the stem cells are used to help create organs, the embryo is destroyed. It should be emphasized that neither of these technologies is yet fully realized nor have the ethical ramifications been resolved. In addition, there are potential problems with cross-species diseases coming from the recombinant animal into humans. These issues will need to be resolved before any of this technology is applied.

Organelles

Mitochondria and chloroplasts have evolved for specialized functions in eukaryotic cells. Mitochondria are thought of as the powerhouse of the cell because many enzymes involved in cellular respiration and ATP production are located there. Chloroplasts serve as the site of photosynthesis in plants. Both mitochondria and chloroplasts share several characteristics with modern prokaryotic cells. All three generally have a circular double-stranded DNA genome (exceptions include some protozoans such as *Paramecium* and *Tetrahymena* that have linear mitochondrial DNA molecules). Their genomes are neither enclosed within a nuclear membrane nor associated with histone proteins (hence no nucleosome organization). They each code for part of their own protein-synthesizing systems (all rRNAs, tRNAs, and at least some of the ribosomal proteins). Many of the enzymes and other proteins that function in these organelles, however, are encoded by nuclear genes, synthesized on 80S ribosomes, and transported into these organelles. Their ribosomes are usually 70S or smaller and are sensitive to antibiotics and other substances that have no effect on the 80S eukaryotic cytoplasmic ribosomes. Protein synthesis is initiated by formyl-methionyl-tRNA. The nucleus, mitochondrion, and chloroplast are each bounded by a double-membrane envelope, but only the nuclear membrane contains pores. Mitochondria and chloroplasts grow in size and then seem to split in two, in a process akin to binary fission in bacteria.

1   MITOCHONDRIA

Mitochondria are organelles found in the cytoplasm of both plants and animals. They contain the enzymes of the electron-transport chain that carry out oxidative phosphorylation in the production of adenosine triphosphate (ATP, the main source for energy-requiring biochemical reactions). Unlike chloroplasts, the mitochondrial genome (**mtDNA**) varies markedly in length between species. For example, in fungi, such as the yeast *Saccharomyces cerevisiae*, the mtDNA is about 86 kb, and in most metazoan (multicellular) animals it is on average 16 kb. Much of the mtDNA of fungi and plants is thought to be noncoding (perhaps "junk" or "selfish" DNA). Animal mitochondrial genomes typically encode the same 37 proteins: 2 rRNA genes, 22 tRNA genes, and 13 protein-coding genes. The proteins encoded are involved in respiration (i.e., cytochrome oxidase), DNA replication, transcription, and translation (e.g., ribosomal proteins). Mitochondrial genomes are also typically AT rich ($\sim 70\%$).

One or more mitochondrial DNA molecules resides within each of the several nucleoid regions within the mitochondrion. If a cell contains 250 mitochondria, each with 5 mtDNA molecules, there will be 1250 mtDNA copies in that cell. Mitochondrial ribosomes are also highly variable between species (e.g., 55S in animals, 73S in yeast). There is also some interspecific variation in mitochondrial tRNAs. Some codons are read differently by mitochondrial tRNAs than by nuclear-encoded tRNAs. For example, AUA codes for methionine (not isoleucine) and UGA codes for tryptophan (not translation termination) in mammalian mitochondria. Mitochondrial mRNAs of fungi and higher plants contain introns, but those of animals lack introns and are transcribed as polycistronic mRNAs that become cut into monocistronic mRNAs before translation. Mitochondria have no DNA repair systems. Hence, the mutation rate of mitochondrial DNA is much higher than that of nuclear DNA.

2   CHLOROPLASTS

Chloroplasts contain the enzymes for photosynthesis and are thus characteristic only of plant cells. Most plant cells contain numerous chloroplasts. A few plants such as the unicellular alga *Chlamydomonas* (see Fig. 5-1) contain a single chloroplast. In most plants, however, each chloroplast genome is usually present in multiple copies. For example, a typical leaf cell of *Euglena* may contain 40–50 chloroplasts. Every chloroplast usually contains several nucleoid regions, each containing 8-10 DNA molecules; thus the entire cell may contain over 500 copies of the chloroplast genome (**ctDNA**). The length of a typical plant chloroplast genome is 120–150 kb of DNA. The number of protein-coding genes contained within ctDNA ranges from 46 to 90. The majority of these proteins are involved in photosynthesis, with the remainder being involved in replication, division, transcription, translation, and biosynthesis. There are also two genes for rRNAs and over 30 tRNA genes. Evidence indicates that ctDNA from liverworts to the higher plants have essentially the same genome (highly conserved). Some of the ctDNA genes (both for tRNAs and mRNAs) are known to contain introns. The RNA polymerase of the liverwort *Marchantia polymorpha* contains α- and β-subunits that are homologous in amino acid sequences to those found in the bacterium *E. coli*.

3   ORIGIN OF ORGANELLES

The most supported idea regarding the origin of organelles is the **endosymbiosis** or **endosymbiont theory.** This theory proposes that eukaryotic organelles arose as a result of symbiotic relationships between early bacterial cells. Several lines of evidence support the theory that mitochondria evolved from Eubacteria, or true bacteria, rather than the Archaea. According to the endosymbiosis theory, a primitive anaerobic-phagocytic type of nucleated cell (called the **urkaryote**) engulfed an aerobic bacterium (the **progenote**) that was able to generate energy by oxidative phosphorylation. The engulfed bacterium somehow escaped digestion and replicated within the cytoplasm. These early symbiotic relationships gradually evolved into a mutualism whereby they could not survive apart from one another. During the evolution of this organelle, the bacteria gave up many of its genes to the nucleus, so that now many of the proteins needed for mitochondrial functions are specified by nuclear genes, made on cytoplasmic ribosomes, and transported into the mitochondria. This is how fully aerobic, nucleated cells (like modern eukaryotic cells) are proposed to have evolved.

At some later time, some of these fully aerobic, nucleated cells may have engulfed photosynthetic **cyanobacteria** (blue-green "algae"). A mutualism gradually developed between these two entities in the evolution of the chloroplasts that characterize the plant kingdom.

There are no real clues in extant organisms as to the evolution of the nucleus. It is thought that the eukaryotic nuclear membrane probably evolved independently of the prokaryotes, possibly by invaginations and coalescences of the cell membrane. The nucleus is a double-membraned organelle like mitochondria and chloroplasts, so any theory regarding the origins of these organelles would have to take this into consideration.

Although the shape of mitochondria is different from that of the bacteria from which they presumably were derived, the mitochondria resemble bacteria in many ways. Both of their genomes are circular and histone-free. Their transcription and translation systems are also similar. On the other hand, some archaebacterial genes (like those in the eukaryote nucleus) have introns. But introns are unknown in modern Eubacteria. Hence, it has been suggested that the progenote may have had introns that were lost during the evolution of the Eubacteria. Interestingly, the mitochondrial DNA of mammalian cells does not contain introns, but many mitochondrial genomes of more primitive eukaryotes do. In addition, Eubacteria and eukaryotes contain ester-linked, unbranched lipids containing L-glycerophosphate, whereas the branched lipids of Archaebacteria are ether-linked and contain D-glycerophosphate. Finally, with the advent of DNA-sequencing technology, mitochondrial gene sequences can be compared with bacterial gene sequences. These investigations are showing that there is a high degree of related-ness between mtDNA genes and bacterial genes.

4   INHERITANCE OF ORGANELLES

In most plants and animals, mitochondria and chloroplasts are strictly inherited only from the female parent (**maternal transmission**) because the male gamete (or that part that enters into fertilization) is essentially devoid of these organelles. It is estimated that in about two-thirds of plant species the inheritance of chloroplasts is strictly maternal. Traits with an extranuclear basis may be identified on the basis of several diagnostic criteria.

1. Differences in reciprocal crosses that cannot be attributed to sex linkage or some other chromosomal basis tend to implicate cytoplasmic factors.

    (*a*) If progeny show only the characteristics of the female parent that can be attributed to unequal cytoplasmic contributions of male and female parents, then plasmagene inheritance is suspect (Example 13.24).
    (*b*) If the uniparental inheritance of a trait cannot be attributed to unequal cytoplasmic contributions from the parents, this does not necessarily rule out cytoplasmic factors (Example 13.25).
2. Extranuclear factors may be detected by either the absence of segregation at meiosis (Example 13.26) or by segregation that fails to follow Mendelian laws (Example 13.27).
3. Repeated backcrossing of progeny to one of the parental types for several generations causes their chromosomal endowment to rapidly approach 100% that of the recurrent parental line. The persistence of a trait in the progeny, when the backcross parent exhibits an alternative character, may be considered evidence for plasmagene inheritance (Example 13.28).

EXAMPLE 13.24

In higher plants, pollen usually contributes very little, if any, cytoplasm to the zygote. Most of the cytoplasmic elements are transmitted through the maternal par-ent. In the plant called "four-o'clock" (*Mirabilis jalapa*), there may be normal green, pale green, and variegated branches due to two types of chloroplasts. Plants grown from seeds that developed on normal green branches (with all normal chlor-oplasts) will all be normal green; those that developed on pale-green branches (with abnormal chloroplasts) will all be pale green; those on variegated branches (with both normal and abnormal chloroplasts) will segregate green, pale green, and variegated in irregular ratios. The type of pollen used has no effect in this system. The irregularity of transmission from variegated branches is understandable if plasmagenes exist in the chloroplasts, because there is no mechanism to ensure the regular distribution of chloroplasts to daughter cells as there is for chromosomes.

EXAMPLE 13.25

The uniting gametes of the single celled alga *Chlamydomonas reinhardi* (Fig. 5-1) are morphologically indistinguishable. One strain of the alga that is streptomycin-resistant (*sr*) and of the "plus" mating type (*mt*$^+$) is crossed to a cell of "negative" mating type (*mt*$^-$) that is streptomycin-sensitive (*ss*). All progeny are resistant, but the nuclear genes for mating type segregate as expected: 1/2*mt*$^+$, 1/2*mt*$^-$. The reciprocal cross *ss mt*$^+$ × *sr mt*$^-$, again shows the expected segregation for mating type, but all progeny are sensitive. Repeated backcrossings of *sr mt*$^+$ to *ss mt*$^-$ fail to show segregation for resistance. It appears as though the plasmagenes of the *mt*$^-$ strain become lost in a zygote of *mt*$^+$. The mechanism that inactivates the plasmagenes of *mt*$^-$ in the zygote is not well understood.

EXAMPLE 13.26

Slow-growing yeast cells called **petites** lack normal activity of the respiratory enzyme cytochrome oxidase associated with the mitochondria. Petites can be maintained indefinitely in vegetative cultures through budding, but can sporulate only if crossed to wild type. When a haploid neutral petite cell fuses with a haploid wild-type cell of opposite mating type, a fertile wild-type diploid cell is produced. Under appropriate conditions, the diploid cell reproduces sexually (sporulates). The four ascospores of the ascus (Fig. 6-4) germinate into cells with a 1 : 1 mating-type ratio (as expected for nuclear genes), but they are all wild type. The petite trait never appears again, even after repeated backcrossings of both mating types to petite. The mitochrondrial factors for petite are able to perpetuate themselves vegetatively, but are "swamped," lost, or permanently altered in the presence of wild-type factors. Neutral petite behaves the same in reciprocal crosses regardless of mating type, and in this respect is different from the streptomycin-resistance factors in *Chlamydomonas* (see Example 13.25).

EXAMPLE 13.27

Another type of petite in yeast, called **suppressive**, may segregate, but in a manner different from chromosomal genes. When haploid suppressive petites are crossed to wild types and each zygote is grown vegetatively as a diploid strain, both petites and wild types may appear, but in frequencies that are hardly Mendelian, varying from 1 to 99% petites. Diploid wild type cells may sporulate producing only wild-type ascospores. By special treatment, all diploid zygotes can be made to sporulate. The majority of the ascospores thus induced germinate into petite clones. Some asci have 4, 3, 2, 1, or 0 petite ascospores, suggesting that environmental factors may alter their segregation pattern. Nuclear genes, such as mating type, maintain a 1 : 1 ratio in all asci.

EXAMPLE 13.28

The protoperithecial parent in *Neurospora* (Fig. 6-5) supplies the bulk of the extra-chromosomal material of the sexually produced ascospores. Very slow spore germination characterizes one strain of this fungus. The trait exhibits differences in reciprocal crosses and maternal inheritance, and fails to segregate at meiosis. When the slow strain acts as protoperithecial parent and the conidial strain has normal spore germination, all the progeny are slow, but possess 50% of the nuclear genes of the conidial parent. Each generation is then backcrossed to the conidial parent, so that the F_2 contains 75%, F_3 contains 87.5%, etc., of nuclear genes of the conidial parent. After the fifth or sixth backcross, the nuclear genes are almost wholly those of the conidial parent, but the slow germination trait persists in all of the progeny.

Exceptions are known to the generalization that cytoplasmic factors are maternally inherited.

EXAMPLE 13.29
When a green strain of geranium (*Pelargonium*) is crossed to a strain with white-margined leaves, the progeny may have green, white, or white-margined leaves. Since the results of reciprocal crosses are the same, it has been hypothesized that plastids can be transmitted to offspring by both the male and female gametes. Cells that contain a mixture of plastid genomes are said to be **heteroplasmic**; those that contain only one type of plastid genome are called **homoplasmic**.

SOLVED PROBLEM 13.2
The direction in which the shell coils in the snail *Limnaea peregra* can be dexteral like a righthand screw or sinistral like a left-hand screw. The maternal genotype organizes the cytoplasm of the egg in such a way that embryological cleavage divisions of the zygote will follow either of these two patterns regardless of the genotype of the zygote. If the mother has the dominant gene s^+, all her progeny will coil dextrally; if she is of genotype ss, all her progeny will coil sinistrally. This coiling pattern persists for the life of the individual. *Limnaea* is a hermaphroditic snail that can reproduce either by crossing or by self-fertilization. A homozygous dextral snail is fertilized with sperm from a homozygous sinistral snail. The heterozygous F_1 undergoes two generations of self-fertilization. (*a*) What are the phenotypes of the parental individuals? (*b*) Diagram the parents, F_1, and two selfing generations, showing phenotypes and genotypes and their expected ratios.

Solution:
(*a*) Although we know the genotypes of the parents, we have no information concerning the genotype of the immediate maternal ancestor that was responsible for the organization of the egg cytoplasm from which our parental individuals developed. Therefore, we are unable to determine what phenotypes these individuals exhibit. Let us assume for the purpose of diagramming part (*b*) that the maternal parent is dextral and the paternal parent is sinistral.
(*b*) (b) Let D = dextrally organized cytoplasm, S = sinistrally organized cytoplasm.

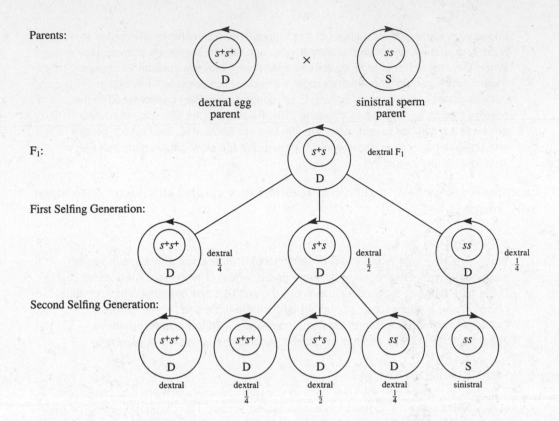

Notice that the F_1 is coiled dextrally, not because its own genotype is s^+/s, but because the maternal parent possessed the dominant dextral gene s^+. Likewise in the first selfing generation, all are phenotypically dextral regardless of their own genotype because the F_1 was s^+/s. In the second selfing generation, we expect the following:

First Selfing Generation		Second Selfing Generation			Summary				
						Genotypes			Phenotypes
$1/4\ s^+\ s^+$	$\times$	All $s^+\ s^+$	$=$	$1/4\ s^+\ s^+$	$s^+\ s^+$	$=$	$1/4 + 1/8$	$= 3/8$	5/8 dextral
$1/2\ s^+\ s$	$\times$	$\begin{cases} 1/4\ s^+\ s^+ \\ 1/2\ s^+\ s \\ 1/4\ ss \end{cases}$	$\begin{matrix}= \\ = \\ =\end{matrix}$	$\begin{matrix}1/8\ s^+\ s^+ \\ 1/4\ s^+\ s \\ 1/8\ ss\end{matrix}$	$\begin{matrix}s^+\ s \\ \\ ss\end{matrix}$	$\begin{matrix}= \\ \\ =\end{matrix}$	$\begin{matrix}1/4 \\ \\ 1/8 + 1/4\end{matrix}$	$= 2/8$	
					ss	$=$	$1/8 + 1/4$	$=$	3/8 sinistral
$1/4\ ss$	$\times$	All ss	$=$	$1/4\ ss$					

SOLVED PROBLEM 13.3

Slow-growing yeast cells called neutral petites lack normal activity of the respiratory enzyme cytochrome oxidase associated with the mitochrondria. Petites can be maintained indefinitely in vegetative cultures through budding, but can sporulate only if crossed to wild type. When a haploid neutral petite cell fuses with a haploid wild-type cell of opposite mating type, a fertile wild-type diploid cell is produced. Under appropriate conditions, the diploid cell reproduces sexually (sporulates). The four ascospores of the ascus (Fig. 6-4) germinate into cells with a 1 : 1 mating type ratio (as expected for nuclear genes), but they are all wild type. The petite trait never appears again, even after repeated backcrossings of both mating types to petite. The mitochondrial factors for petite are able to perpetuate themselves vegetatively, but are "swamped," lost, or permanently altered in the presence of wild-type factors. Neutral petite behaves the same in reciprocal crosses regardless of mating type. Assume that a neutral petite yeast has the chromosomal genes for normally functioning mitochondria, but has structurally defective mitochondria. Another kind of yeast is known, called segregational petite, which has structurally normal mitochondria that cannot function because of inhibition due to a recessive mutant chromosomal gene. What results would be expected among the sexual progeny when the neutral petite crosses with the segregational petite?

Solution: The diploid zygote receives from the segregational petite parent structurally normal mitochondria that should be able to function normally in the presence of the dominant nuclear gene from the neutral petite parent. Sporulation would probably distribute at least some structurally normal mitochondria to each ascospore. The nuclear genes would segregate 1 normal : 1 segregational petite. Let shaded cytoplasm contain defective mitochondria.

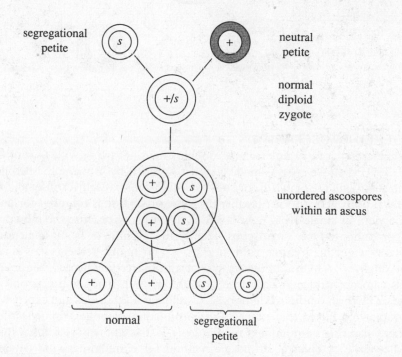

SOLVED PROBLEM 13.4

A condition called "poky" in *Neurospora* is characterized by slow growth due to an abnormal respiratory enzyme system similar to that of petite yeast. The poky trait is transmitted through the maternal (protoperithecial) parent. A chromosomal gene *F* interacts with poky cytoplasm to produce a faster growing culture called "fast-poky," even though the enzyme system is still abnormal. Poky cytoplasm is not permanently modified by transient contact with an *F* genotype in the zygote. It returns to the poky state when the genotype bears the alternative allele *F'*. Gene *F* has no phenotypic expression in the presence of a normal cytoplasm. If the maternal parent is fast-poky and the paternal (conidial) parent is normal, predict the genotypes and phenotypes of the resulting ascospores.

Solution: Let shaded cytoplasm contain poky mitochondria. The chromosomal alleles segregate in a 1 : 1 ratio, but poky cytoplasm follows the maternal (protoperithecial) line. The recovery of poky progeny indicates that poky cytoplasm has not been altered by its exposure to the *F* gene in the diploid zygotic stage.

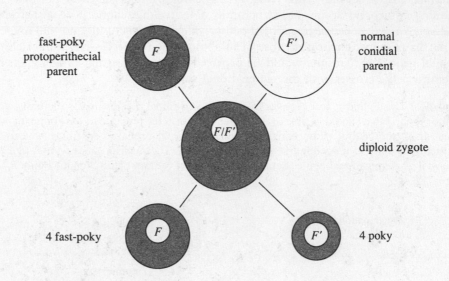

fast-poky protoperithecial parent — *F*

normal conidial parent — *F′*

F/F′ — diploid zygote

4 fast-poky — *F*

F′ — 4 poky

SOLVED PROBLEM 13.5

Commercial corn results from a "double cross." Starting with four inbred lines (A, B, C, D), a single cross is made between A and B by growing the two lines together and removing the tassels from line A so that A cannot self-fertilize, and thus receives only B pollen. In another locality the same procedure is followed for lines C and D. The yield of single-cross hybrid seed is usually low because the inbred parent lacks vigor and produces small cobs. Plants that germinate from single-cross seed are usually vigorous hybrids with large cobs and many kernels. It is undesirable for the single-cross hybrid to self-fertilize, as this inbreeding process commonly produces less vigorous progeny. Therefore, a double cross is made by using only pollen from the CD hybrid on the AB hybrid. Detasseling is a laborious and expensive process. A cytoplasmic factor that prevents the production of pollen (male-sterile) is known. There also exists a dominant nuclear gene *R* that can restore fertility in a plant with male-sterile cytoplasm. Propose a method for eliminating hand detasseling in the production of double-cross hybrid commercial seed.

Solution: Let S = male sterile cytoplasm, F = male fertile cytoplasm. The ABCD double-cross hybrid seed develops on the large ears of the vigorous AB hybrid. When these seeds are planted they will grow into plants, half of which carry the gene for restoring fertility so that ample pollen will be shed to fertilize every plant.

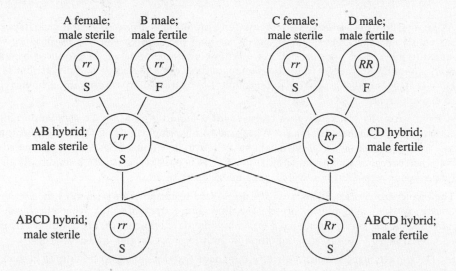

Supplementary Problems

13.6. The haploid genome of *Drosophila melanogaster* contains approximately 1.8×10^8 nucleotide pairs. The chromosomes of a polytene nucleus collectively have about 5000 bands. Assuming that 95% of the DNA is located in these bands and 5% is in the interband regions (*a*) determine the average number of nucleotide pairs in each band and interband region and (*b*) estimate the average number of genes per band and interband if an average gene contains 10^3 nucleotide pairs.

13.7. The puffing pattern of *Drosophila* polytene chromosomes seems to change in a predictable pattern during larva development. (*a*) It has been suggested that these puffs are the sites of genes transcription. How could this hypothesis be tested experimentally? (*b*) The synthesis of a particular protein coincides with the appearance of a specific puff in one of the polytene chromosomes. What inference can be made from this observation?

13.8. A sex-linked mutation in humans results in deficiency of the enzyme glucose-6-phosphate dehydrogenase (G6PD). Some individuals with this enzyme defect are more resistant to malaria than are those without this enzyme defect. Among those parasitized, approximately half of the blood cells of the resistant females contain the causative parasite; the cells of G6PD-deficient males are not parasitized. How can these observations be explained?

13.9. Different mRNA molecules have characteristic half-lives. Propose a method for estimating the half-life of a specific mRNA.

13.10. If a certain protein is found in the Golgi apparatus, how can you explain the fact that its cytoplasmic mRNA transcript contains 24 codons at its 5′ end that are not represented by corresponding amino acids at the amino terminus of the protein?

13.11. The gonads of *Drosophila* develop from material in the posterior end of the oocyte cortex (outer layer) that contains densely staining polar granules. An autosomal recessive mutation *gs*, when homozygous, causes adult females to produce oocytes without polar granules; progeny that develop from such eggs do not develop gonads. If parents are of genotype gs^+/gs, predict the results for the next two generations.

13.12. A mutant gene in *Drosophila* called *antennapedia* causes legs to develop on the head where antennae normally appear. To what class of developmental control genes does *antennapedia* belong? Offer an explanation as to how this mutation might cause abnormal development.

13.13. A female fruit fly is homozygous recessive for a mutation in a maternal effect gene important for embryonic development (m/m). (*a*) What kind of offspring will this fly produce when mated to a wild type male (m^+/m^+)? (*b*) What kind of offspring will be produced by a male that is homozygous recessive for the same mutation when mated to a wild-type female? (*c*) How is it possible to generate this female fly (m/m) from a cross involving a homozygous recessive parent?

13.14. A female fruit fly is heterozygous recessive for a mutation in a zygotic gene important for embryonic development (z^+/z). (*a*) What kind of offspring will this fly produce when mated to a wild-type male (z^+/z^+)? (*b*) What kind of offspring will this fly produce when mated to a male heterozygous for the same mutation? (*c*) Is it possible to isolate a mature fly that is homozygous recessive for a zygotic gene mutation? Explain.

13.15. The bicoid protein (see Example 13.20) is localized to the anterior portion of the *Drosophila* oocyte by the action of two proteins swallow and exuperantia. (*a*) What effect would a mutation in the *swallow* gene have on embryonic development? (*b*) How would this effect be altered in a fly containing a mutation in both the *swallow* and the *nanos* genes?

13.16. (*a*) Design a mutant screen to identify dominant (gain-of-function) mutations in maternal effect genes in *Drosophila*. (*b*) Design a screen to identify recessive zygotic gene mutations.

13.17. The *Drosophila* gene *hunchback* (see Example 13.20) contains five repeats of 5'-TCTAATCCCC-3' in its promoter. Bicoid protein is known to bind to each of these sites. When all five sites are bound with bicoid, maximal transcriptional activation occurs. What would be the phenotype of a fly that contains a deletion of three of the five repeats?

13.18. Why are translation controls less critical for eukaryotic genes than for many prokaryotic genes?

13.19. The *ced-3* gene (see Example 13.22) is required for apoptosis in *C. elegans*. (*a*) How would a loss-of-function mutation in this gene effect *C. elegans* development? (*b*) What about a gain-of-function mutation? (*c*) Is there likely to be a human oncogene counterpart to the *ced-3* gene?

13.20. The N terminus of a polypeptide destined to cross a membrane contains a signal sequence or signal peptide that is removed by a signal peptidase enzyme sometime during the passage of the rest of the polypeptide through the membrane. (*a*) What kinds of amino acids would be expected to predominate in the signal peptide? (*b*) Why is protein translocation through the mitochondrial or chloroplast membranes potentially more complex than that through the endoplasmic reticulum? (*c*) In what major respect does the nuclear membrane differ from the membranes of other organelles?

13.21. The concentration of cytoplasmic mRNA is observed to be higher under one condition when compared with another. Does this observation indicate that transcription control of that gene is operative? Explain.

13.22. Pseudogenes are nontranscribed DNA sequences that are highly homologous in nucleotide sequence to functional genes found elsewhere in the same genome. One class of pseudogenes, known as "processed pseudogenes," is characterized by absence of introns and upstream promoter sequences and presence of 3' terminal poly-A tracts. Propose a mechanism that might account for the origin of processed pseudogenes.

13.23. The diagram below represents a spread of nucleolar chromatin, showing gene transcription on a segment of DNA containing a tandemly arranged series of nucleolar rRNA genes. They give the appearance of a linear series of Christmas-tree-like structures, first identified in electron micrographs by O. L. Miller and B. R. Beatty (1969), and have been subsequently referred to as "Miller trees." Identify the following structures or regions. (*a*) The limits of an rRNA gene for the 38S rRNA precursor molecule. (*b*) A nontranscribed spacer DNA region between the rDNA repeats. (*c*) Promoter or initiator region. (*d*) Terminator of an rDNA gene. (*e*) RNA polymerase molecules. (f) 5' end of an rRNA transcript. (*g*) Do you expect to find ribosomes? (*h*) Is a similar phenomenon expected in *E. coli*? If so, explain any predicted differences.

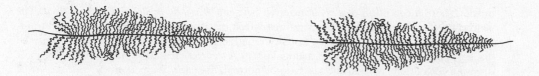

13.24. A recessive chromosomal gene produces green and white stripes in the leaves of maize, a condition called "japonica." This gene behaves normally in monohybrid crosses giving a 3 green : 1 striped ratio. Another striped phenotype was discovered in Iowa, named "iojap" (a contraction of Iowa and japonica), which is produced by a recessive gene *ij* when homozygous. If a plant with iojap striping serves as the seed parent, then the progeny will segregate green, striped, and white in irregular ratios regardless of the genotype of the pollen parent. Backcrossing striped progeny of genotype *Ij/ij* to a green pollinator of genotype *Ij/Ij* produces progeny that continue to segregate green, striped, and white in irregular ratios. White plants die due to lack of functional chloroplasts. Green plants produce only green progeny except when the genotype of the progeny is *ij/ij*; striping then reappears. Interpret this information to explain the inheritance of iojap.

13.25. If a woman contracts German measles during the first trimester of pregnancy, the child may be seriously affected even though the mother herself suffers no permanent physical effects. Such anomalies as heart and liver defects, deafness, cataracts, and blindness often occur in the affected children at birth. Can these phenotypic results be considered hereditary abnormalities?

13.26. A snail produced by a cross between two individuals has a shell with right-hand twist (dextral). This snail produces only left-hand (sinistral) progeny by selfing. Determine the genotype of this snail and its parents. See Solved Problem 13.2.

13.27. Most strains of *Chlamydomonas* (Fig. 5-1) are sensitive to streptomycin (see Example 13.25). A strain is found that requires streptomycin in the culture medium for its survival. How could it be determined whether streptomycin-dependence is due to a chromosomal gene or to a cytoplasmic element?

13.28. A yeast (Fig. 6-4) culture, when grown on medium containing acriflavine, produces numerous minute cells that grow very slowly. How could it be determined whether the slow growth was due to a cytoplasmic factor or to a nuclear gene?

13.29. Determine the genotypes and phenotypes of sexual progeny in *Neurospora* from the following crosses: (*a*) fast-poky male × normal female of genotype *F'*, (*b*) poky female × fast-poky male, (*c*) fast-poky female × poky male (see Solved Problem 13.4).

13.30. The cells of a *Neurospora* mycelium are usually multinucleate. Fusion of hyphae from different strains results in the exchange of nuclei. A mycelium that has genetically different nuclei in a common cytoplasm is called a *heterokaryon*. Moreover, the union results in a mixture of two different cytoplasmic systems called a *heteroplasmon* or a *heterocytosome*. The mycelia of two slow-growing strains, each with an aberrant cytochrome spectrum, fuse to form a heteroplasmon that exhibits normal growth. Abnormal cytochromes *a* and *b* are still produced by the heteroplasmon. Offer an explanation for this phenomenon.

13.31. Male sterile plants (Solved Problem 13.5) in corn may be produced either by a chromosomal gene or by a cytoplasmic factor. (*a*) At least 20 different male-sterile genes are known in maize, all of which are recessive. Why? Predict the F_1 and F_2 results of pollinating (*b*) a genetic male sterile by a normal, and (*c*) a cytoplasmic male sterile by a normal.

13.32. Given seed from a male sterile line of corn (see Solved Problem 13.5), how would you determine if the sterility was genic or cytoplasmic?

13.33. A bacterial spirochaete that is passed to the progeny only from the maternal parent has been found in *Drosophila willistoni*. This microorganism usually kills males during embryonic development but not females. The trait is called "sex ratio" (SR) for obvious reasons. Occasionally, a son of an SR female will survive. This allows reciprocal crosses to be made. The SR condition can be transferred between *D. equinoxialis* and *D. willistoni*. The spirochaete is sensitive to high temperatures, which inactivates them, forming "cured" strains with a normal sex ratio. (*a*) What would you anticipate to be the consequence of repeated backcrossing of SR females to normal males? (*b*) A "cured" female is crossed to a rare male from an SR culture. Would the sex ratio be normal? Explain.

13.34. How will the sequencing of whole genomes change the way that geneticists approach the understanding of gene structure and function?

13.35. The evolution of mitochondrial DNA occurs at a rate much faster than that of nuclear DNA. Hence, there is much greater variation from one person to another in mitochondrial DNA sequences than in nuclear DNA sequences. Of all the existing human populations, there appears to be greater variation in those of Africa than any other place on Earth. Furthermore, all the human mitochondrial DNA sequences can be arranged into a single phylogenetic tree. Assuming that the mitochondrial DNA of our most ancient ancestors had the same amount of individual variation as that of modern mitochondrial DNA, what are the evolutionary implications of these facts?

Review Questions

Vocabulary For each of the following definitions, give the appropriate term and spell it correctly. Terms are single words unless indicated otherwise.

1. The analogue in eukaryotes of the bacterial Pribnow box. (Two words.)
2. Genes involved in metabolism that tend to be active at all times in all nucleated cells. (Two words.)
3. The total complement of proteins in a cell.
4. A DNA sequence in cis position with a structural gene that potentiates the transcriptional activity of a gene on that same DNA molecule even though it may be far distant upstream or downstream from the gene it influences.
5. A genelike DNA sequence bearing close resemblance to a functional gene at a different locus, but rendered nonfunctional by additions or deletions in its structure that prevent its transcription and/or translation.
6. A small protein that, when covalently attached to target proteins, marks them for destruction by proteases.
7. The removal of intron sequences from mRNA and coupling of flanking exons.
8. Descriptive of mutant genes that cause a normal body part to develop in an abnormal location.
9. Programmed cell death.
10. The name of a theory that explains the origin of mitochondria and chloroplasts in eukaryotic cells.

Multiple-Choice Questions Choose the one best answer.

1. Nucleolar organizing regions (NORs) are most closely associated with (*a*) protein-coding genes (*b*) rRNA genes (*c*) replication forks (*d*) cellular differentiation (*e*) homeoboxes
2. Which of the following statements about mitochondria is incorrect? (*a*) Mitochondrial DNA has a higher mutation rate than nuclear DNA. (*b*) Replication of mitochondrial DNA is

synchronized with that of chromosomal DNA. (*c*) The size of mitochondrial DNA varies considerably from one species to another. (*d*) Mitochondria contain a protein-synthesizing system of their own. (*e*) Mitochondria usually follow a maternal inheritance pattern in species with equal-sized gametes.

3. Differentiation of most somatic cells does not appear to involve the loss of genes or recombination of DNA segments. The most striking exception to this rule is found in (*a*) histone genes (*b*) rDNA (*c*) mitochondrial DNA (*d*) immunoglobulin genes (*e*) hemoglobin genes

4. Which of the following differentiates eukaryotic DNA replication from prokaryotic replication? (*a*) multiple origins of replication (*b*) bidirectional replication fork (*c*) no use of an RNA primer (*d*) use of only one DNA polymerase (*e*) none of the above

5. Hormones are thought to regulate gene activity primarily at the level of (*a*) transcription (*b*) mRNA processing (*c*) transport of RNA from nucleus to cytoplasm (*d*) translation (*e*) post-translation processing of protein

6. Which of the following is not characteristic of most mRNA processing in eukaryotes? (*a*) addition of a poly-A tail at the 3′ end (*b*) addition of an unusual guanine to the 5′ end (*c*) removal of exons and splicing together of introns (*d*) removal of leader and trailer sequences (*e*) more than one of the above

7. There are three kinds of RNA polymerases (I, II, III) in eukaryotic cells, each specific for one class of RNA molecule (mRNA, tRNA, rRNA). Which of the following is a correct match? (*a*) I = rRNA, II = tRNA (*b*) II = mRNA, III = rRNA (*c*) I = tRNA, III = rRNA (*d*) I = rRNA, II = mRNA (*e*) none of the above

8. Apoptosis involves which of the following: (*a*) chromatin condensation (*b*) membrane blebbing (*c*) DNA fragmentation (*d*) cell dissolution (*e*) all of the above

9. A cell in its final functional state and form is said to be (*a*) determined (*b*) transdetermined (*c*) differentiated (*d*) activated (*e*) stabilized

10. Which of the following chain combinations could be found in a functional antibody combining site? (*a*) $L_{lambda} H_{Mu}$ (*b*) $L_{lambda} L_{kappa}$ (*c*) $H_{gamma} H_{alpha}$ (*d*) more than one of the above (*e*) none of the above

Answers to Supplementary Problems

13.6. (*a*) $(1.8 \times 10^8$ nucleotide pairs $\times 0.95)/(5 \times 10^3$ bands$) = 3.4 \times 10^4$ nucleotide pairs per band
$(1.8 \times 10^8$ nucleotide pairs $\times 0.05)/(5 \times 10^3$ bands$) = 1.8 \times 10^3$ nucleotide pairs per interband

(*b*) $(3 \times 10^4$ nucleotide pairs per band$)/(10^3$ nucleotide pairs per gene$) = 30$ genes per band
$(1 \times 10^3$ nucleotide pairs per interband$)/(10^3$ nucleotide pairs per gene$) = 1$ gene per interband

13.7. (*a*) Affix the cells containing the polytene chromosomes to a slide and expose them to radioactive uracils. Active genes will synthesize RNAs containing the labeled uracils. Then wash the slide to remove any unincorporated label and cover it with photographic film sensitive to the radiation (autoradiography). If dots on the developed film are concentrated in the puffed areas, the hypothesis would be confirmed. (*b*) It might be tempting to speculate that one or more genes in the puffed regions is actively synthesizing the protein. However, it is also possible that the protein itself might be responsible for the change in the puff pattern; or perhaps some other protein synthesized at the same time might be responsible. Thus, cause and effect cannot be established from this observation.

13.8. In most female mammals, including humans, one of the X chromosomes is inactivated. In females that are heterozygous for the mutation, about half of the cells would be expected to have an inactive X chromosome containing the normal G6PD gene and an active X chromosome containing the mutant gene; these cells would have a deficiency of G6PD and would not be parasitized. In the remainder of these female cells, the other X chromosome bearing the normal gene would be active, would not be G6PD deficient, and therefore would be parasitized. Males do not inactivate their single X chromosome. Males hemizygous for G6PD deficiency would therefore not be parasitized.

13.9. Expose cells to one or more radioactive ribonucleotides for a defined short period of time (a pulse) followed by large amounts of unlabeled ribonucleotides (the chase). After various lengths of chase, expose all of the labeled mRNAs to a single-stranded cDNA that is of a nucleotide sequence complementary to that of the mRNA species under consideration. Measure the amount of radioactivity in the mRNA-cDNA hybrids trapped on a nitrocellulose filter. For any given chase time, the relative stability of an mRNA should be directly related to the amount of radioactivity detected in the hybrids.

13.10. Proteins that are destined to cross the endoplasmic reticulum (ER; which contributes to the Golgi apparatus) contain an N-terminal leader sequence called the signal peptide. This peptide is cleaved after it has performed its job of aiding the polypeptide in passing through the ER membrane.

13.11. Offspring bearing the normal gene gs^+ are fertile and can produce a second generation. F_1 females of genotype gs/gs produce offspring (regardless of the genotype of their mates) that will be sterile; hence, those F_1 females cannot have "grandchildren." The name of this mutation is "grandchildless."

13.12. *Antennapedia* is a homeotic gene in which mutation causes transformation of one body part into another. In other words, it makes the right structure in the wrong place. Homeotic genes contain a regulatory sequence that responds to signals from other control genes. In embryonic cells of the imaginal (imago = adult insect) leg disks, the wild-type antennapedia gene is normally active; in cells of the imaginal antennae disks, the gene is normally inactive (silenced). A mutant *antennapedia* gene might fail to respond to the signals that normally turn off its normal allele in the antennae disks, and thus it is somehow able to direct development of legs instead of antennae.

13.13. (*a*) The female will be sterile, i.e., not able to produce viable offspring. (*b*) The F_1 offspring will be viable; however, 1/4 of F_2 females will be sterile. (*c*) The mother of this fly must be heterozygous in order to produce viable offspring, so the male parent must be homozygous recessive.

13.14. (*a*) The offspring will be viable: $1/2\ z^+/z : 1/2\ z^+/z^+$. (*b*) 1/4 of the offspring (z/z) will not produce viable embryos. (*c*) It is not expected that a homozygous recessive zygotic gene mutation would be able to develop properly, so no flies of this genotype are expected. However, it may depend on the particular gene and its action.

13.15. (*a*) The bicoid protein will not be as concentrated in the anterior portion of the oocyte (i.e., more of it will be located in the posterior than normal); thus, anterior structures may develop in the posterior end. (*b*) The effect would be exaggerated since *nanos* helps suppress anterior development in the posterior end.

13.16. (*a*) Initial crosses should reveal females that are sterile, because a dominant gene will have an effect even in a heterozygous state. (*b*) The F_2 of mutagenized flies would have to be examined for embryonic lethals. The F_2 is necessary to generate homozygous genotypes from the initial heterozygous mutant (a new mutation is expected in only one of two alleles).

13.17. When more bicoid protein is present, there are higher levels of hunchback protein. If three of five bicoid binding sites are mutated, the effect would be similar to having less bicoid protein and so less *hunchback* gene activation would occur. This would result in a reduced concentration of hunchback protein and a less dramatic gradient throughout the embryo. Formation of head, thorax, and other anterior structures would be altered. Posterior structure formation would most likely be unaffected.

13.18. All eukaryotic cytoplasmic mRNAs are monocistronic, whereas many prokaryotic mRNAs are polycistronic. Thus, within a bacterial operon, if the product of one structural gene is needed and another is not, then gene expression must be controlled at the translation level (or perhaps posttranslationally).

13.19. (*a*) Loss of *ced-3* would result in survival of cells that would normally die. (*b*) Gain of *ced-3* function would perhaps result in premature death of cells programmed to die. (*c*) Yes, only loss of function mutations will have a cancerous effect.

13.20. (*a*) Hydrophobic amino acids are expected in the signal peptide so that it can easily be inserted into the hydrophobic lipid membrane bilayer to initiate transport of the attached protein. (*b*) This is because these organelles are surrounded by a double membrane system, whereas the endoplasmic reticulum membrane consists of a single membrane. (*c*) There are pores in the nuclear membrane.

13.21. Transcriptional control may be operative. However, posttranscription regulation may be solely responsible or in conjunction with transcription control. Primary eukaryotic mRNA transcripts are subject to 5′ capping, 3′ polyadenylation, intron removal and splicing of exons, RNA degradation, and transport across the nuclear envelope; each operation presents a potential control point.

13.22. If cytoplasmic mRNA could be copied into DNA, it would have the characteristics described for processed pseudogenes. The retroviral enzyme reverse transcriptase can make DNA from viral RNA, and this cDNA can then become integrated into the host's genome, thus providing a model for the origin of processed pseudogenes.

13.23.

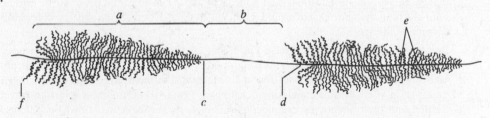

(*g*) No, because ribosomal RNAs and ribosomal proteins are assembled into ribosomes in the cytoplasm, not in the nucleus where Miller trees are formed. (*h*) Yes, but there will be ribosomes present on the transcripts.

13.24. It appears that the chromosomal gene *ij*, when homozygous, induces irreversible changes in normal plastids. The plastids exhibit autonomy in subsequent generations, being insensitive to the presence of *Ij* in single or double dose. Random distribution of plastids to daughter cells could give all normal plastids to some, all defective plastids to others, and a mixture of normal and defective plastids to still others. All plastids are not rendered defective in the presence of *ij/ij* as this would produce only white (lethal) seedlings.

13.25. It is important to distinguish between congenital defects (recognizable at birth) that are acquired from the environment during embryonic development and genetic defects that are produced in response to the baby's own genotype. The former may be produced by infective agents such as the virus of German measles, which is not really a part of the baby's genotype but is acquired through agents external to the developing individual. An active case of this disease usually produces immunity in the mother so that subsequent children of this mother should not be susceptible to the crippling influences of this virus. A hereditary disease is one produced in response to instructions of an abnormal gene belonging to the diseased individual and that can be transmitted in Mendelian fashion from generation to generation.

13.26. Parents: s^+s female $\times$ $s/?$ male; F_1: ss

13.27. Cross $ss\ mt^-$ (male) $\times$ $sd\ mt^+$ (female); if chromosomal, 25% of the sexual progeny should be $ss\ mt^-$, 25% $ss\ mt^+$, 25% $sd\ mt^-$, 25% $sd\ mt^+$; if cytoplasmic, almost all of the progeny should follow the maternal line (streptomycin-dependent) as in Example 13.25, while mating type segregates $1\ mt^-$: $1\ mt^+$.

13.28. From a cross of minute $\times$ normal, a nuclear gene will segregate in the spores in a 1 : 1 ratio (e.g., segregational petite in Problem 13.3). If an extranuclear gene is involved, segregation will not be evident and all spores will be normal (e.g., neutral petite in Example 13.26).

13.29. (a) All phenotypically normal; 1/2 F' (normal cytoplasm) : 1/2 F (normal cytoplasm). (b) and (c) 1/2 fast-poky; F (poky cytoplasm): 1/2 poky; F' (poky cytoplasm)

13.30. One strain may have an abnormal cytochrome a but a normal cytochrome b. The other strain might have an abnormal cytochrome b but a normal cytochrome a. The normal cytochromes in the heteroplasmon complement each other to produce rapid growth.

13.31. (a) A plant in which a dominant genic male sterile gene arose by mutation of a normal gene would be unable to fertilize itself and would be lost unless cross-pollinated by a fertile plant. The gene would be rapidly eliminated from heterozygotes within a few generations by continuous backcrossing to normal pollen parents. (b) F_1: $+/ms$, fertile; F_2: 1/4 $+/+$, 1/2 $+/ms$, 1/4 ms/ms; 3/4 fertile, 1/4 male sterile. (c) Male sterile cytoplasm is transmitted to all F_1 progeny; a selfed F_2 cannot be produced because none of the F_1 plants can make fertile pollen.

13.32. Plant the seeds and pollinate the resulting plants with normal pollen from a strain devoid of male sterility. If the F_1 is sterile, then it is cytoplasmic; if the F_1 is fertile, it is genic.

13.33. (a) If interspecific crosses can transmit the spirochaete, it is probably relatively insensitive to the chromosomal gene complement. Backcrossing would cause no change in the SR trait; indeed, this is how the culture is maintained. (b) The sex ratio would probably be normal. It is unlikely that the spirochaete would be included in the minute amount of cytoplasm that surrounds the sperm nucleus.

13.34. Previously, geneticists did not have knowledge of the location or sequence of genes. Gene function was discovered through the isolation of mutants with particular mutant phenotypes. With the knowledge of gene location and sequence, a different approach can be taken. For example, in yeast, strains containing deletions of genes whose function is not known can be created and their phenotype examined. In addition, gene sequence knowledge allows inference of amino acid sequence and thus allows comparisons of protein sequence with other known genes. This can help geneticists make better educated guesses regarding gene function. Experiments can be tailored to answer specific questions based on these sequence comparisons. For example, if a particular motif is known to be important in a previously identified gene and a newly sequenced gene is found to contain the same motif, mutations can be directed to that region to determine if they have similar functions. This approach is more efficient that screening through thousands of mutant organisms looking for a particular mutant phenotype.

13.35. Africa was probably the place where humans first evolved, and our ancestors emigrated from there to populate the world. Mitochondria are maternally inherited. From this fact and the single family tree for mitochondrial DNA, it has been inferred that there probably was a single ancestral woman (our "mitochondrial Eve") from which all humans derived their mitochondrial DNA.

Answers to Review Questions

Vocabulary

1. Hogness box
2. housekeeping genes
3. proteome
4. enhancer
5. pseudogene
6. ubiquitin
7. splicing
8. homeotic
9. apoptosis
10. endosymbiosis

Multiple-Choice Questions

1. *b*   2. *b*   3. *d*   4. *a*   5. *a*   6. *e* (*c* and *d*)   7. *d*   8. *e*   9. *c*   10. *a*

INDEX

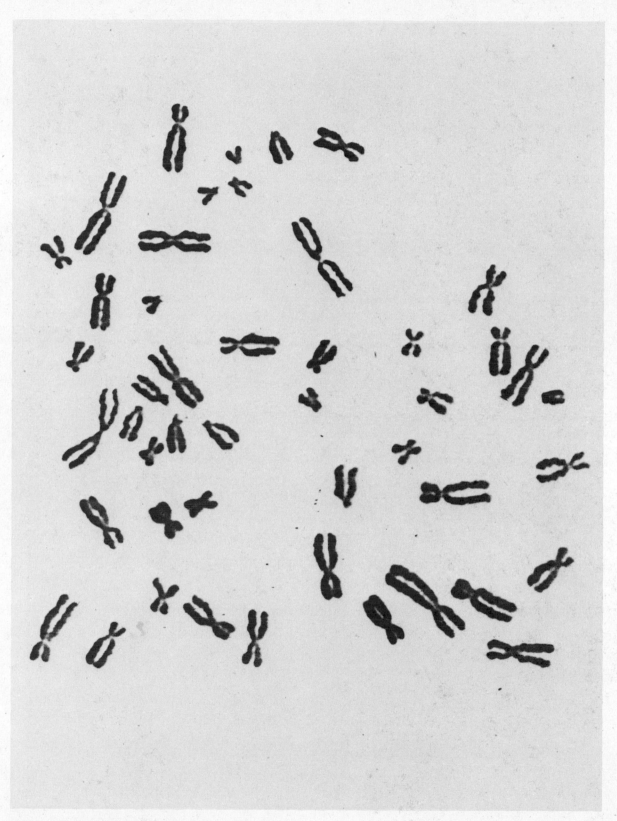

Photograph accompanying Problem 7.36. See p. 249.